T0173046

THREE DIMENSIONAL QSAR

Applications in Pharmacology and Toxicology

QSAR in Environmental and Health Sciences

Series Editor

James Devillers
*CTIS-Centre de Traitement de
l'Information Scientifique
Rillieux La Pape, France*

Aims & Scope

The aim of the book series is to publish cutting-edge research and the latest developments in QSAR modeling applied to environmental and health issues. Its aim is also to publish routinely used QSAR methodologies to provide newcomers to the field with a basic grounding in the correct use of these computer tools. The series is of primary interest to those whose research or professional activity is directly concerned with the development and application of SAR and QSAR models in toxicology and ecotoxicology. It is also intended to provide the graduate and postgraduate students with clear and accessible books covering the different aspects of QSARs.

Published Titles

Endocrine Disruption Modeling, *James Devillers, 2009*

Three dimensional QSAR: Applications in Pharmacology and Toxicology, *Jean Pierre Doucet and Annick Panaye, 2010*

QSAR in Environmental and Health Sciences

THREE DIMENSIONAL QSAR

Applications in Pharmacology and Toxicology

Jean Pierre Doucet
Annick Panaye

CRC Press
Taylor & Francis Group
Boca Raton London New York

CRC Press is an imprint of the
Taylor & Francis Group, an **informa** business

CRC Press
Taylor & Francis Group
6000 Broken Sound Parkway NW, Suite 300
Boca Raton, FL 33487-2742

First issued in paperback 2019

© 2010 by Taylor & Francis Group, LLC
CRC Press is an imprint of Taylor & Francis Group, an Informa business

No claim to original U.S. Government works

ISBN-13: 978-1-4200-9115-1 (hbk)
ISBN-13: 978-0-367-38316-9 (pbk)

This book contains information obtained from authentic and highly regarded sources. Reasonable efforts have been made to publish reliable data and information, but the author and publisher cannot assume responsibility for the validity of all materials or the consequences of their use. The authors and publishers have attempted to trace the copyright holders of all material reproduced in this publication and apologize to copyright holders if permission to publish in this form has not been obtained. If any copyright material has not been acknowledged please write and let us know so we may rectify in any future reprint.

Except as permitted under U.S. Copyright Law, no part of this book may be reprinted, reproduced, transmitted, or utilized in any form by any electronic, mechanical, or other means, now known or hereafter invented, including photocopying, microfilming, and recording, or in any information storage or retrieval system, without written permission from the publishers.

For permission to photocopy or use material electronically from this work, please access www.copyright.com (http://www.copyright.com/) or contact the Copyright Clearance Center, Inc. (CCC), 222 Rosewood Drive, Danvers, MA 01923, 978-750-8400. CCC is a not-for-profit organization that provides licenses and registration for a variety of users. For organizations that have been granted a photocopy license by the CCC, a separate system of payment has been arranged.

Trademark Notice: Product or corporate names may be trademarks or registered trademarks, and are used only for identification and explanation without intent to infringe.

Visit the Taylor & Francis Web site at
http://www.taylorandfrancis.com

and the CRC Press Web site at
http://www.crcpress.com

Contents

PART I   *Actual 3D Models*

PART II   Around the 3D Approaches

PART III   Beyond 3D

PART IV   *Receptor-Related Models*

Series Introduction

The correlation between the toxicity of molecules and their physicochemical properties can be traced back to the nineteenth century. Indeed, in a French thesis entitled *Action de l'alcool amylique sur l'organisme* (Action of amyl alcohol on the body), which was presented in 1863 by A. Cros before the Faculty of Medicine at the University of Strasbourg, an empirical relationship was made between the toxicity of alcohols and their number of carbon atoms as well as their solubility. In 1875, Dujardin-Beaumetz and Audigé were the first to stress the mathematical character of the relationship between the toxicity of alcohols and their chain length and molecular weight. In 1899, Hans Horst Meyer and Fritz Baum, at the University of Marburg, showed that narcosis or hypnotic activity was in fact linked to the affinity of substances to water and lipid sites within the organism. At the same time, at the University of Zurich, Ernest Overton came to the same conclusion, providing the foundation of the lipoid theory of narcosis. The next important step was made in the 1930s in St. Petersburg by Lazarev, who first demonstrated that different physiological and toxicological effects of molecules were correlated with their oil–water partition coefficient through formal mathematical equations in the form $\log C = a \log P_{oil/water} + b$. Thus, the quantitative structure–activity relationship (QSAR) discipline was born. Its foundations were definitively fixed in the early 1960s by the seminal works of C. Hansch and T. Fujita. Since then, the discipline has gained tremendous interest, and QSAR models now represent key tools in the development of drugs as well as in the hazard assessment of chemicals. The new REACH (Registration, Evaluation, Authorization, and Restriction of Chemicals) legislation on substances, which recommends the use of QSARs and other alternative approaches instead of laboratory tests on vertebrates, clearly reveals that this discipline is now well established and is an accepted practice in regulatory systems.

In 1993, the journal *SAR and QSAR in Environmental Research* was launched by Gordon and Breach to focus on all the important works published in the field and to provide an international forum for the rapid publication of structure–activity relationship (SAR) and QSAR models in (eco)toxicology, agrochemistry, and pharmacology. Today, the journal, which is now owned by Taylor & Francis, publishes twice as many issues per year and continues to promote research in the field of QSAR by favoring the publication of new molecular descriptors, statistical techniques, and original SAR and QSAR models. This field continues to grow rapidly, and many subject areas that require larger development are unsuitable for publication in a journal due to space limitations.

This prompted us to develop a series of books entitled *QSAR in Environmental and Health Sciences* to act in synergy with the journal. I am extremely grateful to Colin Bulpitt and Fiona Macdonald for their enthusiasm and invaluable help in making the project a reality.

This book is the second of the series. Its purpose is at least twofold: On one hand, it introduces the theory and practical applications of 3D-QSAR approaches in pharmacology and toxicology to both the neophytes and the experienced scientists; on the other, it provides a clear overview of the strengths and weaknesses of these methods.

At the time of going to press, two other books are in the pipeline. One deals with reproductive and developmental toxicology modeling and the other focuses on the topological description of molecules. I gratefully acknowledge Hilary Rowe for her willingness to assist me in the development of this series.

James Devillers

Preface

Computational chemistry is today playing a major role in the studies of complex processes involved in the design and development of new drugs. Based on structural similarity, data mining from huge chemical databases allows the selection of compounds having the pharmacophore likely to give them an adequate biological activity. The calculation of intermolecular interactions between a drug and its receptor specifies the mechanisms at the molecular scale and may suggest structural modifications that would be able to increase activity.

QSAR models establish relationships between a molecular structure and its activity. Their ability to predict the behavior of untested and even unsynthesized molecules is a valued asset in the quest for new drugs. QSAR models direct research toward the more promising structures from the initial stages of development, avoiding wrong tracks, reducing laboratory tests, and limiting animal experimentation.

Another important field of application is toxicology and ecotoxicology. With the widespread introduction of new chemicals in the market, sometimes with important tonnages, it is crucial to have quantitative models at one's disposal that are able to identify pollutants acting on human health or wildlife and prioritize chemicals to be submitted for in-depth experimental tests.

In the past, QSARs were generally received with skepticism, not always unjustified, on account of being rather crude models and sometimes providing inaccurate results with regard to robustness and applicability range of the models. QSARs today, however, have greatly improved and have made rapid strides in the various fields they are employed in as indicated in the following:

- Introduction of new statistical or mathematical tools for data analysis with nonlinear or nondeterministic approaches such as neural networks and genetic algorithms. These methods also benefit from the increasing power of computers for data processing.
- Improvements of the methods for structural representation and development of large chemical databases.
- Availability of graphical display and interactive visualization tools.

QSARs now constitute, in their own right, an important element of drug design approaches. The newly introduced 3D-QSAR models take into account the spatial characteristics of molecules (geometry, shape, and electron distribution), and even evaluate the fields they create in their surrounding or their interactions with neighboring structures (solvents, or receptors). This results in definite improvement in the field. A more detailed and accurate picture of the molecular behaviors is thus accessible.

Two elements explain the now well-established interest in QSAR models:

- On the one hand, international policies (such as the European REACH project) proposed the use of QSARs (if correctly designed) for hazard identification and risk assessment.
- On the other hand, there has been a progressive fusion of approaches that were previously limited to distinct areas of molecular modeling. 3D-QSARs routinely call for molecular mechanics, quantum mechanics, or molecular dynamics to define the privileged conformations of drugs and their possible interconversions. Docking is currently used to specify the best binding mode of drugs in their receptor pocket. Free energy calculations, reserved in the past to some specific or illuminating examples, can now be performed for a series of molecules and incorporated in 3D models. Conversely, QSARs become an efficient tool for screening chemical databases (possibly after a preliminary filtering process) in the search for new leads.

We would like to express our gratitude to our colleagues and coworkers for their wholehearted support and fruitful discussions and specially to Dr F. Barbault who also designed the cover artwork. At last, we emotionally remember our colleague and friend, Prof. B.T. Fan, who departed prematurely.

Introduction

The term "QSARs" (quantitative structure–activity relationships) encompasses a set of methodologies relating, for a specific process, the biological activity of molecules to some selected features of their physicochemical structure by means of a statistical or mathematical tool. The derived model is then used to analyze the results and to predict the activity of untested compounds.

Property prediction is of paramount importance in Chemistry. From a practical point of view, the interest is not so much on the molecular structure itself but rather on the properties the structure may have. It is therefore not surprising that the search for relationships (more specifically, *quantitative* relationships) between structures and properties or activities presented itself as a major concern several years ago. For example, in the Shanghai Museum it is reported that the "Kaogongji" (roughly, something like the book of the craftsman techniques) proposed in the fifth century BC a qualitative relationship between the composition of bronze and its properties (such as quality of the cutting edge, ease of polishing, and sparkling aspect).

With the boom of combinatorial chemistry, a large number of new chemicals can be readily obtained. It is imperative to have efficient methods for activity prediction not only because such methods save time and resources, but also because they avoid large-scale tests, orienting synthesis toward selected, potentially interesting compounds. Toxicology and ecotoxicology are now faced with the widespread diffusion of many long-life chemicals, for example, polychlorinated aromatics that are able to bind nuclear receptors and disrupt the normal hormonal processes of the endocrine system in humans and animals. The outburst in the number of xenobiotics present in the ecosystem makes such prediction tools a privileged way for prioritizing tests on chemicals the more suspect. Thus, they play an important role in environmental policy and adhere to international regulations for risk assessment and hazard identification. Furthermore, they are in line with policies that recommend a decrease in animal experimentation.

After initial skepticism, justified in part by several "meaningless" models (as quoted by Kubinyi [1]), QSARs are now regarded as valuable, scientifically credible tools in drug discovery and environmental toxicology programs such as the REACH (Registration, Evaluation, Authorization and restriction of CHemicals) policy for chemicals in the EU (http://europa.eu.int/comm/environment/chemicals/reach.html) and the Chemical Assessment and Control Program of EPA (U.S. Environment Protection Agency) (see, e.g., the reviews of Schultz et al. [2] and Schmeider et al. [3,4]).

Trying to relate the properties of a molecule with its structure is an old problem, but models widely evolved in parallel to the advances in chemical knowledge. QSARs also took advantage of the development of calculation methods and graphical display tools. How did we pass from the simple count of a sequence of atoms to the concept of the complementarity ligand–receptor?

BRIEF HISTORICAL EVOLUTION

THE BEGINNINGS

In the following, we will briefly mention some milestones in the development of QSAR models. For a detailed historical presentation, see Kubinyi [1], Rekker [5], Parascandola [6], and Güner [7]. The concept of structure–activity relationship may be traced back to Crum-Brown and Fraser [8,9], who indicated, in 1869, that "there can be no reasonable doubt but that a relation exists between the physiological action of a substance and its composition and constitution" (quoted from [10]). A few years before this, in his thesis at the Faculty of Medicine in Strasbourg (France) in 1863, Cros observed an increase in the toxicity of alcohols to mammals with decreasing water solubility up to a maximum potency (quoted from Kubinyi [1]).

An important notion arose from Langley's work [11] on the antagonism between pilocarpine and saliva. He suggested the formation of a complex between exogenic compounds introduced and a material present on the nervous terminations. This was the concept of the receptor, which later became very useful in choosing the active conformation of a drug and in constructing receptor-based models, as well as in the neighboring field of molecular modeling. The hypothesis of specific interactions was also formulated by Fisher in 1894 [12–14], with the image of "the key and the lock," and was later modified in 1966 by Koshland [15], who envisaged the possibility of receptor deformation on ligand binding. This was the notion of "induced fit."

At nearly the same time, Richet [16] correlated toxicities of narcotics with the inverse of their solubility in water in 1893, and in 1899–1901 Meyer and Overton [17,18] independently found linear relationships between the toxicity of organic compounds and their lipophilicity (ability to partition between water and a lipophilic biophase, the system olive oil–water being proposed as a reference medium).

BASES OF MODERN QSARS

The basic hypothesis is that the structure of a molecule contains features (geometric and/or electronic) responsible for its physical, chemical, or biological properties. Thus, for a given biological process from a set of active molecules assumed to have the same (or very similar) mode of action (MOA), it becomes possible to define a model relating structure and activity provided that the molecular structure can be represented by a set of structural descriptors (numerical values, fragments, etc). This corresponds to "ligand-based" models.

The parameters characterizing the molecular structure may be as follows:

- Descriptors calculated from the 2D molecular formula or the actual 3D geometry.
- Physicochemical quantities (measured or calculated) such as partition coefficients, vapor pressures, ionization constants, and orbital energies.

More precisely, QSARs generally relate, in a series of compounds, the *variations* of the activity to *variations* in the values of computed or experimental characteristics or properties of the structures.

The biological action of a chemical is generally associated from an early stage with (non-covalent) interactions with a specific "receptor" (protein, enzyme, etc.) in the living organism ("receptor-mediated" mechanism). Evaluation of these interactions leads to "receptor-based" models in contrast to the previously mentioned "ligand-based" models that consider only the active molecules and ignore their biological receptors. However, these approaches, which rely more closely on the actual mechanism of action, are, in most cases, more intricate, and to date less frequently developed. Nevertheless, with the increasing number of protein structures (free or bound to a ligand) now available (from X-ray crystallography, NMR, and homology modeling), the number of such applications is also rapidly increasing.

A traditional distinction remains between QSARs and QSPRs (quantitative structure property relationships), which deal with the prediction of physicochemical quantities. Although there are some differences in the nature of the data that are treated (biological data are often "softer" than physicochemical data), this distinction looks rather artificial since mathematical and statistical models are generally the same, and some physicochemical parameters (such as partition coefficient, pK values, orbital energy) may be introduced in the QSAR formulation.

2D MODELS

Linear free energy relationships: The birth of modern QSAR models is generally associated with the pioneering work of Hammett (1937) [19], who defined substituent constants for describing the electronic properties of aromatic compounds, and Taft (1952), who introduced steric substituent constants [20,21]. QSAR models have also benefited from the work of Ferguson [22], who proposed a thermodynamic interpretation of the relationship between nonspecific narcotic effect levels and lipophilicity.

Development was then stalled for some years until a new impetus was provided in the 1960s, when Hansch and Fujita (1964) [23] used the formalism of the linear free energy relationships (also known as "extrathermodynamic relationships") to correlate biological activities with physicochemical properties. We recall here that Hammett proposed to quantify electronic effects in substituted aromatic compounds by σ constants, measured in the dissociation of substituted benzoic acids:

$$\text{For the substituent X, } \sigma_X = \log (K_X/K_H)$$

where K_H and K_X are the dissociation constants of the benzoic acid (the reference) and the X-substituted benzoic acid, respectively. The advantage of this σ scale is that the variations in the equilibrium (K) or rate (k) constants of many reactions can be correlated by equations such as

$$\log (k_X/k_H) \quad \text{or} \quad \log (K_X/K_H) = \rho\sigma_X$$

where k_X (resp. K_X) are the rate (or equilibrium) constant of the X-substituted compound and k_H (resp. K_H) are the corresponding values for the reference (X = H). The reaction parameter ρ characterizes the sensitivity of the process to electronic effects (and may, for instance for rate constants, be related to the partial charge developed in the transition state). This first scale was followed by many other substituent constant scales (σ^-, σ^+, σ_0, σ_R, F, R) for a more precise evaluation of electronic effects, and, later, for aliphatic systems by the σ^*, E_S scales for polar and steric effects, respectively [20,21,24,25]; however, a more precise representation of steric effects (taking into account the shape of the substituent group) was introduced by Verloop et al. [26].

After the definition of the π constants characterizing lipophilicity contribution, Hansch and Fujita proposed a σ-π-analysis on various processes such as activity of benzoic acids on mosquito larvae or of diethylaminoethyl benzoates on guinea pigs [23]. The general form of such a correlations is

$$A = a\sigma + bE_S + c\pi + d\pi^2$$

The biological activity A is expressed as the concentration of the chemical for a given end point (50% mortality or effect, log IC_{50}), the inhibitory power or the dissociation constant of the drug–receptor complex log K_i, etc. Parabolic terms (in π^2) were introduced in Hansch relationships [27] to express the fact that very polar drugs will not reach the receptor site due to their inability to cross lipid membranes, whereas very lipophilic drugs will just stay "trapped" in these membranes and will not pass through aqueous phases. Only compounds with intermediate lipophilicity have a good chance of arriving at the receptor site in a reasonable time frame and with sufficient concentration.

These pioneering studies were followed by several QSARs built with these substituent constants (of experimental origin) [28,29].

Incremental models: Free and Wilson [30] and Fujita and Ban [31] developed a different type of model using additive indicator variables (set to 0 or 1) to encode the presence (or the absence) of certain chemical groups. This method had also been explored earlier by Bruice et al. [32]. In the same vein, structural fragments were used in classification (separation of active, weakly active, or inactive compounds).

In the framework of molecular graphs, the DARC (Description, Acquisition, Restitution and Conception) system of Dubois [33,34] considered contributions of ordered atomic positions (sites) in a hierarchically ordered concentric description of the environment of a focus (the atomic position, or the bond, where the property was localized). For properties not localized on a given site or bond (e.g., ^{13}C shifts or IR frequencies), a "defocalized treatment" is possible. Numerous examples (related to QSPR models) are provided in Ref. [34]. In addition to QSAR/QSPR models, this type of description of the molecular structure has been applied to several other fields of cheminformatics: database management, structural elucidation, spectral simulation, and computer-aided organic synthesis.

Topological indices: Another effective approach took advantage of the similarity between chemical structural formulas and mathematical graphs (atoms corresponding to vertices, and bonds to edges). This led to the definition of topological indices, which started in 1947 with Wiener's work on the boiling point of paraffins, where molecules were described by path counts determined on the molecular graph [35]. This application was followed by the introduction of a deluge of indices (more than 400) encoding, for example, ramification (Randić), shape (Kier and Hall), and cyclization (Balaban). Electronic aspects were also taken into consideration (E state indices of Kier and Hall, electrotopologic indices, or Galvez charge distribution indices). For more details, see Devillers and Balaban [36] or Todeschini and Consoni [37].

3D AND 2.5D MODELS

The preceding topological models are based on the structural formula of molecules (a 2D only representation), and, despite this, led to many satisfactory correlations (predicted vs. observed activities). However, it is clear that the properties of molecules actually depend on their 3D structures. Numerous examples can be found of the influence of geometrical isomerism or conformational preferences in spectroscopy or reactivity even in elementary organic chemistry; hence, the efforts to develop 3D models taking the actual molecular geometry into consideration. This was not without its share of problems. Considering 3D structures implies the choice of a "good conformation." The minimum energy conformation attainable by more or less lengthy geometry optimizations were frequently used but things worked better when the "active conformation" (that bound the receptor) was known (or might be reasonably inferred). Schematically, two avenues were explored:

3D models: A first type of method, exemplified by the well-known, and still widely used, CoMFA method [38] exploits structural information on discrete and individualized points in the neighborhood of the molecules under scrutiny. This corresponds to actual 3D methods.

With the pioneering work of Cramer et al., after the limited success of DYLOMMS (DYnamic lattice-oriented molecular modeling system) [39], the CoMFA methodology [38] asserted itself as the 3D method of reference, and is still widely used 20 years after its inception. The method relies on the calculation of steric and electrostatic potentials on nodes of a lattice surrounding the molecules, all aligned on a common reference in their supposed active conformation. Additional potentials have also been proposed. The steepness of the steric potential caused some problems and a few years later, the CoMSIA approach [40] replaced the evaluation of potentials by similarity calculations with smoother functions.

However, although several programs are now available, alignment (a critical step that must be addressed) remains a problem, especially when considering a series of non-congeneric compounds. But with the increasing number of ligand–receptor complexes solved by X-ray crystallography, NMR, or homology modeling studies (to infer the receptor-binding site structure), docking calculations (determination of the best ligand arrangement in the receptor site) now make the choice of the active conformation of the drug relatively easier.

In addition to CoMFA and CoMSIA, several other approaches were developed. Rather than nodes on a lattice external to the molecular shapes, they considered nodes in the occupied molecular volumes or points scattered on the surface as, for example, molecular shape analysis (MSA) [41], hypothetical active site lattice (HASL) [42], and comparative molecular active site analysis (CoMASA) [43].

2.5D models: The other explored avenue corresponds to what may be called *2.5D approaches* [44]. They take into account descriptors or quantities that obviously depend on the molecular geometry, but the information is condensed in a single numerical value or a vector of a few components (without explicitly specifying the location of the point where information is collected). For example, the delicate alignment phase of CoMFA and CoMSIA is avoided in GRIND [45] by means of an *auto-* and *cross-correlation* transform.

The *similarity concept* (similar molecules have similar property), largely verified but with some exceptions [46], prompted the use of similarity indices (steric, electrostatic, or quantum) as structural descriptors. For a population of N compounds, each molecule is described by its similarity with the other molecules of the set in an N*N symmetrical matrix. One advantage is that only pairwise comparisons are carried out without alignment of the whole set on a common reference [47–50].

Relevant to the same concern, QSDARs (quantitative spectroscopic data activity relationships) took advantage of the sensitivity of spectroscopic data to structural or geometrical modifications to convey molecular information: The underlying idea was that spectra very sharply reflected the molecular structure and so were able to characterize it. Descriptors are here measured or calculated spectral parameters (^{1}H or ^{13}C NMR shifts, IR frequencies, orbital energies) rather than abstract calculated descriptors [51,52].

Alternately, other approaches may be viewed as a direct expansion of the "classical" 2D-QSARs. In addition to constitutional (molecular weight, number of atoms) and topological descriptors (calculated from the structural formula) or substituent constants (σ, π), other descriptors were introduced involving the molecular geometry. HOMO, LUMO energies, for example, have been largely used, as well as molecular surface area and similar quantities. Topographical indices and extrapolation of topological indices, where the actual 3D interatomic Euclidian distances replace the topological distances (number of bonds between atoms), were also introduced, but, seemingly, with a more limited diffusion.

Various packages (e.g., CODESSA [53], DRAGON [54], ALMOND [55], Cerius [56], ADAPT [57]) are now available for the calculation of a large number of such 2D and 2.5D descriptors (nearly 3000). Faced with this deluge, an important problem arose: the selection of the descriptors relevant to the specific biological application looked for (vide infra).

RECEPTOR-BASED METHODS

All these approaches (ligand based) consider only the drug and provide no information on the receptor (there are, of course, exceptions as cited above). On the other hand, "receptor-based" methods try to evaluate interaction energies between the receptor and a potential ligand. The GRID approach developed by Goodford [58], who determined positions for favorable interactions on nodes in the vicinity of a protein, was the starting point. These methods required more sophisticated potential functions than those generally used in ligand-based models and often more refined thermodynamic paths to provide reliable results [59–61]. However, with the ever-increasing sophistication of computers, such thermodynamic-type models, which were until some years limited to specific examples, can now be implemented in QSAR studies.

4D MODELS AND FURTHER

All these models implicitly assume that ligands act under a single conformation and bind in a unique or similar mode. The possibility of several simultaneous binding modes was investigated by Lukakova and Balaz [62] on rigid aromatic hydrocarbons. Another problem is that flexible compounds may exist as a mixture of several coexisting conformations. This problem was addressed in the 4D methodology proposed by Hopfinger [63–65] whereas Funatsu et al. [66,67] used *n-way* PLS analysis to choose the good alignment and the good conformation among various possibilities.

Until recently, the receptor was assumed to be rigid. However, several experimental observations showed that the receptor may undergo some deformation to accept a ligand. In other words, the old "key and lock" image must be replaced by something like "the hand and the glove," which corresponds to "induced fit." In the "fifth-dimension" models, different adaptation protocols, with possible dynamic interchange between them, are considered [68–71]. The possibility of different solvation models was even introduced in the 6D-QSAR approach.

2D- vs. 3D-QSAR Models: What Is the Best Choice?

In the 1990s, with the increasing number of topological indices, this question was the subject of heated debates in several QSAR meetings. However, it seems to us to be more a problem of resources and objective rather than a dilemma. As expressed aptly by a Chinese proverb "no matter a cat is white or black if it catches mice well."

From the various comparisons carried out so far on specific properties, with often (more or less) limited populations of compounds, no definite general conclusion can be drawn. However, on an extended data set, Sutherland et al. [72] indicated a better predictive capability of field-based 3D methods, and emphasized the importance of interpretability of the models. A naive empirical guideline might be as follows:

- If we need only a numerical equation or a mathematical model allowing for reasonably predicting activities, a 2D approach (duly validated) may be sufficient, and, in agreement with Occam's razor principle, it will then be useless to search for a complex 3D model since the calculation of 2D indices is very rapid and requires only knowledge of the structural formula.
- Conversely, if the problem is to get some insight into the main intervening interactions at a physicochemical level, and to access geometrical information as to the spatial areas involved, a 3D approach is far more efficient. It may readily suggest, for example, what structural modifications would be interesting for the synthesis of new active compounds. However, it is more time-consuming and resource-demanding as several processes, such as conformational analysis, choice of conformation, and (sometimes) alignment have to be covered.

QSARs and Related Fields: Database Mining, Molecular Modeling, and Post-3D-QSAR Models

We would like to conclude this short historical survey indicating the synergy arising from QSARs and molecular modeling not only in receptor-based approaches but also in ligand-directed studies. Molecular docking (determining the best conformation and orientation of a ligand in its receptor binding site) is of definite help for aligning compounds in CoMFA and CoMSIA treatments. At the same time, X-ray crystallography and homology modeling of proteins supply additional tools that are of great value. Molecular dynamics, quantifying the conformational flexibility of ligands, has now become an integrated part of numerous QSAR applications.

At the beginning, QSARs were restricted to linear relationships, on strictly congeneric compounds (a common core and some substituent groups), involving standard "substituent constants" in an LFER formalism. With 3D methods such as CoMFA and CoMSIA, the numerical prediction is refined, taking into account the various fields acting on the ligands, and even (with Quasar and Raptor [69,73]) the geometrical adaptation of the ligand, the variation in the dynamic processes, and the solvation modes.

Finally, QSARs, initially devoted to small populations of compounds with a common activity, now constitute new tools for mining chemical databases in view of possibly finding active compounds. Although this operation was mainly treated in the past as a classification problem relying on structure or substructure recognition, the development of a QSAR model, even a crude one, can be efficient to categorize compounds, provided the treatment is automated, for an acceptable speed. Various recent models have exemplified their efficiency [74–76].

BUILDING A QSAR MODEL

The Main Steps

The principal phases in building a QSAR model (Figure 1), that is, establishing a mathematical or statistical relationship that links the biological activity to a description of the molecular structure

$$\text{Biological activity } y = F \text{ (molecular descriptors)}$$

can be summarized as follows:

- *Constitution of the data set*, and splitting it into a training ("learning") set that will be used to adjust the model, and a test set, which the model has never seen and that will be used subsequently to check its predictive capacity. The training set must span, as widely as possible, the structural space with a rather limited number of compounds. The test set must correspond to a "reasonable" extrapolation.
- *Generation of the descriptors* characterizing the molecular structures under scrutiny. The nature and the number of the descriptors depend on the

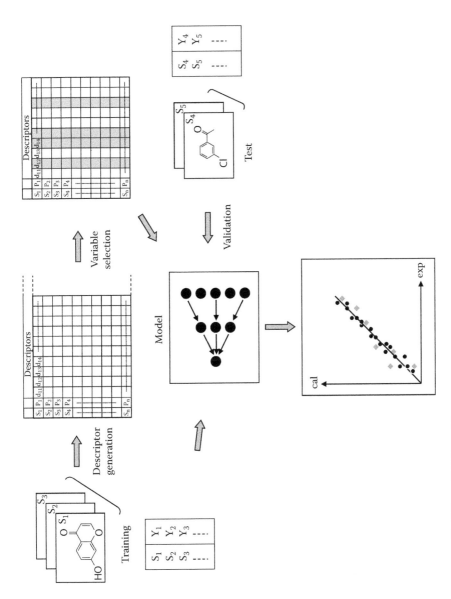

FIGURE 1 Main steps of building a QSAR model.

biological process investigated, the dimensionality (2D, 2.5D, 3D), and the
type of model chosen (atom or group parameters, molecular field-, shape- or
surface-analysis).

- *Development of the model* and evaluation of its quality. This is generally
carried out by examining the statistical criteria of the correlation (calculated
activity vs. experimental values). To avoid overfitting, noise, or "chance
correlation" [77], this step often requires a selection among the numerous
descriptors now available.
- *Validation*: examination of the predictive capacity of the model on the
test set.

After successful validation, the model may be considered as valuable for the prediction of unknowns.

THE MATHEMATICAL OR STATISTICAL MODELS

The continuous advancement of QSAR models was only possible thanks to the introduction of new statistical tools in data analysis, the development of mathematical
methods to represent the chemical structures and treat the information they carry,
and, for 3D applications, the increased resources in computer graphics. Generally
speaking, these models are not specific to certain types of descriptors but rather
determined by the number of variables under consideration, and the linear (or not)
behavior of the response (the biological activity or the physicochemical property
looked for).

The multilinear regression (MLR) that was more or less the default option at the
beginning was completed with principal component regression (PCR) and partial
least squares (PLS) regression to cope with the use of numerous structural descriptors (often many more descriptors than experimental data) [78].

The linear models were then completed with nonlinear approaches. Quadratic
terms (such as the π^2 term in the Hansch relationship) and, more generally, two- or
threefold cross-products were also used in QSAR/QSPR to cope with nonadditive
behaviors due to substituent–substituent interactions [79–81] (see also McFarland
[82]). The DARC correlations, on the other hand, sometimes also involved additive
interaction terms (due to the simultaneous occupation of two sites) [34]. However,
such methods need to know the interaction terms to introduce and fix their expression.

A significant turning point was the introduction of less subjective, nondeterministic
approaches with artificial neural networks (ANN), on the one hand, and genetic algorithms (GA) and evolutionary programming (EP), on the other. Some applications may
be found in the books by J. Devillers [83,84] and in the papers by Luke [85,86].

Artificial neural networks (ANN) were largely used in SAR/QSAR for correlating
activities (or properties) to structural descriptors or categorizing compounds (e.g., actives
vs. inactives). Although this approach is not fully deterministic, ANN offer an attractive,
model-free tool that does not prejudge any physical or biological underlying mechanism
and can accept noisy or incomplete data. These devices are also capable of capturing in
the raw data complex and nonlinear relationships that were often missed by more conventional QSAR approaches [87–91]. Different architectures of ANN were elaborated.

Back propagation neural networks (BNN), among the first proposed, are layered networks generally working with an algorithm of back propagation of the error. As stated by Livingstone and Manallack [92], BNN rapidly produced remarkable results (see, e.g., the works of Aoyama et al. and Andrea and Kallayeh [93,94]) and led to many successful applications. However, settings must be carefully defined, particularly in view of avoiding overfitting or overtraining.

Self-organizing maps (Kohonen maps), largely used as unsupervised classification tools, were also introduced to organize the molecular information derived from points scattered on the molecular surface in the CoMSA method of Polanski et al. [95].

Other network architectures progressively appeared such as radial basis function neural networks (RBFNN), which stood out due to their efficiency and easy implementation [96,97], general regression neural networks (GRNN) [98,99], or probabilistic neural networks (PNN) [90]. More recently, support vector machine (SVM) [100,101] that works in a higher dimension space, where data are projected via a kernel function, was at the basis of several works. This method avoids the local-minima problems, is stable, and is able to build complex models without suffering the "curse of dimensionality" in high-dimension problems. Various non-linear methods (such as projection pursuit regression [PPR]) [102,103] received more limited attention.

Until recently, these methods were mainly used with classical (2D) or 2.5D descriptors, but they can now be used with 3D descriptors: SVM, for example, was able to replace the usual PLS treatment in CoMFA [76].

VARIABLE SELECTION

During the initial stages of QSAR/QSPR studies, the number of available descriptors (the independent variables in the correlation) was somewhat limited, and an intuitive choice was easily made by the user, helped by some presumption about a probable mechanism of action. With the incredible inflation in the number of descriptors now available (about 3000 entries in Ref. [37]), the selection of relevant descriptors became a problem. Among the wide variety of descriptors, some are specifically relevant for studying a particular type of property, but may be irrelevant for other applications; their use can actually be detrimental (unnecessarily complicating the calculation, possibly generating noise, or making interpretation more difficult). Using too many descriptors may also lead to chance correlation, a risk indicated as early as 1972 [77], or create overfitting: When the number of features is large with respect to the number of training samples, a satisfactory model may be easily found for training data but it may perform poorly for a test set [104].

Unfortunately, there is no universal best set of descriptors for any application (*no free lunch* theorem). Various filtering techniques were thus proposed (for a summary, see Ref. [101]). We indicate here, on the one hand, PLS analysis (which carried out some kind of descriptor clustering), the development of which probably allowed for the wide diffusion of CoMFA [38], and on the other hand, evolutionary programming and genetic algorithms [105–107]. These methods proved their efficiency in proposing a valuable optimal model, or, better, a small set of near-optimal solutions in a consensus approach (they can be viewed as different facets of a same reality) [108,109]. Iterative variable elimination was also introduced by Polanski [110,111]. For other methods of variable selection, see [101,112].

MODEL VALIDATION

Finally, it should be mentioned that for a safe prediction, a model must be carefully validated. It is now proven that the correlation coefficient q^2 obtained in leave-one-out cross-validation is indicative of internal consistency (and robust interpolation). But it is not an unfailing criterion of good prediction for unknowns [113,114]. Similarly, s_{pred} values are not a reliable measure of predictive capability [1,115]. New controls were generalized with the use of external validation sets, bootstrapping, or Y scrambling [116]. A rational choice of the training set was also a subject of consideration.

Some Common Sense Remarks

To conclude this introduction, a few points (although obvious) can be reiterated as follows:

- However sophisticated and efficient mathematical or statistical tools may be, a model cannot be better than the weaker of the constituents it was built from, in this instance, the data set, the molecule representation (descriptors), and the computational tool.
- Statistical methods may differ in their ability to establish a model. Similarly, some descriptors are better than others on a given chemical family.
- Biological data are, by their very nature, soft data. Precision on an LD_{50} (lethal dose for 50% of a sample), for example, is by far inferior to that obtained on the wave number of a spectral ray. Thus, it would be a waste of time expecting a prediction with an accuracy of three decimals if the experimental data are only reliable to the first one.
- Prediction, whatever be the field (hazard assessment, risk management, virtual screening, synthesis planning), has value if and only if one has confidence in it. Any predictive model requires a careful preliminary examination to check its robustness and its applicability area (including domain extrapolation). In other words, models derived from a (limited) learning set of known compounds must only be used in reasonable extrapolation. Caution is warranted in interpreting QSAR results for chemical classes that are not well represented in the training set [117,118]. The training set should cover the structural space, with a minimum of patterns, equally shared. However, it was indicated that some redundancy must exist to cope with possible single point errors [119].

Outliers (predictions significantly deviating from the experimental values) are often ignored, which tells nothing more, or discarded, which only gives a better correlation coefficient, but no more. In fact, outliers convey some information. If the experimental datum is right (which must be checked first) and if the descriptors do not show values very different from the remainder of the pool ("exotic" compound or too large extrapolation), it means that the model ignores, for this point, a supplementary inter-action mechanism, which perhaps is worth investigating in detail (see also Ref. [120]).

- Translation of in vitro results to in vivo conditions is not always obvious. Thus, vinclozolin, which induces antiandrogenic effects in vivo, does not bind AR (androgen receptor) in vitro, its action being due to its metab-olites M1 and M2 [121]. In contrast, in the search for SARMs (selective androgen receptor modulators), newly synthesized molecules, mimicking known actives and looking promising due to their binding affinity in vitro, are ineffective in vivo because of rapid metabolization [122]. Desolvation energy differences and H-bond properties may also complicate the com-parison [123].
- Existence of a good relationship between (say) two variables is not a guar-antee that some causality relationship exists between them. This was the subject of numerous jokes. For example, in eastern France, there is a close relationship between the number of births and the number of stork nests, which is obviously a spurious relationship, since everybody knows that boys are born in cabbages and girls in roses (look, however, for a common factor, the number of houses and of chimneys).
- Practitioners have now at their disposal highly sophisticated and efficient programs. For the most usual cases, the default values often give good results but not automatically the best ones. It is better trying to optimize settings. Some applications require more care. Energy minimization may get trapped in a local minimum, convergence of an MD simulation must be checked, free energy perturbation methods require a careful examination of hysteresis or drift, choice of the training and test sets may be well consid-ered, etc. And, of course, if it is pleasant to have at one's disposal plenty of descriptors, it is wise to use as few as possible and to prefer those that have a clear physicochemical meaning.

On this point, in Europe, the REACH legislation stipulates some requirements that reliable QSAR models must meet: definite end point, unambiguous algorithm, defined domain of applicability, appropriate measure of goodness-of-fit, robustness and predictivity, and, if possible, mechanistic interpretation. They are not so far from the "good practice in QSAR" defined by Unger and Hansch as early as 1973 [124], quoted from Kubinyi [1]: "Select independent variables, justify the choice by sta-tistical procedures, apply the principle of parcimony (Occam's razor), have a large number of objects as compared to the number of variables, try to find a qualitative model of physico-chemical or biochemical significance."

Among the diverse possible organizations on account of similarity between the proposed methods, we have divided the book into four main parts. Part I is dedicated to field-based methods; CoMFA and related methods such as CoMSIA, FLUFF, SOMFA; and to shape-, surface-, or volume-based approaches (MSA, excluded vol-ume, LIV, HASL, receptor surface model, COMPASS, and CoMSA).

Part II treats methods using, but not explicitly, 3D information. This part covers autocorrelation methods (GRIND), similarity-based methods (similarity matrices, quantum similarity indices), and quantitative spectroscopic data–activity relationships (QSDARs). Some applications in data mining are also discussed.

Part III deals with post-3D models: adaptation of the receptor and simultaneous presence of several conformers or of several solvation mechanisms.

Part IV presents some "receptor-related" approaches, docking and free energy calculations, which are treated at various levels, extensive sampling of the phase space or approximate methods (linear interaction energy, Poisson–Boltzmann, or generalized Born models). A case study covering several parallel approaches is then developed.

After some concluding remarks, Appendix A provides some mathematical details on the approaches that have been discussed. Appendix B presents data on the steroid benchmark.

Part I

Actual 3D Models

There are various ways to incorporate geometrical information, with more or less details, into QSAR models. The most direct ones calculate values of physicochemical quantities, selected as descriptors of the molecular structure, on discrete positions, the location of which is specified and used to give a data structure for the subsequent statistical analysis.

In the most popular CoMFA [38] and CoMSIA [40] approaches, structural information is provided by the fields a molecule exerts in its neighborhood, whereas receptor surface models (RSM and GERM) consider points scattered on the molecular surface or a substitute lining it for the whole population. Shape analysis [41] or hypothetical active site lattice [42] methods focus interest on elementary elements interior to the molecular volume. At last, search for a common geometrical arrangement of selected atoms or "structural features" is the basis of methods like CoMASA [43] and pharmacophore-based approaches. Since the position of the probes is important in processing data, these methods (generally) imply alignment of the molecules on a common template, which may lead to some difficulties for non-congeneric series of flexible chemicals.

1 Comparative Molecular Field Analysis

No doubt, comparative molecular field analysis (CoMFA) [38] is the most popular and widely used 3D-QSAR method. After the limited success of dynamic lattice-oriented molecular modeling system (DYLOMMS) [39], CoMFA could be widely spread, thanks to the introduction of an efficient statistical tool, partial least squares (PLS) regression, and the increased opportunities to handle molecular structures on computers, giving birth to a turnkey package included in SYBYL and provided by TRIPOS [125]. More than 20 years after its inception in 1988, CoMFA (and the neighboring approach comparative molecular similarity analysis [CoMSIA]) [40] remains a reference to which nearly all newly proposed methods are compared.

CoMFA methodology and salient results have been the subject of extensive reviews and books [119,126–140]. An extensive bibliography may also be found in some theses [141,142]. So, we will briefly present the principle of the method, the major improvements proposed along years, and we will only put stress on the most recent developments. In this chapter, we will not focus too much attention to traditional applications since numerous examples will be found in other chapters where varied 3D methods are presented and their performance compared to that of CoMFA.

1.1 CoMFA: BASIC STEPS AND CAVEATS

The basic premise, common to QSAR models, is that the biological response (in most cases) reflects non-covalent interactions between a receptor and the ligand that will bind to it. It is assumed that all molecules interact with the receptor at the same binding site in the same (or similar) manner. However, Lukacova and Balaz considered the possibility of competitive binding mechanisms [62]. Variations in these interactions are related to the observed variations in the ligand affinity. CoMFA assumes that non-covalent forces dominate these interactions and can be modeled by the fields (electronic and steric) created by the ligand in its vicinity. These fields will be evaluated on regularly spaced nodes of a 3D lattice in which molecules are embedded. So, differences in activity may be correlated with differences in these non-covalent fields. This contrasts with "classical" QSARs, where structural descriptors, encoding relevant features of the molecules under scrutiny, generally correspond to characteristics (local or global) directly attached to the molecular framework such as heat of formation, molecular surface area or volume, atomic charges.

3

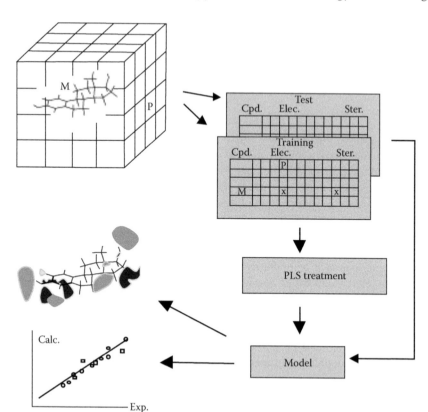

FIGURE 1.1   Main steps of CoMFA method.

The successive steps in a CoMFA analysis are (Figure 1.1)

- *Alignment*: Molecules are superimposed on a common template in a 3D lattice surrounding the molecules.
- *Field calculation*: Interaction energies (steric and electrostatic) are calculated locating a "probe atom" on the nodes of this 3D lattice.
- *Correlation with activity*: Interaction energies are correlated with values of the studied property, using PLS method [78]. Variants and nonlinear methods have been also used (see Refs. [143–145]).
- *Contour plots*: From PLS coefficients, areas may be delineated, in the space surrounding all molecules, where an increase of steric potential (due to increased bulk in the vicinity), respectively, an increase of positive electrostatic potential, contributes to increase (or decrease) binding affinity.

The CoMFA approach was first presented by Cramer [38] on the example of binding affinity of a series of steroids for corticosteroid-binding globulins (CBG) and testosterone-binding globulins (TBG). The structural formulae and activity of these molecules are given in Appendix B: Steroids. Subsequently, this data set was quasi-systematically used, and accepted as a benchmark, to evaluate new proposals of 3D-QSAR models. However, as stressed by Coats [126] and Kubinyi [138], this might not be the best choice. A significant number of erroneous structures were included in this first analysis. They have been now corrected by Gasteiger et al. [146]. With corrected values, Coats [126] led to $q^2 = 0.734$ and in prediction, $r_{ext}^2 = 0.13$, but discarding compound **s-31** (the only one to bear a fluorine atom in position-9), $r_{ext}^2 = 0.71$, with three latent variables (LV). Steroid **s-31** also appeared as an outlier in most of the 3D-QSAR models. Slightly better results have been obtained with optimized settings (vide infra). We specify that, systematically in this text, q^2 refers to the determination coefficient obtained in leave-one-out cross-validation (L-O-O) on the training set, and r_{pred}^2, r_{ext}^2 to the determination coefficient in prediction on the test set (vide infra).

It was also remarked that the reactivity range is somewhat limited (2.9 log units for log K). Furthermore, the splitting into training (compounds **s-1** through **s-21**) and test (compounds **s-22** through **s-31**) may be criticized since it did not correctly span the structural space. Indeed, this training set did not include some specific groups present in the test (2-α Me, 9-α Fluorine, 16-α Me, or 21-acetoxy). A modified choice was then proposed: in training compounds **s-1** through **s-12**, plus **s-23** through **s-31** and in test **s-13** through **s-22**. With this set, $q^2 = 0.713$, $s = 0.50$ and in test $r_{pred}^2 = 0.784$, $s = 0.63$ [115].

CoMFA led to countless successful applications; however, to get significant results, some precautions must be taken at each step. We prefer here to start with the first (default) settings of CoMFA before examining possible variants and discussing the choice for optimal settings.

1.1.1 ALIGNMENT AND CHOICE OF AN ACTIVE CONFORMATION

No doubt, alignment, sometimes time consuming and requiring some chemical and biological knowledge, is the most crucial part of the treatment and often a real bottleneck. Actually, this alignment step is also necessary to several other QSAR approaches, see Chapters 2 and 3. And, on account of these difficulties, efforts were made to get rid of this problem with translation rotation invariant (TRI) models or descriptors, see Chapters 3 and 4.

This problem of alignment: superimposition of the investigated molecules in their active conformation in view of deriving QSAR models to estimate binding affinities is a part of a more general concern. Indeed, structural alignment of ligand molecules, or substructure overlap, has been the method of choice for structure comparison and analysis of molecular similarity/diversity. Evaluation of structural similarity is, for years, a common (and largely used) tool in high-throughput virtual screening of chemical databases. In addition to the derivation of QSAR models, it also intervenes, for example, in pharmacophore elucidation (extracting a small group of atoms or the key structural features necessary for a molecule to be recognized by a specific receptor and so present in active compounds) or in receptor modeling to get some insight into the specific process of binding. Note that these three facets rely on common problems quite present in database virtual screening.

Turning now to the specific problem of alignment in CoMFA approach, it is clear that this step may seem rather straightforward when treating congeneric and rather

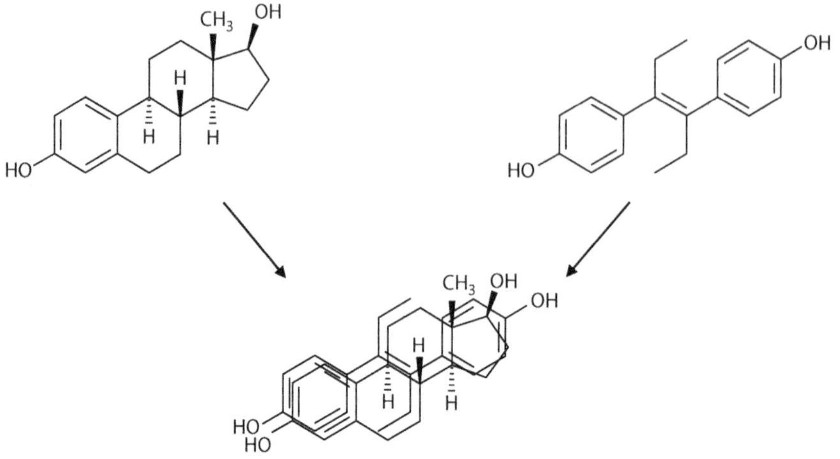

FIGURE 1.2   Superimposition of estradiol and diethyl stilbestrol.

rigid molecules; as, for example, the steroid set, largely used to develop and test 3D-QSARs [38]. A molecule, generally among the most active ones (since it probably corresponds to optimal interactions) is selected as a reference template, and a rigid part, common to all other molecules (if possible), is chosen as a "seed" (Figure 1.2). When the data set encompasses different chemical subfamilies (e.g., steroids and more flexible molecules), it may be useful to choose some "secondary references": in each subfamily, a lead is first aligned with the template and constitutes itself a template for its own group.

In fact, this problem encompasses two facets:

- Establishing the correspondences between the atoms of the seed and some atoms of the studied molecules
- Determining the best geometrical transformation (translation and rotations) to get the best superimposition

In the easiest cases (series of congeneric structures), the first step is straightforward, but things may be harder for other data sets. The simplest way to achieve superimposition in a homogeneous series of compounds (when correspondences are unequivocal) is clearly a rigid atom–atom fit procedure. The quality of the fit is determined by the root mean square deviation (RMSD) on atom locations (after superimposition). This superimposition involves 6 degrees of freedom (3 for translation, 3 for rotation), but a difficulty arises from the possibility of multiple minima giving suboptimal solutions. Since the pioneering work of Kabsch [147], a lot of approaches have been proposed for such distance-based atom–atom superimpositions. Several of them would be a priori usable in CoMFA analyses. Nevertheless, most of the current applications of CoMFA refer to the SYBYL routines directly provided by TRIPOS. For an extensive review, see Lemmen and Langauer [148]. We just go into few details for some commonly used methods.

1.1.1.1 Multifit

However, ligand flexibility is often an important aspect to be taken into account. In the "multifit" procedure allowing for ligand flexibility, a supplementary "spring" is added to the ligand force field in order to force the corresponding atoms to get closer. Bhongade and Gadad [149] examined indole/benzoimidazole-5-carboxamidines as inhibitors of urokinase-type plasminogen activator. This enzyme is largely involved in tumor metastasis and increased interest focused on its inhibitors as possible therapeutic agents in the treatment of cancers.

Different strategies were compared for aligning the molecules under scrutiny on the selected template:

- Atom-based RMS fitting
- Shape-based RMS fitting (on centroids of rings or groups)
- Flexible fitting (multifit) of atoms
- Database RMS fitting to the template

For the investigated data set (training 28 molecules, test 7), the four options gave nearly similar results (q^2 from 0.615 to 0.579, r^2_{pred} from 0.616 to 0.555); the best ones being obtained for the first choice.

1.1.1.2 Field-Fit Approach

Since the basic descriptors used in CoMFA are the field values in the space surrounding the molecules and not the positions of the atoms, a field-based alignment strategy has been proposed. Rather than optimizing the location of atoms, the alignment may be refined in a "field-fit" minimization searching for the best similarity between the fields generated by the template and by the molecules to be aligned.

The differences between the field values (steric or electrostatic) assigned to each node of the lattice are summed up in the objective function to minimize. Weights may be given to put more importance to certain nodes or to eliminate nodes where the fields are weak and meaningless [150]. In the energy minimization process (usually by molecular mechanics), additional penalty terms are added to the force field to reflect the degree of similarity of the steric and electrostatic fields (between template and fitted molecule). Internal coordinates of the fitted molecule are adjusted to ensure optimal field and geometric overlap with the template. This causes an increase of its potential energy and possibly some structural distortions. The fitted molecule is therefore reenergy minimized, without the "field-fit" penalties, to relax to the nearest local minimum-energy structure. It was quoted that such a field-fit alignment may avoid multiple possible solutions due to symmetry or presence of multiple rings [121].

1.1.1.3 Steric Electrostatic Alignment

In another type of approach, steric and electrostatic alignment (SEAL) [151] uses a similarity score to provide an automated method for a rigid body alignment. For each tested molecule, numerous randomly generated starting orientations are tried, with definition of a similarity index with the template, in order to maximize electrostatic and steric overlap. The formalism of SEAL was further used in CoMSIA [40], a parent approach to CoMFA, for the calculation of probe–ligand interactions. It intervenes also in the derivation of similarity matrices (see Kubinyi [115]) and in comparative molecular active site analysis (CoMASA) [43].

The similarity score of SEAL is a sum of a set of atomic terms:

$$A_F = -\sum_{i=1,n}\sum_{j=1,m}(w_e q_i q_j + w_s v_i v_j)\exp(-\alpha r_{ij}^2)$$

where

the summation is running over the atoms i of the first molecule and j of the second one, distant of r_{ij}

q represents the charge

v is a function (generally cubic) of the atomic van der Waals radius

α is an attenuation factor (usually 0.2)

w_e, w_s are user-defined parameters to weight the influence of electrostatic and steric effects

The alignment is obtained via random translations and rotations of one of the structures with respect to the other and minimization of the alignment function. The SEAL methodology provided an objective way to ensure alignment, free from user's subjectivity.

But, curiously, in other applications regarding the same steroid benchmark, an anecdotal remark [115] was that SEAL proposed, for certain pairwise comparisons (e.g., 3-keto, 17-hydroxy and 3-hydroxy, 17-keto), a head-to-tail superposition, with identical functional groups together, rather than superimposing the skeletons. But this induced small variations in the results!

A flexible variant TORSEAL was further proposed with random perturbations and force-field minimization step but an initial assumption for the alignment was still necessary [152]. The MultiSEAL system proposed by Feher and Schmidt [153] carries out a fully automated and systematic conformational overlay for multiple molecules, devoid of the need for initial assumptions. The basic idea is that aligning molecules by pairwise comparisons may not be optimal, and that it is better to examine the overall overlay.

From the similarity score of SEAL, it is easily established that, for a simultaneous alignment of (say) three molecules A, B, C, the total overlay score is simply the sum of all the pairwise scores (A/B, A/C, B/C), and so on. The total score characterizes the overall overlay; individual scores describe the quality of the fit for a given pair. The process can treat multiple conformations. At the beginning, a conformational analysis is performed. In this application, random incremental pulse search (RIPS) of Ferguson et al. [154] was used. The structures were optimized and conformers in a predefined energy range were selected. To limit the number of conformers, they were clustered and a representative of each cluster (that of minimal energy) was selected. The best scoring overlays (say 200) for each step were kept to avoid biasing end results in the process development.

Although potentially applicable in QSAR, only qualitative applications were given. The method was tested on various examples: two angiotensin II receptor antagonists, four 5-HT3 receptor antagonists, and four flexible dopaminergic compounds. The approach was also used to rationalize the activity of highly flexible colchicine analogues, putative inhibitors of tubulin polymerization. It was suggested that the weak antimitotic activity of 2-methoxyestradiol corresponded to an imperfect fit with the active site. One interest of the method is that all conformations of all molecules can be considered simultaneously to obtain all possible binding modes and mutual orientations. On the quoted examples, results competed well with those given by the genetic algorithm (GA) approach of Jones et al. [155].

1.1.1.4 Property Density

An approach somewhat similar to that of SEAL was proposed by Labute et al. [156], on the basis of property densities represented by Gaussian functions. The method is objective and produces a series of alignments based on a scoring function, sum of a similarity function, and an internal energy term (to filter high-energy conformers). It requires neither predefined fit-centers nor pre-orientation of the molecules.

For n points, the property density is

$$f_p(\mathbf{x_1}, \mathbf{x_2}, \ldots, \mathbf{x_n}) = \sum_{i=1,n} \left(\frac{w_i}{n} \right) \left[\frac{a^2}{(2\pi r_i^2)} \right]^{1.5} \exp\left\{ -\left(\frac{a^2}{(2r_i^2)} \right)(\mathbf{x} - \mathbf{x_i})^2 \right\}$$

where
   $\mathbf{x_i}$ denotes the position of atom i
   w_i is the degree to which atom i has the property P
   a and r are tunable parameters

Various properties were considered: volume (in fact, presence of an atom), aromatic character, donor-type, acceptor, or hydrophobic (with binary values). Log P, molar refractivity, and exposed surface were also introduced but they did not improve the results. The similarity function between two molecules is computed as the overlap of their property densities (i.e., also a Gaussian function).

The search is carried out with RIPS [154]. Three atoms are randomly chosen for superposition, and the score function, sum of the similarity function f, and the potential energy U (to filter high-energy conformers) is minimized:

$$F = -kT \, \text{Log} \, f + U$$

Rather than the high-scoring cases, only the best solution is here retained. The method was validated by comparison of the proposed alignment and crystal structures of protein ligand complexes. See the example of dihydrofolic acid (**1.1**) and methotrexate (**1.2**), where, in particular, the X-ray alignment of the cycles is retrieved rather than the "intuitive" alignment [157] (Figures 1.3 and 1.4).

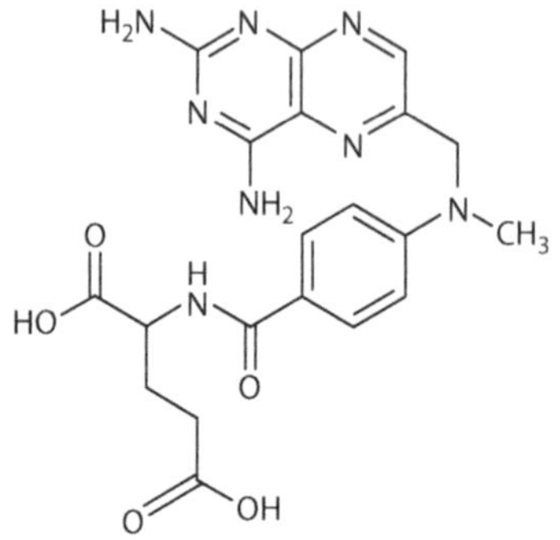

1.1

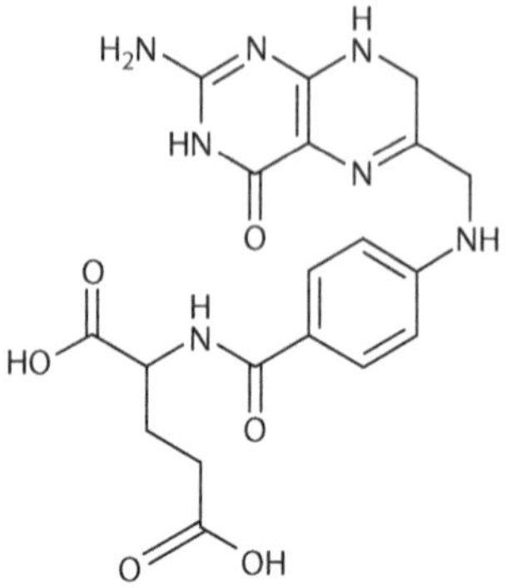

1.2

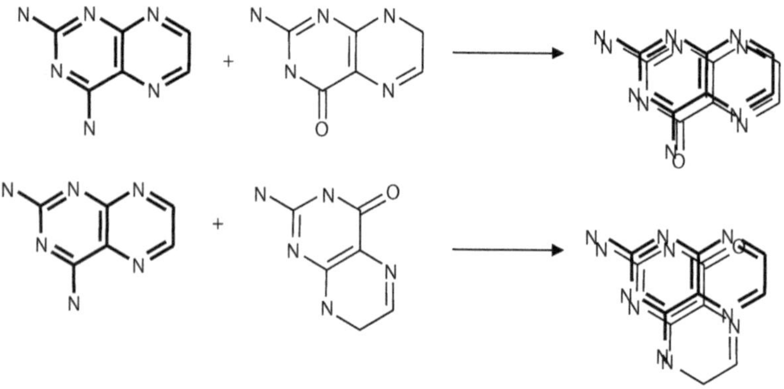

FIGURE 1.3   The intuitive superimposition of the pteridine part differs from the experimental X-ray alignment.

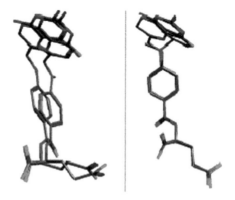

FIGURE 1.4   Overlay of methotrexate and dihydrofolic acid. (Left) Alignment in the crystal; (right) best calculated flexible alignment, obtained at the final set of parameters optimized for all ligand pairs (RMSD of 1.4 Å). The program retrieved the X-ray alignment. (Reproduced from Labute, P. et al., *J. Med. Chem.*, 44, 1483, 2001. With permission.)

However, the authors put some caveats as to limitations of alignment methods. For example, equilibrium geometries obtained by X-ray crystallography may differ from those calculated with a force field. Bad alignment may occur if large regions of a molecule have no equivalent in the other (missing information) or when binding in a spacious receptor pocket (the only actual constraint being that few corresponding groups interact with the same receptor sites whereas programs would tend to overlay all atoms).

A great deal of efforts was then made to propose new tools in order to improve the derived models, particularly for the treatment of non-congeneric series, where a common subset of atoms (the "seed") cannot be easily chosen. The methods also differ in the treatment of flexibility (rigid system or no) and in the search engine (systematic or stochastic conformational exploration), see Labute et al. [156] and references therein. This point will be addressed further. Fortunately, in favored cases, some additional structural information may be gained on the target protein or on a complex ligand–ligand-binding domain (LBD) (via X-ray crystallography, NMR, or homology modeling) providing efficient supplementary tools. For example, docking ligands into the receptor-binding domain is an interesting alternative to classical alignment, see Section 1.3.

1.1.2 FIELD CALCULATION

For field calculation, the probe atom mimics a sp^3 hybridized carbon (radius 1.53 Å, charge +1.0). 6–12 Lennard-Jones and Coulombic potentials are respectively used. Charges are generally derived from the Gasteiger–Marsili model relying on partial electronegativity equalization [158–161] or calculated by semiempirical quantum mechanics programs, such as AM1 or PM3 [162–163], with (often) a distance-dependent dielectric constant ($\varepsilon = r$). The lattice, with a mesh of (usually) 2 Å, extends at least 4 Å in all directions beyond the common volume of the superimposed molecules (nodes internal to the common volume are ignored). At short distances, these potentials may take very large values and a truncation threshold (generally ±30 kcal/mol) is fixed. Electrostatic interactions at "sterically forbidden" points (high steric energy) are ignored, and a mean value for these nodes is retained ("column dropping").

1.1.3 CORRELATION

For the correlation of field values with activity, interaction energies are gathered in a spreadsheet where rows correspond to individual molecules and columns to individual nodes (Figure 1.1). Since there are much more values of potentials (e.g., 1000 nodes × two fields) that constitute the independent variables ($\mathbf{X}$ matrix) than activity values (dependent variable, vector $\mathbf{y}$) for a data set of (say) 50 compounds, analysis is carried out using PLS method [78,164–166]. Basically, PLS extracts from the independent variables a reduced set of composite, abstract axes (linear combination of the original descriptor variables). The process is somewhat similar to the determination of the principal components of principal component analysis (PCA) for correlating the observed activities. In fact, these latent variables (LV) are slightly skewed from the principal components for maximum intercorrelation with the dependent variable $\mathbf{y}$ [78,119].

In such analyses, columns (nodes) with small variations of interaction energies may be discarded to reduce the computational task ("column filtering" makes the calculation faster by about 10 times) [121]. The quality of the model and the optimal number of components (trade-off between the precision and the complexity of the model) are characterized from the cross-validated determination coefficient q^2 in L-O-O process:

$$q^2 = 1 - \frac{\text{PRESS}}{\text{SSD}}$$

where

SSD is the sum of squared deviations between observed values and the mean of property values in the training set

PRESS is the sum of squared deviations between observed and calculated property values $= \Sigma \, (y_{obs} - y_{cal})^2$

The optimal number of LV (or principal components) to retain is determined by the first slight drop in the increase of q^2 or alternatively by the first increase of PRESS, with the model complexity. Introducing more LV may lead to overfitting.

The model is then applied to the whole set to get the conventional determination coefficient r^2. Schematically, r^2 measures the model's goodness of fit and its internal consistency, whereas q^2 evaluates its robustness and its ability to interpolate within the training set. $r^2 > 0.9$ and $q^2 > 0.5$ are usually considered as significant [167]. A negative q^2 would mean that a "no model" (taking the mean of the observations) works better than the proposed model.

Results generally mention SPRESS:

$$\text{SPRESS} = \left(\frac{\text{PRESS}}{n - a - 1} \right)^{0.5}$$

where

n is the number of compounds

a is the number of components in the model, which takes into account the complexity of the model

Otherwise, the standard error of estimate (SEE) and the (cross-validated) standard error of prediction (SDEP) are given:

$$\text{SEE} = \left(\frac{\text{PRESS}}{n} \right)^{0.5} \quad \text{and} \quad \text{SDEP} = \left(\frac{\text{PRESS}}{n} \right)^{0.5}$$

In fact, they rely more on root mean square error (RMSE) than on standard error(s). For prediction,

$$r^2_{pred} = 1 - \frac{\text{PRESS}}{\text{SSD}}$$

is calculated on the sum of squared differences between calculated and experimental values for objects (molecules) in the test set and SSD calculated on the mean of the *training set* objects alternately r^2_{ext} is the usual determination coefficient between observed and predicted values for the test set. Bootstrapping or y-scrambling may be used as other criteria of model validity. In bootstrapping, several random selections of training objects are carried out in order to get some insight about the stability of the model. In y-scrambling (randomization, shuffling) [48,116], the elements of the response vector are shuffled by about 100 random exchanges, and models are sought for with the reordered responses. If satisfactory correlations are obtained, the significance of the QSAR must be suspected: the method is potentially able to model any kind of data. Of course, the point is not to test the model obtained from the original data, but to carry out the entire selection procedure (i.e., entirely rebuild the model) from the permuted data. Scrambling also checks for the risk of chance correlation [77,168].

A variant of PLS, sample distance PLS (SAMPLS) [144] uses a kind of covariance matrix that provides a faster treatment (with equivalent results) particularly attractive in L-O-O cross-validation. Rather than an N*M matrix, SAMPLS works on a smaller N*N matrix (where N is the number of compounds, M is the number of descriptors).

It is important to remark that q^2, which was in the past largely invoked as criterion of quality (high q^2 indicating "good correlations"), is now considered more as a test of good internal interpolation within the training set rather than a sign of good predictivity for unknowns [113]. Predictive ability is better characterized by use of an external prediction test (not used for defining the model). Nevertheless, one advantage of q^2 is to be perfectly reproducible (and for this reason, largely used for comparing models) in contrast to r^2_{cv} obtained in leave-some-out processes with random splitting of the data set into train/test. However, a good estimate of r^2_{pred} can be gained from the mean value of r^2_{cv} averaged on multiple runs. A composite score index was also proposed by Schultz et al. in ecotoxicology studies [169]:

$$SCORE = 5r^2_{pred} + 4q^2 + 2r^2$$

Let us note that a nonlinear PLS method was proposed by Hasegawa et al. [145] in the modeling of phenyl alkylamines with the monoamine oxidase inhibitory activities, and later [170] in a study of inhibitors of *Escherichia coli* DHFR. A quadratic expression was used between the **X** and **y** scores in place of the classical linear dependence. Another variant consists in using the k-nearest-neighbor (k-NN) method to analyze the CoMFA fields [171]. This, in fact, helps in analyzing the similarities between these field values. This point is addressed in Chapter 5.

1.1.4 CONTOUR PLOTS

Besides the usual prediction of activity, one of the interesting points of CoMFA is provided by the contour plots. They delineate areas in the space surrounding aligned molecules where an increase of steric potential (bulk brought about by larger molecules in the training set), respectively an increase of positive electro-static potential contributes to increase (or decrease) inactivity. Schematically, the (very numerous) coefficients of the CoMFA treatment (giving the contribution of each lattice node to the PLS model) are used to define contour plots, correspond-ing to fixed values of the product "SD × Coefficients." For example, for the steric field, a positive value on a node indicates that an increase of the field felt on this node would correspond to an increased activity. An example of contour plots is given in Figure 1.5 for a series of analogues of BMS-806 (**1.3**) that inhibits the first steps of HIV-1 infection by blocking the binding of host cell CD4 with viral gp120 protein [172].

The main biological function of HIV-1 reverse transcriptase is to transcribe the HIV-1 RNA genome into double-stranded DNA, which is subsequently integrated in the host cell genome. Among the inhibitors of HIV-1 reverse transcriptase, nucleo-side inhibitors (NRTIs) bind the same site as the viral DNA, whereas non-nucleoside inhibitors (NNRTIs) bind in an allosteric site, which induces a conformational change interfering with viral DNA binding to HIV-1 RT. This makes NNRTIs par-ticularly attractive because their binding site is unique to the RT of HIV-1, and there-fore less adverse side effects may occur.

1.3

Contour plots may suggest what structural modifications are introduced in these regions to modify activity (e.g., more bulky, more positively charged group). Contour plots may be viewed as representing very roughly a complementary image of the receptor-binding domain. However, the limited amount of information they bring involves extreme caution in such an interpretation. They may even be consid-ered as a "self-fulfilling prophecy since totally dependent on alignment and com-position of training set" [173]. Contour plots are directly related to the changes in structures that lead to changes in activity. Absence of a region in a contour does not mean this region is unimportant; it only indicates it is constant in the data set [174]. Nevertheless, contour plots may give some indices as to the mechanisms interven-ing in binding and may also be used to inspect the necessary complementarity of bulk and charge for possible ligands, for example, to identify potentially active metabolites [121]. Pajeva and Wiese [175], from a study on simulated data, pointed out that contour plots (with the usual settings of 20% for negative or disfavored and 80% for positive or favored regions) are sensitive to the distribution of positive and negative field values, so that some care is needed in their analysis, see also Kroemer et al. [130].

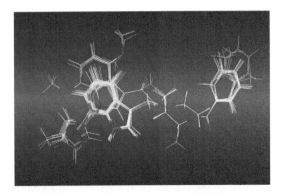

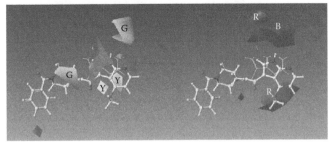

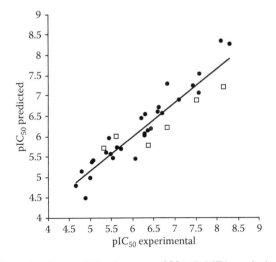

FIGURE 1.5 (See color insert following page 288.) CoMFA analysis for analogues of BMS-806. (Up) Alignment of 36 ligands' 3D structures; (middle) CoMFA steric (left) and electrostatic (right) contours. Green (G) areas indicate regions where bulky groups increase activity whereas yellow (Y) contours indicate where bulky groups decrease activity. Blue (B) areas represent regions where positive groups enhance activity, whereas red (R) contours indicate where negative groups increase activity. BMS-806 compound is shown with ball and stick representation. Its analogue, which possesses a sulfonamide instead of a ketoamide group, is displayed in wireframe. (Down) Plots of predicted vs. observed pIC_{50} of CoMFA model; square points represent the compounds of the test set. (Reproduced from Teixeira, C. et al., *Eur. J. Med. Chem.*, 44, 3524, 2009. With permission.)

1.1.5 GRID: A Precursor, Receptor-Oriented Approach

Few years before CoMFA, a comparable methodology was proposed by Goodford [58] for determining the energetically favorable binding sites to a target macromolecule. Interactions are calculated on a grid throughout and around the protein. The approach has been later extended to small molecule targets and became an alternative method to CoMFA, but it seems, however, less frequently used [176].

The interaction energy on a node is the sum of pairwise interactions between the probe and a single "extended atom" of the protein (or the target molecule). GRID considered a variety of probes: amino (NH_3^+), carbonyl oxygen, carboxy oxygen (O^-), hydroxyl, methyl, and water. Extended atom is taken here in the sense of molecular mechanics: for example, an NH_3^+ group is treated as a single entity with an adapted van der Waals radius of $1.75\,\text{Å}$. In contrast to CoMFA, the different contributions to E_{int} are summed up and not considered separately:

$$E_{int} = \sum E_{lj} + \sum E_{el} + \sum E_{hb}$$

summation running over all extended atoms of the protein (or molecule). The first term corresponds to a 6-12 Lennard-Jones potential. The electrostatic contribution E_{el} is expressed by

$$E_{el} = \frac{pq}{K\zeta}\left\{ \frac{1}{d} + \frac{\left[\frac{(\zeta-\varepsilon)}{(\zeta+\varepsilon)}\right]}{[d^2 + 4s_p s_q]^{0.5}} \right\}$$

where p, q are the electrostatic charges on the probe group and the pairwise protein atom, at distance d. K covers geometrical factors and numerical constants. The term $s_p s_q$, associated to the depth of the probe and atom in the case of probe–protein interaction is set to zero for small target molecules. ε and ζ are, respectively, the dielectric constants of the solution (80) and of the protein (4).

The last term characterizes H-bonding according to

$$E_{hb} = \left[\frac{C}{d^6} - \frac{D}{d^4}\right] \cos^m \theta$$

where
  θ is the angle protein donor …H… acceptor probe atom
  C and D are taken from tables
  m is assigned a value of 4

However, E_{hb} is set to zero if $\theta \leq 90°$. Note that if a hydrogen bond exists (favorable interaction), the corresponding Lennard-Jones repulsion is not considered. If the probe atom is the donor, $\cos\theta$ is set to unity (it is assumed that the probe can orient itself for the most favorable interaction). The corresponding probe is a "neutral" H_2O molecule (radius 1.7 Å) with no dipole, but able to donate up to two hydrogen bonds and also accept up to two. This field is calculated on each lattice node. For each probe, contour surfaces corresponding to energy levels can be displayed (negative values corresponding to regions of attraction between probe and protein).

Some original features of GRID (several probes, directional H-bonds) will be taken up in further improvements of CoMFA. GRID could be used alone (often coupled with the variable-selection program generating optimal linear PLS estimations (GOLPE) [177], but it was also at the origin of two efficient approaches for building QSAR models: GRid INdependent Descriptors (GRIND) [45] and VolSurf [178,179]. GRIND, which characterizes pairs of nodes of favored interactions around the molecules, has the advantage of alignment independence and allows for back-projection onto the original structures (see Chapter 4). VolSurf derives 2D numerical descriptors from the 3D information of GRID and relies on 2.5 approaches (see Chapter 7).

GRID/GOLPE was used, for example, by Ragno et al. [180] on an enlarged heterogeneous set of 103 histone deacetylase inhibitors. Models developed for four congeneric series proved to be highly predictive (outperforming previous 3D-QSARs), and contour maps showed good agreement with the structural features of the binding site.

1.2　IMPROVEMENTS OF CoMFA

About 10 years after the first applications, various reviews examined some proposals for improving performance [127,130]. More recently, Melville and Hirst [181] and Peterson [141,182] carried out in-depth studies of the influence of the parameters proposed in the various possible options of SYBYL and of their interrelations at the performance level.

We just rapidly summarize some proposed modifications.

1.2.1　VARIABLE SELECTION

Calculation of field values on the lattice nodes leads to a huge number of variables compared to the generally smaller number of activity values obtained for about 50–100 compounds. PLS treatment can cope with such data arrays, with the calculation of few significant LV, linear combination of the original ones. But one can hope for better results by elimination of variables bearing nonsignificant or spurious information, not directly related to biological activity.

1.2.1.1　GOLPE

GOLPE [177] aims to select the most informative variables from the data matrix, after a first usual preprocessing that eliminates columns of constant values and those where the SD is too small. Then, a first selection is carried out from the weights of the different variables in a full PLS treatment. Only the most important variables (nodes) are retained, in successive iterations: say about 200 from the 2000 initial node values.

A fractional factorial design (FFD) protocol is then carried out to improve the predictability of the model. A large number of PLS models (more than twice the number of the remaining variables) are carried out, where variables are included or excluded alternately. The performance of these models is evaluated on SDEP in L-O-O. A design matrix is built: rows represent the experiments and columns represent the different variables, encoded in each cell by zero or one depending on the variable being included or excluded in that experiment. The Yates' algorithm [183] estimates the importance of each individual variable on the predictability. Variables can be classified as favorable, uncertain, or detrimental (for model predictability). Some dummy variables are incorporated in the calculation in order to fix a significance threshold. Detrimental variables are discarded for establishing the final model.

1.2.1.2   Region Focusing: q^2-Guided Region Selection

On another hand, it was shown [184] that the quality of the CoMFA models might heavily depend on the overall orientation of the superimposed molecules in the lattice and could vary by as much as 0.5 units on q^2. To get rid of this problem, Cho and Tropsha [184] proposed a "q^2-guided region selection (q^2GRS)." The basic idea was to eliminate regions where fields did not correlate well with changes in the activity. After a conventional CoMFA, the lattice embedding the aligned molecules was divided into 125 small regions where CoMFA was carried out with a smaller mesh (e.g., 1 Å vs. 2 Å for the usual treatment). Only regions where q^2 (from these analyses) were greater than a defined threshold were selected to create a master region file used to perform the final CoMFA treatment.

Binding of nonsteroidal ligands to estrogen receptor (ER) was studied by Sadler et al. [185] on 30 chemicals (such as DES metabolites and indenestrol analogues), with binding affinity measured on mouse cytosol uterus. Various alignments were tested based on the oxygen atoms in 3- and 17-positions of estradiol and either the centroids of rings A and D or carbons of ring A (the best model), with field-fit or not. With region focusing, the best model led to $q^2 = 0.796$ with three principal components, compared to 0.720 (two PC) with traditional CoMFA.

1.2.1.3   All-Orientation Search and All-Placement Search

Another solution is the "all-orientation search and all-placement search" of Wang et al. [186]: the molecular aggregate was rotated or translated systematically within the grid. A CoMFA was performed for each orientation or placement, and the best (highest q^2) was retained, see, for example, Chen et al. [187].

However, in the study of Shi et al. [188] on estrogen activity, neither region focusing nor the "all orientation, all placement search" of Wang et al. [186] or introduction of log P gave significant improvement. The study encompassed 130 chemicals covering a wide range of structural diversity and six orders of magnitude spread of RBAs. On another hand, adding a "phenol indicator" raised q^2 and r^2 to 0.707 and 0.903 (with no significant change on SEE), respectively, an improvement consistent with the well-known importance of H-bonding from C-3 in the steroid family.

1.2.1.4   Smart Region Definition (RD Algorithm)

The treatment of Cho and Tropsha selected regions where the variations of the fields are the most significant. With the same concern, a slightly different solution was proposed by Pastor et al. [189,190] with the "smart region definition" in the neighboring GRID/GOLPE approach.

The process aims to evaluate the effects of regions of variables rather than individual variables. These regions correspond to gathering grid points (variables) containing the same chemical information but spread on several contiguous, although isolated independent variables (column in the field matrix). The treatment adds some continuity constraints preexisting in the 3D lattice but lost on unfolding the matrix in the mathematical PLS treatment. Calculations (with these new, fewer, variables) are made faster and interpretation becomes easier.

The approach encompasses three steps (Figure 1.6):

- Detection of the most informative variables in the initial PLS treatment and selection as "seed" nodes
- Building around the seeds Voronoi polyhedra containing neighboring variables in the 3D space
- Merging polyhedra containing similar information into larger regions

As in GOLPE, region selection is achieved exploiting a combination matrix. But here, each column represents a 3D region and each row (each experiment) a combination of these regions (in and out) in the model tested. The next step evaluates the

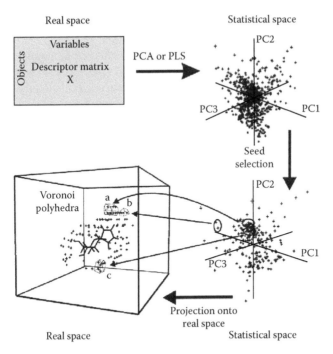

FIGURE 1.6 Scheme of the first and second steps of the RD algorithm. The information flow starts from the descriptor matrix, goes along the statistical space, and finally ends in the real 3D space of the molecule. Seeds close in the statistical space can be variables far away in the real space. Polyhedra a and b can be merged in the third step of the RD algorithm only if they contain the same chemical information. (Reproduced from Cruciani, G. et al., *Perspect. Drug Discov. Design*, 12–14, 71, 1998. With permission.)

effect of the regions on the predictive ability and classifies them as beneficial, uncertain, or detrimental for prediction. Similarly to GOLPE, some dummy variables are incorporated and detrimental regions are discarded in the final model. This is carried out simultaneously whereas it was stressed that the q^2GRS involves separate calculations for the sub-lattices.

1.2.1.5 Domain Mode Selection

The problem of region selection was also approached by Norinder [191] who proposed a domain mode selection gathering in sub-boxes 3^3 or 4^3 elementary cells of an initial CoMFA-type lattice. This approach was compared to a single mode selection (similar to GOLPE), with the difference that a "variable" is now a contiguous domain of variables (sub-box) and no a single variable (a node).

For this study, three sets of steroids were chosen for the rigidity of their skeleton, avoiding possible problems or uncertainties on alignment. However, predictive ability on test sets for the three series investigated was not improved with respect to the original models. Pastor et al. [190] concluded that sub-boxes or other domains defined by geometric criteria may lead to volumes containing differing pieces of information, whereas a same type of information may be split into differing volumes. Furthermore, the predictive ability is evaluated for each individually, without assays for combining a reduced number of them, whereas the RD algorithm gathers contiguous regions with the same type of information.

The efficiency was checked in a study of glucose analogue inhibitors of glycogen phosphorylase b (GPB) (where crystal structures were known for all compounds, avoiding alignment problems). Smart region definition (retaining 459 variables from 7290 nodes) led to better results that variable selection or q^2 region-guided selection.

1.2.2 Introduction of New Fields and Supplementary Descriptors

In countless applications, "standard" CoMFA using only steric and electrostatic fields was quite successful. However, the supplied information might be refined, using other probes than H^+ and C sp^3, supplemented with the definition of new fields, or even introduction of other physicochemical parameters [131]. These adjunctions are relatively easy to implement, but, as stressed by Waller, these options "do not preclude the user for applying physical and chemical sense to the development of the model."

Particular attention was focused on the treatment of hydrogen bonding and hydrophobicity, which may be treated not sharply enough, considering only electrostatic and steric fields.

1.2.2.1 Hydrophobicity Potential

Since the observations of Meyer and Overton, as soon as 1899, hydrophobicity was recognized as an important parameter for understanding drug activity. Inclusion in CoMFA of a hydrophobic field was proposed by Kellogg et al. [192] to provide a new interpretative tool. In HINT (Hydrophobic INTeractions), hydrophobic interactions are evaluated via an empirical potential,

$$A_j = \sum_i a_i S_i R_{ij}$$

where

a$_i$, S$_i$ hydrophobic atom constant and solvent-accessible surface area for ligand atom i

R$_{ij}$ = exp(−r$_{ij}$) Gaussian attenuation factor, function of the distance r$_{ij}$ between atom i and the probe j

Atomic contributions a$_i$ were derived from the fragment constants used in the computation of C Log P [193,194].

On the example of the 21 steroids binding CBG studied by Cramer [38] and Norinder [195], adjunction of HINT to the usual electrostatic and steric fields did not statistically improve the quality of the model. However, the authors stressed that this three-field model made easier the interpretation. Contour plots of the hydrophobic field indicated regions where an increase in hydrophilicity leads to a decreased activity (e.g., presence of a phenolic OH at position-3 or a carbonyl at position 11), whereas an increase in hydrophobicity near the C-18, C-19 positions of side chains is beneficial.

In a study of estrogen activity, Waller et al. [173] examined a population of 55 molecules corresponding to 8 structurally diverse subsets (phenols, DDTs, DESs, PCBs, phtalates, phytoestrogens, steroids, and pesticides), on 6.5 log units of binding affinity (measured on mouse uterine cytosol assays). Rigid body alignment to the template estradiol by SEAL avoids subjective hypotheses. Adding to the usual steric and electrostatic fields, a hydrophobic contribution (HINT) led to a better model, for both internal prediction and robustness, than usual CoMFA or HINT alone (r^2 = 0.881, q^2 = 0.590). Interestingly, external predictive ability was examined excluding a whole structural family and training the model on the other ones. Errors on these "unbiased" predictions were typically on the order 1.0–2.0 log units depending on the structural extrapolation from the training set (whereas in the fitted mode, error was about 1 log unit). L-O-O predictions were not significantly better. This may indicate that the training set already encompassed a sufficient structural diversity to be able to treat new families.

This study presented an example where the binding mode is not the same for all compounds. If phenolic rings were generally superimposed on ring A of estradiol, for the 2,4,6-trichloro-4′-biphenylol, the phenol ring was oriented over the estradiol D ring and its 17-OH group (Figure 1.7). This position provided better hydrophobic interactions in the central part of the molecules (over B and C rings). Similarly for ER binding of Bisphenols A, the intuitive choice, adjusting the two hydroxyl groups on those on the 3- and 17-groups of E2 or DHT, must be discarded; a conformation with a hydrogen bond missing but possessing stronger hydrophobic interactions is preferred [196].

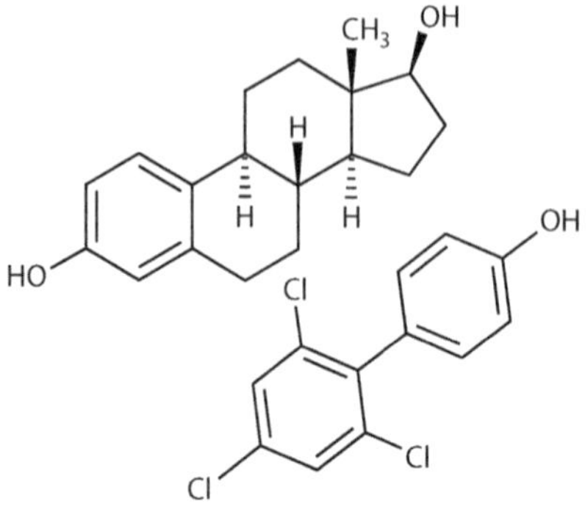

FIGURE 1.7   Preferred orientation of 2,4,6 trichloro-4′-biphenylol superimposed on estradiol scaffold.

Pajeva and Wiese [175] examined on simulated data how the hydrophobic areas (from HINT) were mapped by CoMFA (they often appear widely extended) and how their contribution is different from standard steric and electrostatic fields. They pointed out that log P values alone are predictive only if the target property is linearly dependent on them. They proposed also to evaluate the similarity between fields of different nature by correlations between the **X**-scores of the first components.

1.2.2.2   Molecular Lipophilicity Potentials

Molecular lipophilicity potentials (MLPs) have been also considered [197,198]. The potential proposed by Gaillard et al. [199] is based on the atom lipophilic system of Moreau and Broto [200]. At lattice node j,

$$MLP_j = \sum_{i=1,n} f_i exp\left(\frac{-d_{ij}}{2}\right)$$

where
  f_i is the contribution of atom i
  d_{ij} is the distance between atom i and node j
  summation over all atoms (i) of the ligand

See also Ref. [201] for the derivation of hydrophobic parameters. Note that a steeper distance function (exp[-d]) was used by Fauchere et al. [197]. The main advantage of these models is not a definite improvement of the statistical results but a better description of the interactions, and so a better interpretability. A similar argument was also invoked for the introduction of the five fields of the CoMSIA approach [40].

Gaillard et al. [202] introduced the MLP in an extended study of arylpiperazines, (aryloxy)propanolamines, and tetrahydropyridyl-indoles binding the 5-HT$_{1A}$ receptor (185 ligands). Using a stepwise procedure, they obtained self-consistent alignments suggesting favorable/unfavorable regions for binding, which could not be obtained in single chemical class analyses. In glycine conjugation, including MLPs allowed for grouping together aromatic and aliphatic carboxylic acids in a single model, which was not possible in a MLR treatment [203]. Other examples concerned inhibition of monoamineoxidase [204,205], see also Norinder and Altomare et al. [206,207].

1.2.2.3 Polarizability Field

Polarizability effects could also be introduced as proposed by Bradley and Waller [208]. This field was also more recently included in self-organizing molecular field analysis (SOMFA) method [209]. The underlying idea is that polarizability, through dispersion forces, could be the determining factor for binding. The polarizability field is given by

$$p = \sum_{j=0,m} \sum_{i=0,n} \frac{\alpha_i}{d_{ij}^2} \quad d_{ij} < vdw_i$$

where

$\alpha_i(\text{Å}^3)$ is the atomic polarizability of atom i

d_{ij} is the distance between the grid point (j) and atom (i) of the molecule

In Bradley and Waller study, the data set comprises 99 compounds binding the cytosolic Ah receptor with $(pEC_{50})_{\mu M}$ from 9.3 to 3.8: 14 polychlorinated biphenyls (**1.4**), 25 polychlorinated and polybrominated dibenzo-p-dioxins (**1.5**), 39 polychlorinated dibenzofurans (**1.6**), 5 polybrominated naphthalenes (**1.7**), and 16 indolocarbazoles (**1.8**) and derivatives (**1.9**). Produced in many industrial or combustion processes, these very widespread and long-lived pollutants are responsible for various toxic effects. TCDD (2,3,7,8-tetrachlorodibenzo-p-dioxin) was chosen as template. With 1 Å grid spacing, usual CoMFA (steric plus electrostratic fields) yielded $r^2 = 0.91$, $q^2 = 0.65$, whereas the three-field model (steric, electrostatic, and polarizability) was only slightly superior for q^2 with $r^2 = 0.91$, $q^2 = 0.72$.

1.4   1.5   1.6   1.7

1.8   1.9

Contour plots indicate that some steric bulk is desired on the lateral positions and that, to be good ligands, compounds need not to be halogenated on the medial positions (where positive charges are desired) [210]. Polarizability contour plots also confirm the beneficial contribution of halogenation on the lateral positions. The authors concluded that some information about polarizability was implicitly included in the steric field, but in systems where this contribution is predominant, an explicit evaluation would be desirable.

1.2.2.4   Hydrogen Bonding

To take into account hydrophobic interactions (not well represented by the CH_3 probe) and the specific effects of H-bonds (and particularly their directional character), Kim [211,212] took up the directional 6-4 potential of GRID from Goodford [58], vide supra,

$$E_{hb} = \left[\frac{C}{d^6} - \frac{D}{d^4} \right] \cos^4 \theta$$

where θ denotes the angle (protein donor) –H– (acceptor probe atom). On a series of benzodiazepines, adjunction of a H_2O probe to the standard CH_3 and H^+ probes gave good results, but this field is nearly collinear with the steric field (predominant in this series of compounds) [211]. Bohl et al. [213] added indicators and H-bond fields representing H-bonding acceptor and donor components. See also Carosati et al. [214] for H-bonding interactions with fluorine atom.

In the scheme proposed by Bohacek and McMartin [215], paired H-bonding fields (acceptor and donor) are created. Nodes distant from H-bond donor or acceptor atoms (or if H-bond interactions are forbidden by steric crowding) are given a null energy. Nodes in a sterically allowed region, and close to an acceptor or donor atom, are assigned energy equal to the steric cutoff value. Once an H-bond CoMFA column has been created, the acceptor component is nominally a "steric" field type and the donor component is nominally an "electrostatic" field type. Bohacek and McMartin [215] used the solvent accessible surface of Lee and Richards [216] to quantify and visualize protein–ligand interactions. This surface, pushed away from the van der Waals surface by a water probe radius, is a powerful guide to show where ligand atoms should be located to ensure good complementarity with the binding site. From a large sample of crystallographic data, it was shown that complementarity is best described by using hydrogen-bonding properties of the binding site. This allows for dividing the surface of the binding site into three zones: hydrophobic, H-bond acceptor, and H-bond donor.

1.2.2.5   Desolvation Energy Field

It might be argued that CoMFA with steric and electrostatic fields did not take into account entropic effects and solvent reorganization on ligand–receptor binding (involving a partial desolvation of both entities). Such effects, ignored in the initial implementations of CoMFA, could be quantified using Poisson–Boltzmann or generalized Born models. These aspects are treated in Chapters 11 and 12 dedicated to "receptor-related models." We just mention here that this point was approached by Waller and Marshall [217] on a study of inhibitors of thermolysin (THER) and angiotensin-converting enzyme (ACE), compounds of potential interest for the management of hypertension. But, in this example, inclusion of this contribution had little influence, the desolvation field being correlated with the electrostatic component.

Sulea and Purisima [218] proposed an evaluation of partial desolvation effect on binding by difference in the reaction field energies calculated by the boundary element method plus a nonelectrostatic component between the bound and free states. The solvated system is partitioned into the solute (low dielectric cavity) and the solvent (of high dielectric constant), embedding the solute and represented by an induced charge distribution calculated at the interface by the Poisson–Boltzmann equation. The non-electrostatic component (cavity formation, dispersion effects, and H-bonding) is estimated by a surface area-based term.

The desolvation field, incorporated in CoMFA, is calculated with a probe atom (neutral C sp^3 carbon), simulating the receptor, exploring a 3D lattice embedding the ligands. Ligand charges were calculated at the 6-31G* level and fitted on the electrostatic potential. This field is dissimilar from both the CoMFA electrostatic field, and the molecular electrostatic potential. But it provides a 3D representation of the hydrophilic/hydrophobic character (with ΔG_{desolv} negative near hydrophobic areas and positive near hydrophilic areas such as those surrounding a carbonyl oxygen, a phenolic, or an amide H).

For the steroid benchmark (action on CGB and TBG), addition of this desolvation field to CoMFA electrostatic and steric fields slightly enhanced the performance (and contributed significantly to the model), whereas HINT field [192] did not improve the results. The partial agreement of the contour plots (from HINT and the desolvation model) suggested that these two models encode correlated properties, although HINT is rather a lipophilic than hydrophobic field [219].

A more qualitative rapid evaluation of conformational entropy loss on binding was proposed by Godha et al. [220] with the hypothesis that this entropy loss is small if the ligand is highly flexible and has a high conformational propensity to adopt an active conformation. Conformational propensity was defined as the population ratio of number of energetically stable conformers to the number of active-conformation-like structures. A satisfactory model could be established on a series of imidazoleglycerol phosphate dehydratase inhibitors. However, the population is rather limited (20 compounds in training, one only in test).

It was also suggested that introducing, in CoMFA treatment, new field contributions with HINT [192] or the MLP [199] might, in part, take this desolvation effects into account. A similar remark was also proposed for log P. Let us note that in GRIND (and in the derived approach VolSurf), hydrophobic interactions are specifically taken into account with the DRY probe (vide supra).

1.2.2.6 Electrotopological State Fields

A 3D field was defined by Kellogg et al. [221] from the atomistic electrotopological states of Kier and Hall [222]. In the E-state formalism, each atom is characterized by an intrinsic state (combining valence state, electronegativity, and local topology), perturbed by the inductive influence of every other atom depending on the interatomic distances. The E-state, defined for "heavy atoms," has then been extended to hydrogen atoms with the HE-state.

At grid node t, the field is

$$E_t = \sum_i S_i f(r_{it})$$

where
  summation over atoms i of the molecule
  r_{it} is the distance between probe node t and atom i
  f is a user-defined distance function (various expressions from $1/r$ to $1/r^4$ and even exponential functions were tested)

On the 21 steroids (the traditional training set), a model with E-state alone gave $q^2 = 0.791$. Adding E-states and HE-states to usual CoMFA fields (plus hydropathic field) improved q^2 of about 3%–5% above the standard. It was also noteworthy that, in this application, the E-state field represented an important contribution (50% or more of the total) [223].

1.2.2.7 van der Waals Potentials

Lennard-Jones 6–12 potential gives very steep potentials near the atoms. This causes various problems: a small distance variation may induce large energy changes, implying that a very acute definition of alignment rules is mandatory. Similarly, going from a node to the next one 2 Å apart (the common lattice mesh) may cause (at the proximity of an atom) very important variations of the steric energy. Furthermore, the differences in slopes between electrostatic and steric fields impose arbitrary settings beyond some cutoff values. The largest variance of the steric field (as compared to the electrostatic contribution) may be also detrimental in the PLS treatment. So, it was suggested to choose a lower cutoff value (e.g., 5.0 kcal/mol). Otherwise, it has been proposed to substitute other softer expressions to the usual Lennard-Jones potential ("soft potentials" were used, for instance, in free-energy perturbation models [224], see Chapter 12).

Alternately, Kroemer and Hecht [225] replaced the CoMFA steric field by binary indicator variables: encoding the presence of an atom in a predefined volume element (elementary cube) within the region enclosing the aligned molecules. If the center of an atom is found in an elementary cube, a value of 30 kcal/mol is assigned to the corresponding grid point (otherwise 0). This value (30 kcal/mol) was chosen for the sake of comparison with the "classical" CoMFA. With this approach, improved results have been obtained on five extended sets—training (80 compounds each) and test (60 compounds each)—randomly chosen among 256 dihydrofolate reductase (DHFR) inhibitors. In the same study, it was confirmed that truncating steric energies at a low value or reducing grid mesh did not improve the standard model (smaller node spacing gave more importance to little variations in the positions of molecules, and so introduced some noise in field evaluation), whereas performance was better with indicator variables [130]. A slightly different option was chosen by Floersheim et al. [226] who binary encoded (1 or 0) grid points within or not the van der Waals volume.

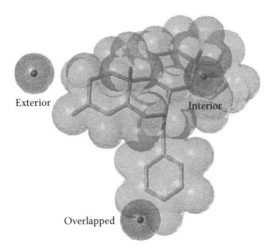

FIGURE 1.8 Schematic representation of the INVOL method. The volume of intersection between the van der Waals envelopes of the ligand molecule (light gray) and probe atom (dark gray) is calculated as steric 3D descriptor in the "overlapped" positions. "Exterior" and "interior" points are usually discarded. (Reproduced from Sulea, T. et al., *J. Chem. Inf. Comput. Sci.*, 37, 1162, 1997. With permission.)

These observations led Sulea et al. [227,228] to propose the replacement of the van der Waals attractive terms by hydrophobicity parameters and repulsive terms by the INtersection VOLume (INVOL) of the van der Waals envelope with the probe atom. This INVOL field, which may be calculated with a Monte Carlo integration technique, varies smoothly with interatomic distances. Points external or interior to the van der Waals surface are usually discarded (Figure 1.8). This model (with only the repulsive component) was tested on 78 steroids binding aromatase and yielded q^2 values comparable to those obtained by the classical CoMFA, but with a limited computational demand. In addition, it was checked that the contour plots were also very similar.

1.2.2.8 Molecular Orbital Fields

Molecular orbital (MO) fields were also proposed for cases where a simple Coulombic representation may be insufficient. A HOMO field (calculated from the electron density at the lattice nodes) was used by Waller and Marshall [217] for characterizing interactions between ionized ligand and Zinc ion in a study of ACE and THER inhibitors. A comparison of CoMFA models incorporating LUMO field was introduced by Poso et al. [229] and Navajas et al. [230]. MO fields were also used for the treatment of metabolic rate constants [231].

1.2.2.9 Introducing Supplementary Descriptors

Additional columns of descriptors (such as HOMO, LUMO, log P, dipole moment) may be added to the usual field values and, in some cases, improve the results. As seen above, in Ref. [188], adding log P was inefficient, whereas a "phenol indicator" increased q^2 up to 0.71 in place of 0.66 with traditional CoMFA. In the study of Ganchev et al. [232]

(on more than 40 halogenated estradiol derivatives binding the ER), incorporation of the dipole moment gave no improvement, presumably because the dipole moment reflects the charge distribution.

Bioactivity (hyperglycemia decrease) and metabolism activity of human glucagon receptor antagonists have been investigated by Chen et al. [233] on extended sets (respectively, 74 and 55 compounds on an activity range of about 3 and 2 log units). A support vector machine treatment was used to select supplementary descriptors (at a 2.5D level) to introduce with the usual CoMFA–CoMSIA fields. Adjunction of log P to the usual fields was meaningful for the activity model. Adding HOMO and LUMO values in the metabolism model also improved performance.

Similarly, in a study of azo dye–fiber affinity, Fumar-Timofei et al. [234,235] observed an improved correlation by adjunction of HOMO and LUMO energies, as independent variables. It was indicated that the observed preponderant importance of HOMO expressed the role of electrostatic interactions in dye–cellulose binding.

1.2.3 INFLUENCE OF PARAMETER SETTINGS

The sensitivity of the CoMFA method to parameters settings in the construction of QSAR models was the subject of numerous studies. A first concern focused on the calculation of charges.

1.2.3.1 Charge Calculation

In most applications, the electrostatic potential is not calculated from the molecular wave function but approximated from atomic charges derived with semiempirical MO methods, in agreement with earlier studies of Kim and Martin [236,237]. As stated by Kroemer et al. [130], this rose a twofold problem. On one hand, how to calculate atomic charges? And, on another hand, is it correct to use atomic charges to represent electrostatic interactions?

The influence of the method used for charge calculation was investigated by Kroemer et al. [238] on 37 ligands of the benzodiazepine receptor (BZR) inverse agonist–antagonist site, and more recently by Hou et al. [239] on 135 inhibitors of the epidermal growth factor receptor, and Hannongbuoa et al. [240] on 101 HEPT derivatives inhibitors of HIV1-RT. The methods, considered at least in one of these papers, encompassed Gasteiger-Marsili [158], Gasteiger-Huckel [159], semiempirical (AM1, PM3, MNDO) [162,163], ab initio (STO-3G, 3-21G*, 6-31G*) with Mulliken population analysis or charges fitting on the electrostatic potential [241]. Charges calculated from the bond increments of MMFF94 based on electronegativity equalization [242] were also included. These studies converge on the conclusion that CoMFA is not very sensitive to the mode of calculation and that semiempirical or Gasteiger's charges give satisfactory results at a reasonable cost. Mapping directly the electrostatic potential onto the nodes (rather than using a Coulombic expression) gives no improvement.

However, an extensive study of Mittal et al. [243] on 30 data sets (more than 2000 compounds) found no significant difference between semiempirical methods and the lighter MMFF94 method that both present more predictive ability than Gasteiger's method.

An important remark of Kroemer et al. [130] is that electrostatic contour plots might be more affected depending on the method used. Scaling steric and electrostatic fields might be another source of difficulty.

1.2.3.2 Field Calculation

The paper of Melville and Hirst [181] more specifically focused on the calculation of the steric and electrostatic fields and on the sensitivity to the lattice position. A particular attention was devoted to "column dropping": Electrostatic descriptors for nodes where steric energy is larger than the cutoff are set to the average of all electrostatic values that were not within the steric envelope. Three series were investigated: the steroid benchmark (31 compounds), 38 D_2 antagonists, and 34 polychlorinated dibenzofurans acting on the Ah receptor. On account of the limited size of the data sets, cross-validation, rather than splitting data into train/test sets, was considered as validation criterion.

The first concern relied on cutoff values. A default value, $E_{cut} = 30$ kcal/mol, is generally applied to the steric and electrostatic fields. It may be applied abruptly or via a smoothing function (the default option):

Field values greater than $E_{up} = 1.2 E_{cutoff}$ are assigned a value equal to E_{cut} and between $E_{lo} = 0.8 E_{cut}$ and $E_{up} = 1.2 E_{cut}$ the original field value E is substituted by a smoothed value

$$E_{smooth} = \frac{E^2 - 2E_{up}E + E_{lo}^2}{2(E_{lo} - E_{up})}$$

Another possibility is the "box option" suggested by Cramer [128] and taken up by Kroemer et al. [244]. In this option, the probe atom is replaced by eight probes at the corner of a box. Its center is the probe atom, and its edge equal two-thirds the normal lattice mesh. The fields assigned to the probe atom are the mean of the values calculated on the eight vertices of the box.

Abrupt cutoff or smoothing gave (on q^2) nearly identical results, larger differences appearing with the box option. The default CoMFA settings (2 Å spacing, AM1 charges, 1/r dielectric function, inclusion of steric and electrostatic fields, column dropping, and smooth cutoff) led to good-quality models, but better results could be obtained. The steric field was found relatively insensitive whereas variations were larger for the electrostatic field. The box method appeared superior, especially if a 1/r dependence of the dielectric constant and column dropping were selected. Filtering lower variance columns reduced the number of variables by about 80% without significant changes of the performance but accelerated the calculation.

Broughton et al. [245] proposed a modification of CoMFA in order to take into account limited conformational mobility. Their study concerned lignan natural products' derivatives related to podophyllotoxin and presenting antineoplastic activity. For each molecule, low-energy conformers were selected (from 1 to 34, depending on the structure) from an energy-related score based upon the strain energy of the conformation and its frequency of occurrence in a stochastic search in the torsion space. An overall field was calculated on each lattice node, with various weighting schemes: simple average, Boltzmann-weighted average, and a variability-weighted average (placing larger values at nodes where field values are high and independent of the conformer). The best results were obtained with the simple average field. The authors concluded that some compounds probably act through more than one mechanism with an importance depending on the cell line used in activity measurement.

1.2.3.3 Coefficient Plots

Another interest of CoMFA is the display of the coefficients plots. Smoothing and box option lead to qualitatively similar results, with positive and negative coefficients distributed in the same region of space. Column dropping reduced the importance of lattice points close (or internal) to the van der Waals surface and significantly improved the model with smoothed field, but variations were smaller with the box technique, which dampened the steep increase of electrostatic field near atomic centers. In this study, the poorest models corresponded to a $1/r$ dielectric constant (whereas field variations in such regions are in $1/r^2$). The display of contour plots was also made clearer and easier to interpret.

Furthermore, it has been shown [128] that the box option improved the stability of the models to small translations of the lattice and generally gave higher q^2. This was in agreement with the study of Melville and Hirst [181] (with a $1/r$ dielectric function and column dropping). These results differed from those of Folkers et al. [129] on HSV1 TK inhibitors, presumably because of variations of electrostatic fields near atomic centers. In line with this observation, the authors remarked that CoMSIA, with smoother fields, also exhibited a small sensitivity to grid position as already quoted [40,115].

It was also noted that, on the studied examples, steric and electrostatic fields encode essentially the same information about drug–receptor interactions, but using the two fields jointly may be interesting for the purpose of interpretation. It was also observed that with a lattice mesh of 1 Å (in place of 2 Å), displacements have a small effect on stability and do not improve the results. This seems surprising, but possibly reflects the fact that PLS may be misled by the high variance of some (rather irrelevant) variables (electrostatic field within the molecular volume) hiding the role of relevant variables but of low variance. Here also, column dropping is beneficial.

The authors concluded that the default CoMFA settings gave good-quality models, but the box approach represents a "simple, immediately applicable method that can potentially increase the quality of the PLS model."

1.2.3.4 Optimization of Settings

Various changes in the parameterization of CoMFA have been attempted, such as extending the grid, imposing ring carbon constraints in field-fit, changing charges on the example of halogenated estradiol derivatives [232].

The influence of data scaling and variable selection in a GRID–CoMFA treatment was investigated by Ortiz et al. [246] on 26 inhibitors of human synovial fluid phospholipase A_2 (HSF-PLA$_2$). Diverse options were examined as to variance scaling (no scaling, auto-scaling, or block scaling) and dielectric scaling considering a constant dielectric, a distance-dependent ε, and the Warshel model, $\varepsilon = 1 + 60[1 - \exp(-0.1r)]$ or the Hingerty model (proposed as giving results comparable to the Poisson–Boltzmann model) $\varepsilon = 78 - 77 \ (r/2.5)^2 \ [(\exp(r/2.5)/(\exp(r/2.5) - 1)]$.

Variable selection was carried out with q^2GRS or with GOLPE [177], relying on the validation of a number of reduced complexity models. In GOLPE, variable combinations were selected according to a FFD strategy, where successive models are generated by elimination of 10% of the variables. In conclusion of this study, it appears that the dielectric constant, energy cutoff, grid spacing, and variable scaling showed a strong influence on the results. With GRID energy function, a constant dielectric ($\varepsilon = 4$) and unscaled data seemed the best choice, yielding with GOLPE $q^2 = 0.754$. It was also indicated that region-guided selection involved a smaller number of variables than GOLPE but might include variables only because they belonged to the same box as important variables. In the same work, a comparison was carried out between CoMFA and COMparative BINding Energy (COMBINE) analysis [247,248], see Section 8.3.

More recently, a systematic study of optimization of the settings, compared to the default values of the SYBYL release, was carried out by Peterson [141,182] on an extended ensemble of data sets. More than 6000 possible combinations of non-default parameters were considered. In addition to the classical steroid benchmark (with corrected valued from Gasteiger [146]), the eight series retained by Sutherland [72] and revisited in that study corresponded to a large diversity in both the chemical structures and the type of activity.

Were examined inhibitors for the following targets: ACE, acetylcholinesterase (AChE), BZR, cyclooxygenase-2 (COX-2), DHFR, GPB, THER, and thrombin (THR) (Figure 1.9). About one-third compounds constituted the test set (they were chosen in each series so as to represent a maximum dissimilarity in order to widely span the structural space) as proposed by Sutherland et al.

Attention was focused on a better understanding of the interdependence of CoMFA parameters monitoring field calculations and their effect on predictive performance. Other settings were left at default values (grid spacing of 2 Å, +1 charged Csp^3 probe atom). Region size, orientation, alignment were also left unchanged with respect to the original publications. Performances were evaluated using r^2_{pred}, and q^2 or the composite score index proposed by Schultz et al. [169]:

$$Score = 5r^2_{pred} + 4q^2 + 2r^2$$

Extensive y-randomization runs also testified that the presented results do not reflect spurious correlations. Peterson's results established that a careful choice of parameter settings in CoMFA might lead to optimal models with improved performance. The gain was substantial if performance with default values was low; less important, as expected, if default parameters yet gave good results. For example, r^2_{pred} and q^2 were improved from −0.01, 0.32 to 0.45, 0.45 for inhibitors of the BZR whereas for DHFR inhibitors, r^2_{pred} and q^2 varied from 0.59, 0.65 to 0.64, 0.68, respectively.

However, there was no unique parameter set optimal for all series. In other words, optimal settings are highly data set dependent, but some general trends might be

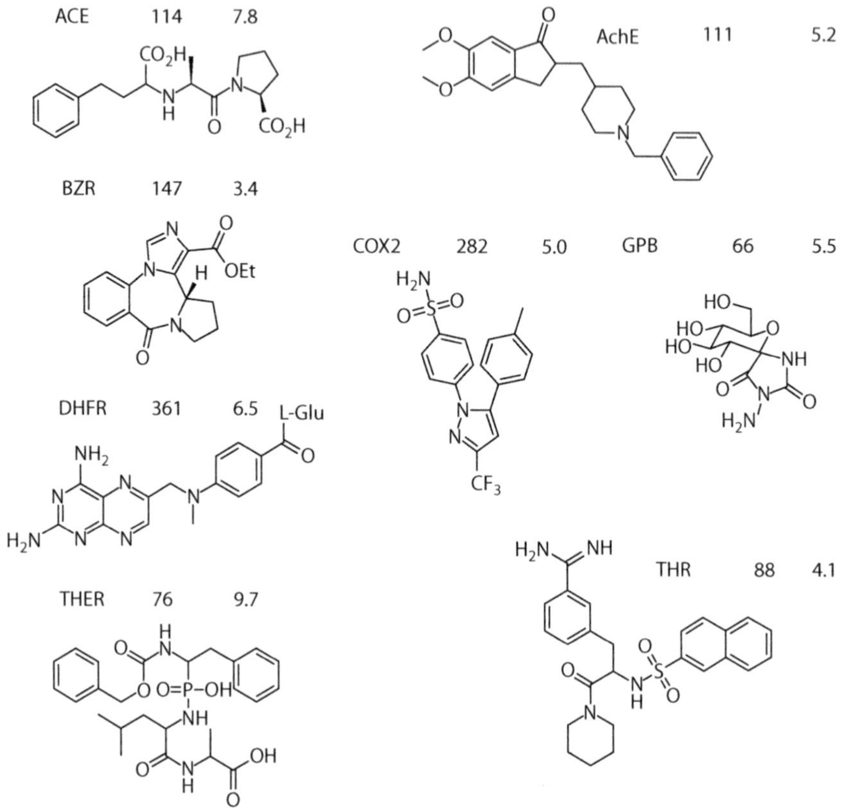

ACE   114   7.8

AchE   111   5.2

BZR   147   3.4

COX2   282   5.0   GPB   66   5.5

DHFR   361   6.5   L-Glu

THER   76   9.7

THR   88   4.1

FIGURE 1.9   The eight inhibitor series of Sutherland's studies. For each series, the target, the number of compounds, and the activity range (log IC_{50} or pK) are specified and a representative compound is given. (Adapted from Sutherland, J.J. et al., *J. Med. Chem.*, 47, 5541, 2004.)

highlighted. All optimal models used both steric and electrostatic fields, never the H-bonding field of Bohacek and McMartin [215]. Abrupt cutoff was always retained (see also [181]). A distance-dependent dielectric function (1/r), box averaging, which limit effects of a bad alignment [181], and a van der Waals repulsive contribution in r^{12}, were largely preferred. But things were less clear for other settings such as cutoff values for the fields or column dropping.

1.2.3.5   Introducing New Probes

Whereas traditional CoMFA only uses as probe a charged C sp^3, we have already seen that the GRID approach of Goodford [58] (vide supra) proposed to use varied probes for a more specific and accurate evaluation of structural effects in the neighborhood of macromolecules. A similar strategy was also introduced in CoMFA.

For example, for a better analysis of H-bonding properties, changing probe to O (charge −1) and H atoms (charge +1) has been proposed [232] in a study of halogenated estradiol derivatives binding the ER. In the same vein, we already indicated that the directional aspect of H-bonds (introduced in GRID) was taken up by Kim et al. [211,212]. Folkers et al. [129] also suggested that introduction of the GRID force field in CoMFA methodology can be beneficial.

1.3 REFINING ALIGNMENT

For nonrigid molecules, the choice of an alignment rule, and beforehand the selection of the active conformation (that which binds the receptor) is of course a major problem. On another hand, although CoMFA (and other ligand-based methods) does not need any knowledge of the receptor, any information on the receptor–ligand interactions may be useful for clarifying and validating the proposed alignments. On this point, help can be gained with docking methods that aim to find the "best" location of a ligand in the receptor pocket, regarding not only geometrical aspects (position, orientation, conformation, and even configuration) but also energetics of the binding process. These methods are also the basis of "receptor-based" (sometimes called "receptor-dependent") QSARs, and of the "de novo design" in the search for new drugs.

Several strategies may be adopted depending upon the structural information available, from methods relying only on known ligands to methods evaluating the interactions of ligands with a receptor the structure of which is known. An extensive review of superposition algorithms (including superposition of small rigid or flexible molecules, our concern for QSAR models) can be found in Lemmen and Lengauer [148]. But it may be remarked that techniques working only from ligands are inherently indirect and only allow for interpolating between known compounds. The influence of a region of space not covered by any ligand cannot be guessed, it may be detrimental in case of unpredicted steric clashes. (A similar remark was made for the contour plots: No prediction is possible for regions where the fields do not change.)

In the absence of structural information about the receptor protein, the common approach is based on the recognition (or the intuitive suggestion) of a pharmacophore model. The functional groups identified (or supposed) to be critical for binding are then superimposed, suggesting an alignment to be used in CoMFA.

At the crudest level, a conformational search is generally first performed by molecular mechanics, to define, for the compounds under scrutiny, their minimum energy conformation that will be used as starting geometry. Systematic grid search or extensive Monte Carlo sampling are often time consuming. If a reasonable ligand structure is known, Molecular Dynamics simulations can provide an efficient conformational sampling.

However, it was stressed that this process relies on a general shape descriptor (common structural features) that does not automatically represent the active site. Conclusions possibly drawn about the active site may be accepted with caution [148]. In addition, using energy-minimized conformations is not a panacea. Sometimes, the best aligned conformation (that would correspond to the "active" form) is not the conformation of minimal energy in the free state, as for bicalutamide binding the androgen receptor [213]. See also Cho et al. [249] on the example of neostigmine. Furthermore, as previously seen, binding modes to a specific receptor may not be the same within a series of chemicals [173,196], vide supra. Consideration of water molecules directly interacting with the ligand may be also important to improve the predictive ability of the models and made easier their interpretation [250].

1.3.1 PHARMACOPHORE ANALYSIS

As we indicated before, the alignment of structures needed by CoMFA and related methods may take advantage of the knowledge of a pharmacophore (a group of atoms that is necessary for a molecule to be recognized by a receptor). It gives some anchoring points for superimposing the investigated structures. Several approaches aimed at a more objective extraction of a pharmacophore from examination of known active (and inactive) compounds. Pharmacophore search can be carried out using an automated analysis [130,251] or active analogue approach [252].

The automated DIStance COmparison (DISCO) method [251] considers the interatomic distances from an ensemble of conformations for a set of molecules. Distances of common features are compared within a defined tolerance value. A clique-detection algorithm identifies the features that could be elements of a pharmacophore model. Considered features are H-bond D- or A-propensity, charges, centers of hydrophobic groups as the most likely location of receptor binding sites. The study of Chen et al. [253] on antitumor molecules encompassed 63 analogues of Colchicine. A common pharmacophore was determined using DISCO, and CoMFA treatment led to $q^2 = 0.797$. In a second phase, 97 virtual compounds were generated from the constraints of the pharmacophore, and their activity predicted from the CoMFA model. Additional structural modifications suggested a new virtual active structure that was able to favorably dock in a telomeric DNA sequence (TTAGGG)3.

An alternative is the Active Analogue Approach of Marshall et al. [252]. The principle is to superimpose active molecules in one of their possible conformations, so that important corresponding groups (the pharmacophore) coincide. The method supposes compounds structurally neighbor enough, so that superimposition is unambiguous. Assuming a given pharmacophore of n points supposed to be present in all molecules, a table of n(n − 1)/2 distances is built. The distance maps of each of the other molecules are compared to that reference. To cope with conformational flexibility, the distances between atoms of the pharmacophore are systematically recorded for each allowed conformation (rotatable bond being rotated by a fixed increment). Sets of interatomic distances that can be found in all active molecules, in one of their possible conformations, represent the possible pharmacophore geometry. During the search, a distance that is not present in one conformation of one molecule cannot be eligible and so is not considered further, progressively reducing the search space.

Another approach to generate a tentative alignment is provided by genetic algorithm superposition program (GASP) [155] that uses a GA to handle flexibility and maps features on a reference molecule without relying on predefined correspondences, see "docking" in Chapter 11.

An appealing alternative solution is also afforded by approaches aiming to identify pharmacophoric patterns [254]. Pharmacophore models are very powerful representations of ligand binding since they capture the concept of bioisosterism and focus on similarity in chemical functionality rather than on topological characteristics. The process also allows for comparing different scaffolds. We just consider here applications where these pharmacophore-oriented approaches helped CoMFA alignment. More direct applications of these techniques for building QSAR models will be detailed in Chapter 3.

The CATALYST [255,256] software proposes a feature-based alignment rather than explicit atom superposition. From a library of low-energy conformers, hypotheses are formulated according to the presence of a minimum number of chemical features encountered in a subset of active molecules. Hypotheses that also match a significant number of inactives are discarded. The surviving hypotheses are then refined adjusting site-point positions by simulated annealing. Excluded volume may be also considered to improve predictive ability [257]. Features encoded are, for example, H-bond donor or acceptor, positive or negative charges, hydrophobic regions. Site points may be atoms or projected positions. The set of features that correlates best with activity (according to a cost evaluation) suggests a pharmacophore model. CATALYST, operating on features and not on atoms, can treat molecules of diverse structures since different groups or functions can belong to the definition of a same feature. Langer and Hoffmann [258], for example, successfully applied CATALYST on a set of 15 highly flexible inhibitors of rat liver squalene epoxidase, whereas atom–atom alignment is difficult. Suhre et al. [259] also used CATALYST to guide alignment before a CoMFA study of renal organic cation transporters.

CATALYST was also applied by Bureau et al. [260], in association with CoMFA, in a study of partial agonist serotonin 5-HT$_3$ ligands; in a case where some discrepancies appeared as to conformational analysis between semiempirical MO methods, empirical force fields, and X-ray crystallography. Using the CATALYST pharmacophore, a CoMFA study led to quantitative activity predictions with a good fit between regions highlighted by CoMFA and features selected by CATALYST. The same group [261] also presented CATALYST as a promising tool in ecotoxicology, with a study of chlorophenols, and a comparison with CoMFA or an evaluation of solvation energies.

A rather similar approach was presented with the grid-based PHASE method of Dixon et al. [262]. Every hypothesis was scored by the geometric alignment of site points in actives and inactives with site points identified in the hypothesis. For establishing a 3D-QSAR model (either with atoms or sites associated to pharmacophore features), molecules were embedded in a grid, with each atom/site represented by a sphere. Each occupied elementary cell gave rise to volume bits (a separate bit being allocated for each different category of atom/site). Thus, a molecule was represented by a string of binary 3D descriptors subsequently treated with PLS. An extensive comparison of CATALYST and PHASE was carried out by Evans et al. [263] on eight series of compounds (more than 2000 chemicals). In most cases, PHASE performed better than CATALYST. For two series, the crystal structures of ligands complexed with their receptor were then used to guide the pharmacophore generation but the resulting models were not improved.

The program SUPERPOSE [264] is based on a pseudomolecule consisting of functional atoms with their properties rather than on real molecules. Hydrophobicity, H-bonding propensity (donor, acceptor, and donor/acceptor) determine the score attributed to each overlay. For 12 pairs of enzyme inhibitors, the best scored overlay reproduced the superposition derived from X-ray crystallography. In another test, for three THR inhibitors, superposition of an ensemble of conformers sampled by high-temperature molecular dynamics led to 13 sets of conformers selected for best common overlay. Among them, one was consistent with active conformation derived from X-ray.

1.3.2 USING SOME CRYSTALLOGRAPHIC INFORMATION

Of course, if the structure of a ligand bound to its receptor has already been solved by X-ray crystallography or NMR, some insights may be gained not only on the important interactions upon binding but also on the preferred conformation/position/ orientation of the ligand in the binding domain, and this is of major interest to guide the alignment. This experimental information on the active conformation (the conformation of the ligand bound to the receptor) is therefore used as a reference template to which the other molecules are fitted. A somewhat similar situation occurs when the known crystal structure only concerns a neighboring protein and not the receptor itself. A putative structure of this receptor active site can then be inferred by homology modeling (vide infra). More and more structures of complexes (ligand– receptor) are now solved, and no doubt, this will impulse the use of such approaches in CoMFA (or other alignment-dependent) QSAR models.

However, using the crystal structure of the complex ligand–LBD is not always without problems, as observed since the earlier applications [265]. Experimental geometries reflect minor geometrical distortions that may strongly modify the fields calculated on neighboring nodes in the lattice (e.g., due to the steepness of the steric potential). Close packing of residue side chains (as seen in the X-ray structure) might suggest that binding in the receptor pocket or H-bond formation could be forbidden [266], see also Refs. [267,268]. The problem of receptor adaptation upon ligand binding, the "induced fit," is examined in Chapter 10.

Earlier approaches: Klebe and Abraham [265], de Priest et al. [269], and Waller et al. [267], for example, used crystallographic information to derive alignment rules. But, in these examples, it was observed that using fitted (calculated) alignments (by multifit or field-fit) gave better results than those based on experimental observations. Let us remark that in Waller's work, field-fit minimization followed the alignment rules derived from crystallographic data.

In a study of 60 inhibitors of AchE, a promising drug in the treatment of Alzheimer's disease, Cho et al. [249] used the X-ray structure of three inhibitors (complexed with AChE) as templates for the superposition of their close analogues in the set. In a similar study on AChE inhibitors [270], building an alignment from the "pose" of only one representative compound gave acceptable models except for some outliers. But better results were obtained in a further study [271] by systematic docking (vide infra).

Subsequent approaches more deeply involve the receptor itself. For example, Waller and Marshall [217] revisited the correlations obtained by de Priest et al. on inhibitors of THER and ACE, compounds of potential interest for the management of hypertension. In addition to the introduction of a new field (HOMO field) and consideration of desolvation energies (with a Poisson model)—vide supra—they proposed a new approach to refine alignment. From observation of some common requirements for successful inhibition on 28 structurally diverse and flexible inhibitors, and systematic conformational search, a unique active site model was defined. But some gaps in the treatment of Zinc–ligand interactions (partly resolved with indicator variables) led the authors to propose a novel method of alignment with the introduction of *complementary receptor fields* (of the THER crystal) to be used as template on which to optimize the various ligands by field-fit. These fields were obtained by probing the residues present in the LBD and reversing the corresponding steric and electrostatic field values. Successful results were obtained for 88 compounds (among them 20 in test).

A semiautomated, flexibility-based procedure NewPred was proposed by Oprea et al. [268] in a study of HIV-1 protease inhibitors, already examined by Waller et al. [267]. The originality of the approach was to consider several conformers per compound (to be relaxed in the binding domain) and to explore alternative binding modes within the limits of the initial alignment rules. For charged ligands, results were significantly improved.

1.3.3 INTEGRATING 3D-QSAR AND MOLECULAR DOCKING

Alternately, docking provides a useful tool for the generation of predictive models and for the identification of the key structural features responsible for the selectivity and biological potency. Docking programs aimed to determine the best "pose" of a ligand into the receptor pocket, that is, the conformation, position, and orientation of the ligand, leading to the most favorable interactions with the receptor.

With the expansion of huge chemical databases and the outburst of combinatorial chemistry, increased interest was devoted to docking programs and their efficiency in high-throughput virtual screening. Owing to the recent interrelations between QSAR models and database treatments, this point will be developed in Chapter 11 devoted to "receptor-related" approaches. Below, we only briefly mention applications of some of the most widely used docking programs.

Suffice it to say that, schematically, docking programs tend to optimize receptor–ligand interactions using some molecular mechanics-type force field. The trouble is that their extensive use in virtual screening led to consider only crude force fields in order to achieve high throughput (thus privileging speed over high accuracy on binding energies). As a consequence, calculated interaction energies cannot accurately reproduce the variations in activity observed in limited series of compounds. In other words, these interaction energies cannot be directly used to derive QSAR models. Fortunately, there is now large evidence, supported by many applications, that the best pose(s) proposed by docking programs do represent realistic binding modes. Thus, a current trend expanded for exploiting the synergy between the accuracy of the structure-based alignment and the computational efficiency of ligand-based methods.

Gamper et al. [272] studied 27 diverse haptens binding the monoclonal antibody IgE(Lb4). The large structural diversity made usual alignment methods (field-fitting or active analogue approach) inefficient. However, a good model could be worked out with AutoDOCK program [273] using, at that time, Metropolis Monte Carlo optimization. A starting set was constituted of nine compounds possessing two or three docked orientations. The possible alignments were subjected to the CoMFA procedure and the alignment that gave the best q^2 value was chosen. The remaining compounds were added successively, choosing always the best q^2. The best model yielded $q^2 = 0.785$. Various assays with modified grids confirmed that tuning the model on one orientation out of several possible for each compound introduced no artifact. The predictive ability of the model was established using in test three new sulfur-containing compounds. However, one may suggest that an enlarged test set would have been desirable.

We already mentioned that in a study of AChE inhibitors [270], a satisfactory model (despite some outliers) was obtained using alignment guided by one pose of a unique compound. In contrast to this "ligand-based" approach, systematically docking all compounds in the receptor pocket ("protein-based" approach) gave better results, with no outliers [271]. It was suggested that the mutual adjustment of side chains and individual ligands in the "protein-based" model is the main factor of this improvement. In these works, it was also shown that a Kohonen self-organizing map allowed for analyzing the diversity of the data set and delineating the structurally similar subgroups where prediction was good, whereas it failed for series of divergent structure.

Sippl [59,274], in a study of 30 ER agonists with GRID/GOLPE, showed that using for alignment, the ligand geometry obtained by docking slightly outperformed a model with flexible alignment of the ligands on the reference template E2. He also observed that direct evaluation of binding energies led to bad results. This point is discussed in Chapter 11. AutoDOCK was also used by Chen et al. in a study of steroids binding the progesterone receptor [275] and of COX-2 inhibitors [187].

Flexibility of the ligands often led to consider an extended set of conformers in the search for the active conformation. However, Mestres et al. [276] showed, on TIBO (**1.10**) derivatives, that different solutions obtained from rigid matching converge when rigidity is relaxed and suggested that a reduced set of conformers per molecule may be sufficient in field-based similarity studies. They also emphasized that it is important to go beyond a pairwise similarity level to obtain consistent solutions in flexible matching. The best multimolecule alignment obtained was in good agreement with the binding geometry and orientation determined from protein–ligand crystal structure. See also Hannongbua et al. [277].

1.10

In a recent application, Guido et al. [278] studied the inhibition of trypanosomatid Glyceraldehyde-3-phosphate dehydrogenase (GADPH). These inhibitors constitute an attractive target in the fight against various parasitic diseases such as sleeping sickness. A large series of adenosine analogues (**1.11**) was explored in order to detect structural differences between human and parasite enzymes, with in view a selective inhibition. Seventy derivatives, with pIC_{50} on a range of 3.5 log units, were split into training (56 compounds) and test (14). For CoMFA and CoMSIA analyses, alignment was carried out with genetic optimization for ligand docking (GOLD) [279] and FlexX [280], using the crystal structure of NMDBA N^6-(1-naphthalene methyl)-2'-deoxy-2'-(3,5-dimethoxybenzamido) (compound 36 in the article).

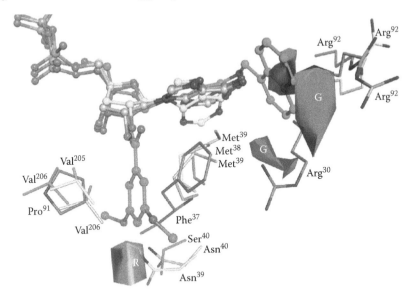

1.11

Both GOLD and FlexX correctly placed the template in the active site (Figure 1.10) and lead to nearly identical statistical results: in CoMFA, $r^2 = 0.93$ and $q^2 = 0.83$ (CoMSIA being slightly inferior). On the test set, a predictive r^2 about 0.8 indicated an excellent predictive power. Whereas for parasite enzyme, the electronegative area is near the NH_2 groups of Asn, or OH of Ser-40, in human GADPH,

FIGURE 1.10 **(See color insert following page 288.)** CoMFA contour plots in selective inhibition of trypanosomatid GADPH. GADPH structures are those of *Leishmania mexicana* (cyan), *Trypanosoma brucei* (yellow), *Trypanosoma cruzi* (gray), and human (magenta). The inhibitor and the cofactor NAD+ are drawn as ball and stick model. In CoMFA contour plots (for *L. mexicana*), the green area (G) corresponds to sterically favored region, and the red one (R) to favorable electronegative area. (Reproduced from Guido, R.V.C. et al., *J. Chem. Inf. Model.*, 48, 918, 2008. With permission.)

it is close to the carbonyl group of Phe-37. Similarly, the sterically favored area is near the Met side chains of parasite GADPH. But in human GADPH, these positions correspond to the more bulky phenyl group of Phe-37. Remark that CoMFA plots are relative to *Leishmania mexicana*. But the other two parasites have close sequence similarity whereas important structural differences are observed with human GADPH.

1.3.4   DOCKING VIA HOMOLOGY MODELING

In the absence of crystal structure of the ligand–receptor complex, some information may be gained from a previously solved complex with a homolog protein (i.e., with a high level of sequence conservation). A model of LBD can be rebuilt using sequence homology with a known receptor structure as template.

As example of homology modeling in CoMFA, we can cite the study of Jalaie and Erickson [281] on two sets of photosystem II inhibitors (benzo- and naphto-quinones on one hand, butenanilides on another hand). A homology model of spinach PSII was derived from the crystal structure of purple bacterium reaction center and used, joint to the DOCK program [282,283], to propose an alignment for CoMFA analysis. This alignment yielded highly predictive models whereas the usual atom–atom procedure failed, see also Refs. [266,270,284].

1.3.5   DOCKING AND PROTEIN FLEXIBILITY

Docking programs give a better view of ligand–receptor interactions and propose consistent schemes for the selection of the active conformation, especially if ligand flexibility is taken into account [283 and references herein]. However, experimental evidence indicated in some cases that in addition to ligand flexibility, the receptor adapted itself to the individual ligands: This corresponds to the so-called induced fit, which was approached with specific methods (Quasar, Raptor) in 5D- and 6D-QSAR models. This point will be developed in Chapter 10.

At a first level, in the framework of CoMFA applications, receptor adaptation was often ignored although some flexibility (e.g., limited movements of side chains) could be taken into account by docking programs [155,285].

In a different way, Knegel et al. [283] introduced receptor flexibility in an extension of the DOCK algorithm [282,286]. Their approach, based on an ensemble of protein structures, did not explore dynamically the protein flexibility but rather considered a sum of experimental information on an ensemble of proteins (derived from crystallography or NMR) and introduced averaging processes. In DOCK, the active site was modeled by a set of overlapping spheres forming a negative image of the active site used to orient the ligands. Interaction energies were calculated with AMBER steric and electrostatic potentials on a scoring grid filling the volume accessible to the ligands. Pre-calculating the protein contribution lowered computer time. A crude initial test discarded orientations where ligand atoms intersected the enzyme, to not consider too many orientations in the subsequent calculation. In this extension, basically, the scoring function was very similar to that of DOCK, with summations implying every atom of every ligand and every receptor atom. But now the sum was extended to all the members of the protein ensemble.

Two methods were proposed: "energy-weighted" average and "geometry-weighted" average. In addition to a scaling factor between steric and electrostatic terms, the trick was to smooth (with a sigmoidal function) or remove repulsive parts of the steric potential for flexible atoms (those with variable positions). The so-determined composite scoring grids allowed for docking differently shaped ligands, with orientations close to those of known ligands.

The implicit hypothesis was that, although a small loss of accuracy, a ligand that fitted well its own receptor still scored favorably with the receptor ensemble. The approach was validated on several examples corresponding to differing contexts: HIV protease where flexibility was related to rotations of side chains, rat p21 oncogenic protein with disorder on side chains and loops, uteroglobulin for which the ensemble of structures was determined by NMR, and retinol with limited flexibility. Binding orientations and computed interaction energies of known ligands were accurately reproduced. In addition, when the structures of conformationally different proteins were known, the use of composite grids made database searches faster.

1.4 PARALLEL DEVELOPMENTS MEET CoMFA

In addition to optimization of parameter settings, introduction of new fields, etc., CoMFA methodology took recently advantage of important advances in the perspective of data mining, dedicated to the search for pharmacophoric features, or recognition of a common substructure. Such new avenues offer possible applications for alignment of multiple ligands in the field of QSARs.

1.4.1 ALIGNMENT VIA HOPFIELD NETWORK

In atom alignment or field-fit alignment (usually used in CoMFA), the location and orientation of the molecule under scrutiny, with respect to the reference template molecule, is guided by an objective function depending on the geometrical position of selected atoms, or on field values on nodes. The novel method proposed by Funatsu et al. [287,288] searches for the subset of atoms (in both the template and the molecule) that has the more similar spatial arrangement, taking into account atomic properties, hydrophobicity, H-bond characteristics, etc. In these aspects, the method presents some analogy with pharmacophore-oriented approaches such as CATALYST [255,256] or PHASE [262,263].

This problem of combinatorial optimization was approached extending a method proposed by Doucet and Panaye [289] for pattern recognition thanks to Hopfield neural networks [290,291]. A brief presentation of Hopfield neural networks is given in the Tool Kit appendix. Suffice it to say that in a Hopfield network, neurons (or "units") are arranged in a single layer and are fully interconnected in symmetrical manner. Connection strengths between units are here fixed, depending on the problem to solve. Activation of neurons only takes binary values 1 or 0 (alternatively 0, −1 may be used). From an initial state (with random activation of units), the network evolves according to Hebb's rule:

$$\text{If} \quad \sum_j w_{ij}S_j - \theta_i > 0 \rightarrow S_i = 1$$

$$\text{Otherwise, } S_i = 0$$

where
the right part of the expression represents the input to neuron i coming from the other neurons j
θ_i is the threshold value
S_i is the activation of neuron i
w_{ij} is the weight of the connection between units i and j

When the network evolves according to Hebb's rule, it can be shown that the associated energy function

$$E = -\frac{1}{2}\sum_i \sum_{j \neq i} w_{ij}S_iS_j + \sum_i \theta_iS_i$$

always decreases or remains stationary. This property can be used to minimize symmetrical quadratic forms of binary variables [292,293]. The problem is mapped onto a Hopfield network, weights corresponding to the coefficients of the quadratic form to treat (Figure 1.11).

Solving the pattern matching problem (or recognition of corresponding atoms in a target molecule and a reference, template, or substructure) enables to establish a correspondence matrix between the atoms of the target and those of the reference. In this form, the problem presents a close analogy with the "traveling salesman problem," which is easily solved by a Hopfield neural network: given a set of cities, how to find the shortest path passing one time and only once by each city? The solution is represented by a correspondence matrix between the cities and their rank in the travel, the objective function to minimize being the length of the travel [294].

In the recognition of an atomic pattern in a molecule, the energy function of the network (the objective function to minimize) is the sum of five terms. The first two impose that there is at most one "1" ("match") in each line and column of the correspondence table. The third term maximizes the number of correspondences. The last two evaluate the distance between properties on associated sites and the consistency of each kind of property (a score matrix penalizes correspondences of atoms with too different properties). From this energy function, weights and bias may be calculated. After an initial randomization, the network evolves and at convergence, it delivers the correspondence table between template and target atoms. Once this best correspondence is found, usual geometrical superposition methods may be used to superimpose the molecules before embedding them in the CoMFA

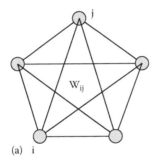

(a) i

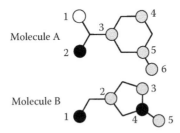

Molecule A

Molecule B

Property criterion						
	A_1	A_2	A_3	A_4	A_5	A_6
B_1	0	1	0	0	0	0
B_2	0	0	1	1	1	1
B_3	0	0	1	1	1	1
B_4	0	1	0	0	0	0
B_5	0	0	1	1	1	1

Solution (property + distance)						
	A_1	A_2	A_3	A_4	A_5	A_6
B_1	0	1	0	0	0	0
B_2	0	0	1	0	0	0
B_3	0	0	0	1	0	0
B_4	0	0	0	0	0	0
B_5	0	0	0	0	0	1

(b)

FIGURE 1.11 Hopfield network (a) and establishing a correspondence table (b).

lattice. See Appendix A for the derivation of the objective function to minimize. One drawback of Hopfield neural networks is that they often remain stuck in a local minimum (see, e.g., Refs. [289,295]). But this problem is relieved using a new algorithm [296] and a Boltzmann machine combining simulated annealing with a Hopfield network.

In a series of molecules, correspondences are determined by successive pairwise comparisons. An advantage of this approach is that matching can include, rather than the traditional steric and electrostatic fields, various features or properties, possibly weighted. The method can also handle suboptimal or partial solutions.

In the work of Funatsu et al. [287], the method was validated on 12 pairs of enzyme inhibitors. It successfully reproduces the real molecular alignments obtained from X-ray crystallography. In this first paper, the purpose was to examine whether the method could successfully reproduce the real molecular alignments. Thus, bounded

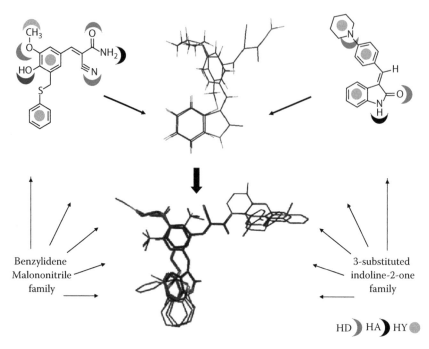

FIGURE 1.12   Alignment of inhibitors of HER2. Alignment of two selected molecules representative of the subfamilies is achieved with a Hopfield network. The other compounds are then built from these selected conformers. Criteria looked for are H-bond donor/acceptor propensity and hydrophobicity. (Adapted from Arakawa, M. et al., *J. Chem. Inf. Comput. Sci.*, 43, 1396, 2003. With permission.)

conformations obtained from X-ray structures were used. In a subsequent publication [288], this novel alignment method was applied to derive 3D-CoMFA-QSARs on two data sets; respectively, on one hand 27 inhibitors of human epidermal growth factor receptor-2 (HER2) and on another hand, 54 inhibitors of COX-2. From sets of conformers of two or three selected molecules, the alignment rule corresponding to the minimal Hopfield energy was selected (Figure 1.12). The other compounds are then built from these selected conformers. Although, each of these two series consists of compounds with different chemical skeletons, good alignments were obtained, leading to satisfactory CoMFA models.

For the 27 HER2 inhibitors (19 benzylidene malononitriles and 8 3-substituted indolines 2-ones): $r^2 = 0.805$, $q^2 = 0.701$ with SEE = 0.18 on 2.7 log unit on pIC_{50}; and for the 54 COX-2 inhibitors (data from Ref. [297]): $r^2 = 0.922$, $q^2 = 0.635$, SEE = 0.36 for a range of 5.2 log units, after elimination of five outliers.

1.4.2   GALAHAD

A new method dedicated to pharmacophore search, but also usable for alignment in CoMFA-type methods was recently proposed with GALAHAD (genetic algorithm with linear assignment of hypermolecular alignment of data sets) [298]. GALAHAD is designed to carry out flexible alignment of ligands that exhibit similar interaction patterns and shape when bound to a target protein; in other words, ligands that share pharmacophoric and pharmacosteric elements.

An important characteristic is the construction of a 3D hypermolecule, merging two superimposed entities (molecules or pre-existing hypermolecules). This may be viewed as a generalization of the concept of 2D hyperstructure, largely used in topological models, for example, in the DARC system [33,34]. Such a hypermolecule aggregates properties of individual molecules (each constituting a subset of the hypermolecule), but without losing information about the properties of these individual molecules.

Basically, GALAHAD, an extension of the LAMDA algorithm [299] adapted from image analysis, carried out pairwise alignment of rigid ligands in a two-step process. First, ligands considered as fully flexible entities are aligned to each other in internal coordinates via a GA operating on a set of models defined by a set of torsional angles. Atoms are characterized by their partial charges and relative positions. Among all possible atom pairs, a preliminary set of intermolecular equivalences minimizing the cost function is defined. It takes into account the neighborhood, evaluated by radial density functions, and the differences in interaction strength of the features (e.g., strong vs. weak H-acceptor).

Then, the best predicted models, on the basis of structural similarity, are filtered for geometric consistency before being aligned as rigid bodies in a common Cartesian space. Each successful overlay of ligand pairs creates a single hypermolecule, and this aggregative process is iterated.

Pharmacophoric and steric multiplets (tuplets), giving independent measures of 3D similarity, plus the total ligand energy, form the input of the fitness function of the GA. This generation of hypermolecules not only helps for alignment but also allows for partial-match queries. This point was addressed by Sheppird and Clark [300] who compared the GALAHAD approach with a single ligand conformation to queries derived (as usual) from an ensemble of randomly selected ligand conformations. The authors emphasized the fact that the GA treatment (with its multiobjective fitness function) generally produced multiple models, each of which representing a different trade-off among the considered criteria (internal strain, number of pharmacophoric features, total energy), hence a high flexibility. Various examples confirmed that the approach was able to generate pharmacophores and retrieve alignments in agreement with crystal structure.

GALAHAD was used by Long et al. [301] as an alignment tool for building CoMFA and CoMSIA models in a study of 24 thiophene–carboxamide analogues, inhibitors of IKK-2 and spanning a 3.1 log unit activity range. These compounds are potentially attractive targets for the development of new drugs against inflammation. This study was interesting since there was no real common pattern for the investigated set (a thiophene ring on one hand and a phenyl on another hand are found in three series but not on the fourth). Eight compounds (among the most active) were chosen to build a GALAHAD master hypermolecule used then as template for the alignment of the other molecules (Figure 1.13). The pharmacophore model generated from this template fitted well a homology model of IKK-2 indicating a rational and successful approach. In parallel to GALAHAD, a usual "common structure

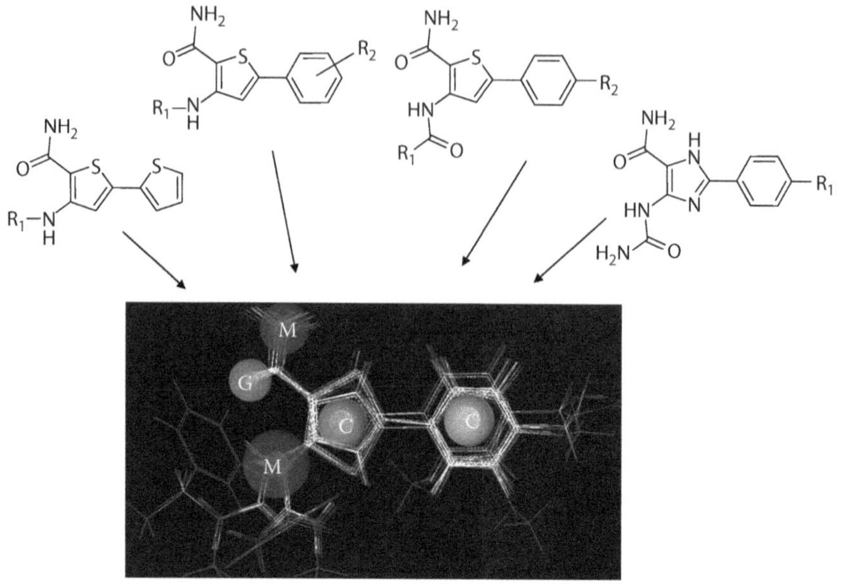

FIGURE 1.13 Structural type of IKK2 inhibitors and generated pharmacophore model. Pharmacophore features are represented as spheres: (C) hydrophobic center; (G) H-bond donor; (M) H-bond acceptor. (Reproduced from Long, W. et al., *QSAR Comb. Sci.*, 27, 1113, 2008. With permission.)

alignment" was performed and the two models compared in CoMFA and CoMSIA treatments. GALAHAD led to better results (q^2 = 0.642 compared to 0.590 for the common structure alignment), a result confirmed for the four compounds selected as test set.

1.4.3   1D TO 3D ALIGNMENT

Another approach to a simultaneous 3D alignment of multiple ligands was provided from Anghelescu et al. [302]. The key feature of the method is that the alignment problem is first solved in a lower dimensional space (1D representation of molecules), and the derived constraints guide the 3D process.

First, conserved pharmacophoric features are identified from a 1D representation (a canonical representation derived from topological atomic distances [303]). Six features are considered (H-bond donor or acceptor, positive or negative charged groups, aromatic and hydrophobic). The 1D alignment is then converted into a 3D one, using three types of constraints:

- Force field constraints to get reasonable geometries
- Alignment constraints based on the conserved features identified
- Knowledge-based constraints (e.g., strong evidence for the active conformation of one member of the series or knowledge of a co-crystal structure)

The method was able to identify common pharmacophoric elements in 10 highly flexible hERG channel blockers (72 rotatable bonds) and to retrieve the X-ray alignment of ER agonists and antagonists.

1.4.4 QUASI

Although not yet directly applied to 3D-QSAR, the QUASI methodology for simultaneous superposition of flexible ligands [304] affords an efficient alternative to the alignment problems inherent to several 3D-QSAR approaches (such as CoMFA and CoMSIA). Its use for proposing a putative receptor site model or screening databases is also foreseeable.

An original aspect is that the possibility for ligands to bind different regions of the receptor, leading to a global partial overlap is also accounted for. This problem was also considered by Mills and Perkins [305–308] with the method of null correspondences (badly matching regions are ignored). Generally speaking, a reference molecule is chosen to perform a series of pairwise superimpositions. Such superimpositions of subsets are time consuming and may be insufficient to generate an optimal simultaneous alignment of all ligands at a time. A possible solution was to perform pairwise comparisons and then analyze the results using consensus match [305,306].

In this new approach, an auxiliary receptor site model is created, which will serve as common reference frame. Similarity is defined by respect to this model, and not by reference to other ligands. The system is dynamic: receptor model and ligands coevolve until optimal self-consistency is achieved. Each ligand is characterized by a set of descriptor points: donor, acceptor, or steric points. H-bond donor points are located on the direction (H-donor atom—Hydrogen), 3 Å apart. H-bond acceptor points are located on the atom itself, and steric points encompass all atoms that are not H-bond acceptors. Two points (one donor, the other acceptor) are associated to amphoteric groups (as OH).

A receptor site model is first created from ligand descriptor points. These points are first extracted from the ligand with the largest number of descriptors, then points from other ligands are added, in a range of 1.0–4.0 or 5.0 Å of any existing point (to avoid disconnected pockets). The number of points is greater than that of individual ligands but less than the sum of all points. So ligands may bind different parts of the receptor model. The program starts with ligands in randomized conformations and orientations but with geometrical centers coincident. This generation of ligand descriptor points bears some analogy with CoMASA of Kotani and Higashiura [43].

1.4.5 N-Way PLS Applications

The success of CoMFA, as well as several alignment-dependent 3D-QSAR analyses, critically depends on a good selection of both the active conformer of every molecule at hand and the alignment rule. In the preceding examples, this choice was guided by user's intuition, pharmacophoric search, or docking experiments.

When no structural data are available, more objective approaches, particularly for flexible molecules, allowing for treating simultaneously several conformations per molecule, and/or various alignment rules were searched for, modifying the initial PLS algorithm [66,67,309–313].

1.4.5.1   From a Matrix to a Wad of Matrices

Schematically, when one alignment rule is defined, and one active conformation is selected for each compound, CoMFA analysis, for example, enables analyzing a data matrix where columns represent the field values at the lattice nodes, and rows the compounds. If one wants to consider more conformations and/or more alignments, the data block is no longer a matrix but an array of dimension higher than 2. Improvements of PLS method, 3-way PLS, 4-way PLS, and more generally N-PLS, address this problem.

1.4.5.2   3-Way PLS

A first method was investigated by Dunn et al. [310] on the example of trimethoprim-like derivatives, inhibitors of *E. coli* DHFR, a good test system since the receptor-bound conformation and alignment for the enzyme inhibitor complex are known. However, it was noted [66] that the algorithm they used came to unfold the data array and use the standard PLS, which may lead to problems. Furthermore, this system was investigated in the 4D formalism with better results [63]. The publication of N-PLS algorithm by Bro in 1996 [314] prompted various studies. We prefer to examine here these works in order of increasing complexity (even if it is not strictly the chronological order), focusing on the correlation results. For mathematical details, see Ref. [314].

Application of 3-way PLS to a rational choice of bioactive conformation was proposed by Hasegawa et al. [66] in a similarity matrix treatment and applied later to a CoMFA data set [312] on the example of 20 styrenes inhibitors of Protein–Tyrosine Kinase. For the most active compound, 12 conformations were selected as possible reference and for each compound of the set, the most similar conformations, within a range of 6.6 kcal/mol from the global energy minimum, were considered (from 4 to 24, depending on the molecule). This led to array $\underline{\mathbf{X}}$ (I*J*K) of dimension (20*1980*12) corresponding to a wad of 12 matrices (one per possible reference template) of size: 20 (number of compounds) * 1980 (field values on the lattice nodes) (Figure 1.14).

Schematically, according to Bro [314], this array $\underline{\mathbf{X}}$ (elements x_{ijk}) may be decomposed into a set of triads, consisting in one score vector, $\mathbf{t}$(I*1) and two weight vectors, $\mathbf{w}^K$(K*1) and $\mathbf{w}^J$(J*1). The score $\mathbf{t}$(I*1) must satisfy two conditions:

- $\mathbf{t}$ has a maximum covariance with $\mathbf{y}$(I*1). In other words, it is highly correlated with the dependent variable $\mathbf{y}$, the biological activity.
- $\mathbf{t}$ models the variance among the independent variables x as much as possible.

More precisely, $\underline{\mathbf{X}}$(I*JK) is expressed as the outer product of score vector $\mathbf{t}$(I*1) and weight vector $\mathbf{w}$(JK*1). $\mathbf{w}$ is further decomposed into two weight vectors, $\mathbf{w}^J$(J*1) and $\mathbf{w}^K$(K*1). So, $\underline{\mathbf{X}}$ and $\mathbf{y}$ are modeled by

$$\underline{\mathbf{X}} = \sum_{h=1,A} \mathbf{t}_h \mathbf{w}_h' + \mathbf{E}$$

$$= \sum_{h=1,A} \mathbf{t}_h \left(\mathbf{w}_h^J \bowtie \mathbf{w}_h^K \right)' + \mathbf{E}$$

and

$$y = \sum_{h=1,A} t_h u_h + f$$

where
- the character $'$ denotes the transpose
- $\mathbb{K}$ is the Kronecker product
- E and f are the residuals
- A is the number of components retained (that which gives the best squared correlation coefficient q^2 in L-O-O)

It is demonstrated that determining $\mathbf{w^J}$ and $\mathbf{w^K}$ comes to a single value decomposition of matrix $\mathbf{Z} = \mathbf{y'X}$, a $J*K$ matrix, where the jk^{th} element is the inner product of $\mathbf{y}$ and the column obtained by fixing the second and third mode of $\underline{\mathbf{X}}$ at j and k:

$$\text{Max}_{(\mathbf{w^J}, \mathbf{w^K})} \left[\sum_{i=1,I} t_i y_i \,\middle|\, t_i = \sum_{j=1,J} \sum_{k=1,K} x_{ijk} w_j^J w_k^K \,\middle|\, \|\mathbf{w^J}\| = \|\mathbf{w^K}\| = 1 \right]$$

That is, the covariance between $\mathbf{y}$ (dependent variable) and the score vector $\mathbf{t}$ is maximized at each component; these models are subtracted from $\underline{\mathbf{X}}$ and $\mathbf{y}$ and the next component is calculated from the residues. From $\underline{\mathbf{X}}$ and $\mathbf{y}$, the weight vectors are determined and then the scores as least squares model of $\underline{\mathbf{X}}$.

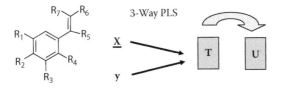

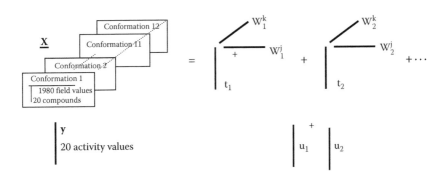

FIGURE 1.14   Decomposition of the wad of matrices by 3-way PLS.

It may be interesting to express the results with a direct relationship between $\mathbf{y}$ and $\underline{\mathbf{X}}$:

$$\mathbf{y} = \underline{\mathbf{X}}\mathbf{b}_{PLS}$$

with $\mathbf{b}_{PLS} = \mathbf{W}\mathbf{b}^*$ where $\mathbf{b}^* = (\mathbf{T}'\mathbf{T})^{-1}\,\mathbf{T}'\mathbf{y}$, $\mathbf{W}$ weight matrix (JK*A), and $\mathbf{b}^*$ modified PLS coefficient vector (A*1). From the correlation coefficient vector $\mathbf{b}_{PLS}$ (loading vector), it was possible to determine what sheet contributes the most to the activity, and so suggests a bioactive conformation. $\mathbf{b}_{PLS}$ is a vector of dimension (jk*1). The absolute sum of elements in each sheet (j*1) determined the importance of the sheet. Alternatively, one can use the score function given as

$$\text{Score function F } = \sum_{\text{(on the h components retained)}} pv_h w_h^K$$

where pv_h and w^K are the partial explained variance of $\mathbf{y}$ and loading value of the second weight vector at the hth component.

In the example at hand, the 3-way PLS analysis of this array $\underline{\mathbf{X}}(I*J*K)$ of dimension $(20*1980*12)$ led to a model with five components yielding $r^2=0.966$ and $q^2=0.491$. Selecting these "active conformations," a traditional PLS could be carried out leading to q^2 and r^2 values of 0.698 and 0.987, respectively. Interestingly, performing a CoMFA analysis on the minimum energy conformations (as it was sometimes done), yielded only $q^2 = 0.396$, highlighting the interest of 3-way PLS in the determination of active conformations.

1.4.5.3   4-Way PLS

In the preceding examples, only one alignment was considered. The method was later extended to select, in a CoMFA analysis, bioactive conformation and alignment rule simultaneously leading to 4-way PLS [311].

The study concerned the antagonist activity of 16 benzodiazepines (**1.12**) to CCK-B (a gastrointestinal hormone), molecules that may be interesting for the treatment of anxiety disorder. For 14 possible template conformers (selected on the most active compound), three different alignment rules were considered: one alignment on the seven-membered ring, another one on the exocyclic urea part, and a third one involving both amide parts.

1.12

With 5096 CoMFA descriptors, the array of independent variables $\underline{\mathbf{X}}(I*J*K*L)$ was of dimensions: $16*5096*14*3$ and the dependent variable $\mathbf{y}$ is a vector $(16*1)$. As in 3-way PLS, the basic formula is

$$\text{Max}_{(w^J,w^K,w^L)}\left[\sum_{i=1,I} t_i y_i \mid t_i = \sum_{j=1,J}\sum_{k=1,K}\sum_{l=1,L} x_{ijkl} w_j^J w_k^K w_l^L \mid \|w^J\| = \|w^K\| = \|w^L\| = 1\right]$$

After determination of the weight vectors in all components, the regression equation can be written as

$$y_i = \sum_{j=1,J}\sum_{k=1,K}\sum_{l=1,L} x_{ij\,kl} b_{jkl} + e_i$$

where $\underline{\mathbf{B}}(J*K*L)$ is the regression coefficient array and e_i the residual error. It is possible to calculate the relative contribution to the model of a particular alignment (resp. conformation) from the regression coefficient array and so, determine the most significant alignment and conformation. For example, for alignment (l),

$$W = \frac{\Sigma_j \Sigma_k B_{jkl}^2}{\left\{\Sigma_l\left[\Sigma_j \Sigma_k B_{jkl}^2\right]^2\right\}^{0.5}}$$

After their selection, a classical 2-way PLS led to good results ($r^2 = 0.973$ and $q^2 = 0.738$), and the treatment was validated on five other compounds. From this final model, contour plots suggested that bulky substituents in the side chain decreased activity whereas a long chain was beneficial. The presence of an *ortho* hydrogen on the pendent benzene ring and H-bond donation from the nitrogen of the urea part were also important.

1.4.6 TOPOMER COMFA

Generally speaking, in CoMFA, alignment is considered as a cumbersome problem. However, according to Cramer et al. [315], the classical approach may overemphasize receptor-bound conformation, minimum energy structure, and field fit. Thus, it makes sense to take advantage of the Topomer methodology for generating an alignment of structural fragments. This approach, presented as "objective and universal," was developed for browsing conventional databases: 3D structures are compared to sets of fragments or "topomers" [315,316]. "Topomer" means "a conformation of a fragment." These fragments (a single conformation, generated from rules) are assembled and oriented by their hanging bonds, and characterized by their steric shape (somewhat similar to CoMFA field).

The ability to reliably generate, for combinatorial library design, 3D models that can be superimposed may also be interesting for automatic 3D-based QSAR analyses. This would allow for treating the enormous quantity of structure-activity data produced by high-throughput screening. The beginnings of this approach can be traced back to the search for efficient method to screen huge combinatorial databases where a common core links up, in a similar spatial position, "side chains" supplied by the reactants. The problem here is to maximize the molecular diversity of candidates in order to discover new leads (avoiding redundant testing) or maximize similarity with known leads [316].

This concept of topomer (and steric field it generates) was first proposed as a molecular diversity descriptor, unrelated to any particular active conformation [316]. The basis of the method is the remark that, in a series of molecules, the variations in activity are only due to the variable part of each structure. These side chains will be clipped off and modeled as topomer fragment, defined by a single conformation with an open valence (attachment bond). An absolute orientation may be provided by overlapping this link to a common core in the series and using a topomeric protocol to select a representative conformation from a standardized 3D model by rule-based adjustments. In other words, a useful alignment may be obtained, simply overlaying the atoms within some selected substructures, the other atoms being arranged with general rules (Figure 1.15).

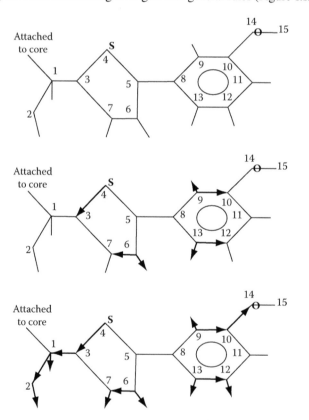

FIGURE 1.15 Example of the generation of a topomeric conformer. (Reproduced from Cramer, R.D. et al., *J. Med. Chem.*, 39, 3060, 1996. With permission.)

Typically, the optimal situation corresponds to series of congeneric compounds sharing a common moiety ("core") and varied side chains. A neighboring case is that of a series of roughly homogeneous molecules where two or several large groups (non identical throughout the series) are connected by acyclic bonds. These groups constitute the topomer fragments (the core being the acyclic bond). Series as steroids (with one, nonidentical, large group, and noncentral acyclic bonds) are less suited to the topomer approach and more easily treated by subgraph superposition. At last, when homology is negligible, identifying more homogeneous subseries is more efficient.

To generate the topomer conformation, the following processes are employed:

- Input structures are each broken into two or more fragments at the central acyclic bond, while removing any core fragment common to the entire series.
- A distinctive "cap" is attached to the open valence and the complete structure is generated by Concord model builder [317,318].
- The model is oriented to superimpose the cap attachment bond onto a vector fixed in a standard topomer grid (typically, 18 Å edge, mesh 2 Å).
- Starting from the root attachment bond, the groups are placed by decreasing importance order. Torsional angles are adjusted. Stereocenters are ignored since chirality is seldom specified.
- Removal of the cap completes the topomer generation.

The approach was developed on a set of 736 commercially available thiols. From similarity in topomer steric fields, 231 clusters were proposed in good agreement with the classical concept of bioisosterism (similarity in shape, although the structures differ). This topomer approach is well suited for the design and screening of neighborhood-based virtual combinatorial libraries created from sets of predefined fragments. But, application to conventional heterogeneous databases requires fragmenting each structure before topomer comparison [319]. Unlike pharmacophore-based 3D searching, topomer comparison considers a combination of fragments rather than a complete structure, with all atoms (and not a small subset), and characterizes their shape by the steric fields and not by some specific geometrical features. Only one topomer conformation is involved rather than a large variety of conformations. With the *dbtop* program, topomer similarity searching appears very efficient on 26 PDE4 inhibitors and 15 serotonin receptor modulators, and outperforms standard Tanimoto similarity coefficients of 2D fingerprints.

In QSARs, the problem is to convert structure-activity observations to sets of mutually comparable fragments. Once standard topomer representations are constructed for each fragment, a set of steric and electrostatic fields is generated for each set of topomers. The "Topomer CoMFA" approach predicts the activity contribution (partial activity) for each side chain. These contributions are supposed to be completely additive and independent (neglecting possible interactions between fragments or with the core).

To test the efficiency of this method, the author reconsidered 15 QSARs previously carried out with the classical CoMFA method (with careful structure alignment) and compared the results with those derived from the topomer alignment (context ignorant). On one hand, a "standard CoMFA" analysis was carried out, the only difference being the topomer model. On another hand, three changes were introduced in the so-called standard topomer CoMFA:

- An attenuation factor limits the importance of distant atoms (which are likely to easily adjust themselves into the receptor).
- Field values are rounded up to a few discretized values. Giving identical values to many lattice points reduces the number of terms in the analysis.
- The number of PLS LV is chosen at the minimum SDEP (and not on the maximum q^2), a more conservative option.

Standard CoMFA leads to weaker q^2 than literature, but the difference is about only 0.1. Standard topomer CoMFA yields still smaller values, but with simpler models (less LV), so that the approaches may be considered as equivalent (in average, variation of q^2 0.134 and 0.6 PLS components). On the 15 investigated series (847 structures), the average q^2 is 0.520 (literature: 0.636) and in prediction, SDEP 0.688 (literature 0.553) for 133 structures. So, "useful prediction does not require manual alignment." It was noted that topomer CoMFA was as efficient as the usual docking procedure that may involve unwanted motions of the common core.

As further development, the concept of topomeric fragment is particularly useful for queries into virtual libraries already composed of topomer structures, in order to automatically search for structures with increased potency [315]. The topomer CoMFA search predicts activity of each side chain that partially satisfies a topomer similarity search criterion. For these similarity searches, two descriptors are used: pharmacophoric features (from a list) and steric field (average value). In this approach, a library was built with 69,751 nucleophilic synthons (19 families) and 89,509 electrophilic synthons (26 families) commercially offered. In 13 of the 15 topomer CoMFA searches (in the series previously investigated), it was possible to retrieve a combination of fragments that would be more potent that any structure described in the original publication (average predicted potency increase = 20×) [315].

1.4.7 CoMFA in Classification

The concept of molecular similarity and its reverse notion dissimilarity (or complementarity) are key concepts in drug design [46]. They are, for example, of prime interest for guiding the design and synthesis of new chemicals. Although primarily designed for QSAR, i.e., correlation, the CoMFA fields can also be considered as structural descriptors able to similarity characterization for some classification problems.

In early screening stages, where noise may be important and activities only roughly assigned, a more qualitative analysis in a rating classification, however, provides interesting suggestions. For such applications, Ohgaru et al. [320] proposed to couple CoMFA with ordinal logistic regression (OLR). This approach was tested on the steroid benchmark and on 31 ACE inhibitors (three classes on a 6.2 log unit range) and appeared robust for both predictive ability and 3D graphical display.

In a subsequent publication [321], a more efficient ordinal CoMFA was developed modifying the preceding logistic CoMFA. First, logistic CoMFA was applied to PLS-generalized linear regression rather than ordinary ordinal linear regression, and second, ridge penalty estimation was introduced. This new model appeared superior.

Principal Properties and similarity calculation: According to Clementi et al. [322], interaction of a system with several probes (e.g., GRID probes) allows for extracting information about the interactions of a molecule with its environment and so defining some of its intrinsic properties. These "principal properties," statistical summary of the behavior of a system, constitute an invaluable tool for selecting the most informative structures in a QSAR.

A first attempt was due to Lin et al. [323], who used CoMFA steric field values in a principal component treatment to define principal properties that can be combined in pairs or triplets in view of shape similarity. GRID probes have also been used by Cocchi et al. [324] in a characterization of the natural amino acids.

However, a drawback was the choice of a conformation and the need for alignment. This difficulty was removed in the work of Langer [325], who considered 72 aromatic moieties (five or six-membered monocycles and benzo-fused bicyclic heteroaromatics with one or two heteroatoms). These groups offered a clear interest in medicinal chemistry as building blocks for bioisosteric replacement. In this application, geometrical constraints were relieved since the systems presented no conformational flexibility and could be aligned by the direction of the dipole moment. Six GRID probes (representative of the main types of interactions) were used and the fields calculated for about 3500 nodes. In a PLS treatment, the first three principal components explained 78% of the total variance and led to a hierarchical clustering in agreement with the common chemical knowledge.

Another application relied on modeling the activity of 16 pyridin-2(1H)-one derivative inhibitors of HIV-1 RT [326]. A satisfactory correlation was obtained with the first two PCs. A neighbor approach was developed by Clementi et al. [322] on 44 heteroaromatic systems. After PCA and clustering, the authors proposed to select 10 heteroaromatics (such as pyrrole, thiophen, indole) as covering the domain of possible structural variations and spanning at best the heteroaromatic space (hydrophobic/hydrophilic character, H-bond D/A propensity, presence of sulfur or oxygen). The approach was developed on 45 heteroaromatics. In addition to the calculation of hydrophobic and hydrophilic surfaces and volume, the best interactions were calculated with six probes (N1, N=, N1$^+$, OH(sp^2), O(carbonyl), O1 (sp^3 OH)) and three multiatom probes (COO$^-$, CONH$_2$, amidine). From a PCA treatment of the scaled data matrix, four principal components were retained. The first component (40% of the variance) separates hydrophobic/hydrophilic behavior, whereas the second one (16% of the variance) relies on the H-bond D/A character. The third PC (16%) expresses shape and hydrophobicity; and the fourth distinguishes systems containing oxygen or sulfur. These four *principal properties* allow for describing the chemical space of the investigated systems into 16 subspaces in a scale of interest in medicinal chemistry for the characterization of building blocks for bioisosteric replacement.

From the CoMFA matrix, it becomes possible to define similarity criteria; however, as well as for classical QSARs, the problems of alignment may limit the generality of the process for extended applications in experimental design.

1.5   COMPETITIVE BINDING

A basic assumption in CoMFA methodology is that all ligands bind the receptor in a same (or similar) manner. However, the question may be posed of different conformations of a same molecule binding a same receptor or, even, of different binding modes. The problem of the action of a conformational mixture was approached by several other ways and will be treated in a separate chapter "4D-QSAR": the simultaneous presence of various conformers constituting the "fourth dimension."

The problem of multiple binding sites was approached by Lukacova and Balaz [62] in a study of the affinity of polychlorinated dibenzofurans (PCDFs) to the aryl hydrocarbon receptor AhR) [327], an intracellular cytosolic protein. This is an important problem since halogenated aromatic hydrocarbons represent an important group of largely widespread and persistent contaminants (particularly owing to their lipophilic character). For these compounds, the structure of the receptor is still unknown, and homology models difficult to build (poor sequence identity). These series have been largely investigated by different 2D or 3D approaches [328], but the possibility of multimode binding in the receptor site was not considered before. Note that for these molecules, no conformational problem occurs and that ambiguity due to symmetry is reduced thanks to IUPAC nomenclature convention.

Such a situation may occur when ligands can bind in different modes or when a ligand binds in an unknown unique mode among several a priori plausible. The approach is similar to that of competitive complexation reactions from a single substrate, where the global association constant is the sum of the individual constants. This competition makes the relationship between the binding energy and the probe interaction energies nonlinear. But a linearized form (from a Taylor expansion) may be iteratively solved until self-consistency. We report here the principle of the method proposed by Lukacova and Balaz [62]. The global association constant (corresponding to the observed activity for molecule i) is

$$K_i = \sum_{j=1,m} K_{ij}$$

where m is the number of binding modes.

Activity is correlated to the CoMFA interaction energy, for molecule i and binding mode j, by

$$K_{ij} = \exp\left[C_0 - C_e * E_{ij} + \sum_{k=1,t} C_k * x_{ijk} \right]$$

k identifies the node of the lattice where the field x is evaluated (such as steric and electrostatic), and t is the total number of such evaluations = number of nodes * number of fields. The second term in the right part corresponds to the influence of conformational energy. In the treated example, where there is no rotatable bond, it may be skipped.

So
$$K_i = \sum_{j=1,m} \exp \left[C_0 + \sum_{k=1,t} C_k * X_{ijk} \right]$$

Remark that for a unique binding mode, the preceding equation is the usual CoMFA relationship, given as

$$\log K = C_0 + \sum_{k=1,t} C_k * X_{ik}]$$

The general equation is nonlinear, and the trouble is that, as generally in CoMFA, the number of coefficients C is larger than the number of compounds. But the trick is to use a limited Taylor expansion:

$$\exp(x) = \exp(M)*(1 - M + x) \qquad \text{if x approaches M. Here, near M} = \ln K_{ij}$$

$$K_i = \sum_{j=1,m} K_{ij} * \left(1 - \ln K_{ij} + C_0 + \sum_{k=1,t} C_k * X_{ijk} \right)$$

or

$$K_i - \sum_{j=1,m} K_{ij} *(1 - \ln K_{ij}) = C_0 * \sum_{j=1,m} K_{ij} + \sum_{k=1,t} C_k * \sum_{j=1,m} K_{ij} * X_{ijk}$$

This equation is solved iteratively, after normalization (i.e., replacing all K_{ij} terms by K_{ij}/K_i since $K_i = \Sigma_{j=1,m} K_{ij}$). This involves settings of the coefficients C, calculation of the K_{ij}, then calculation of the next values of the coefficients C by PLS, and iterating the process until convergence. Details may be found in Lukacova and Balaz [62].

For 34 PCDFs, various binding modes (2, 4, or 16) were investigated, corresponding from an alignment of medial atoms of PCDF (atoms 1, 4, 6, 9) and TCDD (2,3,7,8-tetrachlorodibenzo-p-dioxin) to flip- (left/right or up/down) and shift-motions toward C-7, C-8 and H-1, H-9 (for better overlap on the edges) (Figure 1.16). Then, GOLPE selected the variables before PLS optimization. Assuming a unique binding mode for the 22 compounds of the training test, $r^2 = 0.963$ ($q^2 = 0.786$); whereas for 16 modes, $r^2 = 0.999$, $q^2 = 0.961$, with even a significantly limited number of variables as compared to "usual" CoMFA. The prevalence of the binding modes was specified for each compound. Most of them exhibited one or two modes, but (1, 2, 3, 7) PCDFs would have four significant binding modes. Of the 16 potential

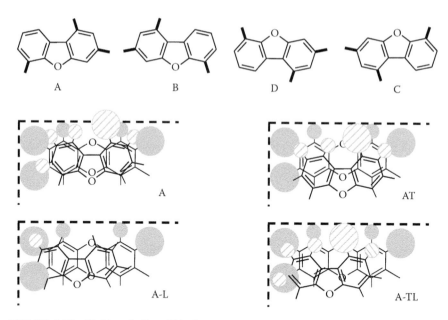

FIGURE 1.16 Multimode ligand binding. Upper part: four modes (A–D) corresponding to flipping up and down, right and left for the 1,3,6 PCDF. Lower part: binding modes generated from mode A for this molecule (hatched spheres) modifying the alignment by shifts, in the plane of the skeleton, to the top (T), the left side (L), the top left corner (TL) of the box enclosing the TCDD molecule (in gray). The binding cavity is represented as a rectangular box (only two walls are indicated). Similar schemes hold for the other modes B–D. (Adapted from Lukacova, V. and Balaz, S., *J. Chem. Inf. Comput. Sci.*, 43, 2093, 2003. With permission.)

modes, three represented about 60% of the whole. And surprisingly, they correspond to edge-shifted positions (in the hypothesis of a "rectangular-box" receptor).

1.6 CONCLUDING REMARKS

Since its inception 20 years ago, CoMFA has been extensively used as the archetype of 3D-QSAR. But in addition to simple correlation equations,

$$Activity = F(steric\ and\ electrostatic\ fields)$$

diverse and efficient extensions appeared; the simplified picture of steric and electrostatic fields can now be supplemented by other fields (polarizability, hydropathic, H-bonding) by means of selected probes. Although the method relies only on ligands (without any compulsory knowledge of the receptor), the contour plots, which, however, must be analyzed with some caution, give some insight about the interaction mechanisms between ligand and receptor, and may suggest structural changes for a better activity.

One of the major limitations of this approach, the alignment of the molecules under scrutiny on a common template, has been now largely relieved by different methods: pharmacophore identification specifies anchoring points, analysis of complexes between a ligand and a receptor, solved by X-rays, provides information on important interactions, interactions that can be also approached by docking methods. Topomer CoMFA also proposed efficient way to generate alignment of structural fragments. At last, the initial necessary hypothesis on an active conformation and a common binding mode (in fact, the alignment) can now be avoided by the N-way approach.

2 CoMFA-Type Grid Methods: CoMSIA, FLUFF, SAMFA, and SOMFA

On one hand, comparative molecular surface analysis (CoMFA) implies calculations of field values on the numerous nodes of the lattice. On the other hand, the expression chosen for the potentials is not without problems. Lennard-Jones potentials are very steep near the atomic positions. Hence, the need for very precise alignment rules (a small variation in position leading to largely different values of the steric field) and the choice, in order to treat simultaneously the steric and electrostatic components, of (somewhat arbitrary) scaling factors, which leads to some information loss. As a consequence, contour maps sometimes present singularities, making their interpretation unclear. Furthermore, limiting structural influences to steric and electrostatic contributions may seem a rather crude approximation (although introduction in CoMFA of additional variables and other fields was proposed).

So, new approaches were developed to alleviate these problems. SAMFA and SOMFA were proposed avoiding these somewhat heavy calculations, whereas CoMSIA or FLUFF-BALL relied on similarity calculations. In addition, an enlarged panel of fields (5) is provided by CoMSIA. Although implying similarity evaluation (developed in Chapter 5), these methods bear strong analogy with CoMFA. Furthermore, numerous studies simultaneously involve both CoMFA and CoMSIA. For these reasons, we prefer to present these approaches in connection with CoMFA rather than in Chapter 5, which is mainly dedicated to similarity matrices.

2.1 COMPARATIVE MOLECULAR SIMILARITY ANALYSIS

The interest for introducing different fields in order to have a more precise analysis of the diverse influences intervening is one of the raisons d'être of CoMSIA proposed by Klebe et al. [40,329]. For example, entropic contributions, solvation loss (by immobilization of the ligand on receptor binding), and hydrogen bonding may be insufficiently covered by van der Waals and Coulomb potentials. These problems are also addressed by CoMSIA.

2.1.1 BUILDING A CoMSIA MODEL—SIMILARITY EVALUATION

After alignment and embedding of the molecules in a 3D lattice (as in CoMFA), the similarity between the atoms of the studied molecule and a probe is evaluated for five properties related to steric, electrostatic, hydrophobic, H-bond donor, or acceptor fields. For property k, the similarity index at grid point q for molecule j is calculated according to

$$A_{F,k}^{q}(j) = -\sum_{i}\left(W_{probe,k} \cdot W_{ik}\right)\exp\left(-\alpha r_{iq}^{2}\right)$$

where
   summation over all atoms i of molecule j
   r_{iq} is the distance between the probe and atom i of the tested molecule
   W_{ik} is the actual value of property k at atom i (volume for the steric part, AM1
      charges, experiment-derived rules for H-bonds, and atom-based parameters for
      hydrophobicity [330])

For the probe, radius r is 1 Å, and other properties are set to arbitrary value +1 (charge, hydrophobicity, H-bond donating, and H-bond accepting). A Gaussian-type attenuation factor generally set at $\alpha = 0.3$ gives smoother variations. With $\alpha = 0.3$, at a distance of 1 Å, attenuation is 26%, at 2 Å distance, 70% of the initial value [43]. Introduction of this attenuation factor avoids singularities on the contour plots and allows for calculation inside the molecular surface. It also makes the analysis less sensitive to grid spacing and changes in lattice position. Data treatment is then carried out with partial least squares (PLS) [78,165,166] or with a variant, SAMPLS [144] more efficient in leave-one-out cross-validation (L-O-O). Neighboring expressions of similarity were also used in steric electrostatic alignment (SEAL) [151] and Labute's treatment [156] or in the similarity matrix of Kubinyi [115].

According to Klebe et al. [40], using more fields than CoMFA does not necessarily increase prediction accuracy (fields may be intercorrelated). But it allows for a sharper analysis displaying space regions where the contributions of the different fields are important for the biological activity and indicating particular characteristics that may be useful for the design of new improved ligands. Particularly, similarity indices may be calculated on nodes within the molecular volume (which is not possible with CoMFA). In CoMFA, contour plots indicate regions of space where molecules would favorably (or not) interact with a possible environment. In contrast, CoMSIA field contributions indicate nodes where a particular property may be given a higher weight to enhance an activity, making contour maps easier to interpret. Thus, contour plots may be more easily used to suggest structural modifications in order to design more active derivatives [331]. However, as in CoMFA, contour plots only reflect structural variations present in the selected population. Difference maps may indicate discriminating regions monitoring selectivity [331,332].

It was also shown that CoMSIA was less sensitive than CoMFA to changes in the orientation of the superimposed molecules, variations of q^2 on translations, and rotations of the superimposed molecules remaining inferior to 0.10 in log RBA. This was attributed to the softer variations of the fields compared to the steepness of the CoMFA fields (especially for van der Waals interactions). As CoMFA, CoMSIA often constituted a reference for establishing the reliability of new QSAR models. So, it will be largely cited in the following chapters. Generally speaking, CoMSIA results are often better than those of CoMFA (but with a greater number of components, which is not surprising on account of the more numerous fields considered). We will just give here few examples and put the emphasis on applications regarding selectivity toward different receptors.

2.1.2   THE CoMSIA FIELDS AND THEIR RELATIVE IMPORTANCE

The first example [40] concerns

- The steroid benchmark for which $q^2 = 0.662$; $r^2 = 0.941$ with steric, electrostatic, and hydrophobic fields. These results are quite comparable to those of CoMFA. Alignment based on SEAL gave similar results.
- Sixty-one thermolysin inhibitors (plus 15 compounds in test). Here also, statistical significance was comparable to a CoMFA treatment.

In a further publication, Böhm et al. [331] studied the activity of 72 derivatives of 3-amidinophenylalanine (**2.1**), inhibitors of trypsin, thrombin, and factor Xα, on an activity range from 3 to 4.7 log units. Sixteen other molecules constituted a test set. Specific inhibition of serine proteinases is of great interest for problems related to the formation of blood clots, a process involving a cascade of reactions. See also Robert et al. [333].

2.1

Ligands were placed from crystal structures of thrombin, trypsin, and factor Xα. Good correlations between predicted and observed affinities were obtained. Factor Xα was an exception with a low r^2_{pred} although an acceptable q^2 in training. To get some insight about the criteria governing selectivity between trypsin and thrombin inhibitors, a CoMSIA analysis was also carried out on the affinity differences. Although this was a greater experimental uncertainty, consistent results were obtained (Table 2.1).

It was suspected that the five fields introduced in CoMSIA may present some collinearity. This point was considered in the study of Böhm et al. [331]. With 5 fields,

TABLE 2.1

Statistical Results of CoMFA and CoMSIA Models for 3-Amidinophenylalanines Inhibitors of Serine Proteinases

	Thrombin		Trypsin		Factor Xα	
	CoMFA	CoMSIA	CoMFA	CoMSIA	CoMFA	CoMSIA
q^2	0.687	0.757	0.629	0.752	0.374	0.594
r^2	0.881	0.950	0.916	0.972	0.680	0.915
LV	4	6	5	9	3	6
r^2_{pred}	0.470	0.432	0.650	0.842	0.384	0.164

Grid mesh = 2 Å

Source: Excerpt from Böhm, M. et al., *J. Med. Chem.*, 42, 458, 1999. With permission.

31 combinations were possible (from only a unique field among the 5 considered to all fields together). For thrombin, trypsin, and factor Xα inhibitors, the various possible combinations gave rather similar results with q^2 values between 0.6 and 0.8. For another example, see Hu et al. [334] and Dias et al. [335].

It is clearly admitted that adding new fields in CoMSIA, compared to CoMFA, gives more information about the important chemical features monitoring activity. However, complicating the model may increase the difficulty of analyses and even adding new explanatory variables, with perhaps some redundancy, may cause overfitting and decrease the predictive ability. In an extensive study, Dias et al. [335] collected 23 data sets (more than 2000 data points) previously studied in the literature. They carefully examined the relative importance of the five fields classically proposed in CoMSIA (steric, electrostatic, hydrophobic, H-bond donor, or acceptor), and the predictive ability of the so-derived models. Such a study might give some insight as to the best choice of descriptors for analyzing a given type of data set. It might also be hoped that the number and the diversity of the collected data would allow for a better understanding of the effects of methodological variations in performing CoMSIA treatments.

The quality of the different models was examined with the cross-validated L-O-O correlation coefficient q^2. For the 31 possible combinations, depending on the model, q^2 significantly varied; and, generally speaking, models with more fields were (as expected) more predictive. Hydrophobic field was more predictive than steric or H-bonding fields, if used alone. But, with three fields, all models were nearly equivalent. The usefulness of the hydrophobic field, already indicated in various studies, was confirmed in this study. It was the best in the unique-field models, and it had the equal-highest relative contribution when all fields were used.

The most commonly used fields in CoMSIA are steric and electrostatic followed by hydrophobic. However, hydrophobic and electrostatic fields had statistically a greater relative contribution over steric field. It was also observed that with steric field only, results were inferior to those of models using steric plus electrostatic fields and these were also inferior to models with steric plus electrostatic plus hydrophobic fields. Although less conclusive, it was noticed that H-bonding fields generally showed a secondary importance compared to electrostatic and hydrophobic fields.

Results indicated great redundancy between the fields (already noted by Böhm et al. [331]). All fields were (more or less) predictive if used alone. But adding a new field to others was of decreasing interest since an important proportion of information was already contained in the fields considered. However, on the basis of the examined results, increasing the number of explanatory variables did not induce overfitting, as it might be suspected.

2.1.3 NEW HYDROGEN-BOND DESCRIPTORS

The importance of hydrogen-bonding interactions in drug–receptor binding prompted Böhm and Klebe [336] to refine the description of the hydrogen-bonding fields proposed in CoMSIA [330]. Putative hydrogen-bonding sites were generated around functional groups according to a limited set of rules determined from crystallographic data on known examples. For example, such sites were placed for an NH group, 1.9 Å from the H atom, along the bond, and for a carbonyl, 1.8 Å from the oxygen atom, in the direction of the lone pairs (Figure 2.1). These binding sites are the centers of the Gaussian functions used to evaluate the similarity with a probe located on the nodes of the lattice surrounding the molecules.

In this new approach, more sophisticated functions are extracted from the numerous crystallographic data collected in IsoStar. This knowledge base of nonbonded interactions gathers about 18,000 scatter plots showing the experimentally observed distribution of one functional group (the "contact group" or "probe") around a binding site. From these data, propensity maps could be derived by the SuperStar program [337] indicating the probability (by respect to a random distribution) to find a putative H-bonding site in a given location. The authors applied SuperStar maps (generated around small ligands) as fields in a 3D-QSAR model. Forty-six central groups were selected (such as methyl, phenyl, formyl, amide, nitro, aromatic, and chloro) to generate the hydrogen-bonding fields to be used in place of the original CoMSIA treatment. The authors also developed an efficient, faster, evaluation of the SuperStar descriptors based on a summation of Gaussian functions with parameters (width and centers) adjusted on experimental data.

The method was tested on a set of thermolysin inhibitors previously studied [40], comparing standard CoMSIA, GRID, and the so-modified CoMSIA (using the SuperStar H-bond donor and acceptor fields). The data set was split into training and test in the ratio 61/15. CoMSIA (with five fields) yielded $r^2 = 0.860$,

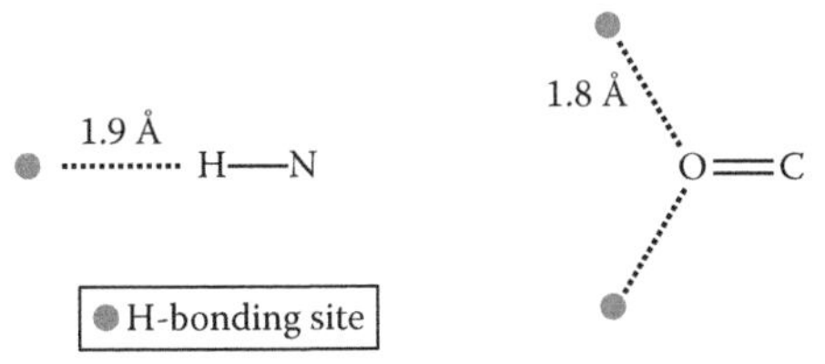

FIGURE 2.1 Location of hydrogen-bonding sites.

$q^2 = 0.536$, $r^2_{pred} = 0.308$, with five latent variables [LV]. GRID, gave similar results ($r^2 = 0.843$, $q^2 = 0.562$, $r^2_{pred} = 0.304$ with four LV) with five fields: acceptor (carbonyl oxygen), donor (NH with lone pair), alkylhydroxy OH (acceptor/donor), sp^3 carbon (steric/hydrophobic), and sp^2 CH aromatic or vinyl (steric/hydrophobic). Comparison was carried out with the seven fields associated with the contact groups in modified CoMSIA, three acceptors (carbonyl oxygen, alcohol oxygen, and terminal oxygen atom), two donors (amino hydrogen and alcohol hydrogen), one steric (aliphatic carbon), and one hydrophobic (aromatic carbon) and gave $r^2 = 0.951$, $q^2 = 0.573$, $r^2_{pred} = 0.492$. Disregarding CH-containing groups led to less satisfactory results, indicating that steric and hydrophobic properties were then less correctly treated.

Mixed models, where CoMSIA fields were complemented by SuperStar hydrogen-bonding fields (neglecting CH-containing fragments), gave the best results, $q^2 = 0.592$, indicating that steric electrostatic and hydrophobic fields were better represented by CoMSIA whereas H-bonding properties were better described by SuperStar fields (using two probes, a carbonyl oxygen and an amino hydrogen). Contour plots from SuperStar gave an information similar (but with more detailed features) to that obtained from classical CoMSIA.

2.1.4 SELECTIVITY PREDICTIONS VIA COMSIA

Beside activity prediction in individual processes, field analysis methods have been increasingly used for selectivity prediction, a particularly important concern for therapeutic applications where detrimental side effects must be avoided.

2.1.4.1 Dopamine D$_4$ Receptor

Dopamine D$_4$ receptor is a potential target for antipsychotic drugs used in the treatment of schizophrenia. However, unwanted side effects (such as movement disorders) have been attributed to the blockade of receptor D$_2$. Highly selective D$_4$ receptor antagonists are therefore needed, all the more so as the D$_4$ selective drug, clozapine may induce blood problems. The study of Boström et al. [332] developed CoMFA and CoMSIA models for structurally diverse antagonists of D$_2$ and D$_4$ receptors (see also Ref. [338]) with a special concern about selectivity).

The starting point was a previous determination of the D$_4$ receptor pharmacophore [339] with three phenyl rings (A, B, G), an ammonium nitrogen (C), and a site point (D) representing an H-bond acceptor, in the N–H direction, within the receptor cavity (Figure 2.2a). It was also shown that dopamine ligands can adopt three binding modes by respect to the aromatic rings (respectively, interactions with A, B, C, D or A, C, D, G, or B, C, D, G).

A data set of 32 structurally diverse D$_2$ and D$_4$ receptor antagonists (representing the three binding modes) was selected with activities spread over nearly 5 and 3 log units for the D$_4$ and D$_2$ receptors, respectively. Nine ligands were chosen as test set. For CoMFA and CoMSIA, alignment was performed on the pharmacophore elements with least-squares rigid-body superimposition. Calculations were made on protonated species. Hydrophobicities were calculated with the atom-based

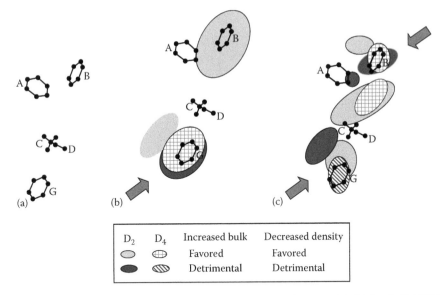

D$_2$	D$_4$	Increased bulk	Decreased density
⬭	⊕	Favored	Favored
⬤	▧	Detrimental	Detrimental

FIGURE 2.2 D$_2$ and D$_4$ receptors. (a) Basic D$_4$ pharmacophore model. (b) Steric CoMSIA contour plots for the D$_2$ and D$_4$ receptors. (c) Idem for electron density. Arrows indicate regions important for D$_2$–D$_4$ selectivity. (Adapted from Boström, J. et al., *J. Chem. Inf. Comput. Sci.*, 43, 1020, 2003. With permission.)

parameters of Viswanadhan and coworkers [340]. Potential locations of H-bonds are derived from experimental information [330,341].

Good results were obtained for both D$_2$ and D$_4$ receptor affinity (respectively, in CoMFA q^2 = 0.68 and 0.49, and in CoMSIA 0.75 and 0.51). Contour plots for the different fields gave some insight about the structures able to possess increased selectivity, CoMSIA contours being the more informative, since using more specific fields, and including regions inside the molecular volume. For steric contour plots, the most striking differences concerned a zone indicating detrimental interactions with the D$_2$ receptor of bulky ligands near the G ring, and an area near the B ring, where steric bulk favored activity. Conversely for D$_4$, steric bulk near G favors binding (Figure 2.2b). Electrostatics contours showed an opposite behavior near ring B, where an increase in electron density would improve D$_2$ receptor affinity on one hand and also near ring G, where electron-rich groups enhanced D$_4$ affinity on another hand (Figure 2.2c).

Similarly, in Figures 2.3 and 2.4 the H-bond acceptor contour plots represented with two compounds are indicated: pyridobenzodiazepine (**2.2**), a selective D$_4$ antagonist, and 2-phenyl-4(5)-[[4-(pyrimidine-2-yl)piperazine-yl]methyl]imidazole (**2.3**), a selective D$_2$ antagonist. The pyridine-type and amidine-type nitrogens of (**2.2**) face the contour indicating for D$_2$ (but not for D$_4$) a detrimental interaction for an H-bond donor of the putative receptor. For compound **2.3**, the pyridine-type and unprotonated imidazole nitrogens face, for D$_2$ and not in D$_4$, two contours where H-acceptor properties of the ligand would be detrimental. H-donor and hydrophobic contour plots are less distinctive.

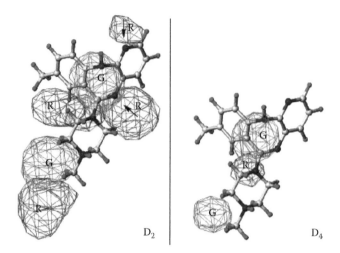

FIGURE 2.3 (**See color insert following page 288.**) H-bond acceptor contour plots for pyridobenzodiazepine (**2.2**). Red (R) contours enclose regions in which the presence of a ligand H-bond acceptor decreases (D_2 or D_4) receptor affinities. (Reproduced from Boström, J. et al., *J. Chem. Inf. Comput. Sci.*, 43, 1020, 2003. With permission.)

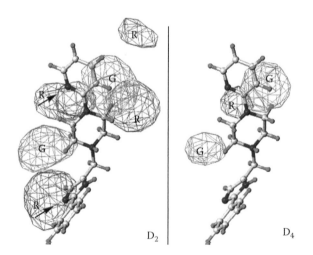

FIGURE 2.4 H-bond acceptor contour plots for 2-phenyl-4(5)-[[4-(pyrimidine-2-yl) piperazine-yl]methyl]imidazole (**2.3**). Area encoding (R, G) is the same as in Figure 2.3. (Reproduced from Boström, J. et al., *J. Chem. Inf. Comput. Sci.*, 43, 1020, 2003. With permission.)

2.2   **2.3**

2.1.4.2 Carbonic Anhydrase Inhibitors

Weber et al. [342] developed with CoMFA and CoMSIA 3D models for selectivity analysis of carbonic anhydrase (CA) inhibitors. This enzyme catalyzes the hydration of carbon dioxide to hydrogen carbonate plus a proton. CA is involved in very varied biological processes, and its inhibitors are used as clinical agents (against glaucoma, tumors, gastric ulcers, etc.). Unfortunately, 16 CA isozymes have been found in different tissues. Their widespread distribution poses important problems of side effects when nonselective inhibitors are administered. Various studies have demonstrated the importance of an unsubstituted sulfonamide group coordinated to a zinc ion in the binding pocket. Several QSAR studies were performed with classical descriptors, but the development of selectivity models was still challenging.

The study of Weber et al. focused on inhibitory activity of 87 chemicals toward three isozymes CA I, CA II, and CA IV. Binding affinities (pK) varied on a range of more than 4 log units. The compounds of the data set fell into six categories with different scaffolds: thiadiazole (**2.4**), thienothiopyrane (**2.5**), benzothiazole (**2.6**), benzenesulfonamide (**2.7**), hydroxamate (**2.8**), and hydroxysulfonamide (**2.9**).

2.4   **2.5**   **2.6**

2.7   **2.8**   **2.9**

In a first step, individual QSARs were established for inhibitors of CA I, CA II, and CA IV. X-ray data indicated that, for all inhibitors, essential interactions are coordination with a Zinc ion and H-bonds with Thr-199. Inhibitors were energy mini-mized in each of the three binding pockets leading to three distinct alignments. For each of these three alignments, three models were built with CoMFA and CoMSIA, with two grid meshes (1 and 2 Å) amounting 36 different models. All, but few excep-tions, were judged satisfactory, and their reliability was confirmed using a separate test set or y-randomization.

In the construction of selectivity models, the affinity differences (ΔpK) for each compound between two CAs were considered. The more robust models were obtained in alignment with CA II (perhaps owing to a greater number of crystal structures available). It is noteworthy that good selectivity predictions (especially for CA I/CA II) are obtained without any protein information.

Isocontours associated to CoMFA steric field or CoMSIA H-bond acceptor field were the more readily interpretable. Figure 2.5 displays the steric isocontours

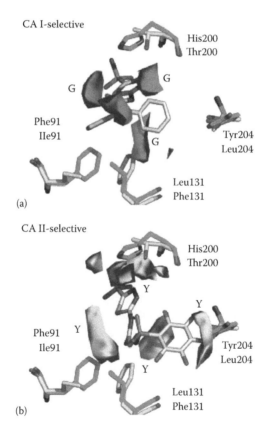

(a)

(b)

FIGURE 2.5   (**See color insert following page 288.**) CoMSIA contour plots and interacting receptor side chains. (a) Green (G) contours correspond to regions where bulky ligand groups increase selectivity toward CA I. Conversely (b) yellow (Y) areas indicate regions where ste-ric bulk increases selectivity toward CA II. (Reproduced from Weber, A. et al., *J. Chem. Inf. Model.*, 46, 2737, 2006. With permission.)

observed for two inhibitors: above, very active CA I inhibitor (**2.10**) and below, selective CA II inhibitor (**2.11**). In Figure 2.5a, dark gray areas correspond to regions where steric occupancy with bulky ligands increases selectivity toward CA I and in Figure 2.5b, light gray contours indicate regions where steric occupancy increases CA II selectivity.

2.10   **2.11**

Phe-131 of CA II penetrates the CA I selectivity enhancing area, whereas in CA II, Leu-131 is more distant. Similarly, several residues of CA I, Hist-200, Tyr-204, and Phe-91 interfere with the CA II-selectivity areas (whereas such interactions are avoided with the less bulky corresponding residues of CA II active site). Hist-200 in CA I restricts the space for bulky chains, whereas Thr-200 present at that position in CA II is sterically less demanding.

However, no good model can be obtained for the C II/C IV selectivity. Reasons may be the high intercorrelation between C II and C IV data, and the small affinity differences observed. The preceding QSAR models were further validated by examination of their prediction of binding affinity. Twelve known ligand inhibitors of CA were docked in the pocket of CA II with AutoDOCK [343] and treated with the CoMFA and CoMSIA models just established. Whereas CoMSIA model gave $r^2_{pred} = 0.497$, the AutoDOCK scoring function failed, giving negative r^2 values.

The efficiency of docking programs to propose the best "poses" (position, orientation, and conformation of a ligand in the receptor-binding pocket) and their relative difficulty to correctly predict affinity variations in QSAR models is a frequently encountered problem that will be detailed in Chapter 10.

2.2 SELF-ORGANIZING MOLECULAR FIELD ANALYSIS

The SOMFA method was proposed by Robinson et al. [344] as an intuitive 3D-QSAR method avoiding complex statistical tools and variable selection procedure.

2.2.1 SOMFA PRINCIPLE

SOMFA is a grid-based, alignment-dependent method. Molecules are embedded in a lattice of nodes as in CoMFA, but the main difference is that SOMFA does not need calculation of fields on these nodes and rather considers intrinsic molecular properties (such as shape and electrostatic potential) and so shows some resemblance with similarity-based methods. It was even noted that this inherent simplicity allows for directly integrating alignment in the model derivation [345].

To each grid node, values of a shape indicator (1 inside the van der Waals volume, 0 otherwise) and of electrostatic potential, calculated from partial atomic charges are given. Note that replacing the van der Waals potential by indicator values were also used by Kroemer and Hecht [225] and by Floersheim et al. [226]. In fact, a SOMFA grid might be trained on any calculable molecular property. But the important point is that, at every node, these values, for a given molecule, are multiplied by the mean-centered activity for that molecule (so as to give less interest to molecules close to the mean activity). Mean-centered activity is represented on a logarithmic scale. The QSAR model relating activity to a property (such as shape and potential) is then derived by linear regression.

From the individual grids for compounds i, a master grid is built. For a master point (x, y, z) on this grid, the SOMFA value is

$$SOMFA_{x,y,z} = \sum property_i(x, y, z) * Mean\_Centred\_Activity_i$$

(summation on the i molecules of the training set). The master grid contains the relative weights of each descriptor variable. In the next phase, the original descriptor grid is reduced to a single number. For molecule i, an estimate for the activity, as defined by a certain property is given by

$$SOMFA_{property,i} = \sum \sum \sum property_i(x, y, z) * SOMFA_{x,y,z}$$

(triple summation on x, y, and z); and activity is calculated by multilinear regression (MLR). It is possible to give different weights to shape and electrostatic potential. The calculated activity becomes a weighted sum of activities calculated separately for shape and electrostatic potential. See Appendix A, Section A.4.2.6.

Another aspect is that it is very easy to visualize the grid points corresponding to larger minimal and maximal values for shape and ESP, indicating the regions more likely to control the biological activity under scrutiny.

2.2.2 Applications of SOMFA

- A first example concerned the steroid data set [344]. On the classical training set (21 molecules), r=0.83. In test, adding 12 other steroids (with activities measured in the same conditions) to the traditional set (molecules **s-22** through **s-31**), SOMFA led to r=0.76 (and 0.83 without compound 31) with a ratio Shape/ESP=6.4. For the traditional test set (molecules **s-22** through **s-31**), standard error of prediction (SDEP) yielded 0.584, which was better than CoMFA (0.84), refined CoMFA (0.72), or (vide infra) similarity matrix analysis (0.64), COMPASS (0.71), or MS-WHIM-FFD (0.66).

  Interestingly, a compound bearing a Fluorine atom at the same position as the famous outlier steroid 31 was correctly predicted, suggesting, according to the authors, that the anomaly observed for the latter compound might be attributed to a different experimental technique for determining activity.

- Another data set corresponded to 35 sulfonamides, endothelin inhibitors (**2.12**) [344]. One-half compounds were selected for test. With a ratio Shape/ESP = 7.3, r = 0.82 in training, and in test 0.73, but in test, the slope (Obs. Activity vs. Pred. Activity) was only about 0.5 indicating a reduced predicted activity range. This was attributed to an alignment problem. Indeed, for the sake of comparison, Robinson et al. [344] used the alignment proposed by Krystek et al. [346], which was not the best one. These molecules have been also investigated with several other approaches, as indicated in the paragraph dedicated to the distance profile (DiP) method from Baumann [116], see Section 4.2.4.

Ar = substituted phenyl or naphthyl

2.12

Visualization of the areas where "important" nodes were concentrated indicated, for example (in agreement with previous CoMFA results), that steric bulk was tolerated in a narrow channel from the fifth position of the 1-substituted naphthyl fragment. Electrostatic build up positive charge is favorable between the 4 and 5 positions of 1-substituted naphthyl structure. A region where additional negative charge is expected to increase activity appears above the naphthyl system and the four and five substituents (Figure 2.6).

- The performance of SOMFA was investigated by Korhonen et al. [209] on two benchmarks, the steroid data set (TBG) and the Sadler estrogen data set (36 compounds binding estrogen receptor [ER]) [185]. In addition, five large xenoestrogens data sets were extracted from the Endocrine Disruptor Knowledge Base (EDKB, http://edkb.fda.gov) amounting 245 molecules for five different ERs.

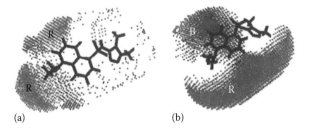

(a)   (b)

FIGURE 2.6   SOMFA master grids for sulfonamides endothelin inhibitors. (a) Shape master grid. (R) represents areas of favorable steric interactions. (b) Electrostatic master grid. The (R) area of favorable positive potential build lies between the 4 and 5 positions of the 1-substituted naphthyl compounds. The (B) area of favorable negative potential lies above the naphthyl system and the four and five substituents. Molecule inside: formula **2.12** with Ar = 5-dimethylamino (1-naphthyl). (Reproduced from Robinson, D.D. et al., *J. Med. Chem.*, 42, 573, 1999. With permission.)

On the Sadler data set, the best SOMFA model [347] obtained using only the molecular shape led to $q^2 = 0.76$ (SPRESS = 0.63). The (somewhat modest) results of SOMFA in this study, as in others, have been attributed to the built-in regression tool, self-organizing regression (SOR) that was shown to be equivalent to SIMPLS [143] and NIPALS with one principal component (whereas more than one component is obviously necessary for mapping complex data sets). Replacing SOR by external tools as multicomponent self-organizing regression (MCSOR) or SIMPLS (in a multicomponent treatment) improved the performance that came comparable to that of CoMFA [209,348]. For details on MCSOR, see Korhonen [348]. For the Sadler data set, SOMFA became better than receptor interaction energies ($q^2 = 0.570$) and comparable to basic CoMFA ($q^2 = 0.720$), but still lower than "sophisticated" CoMFA with region focusing ($q^2 = 0.796$) or GRID with receptor alignment and region focusing ($q^2 = 0.921$) [59,274]. But the authors stressed that some of these studies lacked extended validation.

From a detailed investigation, the authors concluded that MCSOR and SIMPLS gave nearly identical performance. But on the other hand, SOMFA is not very sensitive to the superposition technique as evidenced in a comparison to FLUFF (vide infra) with SEAL. An interesting point is also that the polarizability descriptor of Bradley and Waller [208], either combined with the electrostatic field descriptor or stand-alone, gave valid and predictive models. This descriptor is

$$\mathrm{Polar(p)} = \sum_{\mathrm{on\ atoms\ a}} \frac{\mathrm{Pa}}{\mathrm{r}^3}$$

where
r is the distance between grid point p and atom a
Pa the atomic polarizability of a

- Other applications of SOMFA concerned the activity of 63 dihydropyridines derivatives (**2.13**) [345] and 22 aryl-piperazines (**2.14**) [349] antagonists of α_1-adrenoceptor. These compounds offer potential interest in the treatment of hypertension and benign prostatic hyperplasia. But to avoid undesirable side effects (dizziness and muscle fatigue), selectivity toward the α_1 receptor is needed. A first study [345] concerned 63 dihydropyridines on 3.1 log units in pK_i. One of the most active molecule ($R_1 = p$-NO_2, $R_2 = CO_2Me$, R_3, $R_4 = Me$, Me, X = Phenyl) is chosen as template. Various models were developed with three alignments, charges from AM1 or PM3, and two different grid meshes (0.5 and 1 Å). The best model (AM1 charges, mesh 0.5 Å, and alignment indicated by stars in **2.13**) led to $r^2 = 0.704$ and $q^2 = 0.690$, the shape descriptor intervening for 30%.

2.13

2.14

Visualization of the important points indicated that electropositive groups are favorable around R_3 and R_4 and at positions *o*-, *m*- of the phenyl ring. Electronegative groups are preferred around R_2 and in *para* position of the phenyl. For shape, around R_2 and *para* position, steric interactions are favorable, but they are detrimental around R_3, R_4 and in *o*-, *m*-.

A subsequent study investigated the activity of 22 aryl piperazines, designed from a pharmacophore model [349] involving an aromatic part, a positive ionizable group, and an H-bond donor. The SOMFA model yielded $q^2 = 0.708$ and $r^2 = 0.743$, shape being largely predominant (90%). Steric interactions are favorable near positions 2, 5 of the phenyl ring, but detrimental in 3, 4, and around the aromatic part.

- SOMFA was also applied to a series of 29 1,5-diarylimidazoles (**2.15**) [350] selective inhibitors of COX-2. Selective inhibitors possess the same anti-inflammatory, anti-pyretic, and analgesic activities as nonselective NSAID inhibitors but with fewer gastrointestinal side effects. The best model yielded $r^2 = 0.546$, $q^2 = 0.507$ (on an activity range of 1.7 log units), and indicated a favorable influence of electronegative substituents on the *m*-, *p*-positions of the second phenyl ring (that without the SO_2Me group). Steric interactions also favored medium size substituents on these positions. (For other models on COX-2 inhibitors, see also Refs. [72,187,351].

1.15

SOMFA was also compared to WHIM or 3D-HoVAIF on the activity of 32 artemisinin derivatives [352]. SOMFA and 3D-HoVAIF led to similar satisfactory results, slightly better than WHIM, see Chapter 7. SOMFA was also involved in the 3 + 3D-QSAR models of Martinek et al. [353]. This approach aimed to evaluate the influence, on the variations in biological activity, of molecular flexibility. The relative population of the active conformation is proportional (in log units) to the conformational free energy loss upon receptor binding of a conformational ensemble. The proposed flexibility descriptors furnished predictive QSAR models and suggested active conformation in agreement with a receptor-based prediction on a series of endomorphin analogues. On a data set of 38 PGF2α prostaglandins with antinidatory activity they were used in conjunction to SOMFA descriptors to propose a model with a good predictive level ($q^2 = 0.52$, $r^2 = 0.58$ on a range of 3.5 log units on log ED_{50}). Smith et al. [354] carried out a comparison between SOMFA and phamacophore-based 2D- and 3D-QSAR models from CATALYST, on the binding affinity of 24 structurally diverse human UDP-glucuronosyltransferase 1A4 substrates. In this example, SOMFA ($r^2 = 0.73$) was outperformed by the pharmacophore-based 2D models ($r^2 = 0.80$) or 3D models ($r^2 = 0.88$). SOMFA was also used in a study of thiazolide derivatives, a class of broad-spectrum antiparasitic compounds, studied for their in vitro efficacy against the intracellular apicomplexan protozoan *Neospora caninum* [355].

2.3   FLUFF-BALL

With flexible ligand unified force-field (FLUFF) and boundless adaptive localized ligand (BALL), Korhonen et al. [142,356,357] proposed a semiautomatic, template-based package for superposition of a molecule onto a template and prediction of its biological activity. The method aims to incorporate the dynamic and fuzzy nature of molecules (on account of conformational changes or small movements). The performance appears comparable to that of other 3D-QSAR methods and the approach seems promising for screening large libraries due to its high level of automation and high throughput.

2.3.1   METHODOLOGY

FLUFF carries out the superposition by maximization of the similarity of van der Waals and electrostatic fields. A superimposition force field, based on a modified Merk Molecular Force Field (MMFF94) [242], evaluates the overlap between the template and the molecule under scrutiny. The energy equation encompasses the bonded and nonbonded interactions, as usual in molecular mechanics force field plus a term describing the similarity of the van der Waals and electrostatic fields between the ligand and the template. (Gaussian functions similar to those introduced in SEAL are used for this similarity term.)

The solution is obtained by geometry optimization of this composite energy, but it is mentioned that molecular dynamics or Monte Carlo search can be used. Negative superimposition ("no like that …") may be considered. One original characteristic is also the definition of "logical" molecules (such as one or several molecules, part of real molecules, and arbitrary set of atoms). Groups (such as C_{18}-methyl, H) may be omitted to avoid barrier effects hindering superimposition. Flexibility can be introduced on the ligand or on both ligand and template. Weights may be added on atoms, when alignment is not trivial (allowing for the generation of diverse possible solutions). In Ref. [357], FLUFF gave better superimpositions than SEAL, "presumably thanks to a priori information on weights of the template figures."

In connection with the superposition tool, FLUFF-BALL constitutes a new QSAR approach, based on a local coordinate system tied to the template (and so independent to the global coordinate system). One advantage is that the ligands and the template can choose the best common conformation. To establish a QSAR model, BALL computes the similarity between the template (given as a logical molecule) and the molecules on both van der Waals and electrostatic components. A sparse localized grid, tied to the template, is created (with vertices placed at the atoms of the template, which allows for adaptation to conformational changes). Ligands are described as soft functions: Gaussians primitives and electrostatic potentials (so that a "molecule does not end brutally but slowly fades away").

In the generation of descriptors, the van der Waals similarity is evaluated by three parameters for each template atom: the template atom's own volume, the common volume of the template atom and ligand atoms, and the residual volume of the ligand atom. In a same way, for the charge similarity, three parameters are also evaluated: the electrostatic field projected at the center of the template atom by the template molecule, the difference of projected fields by the template and the ligand at the center of the template atom, and the residual difference of electrostatic fields at the ligand atom. Owing to the asymptotic value at the atom center, a limiting value is set at the van der Waals radius of the atom used as the point of origin. This amounts $6N$ descriptors, if the ligand possesses N atoms. The template atom's own volume and field at center are introduced for scaling purposes. Then, usual statistical methods (as PLS) may be used.

The coarse grid used in BALL has been compared to a sort of region focusing as in Cho and Tropsha [184]. The scarcity of the grid points (and the "fuzziness" of molecules) does not allow for drawing contour plots. However, as the grid is tied to the template, the importance of each site of the template can be estimated.

2.3.2 APPLICATIONS

Beside a test on the "classical" steroid benchmark [356], the FLUFF-BALL software was tested on 245 xenoestrogen molecules from EDKB (http://edkb.fda.gov) with five different ERs (calf, human-α, human-β, mouse, and rat) [357]. This data set, amounting to a total of 374 RBA values [358], recovered six diverse chemical subfamilies (biphenyls, phenols, other phenyl compounds, steroids, indolines, and others). 17-β estradiol was chosen as template molecule (possibly with a preferential weight on its A-ring).

The performance of different pairs associating a superimposition tool (FLUFF or SEAL) and a QSAR model (CoMFA or BALL) was investigated on the five data sets. In all cases, but one, FLUFF-BALL worked the best, the last one being SEAL-CoMFA. "y-Scrambling" and numerous partition (training/external validation) sets confirmed the robustness of the approach. As a typical example, with calf ER (53 compounds), the best q^2 obtained was 0.824 (FLUFF-BALL) compared to SEAL-CoMFA (0.117); FLUFF-CoMFA and SEAL-BALL giving intermediate values

(0.530 and 0.223, respectively). However, these results contrast with the conclusions of Tong et al. [359] that gave for CoMFA, $q^2 = 0.60$. Similar remarks may be made for the other populations investigated.

2.4   SIMPLE ATOM MAPPING FOLLOWING ALIGNMENT

The simple atom mapping following alignment (SAMFA) approach, recently proposed by Manchester and Czermínski [360], constitutes a simplified but efficient alternative to CoMFA methodology, relying on atom-type characterization rather than explicit calculation of molecular fields. As in CoMFA and CoMSIA, an alignment rule is first applied, but instead of a regular grid, an *irregular template* is built from the aligned compound. The largest compound in the set is chosen as template and its atoms are represented as dummy atoms. Each remaining compound is then considered in turn. If any of its atoms lie beyond a cutoff distance (1.2 Å in the application here reviewed) from all dummy atoms, a new dummy atom is added. These dummy atoms constitute the "grid." For each molecule, mapping is performed onto the closest dummy atom and a fingerprint is built, combining dummy atom identifiers and atom types. Allowed atom types are C, N, O, S, P, H, and X (the four halogens). Supplementary details were then added: aromatic C, aliphatic C, HBA (H-bond acceptor), HBD (H-bond donor), and EWG (electron withdrawing group). Another originality of the approach is that data analysis was performed not only with the usual PLS treatment but also with random forest regression [361] or support vector machine (SVM) [100,101].

The efficiency of the method was tested on the steroid benchmark and on the eight other data sets compiled by Sutherlands et al. [72]. The predictive ability was established by Monte Carlo cross-validation (MCCV) also known as random sub-sampling or multiple hold-out [362]. On numerous trials, the whole data set was randomly partitioned into two parts (training and test). After training, the predictive ability was evaluated on the test set; and the draw was iterated on a large number of runs to ascertain that results were stable (convergence was here reached for about 2^{12} repetitions). The quality of the model might be judged by the q^2 probability density (p) and cumulative distribution function Pr:

$$\Pr(q^2 < x) = \int_{-\infty, x} p(t)dt$$

This density distribution was characterized by the median (q^2), by $\Pr(q^2 > 0.5)$ and integral q^2 (an overall measure of model quality)

$$q_{int}^2 = \int_{0,1} [1 - \Pr(q^2 < x)]dx$$

But it was established that median q^2 correlated well with the "classical" L-O-O q^2.

Generally speaking, on the nine investigated data sets, CoMFA and SAMFA were "equivalent for all practical purposes" [360]. It is noteworthy that the number of

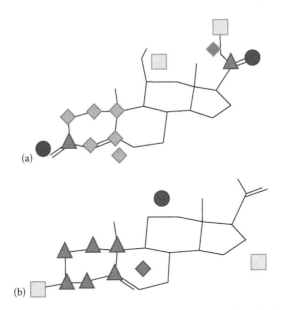

FIGURE 2.7 Interpretation of SAMFA descriptors: visualization of dummy atom/atom types. (a) Features correlated with enhanced CBG binding in the steroid data set; the high-affinity cortisol is shown as a reference compound. (b) Features anti-correlated with CBG binding. Pregnenolone is shown as the reference. Circles are H-bond acceptors; squares are H-bond donors; diamonds are sp³-carbons; and triangles are sp²-carbons. Note that symbols indicate features overall correlated with activity and do not necessarily correspond to the atom types in any particular molecule. (Reproduced from Manchester, J. and Czermínski, R., *J. Chem. Inf. Model.*, 48, 1167, 2008. With permission.)

descriptors (about 100–800) was about 5–10 times inferior to the number of CoMFA points. As to data treatment, random forest regression gave slightly better performance than PLS or SVM.

The interpretation of SAMFA results was rather straightforward, the dummy atoms being directly classified as correlated (or anti-correlated) to activity. On the example of steroid activity to corticosteroid-binding globulin (CBG), as expected, a 17 β-OH, an aromatic A ring, and a sp²-hybridized C-6 corresponded to decreased activity. On C-3, a carbonyl favored activity, whereas a hydroxyl group was detrimental. The approach allowed for easily and directly identifying the sites necessary for an increased activity. This constituted a definite advantage in relation with the problem of the "inverse QSARs" (Figure 2.7).

To explain the (surprisingly) good results obtained with a so simplified model; the authors suggested that QSAR models already implied so many simplifications and approximations that introducing another simplification was not detrimental.

3 Shape-, Surface-, and Volume-Based QSAR Models

While CoMFA and CoMSIA evaluate fields in the space surrounding the superimposed molecules under scrutiny, some other methods focus interest on explicit shape information. As indicated by Polanski et al. [363], shape is one of the fundamental categories used by the human brain for the perception and description of 3D objects. Molecular surfaces and volumes, although constituting only a conventional representation of molecular shapes, proved to be very useful to explain pharmacological effects such as drug–receptor interactions or molecular recognition processes. The influence of the steric interactions was long ago recognized as a determinant factor. Various approaches have been proposed.

As early as 1980, with the general assumption that the difference in activity between any two molecules is a function of the difference of their corresponding molecular properties, Hopfinger [41] proposed a quantitative approach—the molecular shape analysis (MSA)—that stressed the importance of molecular shapes. Later, other models, also based on the examination of the ligand volumes were proposed, with the local intersection volume (LIV) [364] or the excluded volume [365], this method being alignment free. On the other hand, surface-based models were proposed with the receptor surface model (RSM) [366], COMPASS [367], or comparative molecular surface analysis (CoMSA) [95].

3.1 MOLECULAR SHAPE ANALYSIS

The MSA method [41,368] differs from field analysis methods such as CoMFA, CoMSIA, and others since it puts the emphasis only on volumes and ignores electronic aspects (although some "classical" 2D parameters may be included). MSA addresses two major problems appearing in drug design:

- The determination of the "active conformation" of a drug on the one hand
- The characterization of the molecular shape and its influence on the biological activity on the other hand

An important difference with CoMFA or CoMSIA is that, in MSA analyses, the reference compound and its "good" conformation are not chosen a priori but determined in the final phase of the treatment, as those optimizing the correlation.

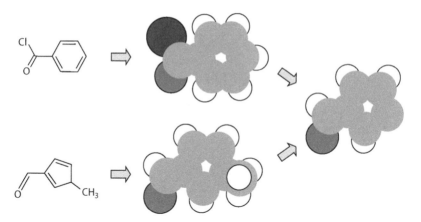

FIGURE 3.1   Common overlap volume.

3.1.1   COMMON OVERLAP VOLUME

In MSA, interest is focused on the evaluation of the common overlap volume with the active conformation of a "shape reference" compound (Figure 3.1). So the derived models encompass (possibly in addition to classical structural variables) pairwise shape descriptors measuring the differences in shape between the molecules of the data set and the shape reference compound.

The common overlap steric volume between two molecules i and j is defined by

$$V_0(i,j) = \sum_k \sum_l (v_{i,k} \cap v_{j,l})$$

(summation over the k atoms of molecule i and the l atoms of molecule j, v atomic volume using van der Waals radii). The union is calculated between atom k (of i) and its nearest neighbor l (of j) after superimposing the two molecules. Rather than V_0, Hopfinger preferred to introduce two alternative functions:

$$S_0(i,j) = [V_0(i,j)]^{2/3} \quad \text{and}$$

$$L_0(i,j) = [V_0(i,j)]^{1/3}$$

(they were not really any surface or distance measurements only clever representations of V_0).

3.1.2   CONSTRUCTING A QSAR: DHFR INHIBITION BY BAKER'S TRIAZINES

The problem is just (?) to determine, among all the molecules of the data set, the shape reference compound, which is a selected conformation of a selected molecule. In some previous models, the active conformation was defined as a conformation (near the energy minimum) common to active compounds and that inactive compounds cannot adopt. In the present approach, the active reference conformation is a "biologically relevant" conformation, which is a specific conformation (to which all molecules will be compared) used to build the MSA-QSAR model. Its selection is dictated by optimizing the resulting QSAR.

This is carried out in a multistep process:

- Conformational analysis
- Hypothesis about an active conformation
- Selection of a candidate shape reference compound
- Pairwise molecular superimpositions
- Measurement of molecular shape descriptors
- Determination of other molecular features
- Construction of trial QSARs

We will now look at these steps in more detail using the example of dihydrofolate reductase (DHFR) inhibition of 27 Baker's triazines (**3.1**), on a 5.1 log unit activity range [41].

3.1

In the conformational analysis (by molecular mechanics) of these molecules, the main parameter was the torsion angle between the two rings. Conformational maps of the most active compound (3,4-Cl$_2$) showed an energy profile with two wells, each with two sub-wells, giving four energy minima ($\theta = 70°$, $130°$, $250°$, and $310°$) (Figure 3.2). From a comparison with some other "interesting" compounds, it appeared that discussion can be focused on the energy profile near the second well (around $270°$) where important variations appeared, whereas the first well did not seem really modified. For the unsubstituted compound (R = H), no change appeared as to the energy well. So, the 1.6 log units decrease in activity (with respect to the 3,4-Cl$_2$ derivative) could not be attributed to conformational changes. However, for the 2,5-Cl$_2$ molecule, this well was reduced to a single minimum destabilized by about 3.1 kcal/mol higher than the global energy minimum. With the 2-F compound (1.3 log unit more active than 2,5-Cl$_2$ but 2.18 less active than the unsubstituted compound), a minimum was found at $260°$ (0.2 kcal/mol. above the global minimum) and an inflection at $300°$ (destabilization about 1.60 kcal/mol). These observations could be easily understood on the basis of steric repulsion of the phenyl *ortho* substituent with the (quasi) axial methyl group on the puckered triazine ring. Hence, the hypothesis that the active conformation corresponded to $\theta = 310°$, or in case of steric clash, to the energy minimum in the $270°$ well.

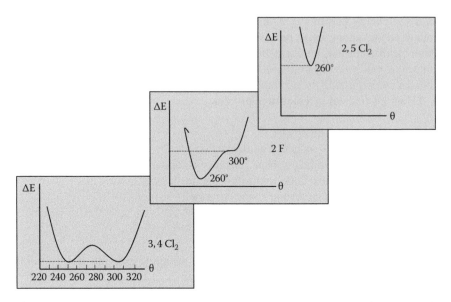

FIGURE 3.2   Relative intramolecular energy ΔE vs. torsional rotation angle θ near the active conformation. The shape of the curves is only indicative. (Adapted from Hopfinger, A.J., *J. Am. Chem. Soc.*, 102, 7196, 1980.)

Then the following steps are sequentially carried out:

- Choice of a tentative shape reference compound.
- Superimposition of the other molecules to this reference (and optimization of the alignment).
- Evaluation of the common overlap volume V_0 and the related quantities S_0, L_0.
- Determination of other classical structural variables, such as molecular refraction (MR), and π values [28,193] for substituents in positions 3 or 4 (that according to previous observations were somewhat related to activity). The dipole moment μ and the difference in energy between the active conformation and the global energy minimum were also considered.

Each active conformation for every molecule of the data set was, in turn, chosen as a shape reference compound and a tentative QSAR was built. In this case, among all these tentative QSARs, the best one was found using the 3,4-Cl$_2$ derivative as the reference. It corresponded here to the most active compound (but this might not be systematically true in other studies). The optimal correlation was therefore

$$-\log IC_{50} = 1.474\,[S_0] - 0.0224\,[S_0]^2 + 0.378\,(\pi_3 + \pi_4) - 17.15$$

$$n = 25, \quad r = 0.960, \quad s = 0.40$$

where IC_{50} is the molar concentration of the inhibitor that produces 50% inhibition. The two outliers detected with respect to this correlation (4-CN and 4-Ph) are interpreted on the basis of steric overlap with the receptor, since they were the two unique compounds with a group largely extending from the C1–C4 segment. Introducing a distance term D_4 (related to the length of a substituent along the C1–C4 direction), a new QSAR was obtained, without outliers:

$$-\log IC_{50} = 1.384 \, [S_0] - 0.0213 \, [S_0]^2 + 0.434 \, (\pi_3 + \pi_4)$$

$$- 0.574 \, [D_4] - 0.294 \, [D_4]^2 - 15.66$$

$$n = 27, \quad r = 0.953, \quad s = 0.44$$

This model outperformed the previous correlation proposed by Silipo and Hansch [369]. It was validated on a test set of seven compounds for which the predicted activity fell within the standard deviation of the correlation, but for a unique compound.

3.1.3 OTHER APPLICATIONS

Nearly the same procedure was followed in the study of 52 1-(substituted-benzyl) imidazole-2(3H)-thiones (**3.2**), inhibitors of dopamine β-hydroxylase (DβH), that might reduce blood pressure [368]. In fact, five highly flexible compounds were discarded, leaving 47 compounds on an activity range of 4.13 log units. After energy minimization, conformers of low conformational energy (few kcal/mol above the global minimum) were selected. The reference compound and the structures to be compared with it were chosen from this pool of conformers. The method implied that, one at a time, a compound was chosen as the reference compound (r) if one of its conformations were retained as the active one. At the same time, one among the conformations of each of the other molecules at hand (u) was selected for comparison, and these two conformations were superimposed. Similarity was measured by the common overlap steric volume $(V_u \cap V_r)$ determined by numerical integration or a relative measure $(V_u \cap V_r)/V_r$.

3.2

To take into account the intramolecular stability of each compound, a *shape commonality index* was defined which decreased the similarity index by a quantity depending on the difference in energy between the absolute conformational energy minimum and the energy of the conformations compared. A weighting factor allowed for giving more or less importance to that term. Other classical molecular descriptors might be added, such as lipophilicity, partial atomic charges, dipole descriptors, and conformational entropy. The construction of trial QSARs was carried out by MLR (after elimination of pairs of strongly intercorrelated descriptors) and the process was iterated to select the best shape reference compound—that which optimized the QSAR.

However, it must be recalled that, as in all multidimensional optimizations, there might exist several local maxima of similar quality. So it might be interesting, for speeding up the calculation, to first select a subset of compounds spanning the entire activity range (including very active, moderately active, and inactive derivatives) and then extend the QSAR to the whole data set.

In this study, the reference compound was the second most active. We have here an example where the active conformation is not that of the crystal packing ($\theta = 76°$). The best QSAR involved V_0, V_0^2, and the sum of partial charges on carbons 3, 4, and 5 of the phenyl ring, and led to $r = 0.90$. Unfortunately, no validation was carried out.

$$-\log IC_{50} = -117.4\ V_0 + 70.4\ V_0^2 + 2.33\ Q_{3,4,5} + 52.12$$

$$n = 45,\quad r = 0.90,\quad s = 0.41$$

This model yielded performance similar to that of a previous classical QSAR [370], but with fewer terms and covering a larger range (47 compounds vs. 25).

3.2 FURTHER DEVELOPMENTS

3.2.1 OTHER SHAPE DESCRIPTORS

The original MSA approach considered the common overlap steric volume, calculated on the basis of atoms represented by spheres. But it was proposed [371] that other molecular shape descriptors might be introduced, derived from the potential energy fields of molecules. This was presented as a "second generation of MSA descriptors." In the work reported here, three probes were investigated: H^+ and O^- as models of a positive and negative charge and a CH_3 unit to characterize the dispersion component of the field. The molecular potential energy field P (for molecule u) is calculated as the sum of a Lennard-Jones component (in r^{-6}, r^{-12}) and a Coulombic one, as usual in MM calculations. Calculations are easier in a spherical intermolecular coordinate system (located on the molecular center of masses) (Figure 3.3).

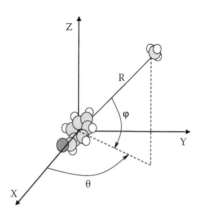

FIGURE 3.3 Polar coordinates.

Various indices were proposed:

- Extrema of $P_u(R, \theta, \varphi)$ that locate the most and least important interaction sites in molecule u
- Radial distribution function, giving a profile of the potential energy field at any distance

$$P_u(R)dR = \int_\theta \int_\varphi P_u(R,\theta, \varphi) R^2 \sin\theta \, d\theta \, d\varphi \, dR$$

- Total energy field, a scalar measure of the overall potential energy field

$$P_u = \int_R P_u(R)dR$$

- Difference of radial distribution functions between molecule u and a reference r (the coordinate frame is centered on it)

$$\Delta P_{u,r}(R,m)dR = \int_\theta \int_\varphi [P_u(R,\theta,\varphi) - P_r(R,\theta,\varphi)]^m R^2 \sin\theta \, d\theta \, d\varphi \, dR$$

with, usually, $m = 2$

- Total difference in potential energy fields

$$\Delta P_{u,r} = \int_R \Delta P_{u,r}(R,m)dR$$

Integration was carried out numerically with resolution of $10°$ for angles and 0.2 Å for R. The last index may be considered as a generalization of the common overlap steric volume V_0. It gives an energetic representation of molecular shape.

This model was worked out on the inhibitory activity of a set of substituted 2,4-diamino-5-benzylpyrimidines (**3.3**) (to bovine dihydrofolate reductase), a series previously studied by MSA [372] or classical Hansch relationships [373]. $P_u(R)$ showed the expected behavior: more or less erratic variations at small R (where dispersion and electrostatic forces appeared), and, after a minimum (reflecting a stable intermolecular binding), convergence to zero, slower for the charged probes H^+ and O^- (the rate depending of the chosen dielectric constant, here 3.5).

3.3

According to the MSA approach, the total difference $\Delta P_{u,r}(R, 2)$ had to be calculated for the 23 molecules of the series with respect to the (active) *cis*-4-NHCOCH$_3$ substituted derivative. This posed some problems: variations looked fallacious at $R < 8\,\text{Å}$, but integration had to be extended up to R about $40\,\text{Å}$ due to slow convergence for charged probes. Fortunately, at $R > 8\,\text{Å}$, curves describing $\Delta P_{u,r}(R, 2)$ were a set of parallel lines, gently slopping to zero. So Hopfinger proposed to replace the evaluation of $\Delta P_{u,r}$ by the relative heights of these curves at a given distance D. For the series under scrutiny, the best correlation was obtained for $D = 12\,\text{Å}$, leading to:

$$-\log \text{IC}_{50} = -2.34\ \text{F} + 0.29\ \text{F}^2 + 0.37\ (\pi_3 + \pi_4) + 9.39$$

$$n = 22, \quad r = 0.961, \quad s = 0.105$$

where $\text{F} = \log\ [\Delta P_{u,r}(12,2)^{0.5}]$. This model compared well with the MSA correlation previously proposed by Hopfinger [372]:

$$-\log \text{IC}_{50} = -21.31\ \text{V}_0 + 2.39\ \text{V}_0^2 + 0.44\ (\pi_3 + \pi_4) + 52.23$$

$$n = 23, \quad r = 0.931, \quad s = 0.137$$

and also with the Hansch model [373]:

$$-\log \text{IC}_{50} = 0.622\ \pi_3 + 0.322\ \Sigma\ (\sigma_3 + \sigma_4 + \sigma_5) + 4.99$$

$$n = 23, \quad r = 0.931, \quad s = 0.146$$

where electronic effects also intervened (expressed by Hammett's σ). In the studied series, it was also shown that the best correlations are obtained with H$^+$ or O$^-$ probes (but, as indicated, the choice of the active conformation perhaps implicitly reflected steric influences). It was also noted that, despite the good results obtained, the local dielectric behavior (and the choice of a macroscopic dielectric constant) was not quite clear.

3.2.2   A MUTANT AS REFERENCE STRUCTURE

Cholecystokinin (CCK) (**3.4**) is a family of peptides that functions as digestive hormones in the periphery and as a neuromodulator in the central nervous system. CCK receptor antagonists have been proposed as potential drugs for the treatment of various diseases from appetite disorders to pancreatic carcinoma or psychiatric problems. The study of Tokarski and Hopfinger [374] concerned the binding affinities for rat pancreas receptor of 53 chemicals, on an activity range of 5.4 log units (CCK-A receptor antagonists), and focused on racemates.

3.4

Energy minimization was carried out with MNDO, which gave results quite consistent with the crystal structure. Then the active conformation (that biologically relevant) was sought for. It was suggested that active analogues were able to adopt an active conformation that was not stable for inactive compounds. This was the "Loss in biological activity-loss in conformational stability" (LBA-LCS) principle. This search was first performed on a subset of five compounds of largely varied activity but with a same substitution pattern on the benzodiazepine ring, so that the differences in activity could be attributed to the substituents on the 3-amido aromatic ring. Superimposing the amide nitrogen, the amide carbonyl and the first carbon atom bonded to the amide carbonyl guided the pairwise comparison.

Various descriptors were added to the shape descriptors: lipophilicity (log P), π constants, σ Hammett's constants, HOMO and LUMO energies. Conformational energy (relative to the global minimum) and the distance from the carbonyl carbon to the furthermost atom in the direction of the p-phenyl substituent were also considered.

All compounds in the proposed active conformation were candidates for the shape reference compound. The choice might be guided by the regression of the common overlap volume and the nonoverlap volume (V_{non}) with activity, but it appeared that no single analogue was the best shape reference compound: different antagonists mapped nonidentical volumes of the allowed receptor space. So an artificial shape reference compound, the "mutant" (Figure 3.4), was built from the overlapped structures of few analogues and the accessible receptor space was expanded.

Various 3D-MSA-QSARs were proposed, depending on the considered subsets of compounds (phenyl analogues, p-substituted phenyl analogues, indole analogues, and phenyl and indole analogues). For example, for all RS-compounds (minus 3)

$$-\log IC_{50} = 0.042\, V_{ov,m} - 0.045\, V_{non,m} - 0.364\, \Delta E + 0.560 \left(\sum \pi 1.5 \sum \text{Å} \right)^2$$

$$+ 0.788\, I_M + 0.487\, I_F + 2.483$$

$$n = 50, \quad r = 0.93, \quad s = 0.47$$

where $V_{ov,m}$ and $V_{non,m}$ are the common overlap and nonoverlap volumes with respect to the mutant. ΔE is the conformational energy difference with the absolute minimum, since in addition to molecular shape (characterized by the overlap and nonoverlap volumes), the conformational stability is also important. I_M and I_F are

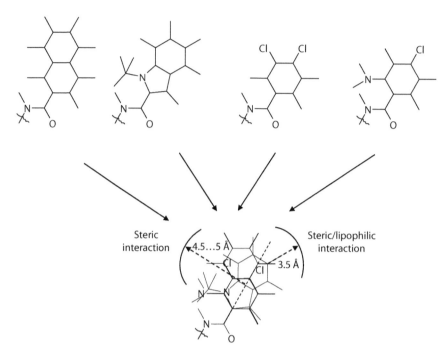

FIGURE 3.4 Binding to the CCK-A receptor. Planar view of the 3-amido part of the mutant in the proposed binding site, and the four skeletons used to build it. Possible steric–lipophilic interaction sites are indicated. (Adapted from Tokarski, J.S. and Hopfinger, A.J., *J. Med. Chem.*, 37, 3639, 1994. With permission.)

indicator variables encoding the presence of a methyl group on the endocyclic amide nitrogen and a fluorine atom on the 5-phenyl ring of the BZD ring. ($\Sigma \pi$ 1.5 Å^2) is a lipophilic contribution estimated from spheres of varying radii and centered on the phenyl *p*-Iodo atom of the most active 3-(benzoylamino)benzodiazepine, the best value corresponding to a radius of 1.5 Å. The intervention of this term might be an indication for a hydrophobic pocket of the receptor.

From these 3D-QSARs that allow for characterizing the substructures with the highest probability to interact with the receptor [374], a 3D pharmacophore for the 3-amido substituents might be proposed from the LBA-LCS principle, and some guidelines might be suggested as to structural reasons of inhibitory activity variations. Ideally, the best conformation would correspond to $\Phi_1 = -130°$, $\Phi_2 = 130–140°$ (**3.4**), corresponding to the maximum common overlap and the minimum nonoverlap. The mutant structure (selected overlapped structures) essentially depicted the allowed space of the binding site, groups extending outside this volume being detrimental.

3.2.3 APPLICATION OF MSA TO FLEXIBLE STRUCTURES

The study of 19 pyridobenzodiazepinones (**3.5**) inhibitors of muscarinic M2 and M3 receptors illustrated the application of MSA to highly flexible compounds [375,376]. These systems comprised several torsional angles, generating many conformers in a 6kcal/mol energy range above the global minimum (from 9 to more than 700). According to the general scheme of MSA, the pairwise superimpositions were carried out using the tricyclic nitrogen N1 (near the carbonyl group) and the two adjacent ring carbons.

3.5

In agreement with observations in other series, the best reference compound was here both the most active and the largest. This corresponded to the fact that such a compound "had the pharmacophoric features necessary to activity and a conformational profile that could adopt the active conformation" [375,376]. An important parameter was the steric molecular shape COSV

$$COSV = \frac{V_{test} \cap V_{ref}}{V_{ref} \cap V_{ref}}$$

where V_{test} and V_{ref} correspond (for a given alignment) to the current conformation of the examined compound and the reference conformation of the reference compound [310]. But other parameters were necessary for deriving satisfactory QSARs: the presence of a lipophilic binding site and the contributions to solvent accessible surface area (SASA) of atoms near the lipophilic pocket. These contributions were calculated for atoms within a sphere of variable radius at the center of the 7-membered ring (according to a scheme already seen [374]). Indicator variables for specific nonallowed steric receptor sites (NAS), affecting M2/M3 selectivity, were also introduced.

Repetitive use of PLS method allowed for extracting the most significant structural features subsequently used in classical MLR for which r^2 was about 0.8, and SD about 0.3 for 19 compounds on a 2.5 log unit range.

3.2.4 RELAXING CONSTRAINTS ON ALIGNMENT AND ACTIVE CONFORMATION

MSA seeks the active conformation of a molecule expressing shape similarity in terms of scalar descriptors, such as common overlap volume with some reference compound, and possibly other classical descriptors. In a previous study on flexible muscarinic receptor ligands, a common bound conformation and receptor alignment were assumed [375,376]. These constraints of fixed conformation and fixed alignment have been relaxed in a subsequent study of the inhibitory action of Trimethoprim-like derivatives (**3.3**) on *Escherichia coli* DHFR [310]. In addition, it was possible to test the proposed approach since the receptor-bound conformation and the alignment were known from the enzyme–inhibitor complex.

For 20 derivatives, 10 conformations and 4 alignments were selected. Conformations were chosen on the basis of the bound conformation of Trimethoprim (i.e., not the free molecule minimum energy), of its eight local minimum-energy conformations (within 5 kcal/mol of its global minimum) and a secondary minimum found for compounds with restricted conformational freedom. Alignment 1 was based on atoms of the 2,4-diaminopyrimidine group, alignment 2 fixed the methylene bridge, the third one the carbons bearing the 3-methoxy groups, and the last one the central carbon and the opposite C and N on the two phenyl rings.

Descriptors were constituted by 2D substituent constants (π, $\pi_{3,5}$, molar refractivity) and the COSV values (vide supra). 3-way PLS led to a model with two latent variables accounting for 61% of the variance of the activity data. The best set conformation/alignment was determined with the scoring function $CAW_{m,n}$ where m and n encoded conformation m and alignment n:

$$CAW_{m,n} = \sum_{a=1,A} Var_a W^2_{a,m,n}$$

Var_a = variance of Y(activity) explained by the **X** data (independent variables) in component a

W is the corresponding scaled **X** loading

This analysis proposed two nearly equivalent models (in COSV or its square) with alignment 2 and two different reference conformations, with r^2 about 0.50 and q^2 about 0.42. One was in agreement with the receptor-bound conformation. For this geometry, conformational flexibility could be taken into account with the torsion angle entropy S (for the phenyl rings) evaluated according to the TAU theory [377]. Using genetic function approximation (GFA) [108], a one-component PLS model was generated, involving COSV, NOV, S^2, and MR^2 with $r^2 = 0.913$, $q^2 = 0.816$. However, in a subsequent treatment, in the framework of 4D-QSAR [63], better results were obtained: $r^2 = 0.957$, $q^2 = 0.885$, SD = 0.34 on a range of 5.7 log units.

3.2.5 Tensor Representation of QSAR Analyses

Considered in its full generality the QSAR problem is quite complex: each molecule of the data set may exist in several conformations, and various alignments may be hypothesized. Furthermore, the various variables describing the system often are of quite different nature. A general formalism was proposed by Hopfinger et al. [375] to express, on a synthetic condensed presentation, the relationships between dependent properties (e.g., biological activity) and independent molecular feature measures (descriptors). It used a tensor representation of multidimensional blocks associated to the various variables describing the system, taking into account the fact that each molecule might exist in several conformations and that several alignments might be hypothesized. See also Section 1.4.5.

So the general representation of a QSAR is

$$\text{Absolute}: \mathbf{P}_u = \mathbf{T}_u \mathbb{K} \left[\mathbf{V}_u(s,\alpha,\beta), \mathbf{F}_u(p,\mathbf{r}_{i,j,k},f,\alpha,\beta), \mathbf{H}_u(h_p,\alpha,\beta), \mathbf{E}_u(e_p,\alpha,\beta) \right]$$

$$\text{Relative}: \mathbf{P}_{u,v} = \mathbf{T}_{u,v} \, \mathbb{K} \left[\mathbf{V}_{u,v}(s,\alpha,\beta), \, \mathbf{F}_{u,v}(p,\mathbf{r}_{i,j,k},f,\alpha,\beta), \, \mathbf{H}_{u,v}(h_p,\alpha,\beta), \, \mathbf{E}_{u,v}(e_p,\alpha,\beta) \right]$$

$\mathbb{K}$ representing the tensor product operator, and u and v corresponding to any molecule of the training set and a reference molecule, respectively. In this formalism:

- **P** is the property looked for, the dependent variable, either a vector $\mathbf{P}_u$, property value for compound u, or an array $\mathbf{P}_{u,m}$ value of property P for compound u in biological assay m (in cases where the activity of a series of compound is measured with respect to several targets: e.g., human ER, mouse ER, calf ER).
- $\mathbf{T}_{u,v}$ is the transformation tensor (to be determined) that optimally maps the composite tensor **[VFHE]** onto $\mathbf{P}_u$.
- **V** is the tensor of intrinsic molecular shape (IMS) features. Their set, possibly depending on conformations (set α) and alignments (set β), gives information about shape.
- **F** is the molecular field feature tensor (where p is the set of field probes and $\mathbf{r}_{i,j,k}$ the points where fields are evaluated), bearing information on shape beyond the molecular volume. In the same tensor, f corresponds to field-related molecular features not derived from (p, $\mathbf{r}_{i,j,k}$), such as the dipole moment.
- **H** (with set h_p of features not derived from IMS or molecular field) corresponds to calculated physicochemical features (lipophilicity and solubility).
- **E** corresponds to experimental physicochemical features (set e_p).

This formalism includes current QSAR/QSPR models as constrained subtypes [375]. For example, Hansch-type relationships may be expressed as

$$\mathbf{P}_u = \mathbf{T}_u \, \mathbb{K} \, [\mathbf{H}_u(h_p,\alpha',\beta')]$$

the prime (on α and β) indicating that conformation and alignment are not needed. For CoMFA:

$$\mathbf{P}_u = \mathbf{T}_u \, \mathbb{K} \, [\mathbf{F}_u(p,\mathbf{r}_{i,j,k},f^*,\alpha^*,\beta^*), \mathbf{H}_u(h_p,\alpha^*,\beta^*)]$$

the asterisk indicating a single selection of conformation and alignment. And for MSA:

$$\mathbf{P}_{u,v} = \mathbf{T}_{u,v} \, \mathbb{K} \, [\mathbf{V}_{u,v}(s,\alpha,\beta^*), \mathbf{H}_{u,v}(h_p,\alpha,\beta^*)]$$

with various conformations (α) but a single alignment selected. A good account of this formalism was presented in the case of multiple possible alignments of active conformation.

3.2.6   MSA ANALYSIS OF ALLOSTERIC MODULATORS
OF THE MUSCARINC RECEPTOR

To get more insight about allosteric modulators of muscarinic receptors, a detailed conformational MSA was carried out by Holzgrabe and Hopfinger [378]. An allosteric effect corresponds to a retarded dissociation of the antagonist–receptor complex in the presence of allosteric substances. This, for example, may be beneficial for over-additive antidote action. This effect is found with very diverse structures (alcuronium, bispyridinium, or hexamethonium derivatives) that stabilize antagonists binding to M2-cholinoceptors.

The study concerned seven flexible bispyridinium or hexamethonium derivatives: DUO (**3.6**) (only the left part of the molecule is represented, the right one is symmetrical; bold lines correspond to possible torsional motions) and IWDUO compounds of very diverse structures. Activity was measured by minus the log of the concentration of a modulator at which the rate of dissociation of a radio-labeled antagonist is reduced by 50%. The goals were to find an active conformation for each molecule and to identify the key physicochemical properties of the ligand bioactive form in order to propose a spatial pharmacophoric hypothesis. Alcuronium (**3.7**) (the most active compound, with a rigid structure) was arbitrarily chosen as reference. For each compound, a starting conformation (from energy minimization) was submitted to molecular dynamics simulations. But none of the derivatives adopted a conformation where the positive charges and the aromatic rings were in a position comparable to that found in alcuronium. So the authors used, for six selected atoms, a torsional flexible-fit procedure to propose tentative aligned geometries. These geometries were re-optimized and aligned in a rigid-body fit. All compounds adopted an S-shape conformation (distorted sandwich) (Figure 3.5).

Duo O   X = O
Duo N   X = N
Duo C   X = CH₂

3.6   **3.7**

Classical MSA was then carried out, but besides overlap (V_o) and nonoverlap (V_{non}) volumes, integrated potential field differences ΔF [371] were calculated with three probes (O anion, methyl group, proton) on account of previous indications of the importance of electrostatic effects in the activity of such compounds. It was indicated that these differences between pairs of analogues gave some indication of "how differently these two molecules were seen by the receptor" [378].

From GFA analysis, the best models highlighted the role of V_{non} and the electrostatic field parameter ΔF_{O-} or ΔF_{H+} (that are interchangeable), and satisfactory

FIGURE 3.5 Superimposition of all compounds using the best alignment for each molecule. (Reproduced from Holzgrabe, U. and Hopfinger, A.J., *J. Chem. Inf. Comput. Sci.*, 36, 1018, 1996. With permission.)

linear correlations (r^2 about 0.9) were obtained. However, the authors emphasized that the small number of compounds, their large structural diversity, and the relatively restricted activity range (1.4 log unit) limited predictive capacity. Similarly, the choice of the large molecule alcuronium as a negative image of the allosteric binding site of the muscarinic receptor protein might also be questioned. But this study allowed for proposing a hypothesis about a spatial pharmacophore. Molecular volumes did not govern activity that was controlled by electrostatic interactions (two positive charges and a terminal aromatic ring separated by a distinct distance), in agreement with experimental observations.

3.3 VOLUME-BASED MODELS

3.3.1 EXCLUDED VOLUME

Excluded volume descriptors were proposed by Tominaga and Fujiwara [365] to provide a model relatively insensitive to the relative orientation of molecules and avoiding the delicate phase of alignment. The model is based on the excluded volume between specific type of atoms in the molecule and a series of probes. These probes are layered like an onion (Figure 3.6). Each is constructed by the excluded volume of two spheres with different radii and a common center located at the center of gravity of each molecule. The descriptors may be viewed as representing, for each molecule, the expansion of a specific type of atom in 3D space.

The first probe has a radius of 2.5 Å (like an iodine atom). The second probe is constituted by 60 atoms (I) arranged to form a quasi sphere surface as a fullerene. The subsequent probes were defined similarly. The distance between the centers of the atoms of each layer is 0.5 Å. This value (as the iodine van der Waals radius) has

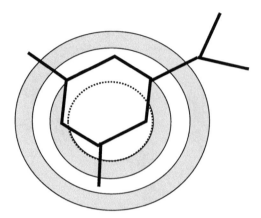

FIGURE 3.6 Novel 3D descriptors using excluded volume: application to 3D-QSAR. For the sake of clarity, the molecule has been shifted with respect to the system of spheres. Centers of gravity of the two systems must be coincident. (Adapted from Tominaga, Y. and Fujiwara, I., *J. Chem. Inf. Comput. Sci.*, 37, 1158, 1997.)

been determined as the best one. The following probes are similarly built. Fifteen types of atoms are considered [379] (the other atoms being removed for the calculation). The descriptors (to be treated by PLS) are constituted by 21 values (one per layer) when considering the whole molecule (EV_{whole}), 315 when specifying the atom type (EV_{type}), and 336 using both (EV_{both}).

The model was validated on the classical 21 steroids binding corticosteroid-binding globulin (CBG). Sensitivity to orientation was first examined on 10 runs starting from randomly positioned molecules. The best results were obtained with EV_{type} with a mean q^2 of 0.766 and a standard deviation of 0.07 (the extremes being 0.856 and 0.658), indicating a predictability superior to that of CoMFA ($q^2 = 0.662$) [38].

3.3.2 LOCAL INTERSECTION VOLUME

Verli et al. [364] proposed new descriptors, such as LIV, to construct predictive 3D-QSAR pharmacophore models. The model was based on a local 3D shape descriptor: LIV was defined as the intersection volume between the atoms of a molecule and a set of spheres (of carbon atom size), which composes a 3D box, in analogy with grid methods. Molecules are embedded in the box according to a previous alignment on a template (best steric and electrostatic fit determined by the method of Good et al. [380,381]).

The approach was validated on the example of benzodiazepine receptor ligands. Benzodiazepine exerts an allosteric effect on γ-aminobutyric acid (GABA) action at $GABA_A$ receptors, potentiating the inhibitory action of GABA (a major inhibitory neurotransmitter in mammalians) in the brain. This results in a continuum spectrum of effects, from agonist effects (anxiolytic, anticonvulsant) to null effect (antagonists) or even inverse agonist effect (anxiogenic, convulsant). Fifty-eight benzodiazepine receptor ligands (**3.8** through **3.10**) (with a non-benzodiazepine structure) corresponding to various series of dihydroindolo-β-carbolines, dihydropyrazolo-quinolinones, and β-carbolines were investigated. Binding affinities pIC_{50} were determined in replacement of [³H]-diazepan on binding $GABA_A/BzR$. For the nonrigid structures, the 20 most diverse structures (highest RMS distance) were used to find the best superposition on the template (there was no guarantee that the active conformation was that of minimum energy). Alignment was first carried out with the atoms of the rigid part, and then refined using similarity indices for steric and electrostatic fields [380,381].

3.8   3.9

3.10

A grid matrix of hard spheres was built from unitary cells corresponding to a quad-rangular base pyramid with edge lengths of 3.08 Å (twice the van der Waals radius of a carbon atom). The intersection of the molecular volume (using van der Waals radii) with each of these spheres constituted the LIV descriptors. In the elaboration of the 3D-QSAR model, a first data reduction was performed, eliminating descriptors of null variance and those corresponding to details specific to few compounds (very low occurrence). A second level was carried out in a GA-PLS treatment (optimiza-tion, using GFA [108,109]). In the quoted example, this led to a linear model involv-ing eight LIV values on some points of the grid matrix. The best model yielded $r^2 = 0.802$, $q^2 = 0.722$.

A graphical representation of the active LIV could be compared to the phar-macophoric model of Cook et al. [382] (Figure 3.7) proposing H-bonding sites at position A2 (acceptor), H1 (donor), H2/A3 (donor acceptor), three lipophilic pockets (L1, L2, L3), and a region of steric repulsion (S). In the LIV treatment, positive con-tributions were associated to sites LIV-139 corresponding to A2, LIV-110 (H1), and LIV-130 (lipophilic region L3). Negative contributions were associated to LIV-065, LIV-074, and LIV-111, at the east and northeast of the molecule, correlated to the area S1 of steric limitation. Occupation of LIV-140 corresponded to a detrimental situation since it prevented H-bonding to the receptor (H1). A similar influence was

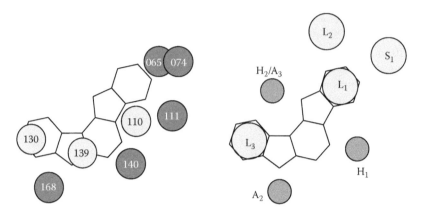

FIGURE 3.7 Comparison between the graphical representation of the LIV 3D-QSAR pharmacophoric model for ligands of $GABA_A$/BzR and schematic pharmacophoric model proposed by Cook and coworkers. (Adapted from Verli, H. et al., *Eur. J. Med. Chem.*, 37, 219, 2002.)

observed for LIV-168 (and A2). Despite this good agreement, the subsite H2/A3 was not identified in the LIV analysis.

For the six outliers appearing in this treatment, it was suggested that some substituents, although not directly interacting with the receptor, might affect the alignment (based on steric and electrostatic characteristics) and so modify the interactions with the receptor.

Martins et al. [383] also developed a LIV model for a series of aromatic heterocyclic compounds known as ligands of the IP receptor. Prostacyclin I2 (PGI2) is an endogenous chemical mediator acting as a potent inhibitor of platelet aggregation. For 42 aromatics, on a range of 3.3 log units on pIC_{50}, the best model retained yielded $r^2 = 0.92$, $q^2 = 0.84$ with six LIV. As in the preceding example, most of the important LIV (with positive or negative contribution) could be associated with previously proposed pharmacophoric sites [384].

3.4   HYPOTHETICAL ACTIVE SITE LATTICE

The hypothetical active site lattice (HASL) approach of Doweyko [42] is a grid-based, alignment-based method relying on a 4D molecular representation. The fourth dimension here represents a user-selected physicochemical property. It should not be confused with the fourth dimension defined by Hopfinger [63] that corresponds to the simultaneous existence of several conformers of a same molecule. Unlike CoMFA or CoMSIA, it works on selected nodes inside the molecular volume and not on field evaluation outside the molecular shape.

3.4.1   CONSTRUCTION OF THE HASL AND DERIVATION OF A QSAR MODEL

In the HASL model, the shape and physicochemical characteristics are captured on a set of regularly spaced "colored" points. From a 3D orthogonal lattice, only points internal to the van der Waals volume are retained, and encoded according to some property value, for example, electron-rich or electron-poor (in a more recent version,

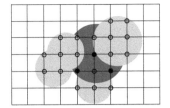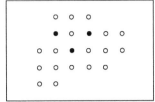

FIGURE 3.8   Building a HASL model. Schematization in 2D.

two attributes may be given among H-bond acceptor/donor, atomic charges, hydrophobic constants, electrotopological parameters) (Figure 3.8). With a resolution of about 3 Å, a 20-atom molecule is represented by about 30 lattice points.

To build an HASL model, molecules are first energy minimized (several conformers being possibly retained). Then in the series of molecules under scrutiny, one is chosen as a stationary reference and others with their own lattices are compared to it, in pairwise comparisons. The best fit is searched for by stepped translations and rotations so as to maximize a FIT index calculated from the fraction of common lattice points (each step generating an intermediate lattice):

$$FIT = \frac{L_{common}}{L_{ref}} + \frac{L_{common}}{L_{molecule}}$$

Thanks to the encoding of the lattice points, the best alignment gives a maximum complementarity of volume and physicochemical characteristics. To derive a QSAR model, the value of the investigated property (e.g., pK for an inhibitor) is distributed equally (in the first step) over all the data points occupied by the structure, so that (at this step) the property value may be reconstructed by summation of these point values.

Merging all the aligned lattice points representing the series of molecules leads to a composite construction containing all points of either lattice with the property values (associated with lattice points) averaged. The objective is to be able, for every compound, to retrieve the associated property value by summation of the partial values on the nodes occupied by the molecule. So the initial distribution has to be corrected iteratively for each molecule, up to global convergence. Corrective terms are derived from the current error (calculated pK – measured pK): **IN** for points common to the molecule and HASL, **OUT** for points not used. In fact, it is a redistribution since **IN + OUT** = 0, increments **IN** and **OUT** being, each one, equally distributed on the corresponding occupied sites. As in CoMFA, the model is validated by cross validation and a final model is built on the all population (Figure 3.9).

Once this HASL model is built, activity can be predicted for a new structure. After alignment with the reference molecule, summation of point values on the occupied nodes yields the predicted activity. Let us note that if the target molecule presents functional groups not sampled, or lying outside of the HASL, results may be unsatisfactory.

Beyond QSAR, HASL represents a kind of composite map of the active binding site, consistent with both structures and activities. These point values give some idea

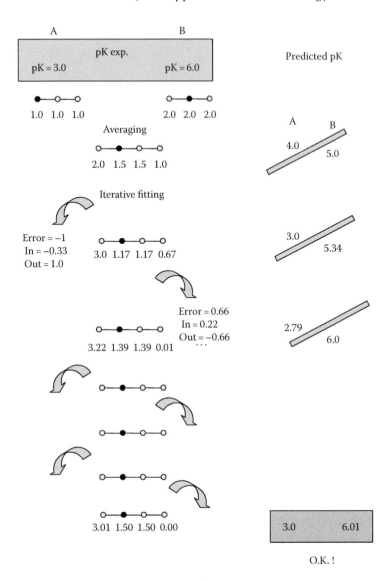

FIGURE 3.9 Establishing a HASL model. pK estimation by an averaging method. The right part gives the predicted pK during the successive iterations. (Adapted from Doweyko, A.M., *J. Med. Chem.*, 31, 1396, 1988.)

of the spatial location of points favorable or not to activity (possibly depending on their electron-rich or electron-poor character) and delineate regions where a given feature (e.g., hydrophobicity) is beneficial. To avoid overfitting of the model, a trimming process was developed to eliminate the less informative points. The partial activity values were redistributed while verifying that relevant points have not been removed [385]. Iteratively eliminating 2.5% of least contributing points led to simplified models limited to about 20 points but still containing relevant information.

Such a presentation highlighted the nodes that were important in the prediction of activity. These schemes somewhat remind us of the contour plots of CoMFA or CoMSIA. However, in CoMFA this diagnosis came from points outside of the molecule, whereas in HASL, lattice points were located inside the molecular volume.

3.4.2 APPLICATIONS OF THE HASL MODEL

Among the examples of HASL treatment, 41 glutathione (**3.11**) and flavonoid (**3.12**) analogues, inhibitors of yeast glyoxalase-I were examined [42]. At an arbitrary lattice resolution of 3 Å, the HASL model consisted of 88 lattice points. Point encoding (1, 0, −1) roughly indicated atoms with low, medium, or high electron density (−1 for H of alcohols, amines, S of sulfoxides, sulfones, carbonyl carbons; 0 for carbon of alkanes, alkenes, alkynes; +1 for O, N, halogens). This encoding tended to overlay atoms with equivalent character and induced a molecular alignment with electronic consistency.

3.11   **3.12**

After iteration on the inhibitory data of each compound distributed on the occupied nodes, the model converged to an average error of prediction 0.1 pK unit on a 3 log unit range (r = 0.981).

To examine the predictive ability of HASL, a second study [42] was carried out on DHFR inhibitors analogues of **3.3**, an interesting example since numerous inhibitors of a wide structural variety were known, and the crystal structure of methotrexate in the active site of *E. coli* dHFR was resolved. Seventy-two inhibitors (on 7 log units in pK) were split into learning (37 compounds) and test (35) sets. At 2.8 Å resolution (a compromise between the precision of the representation and the number of points to treat), an excellent result was obtained in learning (r = 0.999), but in test r was only 0.753. As an example of these qualified results, one can note that more than 20 compounds with actual pK values of about 6 are predicted in a range between 5 and 8 (in log units).

Eighty-four analogues of statin **3.13** inhibitors of HIV-1 protease (HIV-1PR), a critical link in the activation of immunodeficiency virus, were studied in Ref. [385]. Alignment was carried out by reference to the X-ray structure of two bound inhibitors, and an 899 lattice points HSAL was built at 2 Å resolution (construction guided by atom encoding as electron-rich, electron-poor, or electron-neutral). Very good results were obtained (r² = 0.988 on a 6.5 log unit range of activity). Using the trimming process led to an 11-point model of good performance ($r^2 = 0.827$). This set of 11 points might be considered as a pharmacophore. In a further step, comparing the atoms positions (in the superimposed molecules) to the location of these points gave some insight about the inhibitor–receptor interactions, particularly regarding several critical sites for H-bonding.

3.13

Another study [386] compared CoMFA and HASL treatments of antiulcer agents: a set of substituted imidazo[1,2-a]pyridines (**3.14**) and related analogues, for which both in vivo gastric antisecretory activity and in vitro inhibitory potency on the gastric proton pump enzyme H^+/K^+ ATPase (either purified or in gastric glands) were known. After energy minimization, for each compound the minimum energy conformation was retained plus a conformation similar to that of the most active compound chosen as template (phenyl group in a pseudo-equatorial conformation) (**3.15**). Alignment to the template was based on optimization of steric volumes (25%) and electrostatic potential (75%).

3.14   **3.15**

Using template-based conformations (and correcting experimental data for the percentage of compounds protonated at pH = 7.4) CoMFA leads to $q^2 = 0.529$ and 0.508 respectively for in vitro (pIC_{50}) and in vivo (pED_{50}) gastric gland activities; $r^2 = 0.964$, SEP = 1.23, SEE = 0.34 and $r^2 = 0.938$, SEP = 1.31, SEE = 0.47, respectively. Prediction for a test set of six compounds leads to errors within 1 log unit.

With the same conformations, an HASL model was built with 265 points (resolution 1.5 Å). After 36 cycles of a trimming process, eliminating at a time 2.5% of points contributing to the least information, a final 15-point model was proposed, yielding to $q^2 = 0.853$ for in vitro purified H^+/K^+-ATPase activity (in CoMFA 0.539).

Electron-neutral atom type (in HSAL) may be considered as exerting a steric effect. Both CoMFA and HSAL indicate a sterically favored region near 2-, 3-, 8-positions of the imidazo[1,2-a]pyridine ring, and a sterically disfavored region near the 7-position. However, interpretation is more difficult for electron-rich or electron-poor points, protonation and solvation effects possibly obscuring the relationship between the atom type and the field it creates.

Receptor modeling and HASL methodology have been combined in the study of Dastmalchi and Hamzeh-Miverhrod [387] on aldehyde oxidase (AO) (a member of non-cytochrome P-450 enzymes). AO is involved in the metabolism of very diverse endogenous compounds and drugs. A better knowledge of the enzyme–substrate interactions is therefore of prime importance for drug development. The study aimed to generate a 3D model of human AO and clarify the interactions between this enzyme and a set of phthalazine/quinazoline derivatives (**3.16** through **3.18**).

3.16   **3.17**   **3.18**

A model of AO was first built by comparison with the crystal structure of bovine xanthine dehydrogenase, followed by MD simulations and energy minimization. The consistency of the proposed solution was carefully checked and an HASL model was elaborated with the ligands docked into the active site. After training on 22 compounds, for a test set of seven derivatives, $r^2_{pred} = 0.65$ on about 2 log units. In this paper, some information about the intervening interactions with the receptor was given on molecule (**3.18**) chosen as example: a staking π–π interaction between the benzene ring of the ligand and the phenyl side chain of residue Phe-923, an H-bond between N3 and the amide proton of Ala-109, and between the carboxyl group of Glu-882 and the terminal OH of the R_1 substituent. Hydrophobic interactions with the CH_2–CH_2 fragment and the pyridine have also been suggested.

A multiconformer HASL technique has also been proposed [388] in the analysis of five HT1 thienopyrimidinone ligands. The lowest energy conformer of the most active molecule was chosen as the template, and then each molecule was represented by five separate low-energy conformers. All of them were used in the generation of the HASL 3D-QSAR models in a search for maximum similarity with the conformations of active molecules.

3.5 RECEPTOR SURFACE MODEL

A receptor model aims to postulate and represent the essential features of the putative active site of a receptor. In that sense, it is somewhat a complementary notion to a pharmacophore model, small subset of atoms necessary for a molecule to be recognized by a receptor, and common to all molecules binding to it. But, a receptor model can contain detailed information about binding of molecules with varied features or topologies which cannot be easily included in a pharmacophore model. For example, excluded areas or hydrophobic regions are difficult to represent in a pharmacophore model, whereas they can be incorporated into an RSM [366].

Once a reasonable RSM has been generated, a series of structures can be evaluated against the model and a QSAR can be built from the receptor–structure interactions. It must be noted that such an approach takes information from ligands only, and so it is quite separate from the "receptor-related" methods (Chapters 11 and 12) where the structure of the true receptor is explicitly known, or must be reasonably inferred.

3.5.1 CONSTRUCTION OF A RECEPTOR SURFACE MODEL

The RSM proposed by Hahn and Rogers [366,389,390] is basically a nonatomistic, explicit 3D surface construction with associated properties mapped on it. The surface encloses a volume common to all the aligned molecules, which is assumed to be complementary to the real receptor site. The surface itself, built from a set of points organized in a network of triangles, represents information about steric factors, the associated properties (mapped from the active structures) being hydrophobicity, partial charges, and H-bonding propensity, for example. This surface is the tightest surface that lines all training compounds. This may be more severe than the actual shape of the receptor. So an original point is that the model may be opened with holes cut out in the surface, indicating unknown regions or parts able to accommodate bulkier groups than those used in the construction of the model (solvent accessible areas).

The approach starts with the selection of some subsets of the compounds among the most active ones, since it may be thought that the most active compounds present the most favorable interactions with the receptor, whereas for inactive molecules, some interactions tend to be detrimental. The first step is alignment, to build an aggregate that preferably reflects the bound conformations. Various methods are available (as already seen), but as always, a bad choice may lead to poorly predictive models [390 and reference therein].

Shape fields are constructed for the aligned molecules and then combined to give the final shape field from which the surface is generated. A set of field sources are placed on the atoms of the aligned molecules and the fields they generate are calculated and summed up on a grid of points. A grid spacing of 0.5 Å leads to an average density of 6 points per Å². Then an isosurface is created, characterizing the molecular shape (e.g., using the "marching cube" algorithm).

For the shape field two functions were considered:

- A van der Waals field $V(r) = r - VDWr$, r distance from the point to the atom and VDWr van der Waals radius of the atom
- Another distance function was also used, giving a smoother representation and hiding some details: the Wyvill function (a polynomial expression in R^6, R^4, and R^2, so that $V(0) = 1$, $V(R) = 0$, and $V(R/2) = 1/2$; where R is twice the van der Waals radius

$$V(r) = -\frac{4}{9}\left(\frac{r}{R}\right)^6 + \frac{17}{9}\left(\frac{r}{R}\right)^4 - \frac{22}{9}\left(\frac{r}{R}\right)^2 + 1 \quad \text{for } 0 < r < R$$

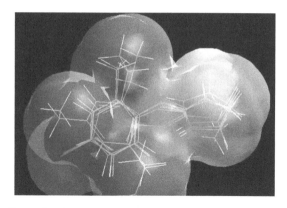

FIGURE 3.10   (See color insert following page 288.) RSM built (with van der Waals function) from the five more active octopamine agonists (see Section 3.5.3) and colored by hydrophobicity. A (dark) brown area corresponds to a positive contribution of hydrophobicity, (light) gray areas to a negative contribution of hydrophobicity. Note that the phenyl ring and its substituents are hydrophobic whereas the heterorings (imidazolidine and oxazolidine) are hydrophilic. (Reproduced from Hirashima, A. et al., *Internet Electron. J. Mol. Des.*, 2, 274, 2003. With permission.)

From these functions, surfaces may be created at a given distance of the van der Waals surface. This allows for a certain distance tolerance when the ligand binds the receptor.

Points are attributed information about possibly interesting properties (charge, electrostatic potential, H-bonding propensity, hydrophobicity). Since the surface is built over a set of molecules, the property associated to a point is an averaged value. The basic idea is then that the receptor may be viewed as the complement of this surface: charge is minus the average of partial atomic charges on the closest atom in all molecules. Similarly, electrostatic potential is the opposite of the potential at that point created by all atoms. H-bond is binary encoded (-1, 1) on points where an H-bond donor or acceptor is favorable. These points are found by projecting away a cone from the H-bond acceptor (O or N atoms) or donor atom (H attached to O or N). Hydrophobicity is encoded 1 for points in a hydrophobic region, 0 otherwise. A hydrophobic point has a low partial charge, a low absolute value of electrostatic potential, and low H-bond donor or acceptor characteristics.

Color-encoding these properties (with intensity depending on their magnitude) offers visual intuitive information on the characteristics of the active site (Figure 3.10). Only one type of property may be visualized at a time but it is quite easy (and rapid) to switch from one property to another.

3.5.2   CALCULATION OF INTERACTION ENERGIES AND RECEPTOR SURFACE ANALYSIS

Property values may be also used for calculating interaction energies. Our specific concern being QSAR, we now focus on these calculations. An interesting feature of this approach is that a molecule may be energy minimized in the presence of the receptor (modeled by its surface) thanks to a simplified force field *Clean*. In Clean, the usual terms, such as bond stretching, angle bending, and torsion, are expressed by simple functions (often harmonic) without the corrective terms intervening in more sophisticated MM programs. Interaction energies are summed up on all pairs atom/receptor point. But, in addition to the usual Coulombic electrostatic term and van der Waals term (Lennard-Jones potential), a desolvation contribution is added to the total energy, after minimization is performed. This term may introduce, for example, a penalty when a polar atom is placed in a hydrophobic region of the receptor surface. Once a receptor model has been generated, a QSAR can be built using the receptor–ligand interaction energies as descriptors.

Several energy descriptors are calculated:

- $E_{interac}$: energy of interaction between the structure and the surface representing the receptor (van der Waals and electrostatic terms)
- E_{inside}: intramolecular strain energy of the ligand, inside the receptor (receptor frozen, ligand free)
- E_{relax}: the bound conformation of the ligand is minimized again in the absence of the receptor and relaxed to the closest energy minimum ($E_{relax} \leq E_{inside}$)
- $E_{strain} = E_{inside} - E_{relax}$ represents the variation of strain energy between the bound and the closest unbound conformations (and not to the global energy minimum), the receptor being considered as rigid

The RSM is composed of a huge number of points with their corresponding energy descriptors. The interesting points (those associated to variations of these descriptors within the series of molecules) are selected using the GFA [108]. Typically, about 300 models involving 15 descriptors are submitted to 5000 evolution steps before the final PLS analysis [366]. Note that, when the model so constructed is applied to new molecules (in test), these need not be aligned precisely, since this will be carried out in the minimization step for the calculation of energy descriptors.

The RSM was applied [390] to the traditional and largely studied steroid benchmark (affinity to CBGs). The receptor surface was deduced from the six more active molecules. Considering only $E_{interac}$, the model gave a correlation with $q^2 = 0.646$ on the 21 compounds of the training set, but for the other 10 compounds (test set) $r^2 = 0.006$ due to one outlier **s-23** (but, if excluded, r^2 increased to 0.696). The outlier possessed an acetoxy group on carbon C-21 and exhibited the same affinity as its (well-predicted) homologue with an OH on C-21. So the authors suggested that the acetoxy group might be located outside the receptor surface in a solvent accessible part. Indeed, opening the receptor surface in that region made r^2_{ext} reach 0.652.

Another puzzling point is that compound **s-31** (the only one with a fluorine atom on C-9) was in some models found as an outlier, whereas here it was correctly predicted. This was due to an empirical solvation correction (in Clean) that penalized polar groups in a hydrophobic region. This example showed that incorporation of some qualitative knowledge (and the help of visualization) allowed refining the model and getting better predictions than CoMFA and COMPASS. However, the experimental activity value of **s-31** has been questioned [344].

The other example concerned 47 dopamine β-hydroxylase inhibitors (see compound **3.2**) (reducing blood pressure) previously studied by Burke and Hopfinger [368] in MSA. To build the receptor surface, 12 models were examined, from the most active compound alone to all compounds considered together, with alignment on the most active molecule and the shape reference conformation of Ref. [368]. Two tolerance values (0.1 and 0.2 Å outside the van der Waals surface) were tried. The selection was carried out with GFA. The most useful (four) descriptors were derived with a model built from the three more active molecules, at 0.1 Å tolerance.

These descriptors were then used together with the Hopfinger–Burke's descriptors. But GFA selects only $E_{interac}$ and E_{inside} incorporated in spline functions. q^2 reached 0.788 in L-O-O and 0.669 in 16 Leave-Some-Out processes. Curiously, the V_0 volume, although proposed for the correlation, was not retained, but it was somewhat correlated to $E_{interac}$.

3.5.3 OTHER APPLICATIONS

The RSM approach was used by Hirashima et al. [391] in a study of 60 octopamine (OA) agonists binding receptor OAR3 in Locust nervous tissue (on a range of 5 pK units). Ten other structures constituted a test set. The aim of the study was to get useful information about characterization and differentiation of the OA receptor. Octopamine (**3.19**), 2-amino-1-(4-hydroxyphenyl) ethanol, is the monohydroxylic analogue of the vertebrate hormone adrenaline, and the octopaminergic system was recognized as an interesting target for the development of selective, safer pesticides. The structures under scrutiny mainly encompassed 2-(aralkylamino)-2-thiazolines (**3.20**), 2-(aralkylmercapto)-2-thiazolines (**3.21**), 3-(substituted phenyl)imidazoline-2-thiones (**3.22**), and 2-(arylimino)imidazolines (**3.23**).

Multiple conformations were generated by Boltzmann jump followed by energy minimization to determine the lowest energy conformers. A conformer of the most active OA agonists was chosen as reference for alignment. This was performed using the maximum common subgroup (MCSG) method and a rigid-atom pairwise superimposition. In this approach, a set of different RSMs were built from different combinations of the most active analogues. Then a GA-PLS was used to select the RSM that led to the best QSARs (nonlinearity is included thanks to spline-based terms). From 1730 initial descriptors, the final QSAR involves a linear sum of 14 interaction energy terms (or their squares) on identified positions. Satisfactory results were obtained with $r^2 = 0.877$, $q^2 = 0.766$ (PRESS = 21.2) for the 60 compounds.

3.5.4   COMPARING DOCKING AND RSM MODELS

The RSM approach was dedicated to situations where the active site of the receptor is not known. However, in favorable cases (known receptor), it is interesting to compare the conclusions of the (blind) RSM approach to those of docking (a receptor-based approach), as was done in the study of Shaikh et al. [392]. Among the c-Jun-N-terminal kinases (also called "stress-activated protein kinases"), JNK3 is largely selectively expressed in the brain. JNK activity is crucial in the immune response and cell apoptose. Inhibition of JNK may be useful in the treatment of diseases related to apoptose, such as rheumatoid arthritis or neurodegenerative diseases. The study concerned 44 benzothiazole derivatives (**3.24**) inhibiting the JNK3 kinase activity, on a range of 2.37 pIC_{50} units. Thirty-four compounds were selected as a training set. Alignment was performed with the maximum common substructure graph method, followed by rigid fit of atom pairings.

3.24

Molecular field analysis (MFA) was carried out with steric and electrostatic components on grid nodes, plus additional descriptors (dipole moment, polarizability). Regression analysis was performed by GA-PLS (50,000 generations of 100 models) leading to an optimum model with five latent variables [108]. From the PLS equation, it was possible to trace back to the nodes where a probe (Me or H$^+$) induced important contributions in the MFA analysis. In a second step, a RSM was built from the five most active molecules (aligned as previously) with the van der Waals field function.

These models led to linear relationships highlighting discrete positions. In this study, the field analysis method MFA ($r^2 = 0.849$, $q^2 = 0.616$, $r^2_{pred} = 0.721$) slightly outperformed RSM analysis ($r^2 = 0.766$, $q^2 = 0.605$, $r^2_{pred} = 0.535$). MFA emphasized the fact that bulky groups near the N-ethyl-3 pyridine position of some of the most active compounds (see compound **3.25**) were detrimental, and that a pyrimidine nitrogen was necessary. Other terms of the equation also indicated the beneficial effect of steric and e-withdrawing interactions at the cyanide group (that may be replaced by a bulkier e-withdrawing group).

3.25

In agreement with MFA, the RSM approach confirmed that an NH group was required in position of the pyrimidine amino nitrogen group (whereas steric effects at this position would decrease activity) and that a pyrimidine nitrogen was necessary for activity. Other important favorable elements were the NH of benzothiazole and electrostatic interactions near the tail part of the pyridine ring for some of the most active compounds.

Mapping charges, H-bond propensity, or hydrophobic character on the molecular surface comforted the preceding observations (Figures 3.11 and 3.12), and accounted for the changes in activity observed when substituting, for example, hetero-atoms (N, S) by carbons:

- For the benzothiazole part, an H-donor group at the NH position, a hydrophilic area nearby and a hydrophobic group at S-position were important for activity.
- Negative charge and H-acceptor groups at the tail of the N-Et-3-pyridine ring (when present) played an essential role at this position.

FIGURE 3.11   (See color insert following page 288.) RSM: mapping of H-bonding propensity. H-bond donor regions are shown in (P) purple while H-bond acceptor regions are shown in (G) cyan color. The compound **3.25** is represented inside. (Reproduced from Shaikh, A.R. et al., *Bioorg. Med. Chem. Lett.*, 16, 5917, 2006. With permission.)

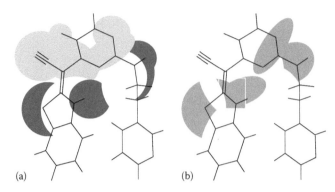

FIGURE 3.12   Schematic representation of RSM: (a) mapping of charges, (b) hydrophobic regions. (a) Positive regions are shown in dark gray and negative regions are shown in light gray. (b) Hydrophilic regions are shown in light gray. Hydrophobic regions are all around the skeleton. The compound **3.25** is represented inside. (Adapted from Shaikh, A.R. et al., *Bioorg. Med. Chem. Lett.*, 16, 5917, 2006.)

- For the pyrimidine ring, nitrogen negatively charged, H-acceptor group (N), and hydrophilic areas are important. An H-donor group was also required at the NH group attached to the pyrimidine ring, and an H-acceptor at the cyanide position. An intramolecular H-bond was also indicated between the benzothiazole NH and pyrimidine nitrogen.

More information on drug–receptor interactions was provided by ligand docking in the crystal structure of JNK3. The active site was defined as a sphere of radius 6.5 Å surrounding the bound ligand. Minimized structures of docked ligands were then submitted to MD simulations to get stable drug–receptor complexes.

Ligands bound in an ATP-binding pocket (Figure 3.13). Among the principal interactions, one could note the H-bonds formed between the pyrimidine nitrogen atom and the amino group attached to the pyrimidine ring with the NH and CO groups respectively of Met-149. Such a situation mimicked the bidentate interaction in ATP. The intramolecular H-bond (benzothiazole-pyrimidine nitrogen) was also confirmed. Important H-bonds implied the pyridine tail chain (borne by the pyrimidino amino group) and receptor residues, Ile-70, Asn-152, Gly-71, and particularly Gln-155. This interaction might be important for the selectivity, since this residue (common in JNK 1, 2, or 3) was not conserved in other kinases. These observations were quite consistent with the conclusions of MFA or RSM analysis as to the importance of the H-bonds involving the amino pyrimidine ring and the role of the pyridine tail.

3.5.5   Genetically Evolved Receptor Model

A little different from Hahn's method, the genetically evolved receptor model [393–395] aims to produce, at atomic level, models of receptor sites based on a small set of known structure–activity relationships. But it may be also used to derive QSAR from a set of ligands of known activity. The principle is to distribute a number of explicit model atoms (40–60 atoms) in the space around a series of aligned ligands

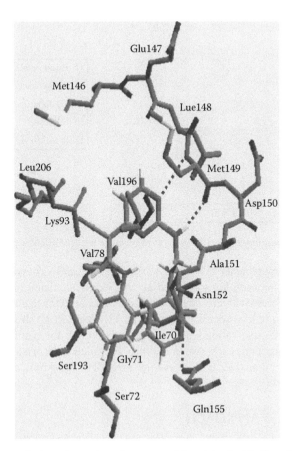

FIGURE 3.13 Docked conformation of compound (gray) in the binding site of JNK3. All the hydrogen atoms of the protein are removed for clarity. H-bonds are shown with dotted lines. (Reproduced from Shaikh, A.R. et al., *Bioorg. Med. Chem. Lett.*, 16, 5917, 2006. With permission.)

and calculate interaction energies between the ligands and this model of receptor. By changing the nature of the atoms at the various positions, models can be constructed that correctly reproduce the experimental activity.

Fourteen atom types are defined, with associated partial atomic charge (approximating the values found in the standard 20 amino acids), E_{min} and R_{min} values for the van der Waals interactions. Atom types correspond, for example, to C (of CH, resp. CH_2 or CH_3), O carbonyl, O hydroxyl, etc. To generate the receptor model, points are distributed evenly on a sphere surrounding the ligands. An aliphatic carbon atom is placed at each point and its position is adjusted by optimizing its radius so as to get maximal van der Waals attraction to the ligand molecules. A "cushion" of 0.1–1 Å allows for compensating the lack of flexibility of receptor and ligands (Figure 3.14).

Then a population of models (typically 500–1000) is created where these atomic positions are occupied by atoms identified by their type. Intermolecular interaction energies between the current model and each ligand are calculated (electrostatic and

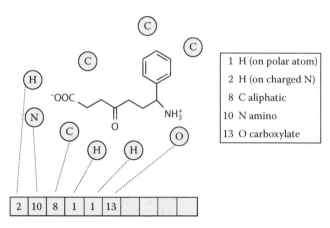

1	H (on polar atom)
2	H (on charged N)
8	C aliphatic
10	N amino
13	O carboxylate

2	10	8	1	1	13				

FIGURE 3.14   Building a genetically evolved receptor model GERM.

van der Waals components). The fitness criterion for the GA is given by the correlation coefficient between log (observed activity) and calculated interaction energy. Then the GA evolves with crossover and mutation operations to convergence.

The efficiency of the process was tested on 22 sweet-tasting structures, on 3.5 log unit activity. Extensive validation tests were carried out: for example, in three subsets of eight compounds each, two are in turn discarded to form a test set: in training r^2 is always better than 0.89, with an average error of 0.08 log unit, and in test average error is about 0.4.

3.6   COMPASS APPROACH

The search of the "bioactive conformation" and the major role of steric and polar influences were also at the heart of COMPASS developed by Jain et al. [367]. COMPASS aims to construct a quantitative binding site model that will be able to predict affinities. It also characterizes the bioactive conformation and alignment of ligands. But this approach sets itself apart by three main innovations:

- COMPASS stresses the importance of the molecular surface in the drug–receptor interaction, and so uses a surface-only representation of molecules, avoiding examination of internal zones.
- Nonlinear models are built thanks to an artificial neural network (ANN).
- Selection of a conformation and a relative alignment for each molecule are automatically performed (let us recall that the bioactive conformation may be quite different from vacuum minimum energy or crystal structure).

The first step is an initial guess for each molecule of a "pose" (the conformation, supposed to be the bioactive one, in an approximate alignment). This implies the construction of low-energy conformers (e.g., a Monte Carlo search selects conformations within 5 kcal/mol from the energy minimum) and the identification (by the user) of a substructure common to all molecules (like a pharmacophore) to assign an approximate alignment with maximum similarity with the conformations of active molecules.

Then (second step), from these guesses, the structural features are extracted at points (about 300) scattered near the van der Waals surfaces and fixed in an invariant, unique reference frame. A single set of sampling points is used for all poses of all molecules. Shape is characterized by the distance of the sampling point to the molecular surface, and H-donor or H-acceptor features by the distance to H-bonding groups. In the third step, a statistical model is built, thanks to a three-layer back propagation neural network (in the example, architecture 265/3/1). The network is fed with the features extracted from the sampling points and delivers at output the calculated activity. An original point is that the inputs of the ANN are not directly the features selected as descriptors (the distances x) but a Gaussian function (centered on the feature value) with a dampening factor,

$$G(x) = z \exp \left[-\frac{(x - \mu)^2}{2\sigma^2} \right]$$

to take into account the fact that there is an optimal distance (μ) for drug–receptor interaction (too small leads to unfavorable contacts, too large to weakened interactions). A dampening factor σ allows for some tolerance of ligand variation, and according to the authors, this (plus the automated pose alignment) may be viewed as somewhat addressing the receptor-induced fit (more sharply characterized with the 5D-QSAR models, see Chapter 10). The coefficient z specifies the weight of feature x in binding affinity.

Another important aspect is that the system will tend iteratively to realign molecules by successive translations and rotations to derive better poses. The basic idea is that the pose predicted as the most active is likely to be the bioactive one. At a time the system may choose among the poses previously generated or may generate new ones in the optimization process. So it adjusts its prediction with respect to the observed activities. Convergence is reached when the predictions for all molecules are close to the experimental values. At convergence, COMPASS gives the "best" model, and is able to predict the activity of an unknown molecule.

On the steroid benchmark (31 compounds binding CBG and 21 binding testosterone-binding globulin (TBG)) COMPASS with $q^2 = 0.89$ and 0.88 for CBG and TBG binding respectively outperformed the results of Good et al. [396] in a similarity treatment (0.74 and 0.53) or CoMFA (0.69 and 0.4) with two fields (electrostatic, steric).

Using points near the surface largely reduced the number of points compared to a CoMFA lattice and gave for the steep changes of the steric field more precise information than that gained from the usual 2 Å sampling of CoMFA. To best understand reasons for the good results of COMPASS, the authors also ran various assays, confirming that the improved performance resulted from the three innovations introduced, and particularly the automated choice of conformations and alignment.

In a further publication, Jain [397] examined 20 molecules of 2 chemotypes (**3.26** and **3.27**) binding 5-HT$_{1A}$ receptor (affinity range about 4 log units). The model was tested on 35 new structures (**3.26** through **3.31**), with often scaffolds or functionalities not present in learning. With two different chemotypes no obvious alignment appears. So a new procedure was proposed. For the surface-feature calculation, reference points were uniformly scattered on two spheres of radius 6 and 9 Å, centered at the origin of a Cartesian reference frame. At each point, three distances were computed (to the van der Waals surface, and to the nearest H-donor and H-acceptor atoms) leading to 42 feature points per sphere for each of the three distance types. A scalar strength (from 0 to 1) indicated the angle between the radius to the point and the H-bond directions.

3.26   3.28   3.30

3.27   3.29   3.31

The next step was the choice of the set of bioactive poses. A similarity measure evaluated the differences in features between the reference and the current pose and also took into account the directionality constraints on potential H-bonds. Let us remark that, due to the Gaussian form of the similarity measure, if the differences in shape were large, the contribution to the similarity function would be nearly null, so that parts of molecules might be allowed to protrude if other areas corresponded well.

Starting from multiple initial alignments, one pose per molecule (among about 50) was selected on the basis of maximum joint similarity of all molecules to the most active one. Once a pose was obtained for each molecule, it was submitted to the ANN for initial training. Then the entire pool of possible poses was considered for the final training. In this study, molecular scaffolds were markedly different (six types as illustrated), and experimental data concerned either resolved compounds or racemic mixtures.

However, the approach proved to be very efficient. After training on 20 resolved compounds, the model was used to predict activity of 35 new compounds. For racemic mixtures, the two isomers were calculated and the mean (in log space) retained. Discarding one outlier, q^2 reached 0.81 in 10-fold cross validation (mean error of prediction 0.53). In test, on 35 molecules, the overall mean error was 0.55 log unit on a 3.5 log unit range. In this process, the method tended to find, for all actives, bioactive poses similar to each other and different from those that inactive compounds could adopt. Painting points depending on favorable (or not) steric and polar interactions (from their calculated contribution) made easier the visual inspection of what might be the binding site model.

With relatively sparse data, COMPASS constructed accurate quantitative binding site models, and deduced binding constraints from structure–activity data, without extensive conformational and crystallographic studies and avoiding the problems often generated by molecular flexibility.

The 3D-QSAR (TDQ) approach of Norinder [398] presents some analogy with COMPASS proposed by Jain [367]. It involves a conformational analysis model with rigid object optimization but uses PLS as statistical tool. Space was structured either by a grid (as in classical CoMFA) or by three surface layers at 1.0, 1.5, and 2.0 times the atomic van der Waals radii. Two sets of molecular fields were considered: on the one hand, the usual nonbonded and electrostatic potentials, and on the other, a shape and H-bonding description as in COMPASS according to a distance-dependent formula:

$$1 - \left(\frac{d_{ij}}{3}\right)^2$$

where d_{ij} denotes the distance (in Å) of the current node to the van der Waals surface or to the closest (non-H) atom of the acceptor or donor group.

For each conformation, molecules, supposed rigid, are iteratively oriented by translations–rotations in order to maximize the predicted biological activity. The model was tested on the steroid benchmark and on styrene-type inhibitors of protein-tyrosine kinase. For the latter series, better results are obtained with the COMPASS-type description than with the usual MFA protocol involving a fixed, single-conformer orientation. It was also noted that there was no need for an exact alignment and that allowing the structures to move during optimization limited the influence of the positioning of descriptor points.

3.7 COMPARATIVE MOLECULAR SURFACE ANALYSIS

In the CoMSA approach developed by Polanski et al. [95,399], the compared properties do not describe a discrete set of points (as in CoMFA) but rather averaged property values calculated on certain areas of the molecular surface. So the approach combines a Kohonen network to determine these averaged values and PLS method to derive a QSAR model.

3.7.1 DERIVING MOLECULAR DESCRIPTORS FROM A KOHONEN MAP

Kohonen network and self-organizing maps (SOMs) are nonsupervised classification tools, allowing for reducing the dimensionality of data, although respecting topological relations between them (proximity relations) [400–403]. SOMs have been used to transform 3D molecular surfaces (possibly encoded with property values) into 2D maps, making their comparison easier [400] but they may also be proposed to define quantitative SARs. Basically, in these QSAR applications, a molecular property (here the molecular electrostatic potential [MEP]) is calculated on points randomly scattered on the molecular surface. These points are projected onto a 2D feature map formed from Kohonen neurons. The training process minimizes the differences between the Cartesian coordinates x_{si} of the surface point s and the weights

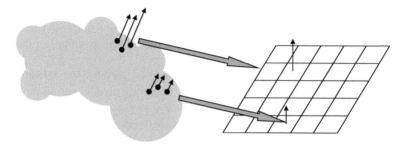

FIGURE 3.15   Kohonen map. Property values (MEP) on surface points are projected on a Kohonen map. For each cell a mean value is calculated.

of the winning neuron (the neuron where the point will be projected). Each point of the molecular surface is projected on neuron sc that have weights w_{ci} the closest to x_{si}

$$Out_{sc} \leftarrow min\left[\sum_{i=1-3}(x_{si} - w_{ci})^2\right]$$

After training, neighboring points on the molecular surface are projected on neighbor neurons (cells) of the Kohonen map. For the Kohonen level (the map) the terms "neuron" or "cell" will be equally used, to remind the structure of this level. The property calculated on a point of the molecular surface is attributed to the corresponding neuron. As the number of neurons of the map is significantly smaller than that of surface points, some surface points may be projected on the same cell of the Kohonen map. So, on each cell, a mean value of the property on these points is calculated and attributed to the cell (Figure 3.15).

One map is obtained per compound. In CoMSA, these maps are not directly compared. A template structure is chosen for training the network. Then, this network is used to process (with its own weights) the other molecules of the data set (counter-templates), giving for each compound a *comparative map* that characterizes the differences in the molecular surfaces between the template and the molecule under scrutiny (Figure 3.16). For each compound, cells on these maps are encoded with the (mean) value of the MEP for the corresponding region of the molecular surface. The data matrix so constituted for the whole series of molecules is then analyzed by PLS.

Two parameters have to be considered: the number of neurons of the Kohonen map (something like the "resolution" of the map) and the "winning distance" parameter (MD): the maximum distance between the weights of the neuron and the coordinates of the points projected on it. If MD is too small, minute differences in the shape of the surface between a compound and the template make some points "unseen" by the neurons and result in empty neurons on the map of the compound. With a larger tolerance, fewer neurons are empty. If the surface considered cannot be superimposed on the template, the corresponding output neurons remain empty, getting no signal from the

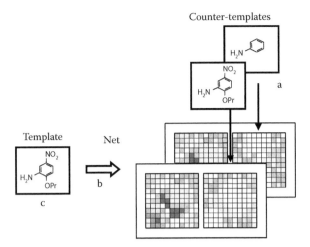

FIGURE 3.16 A scheme for the comparison of a series of molecules with a single template. A series of molecules (a) is consecutively processed by a single neural network (b) trained with a reference molecule (c). This gives a series of pairs of maps in which the first one is always the same map of the reference template molecule. (Reproduced from Polanski, J., *Acta Biochim. Polo.*, 47, 37, 2000. With permission.)

molecule processed. The projection is controlled by the atom coordinates. So, if empty neurons are considered, information contains both shape and electrostatic features.

3.7.2 Applications of CoMSA

In Ref. [95] two examples were presented: the traditional benchmark of steroids binding CBG and TBG and the ionization constants of benzoic acids (the reference process for the definition of Hammett's σ constants). Let us first consider the example of 49 benzoic acids (substituted in *meta*- or *para*-position) previously studied by Kim and Martin [237] on a range of 1.59 σ units, from p-NH_2 to p-SO_2CF_3. The first step was to choose a template molecule. A parallel CoMFA study evidenced that the region of the carboxylic group best modeled the σ. So, in the CoMSA analysis, formic acid was selected as the template involved in three steps: training the network, superimposing the molecules (alignment), and processing them.

Kohonen maps of 20×20 or 30×30 neurons were considered, mapping the MEP on the regions of the carboxyl group only. The unsolvated neutral form only was taken into account, and charges calculated from the PEOE (partial equalization of orbital electronegativity) approximation [158,159]. These MEP values are gathered in a matrix of 49 rows (molecules) and 900 (or 400) columns (averaged MEP values on the neurons) submitted to a PLS analysis. Since the studied phenomenon is (exclusively) monitored by electrostatic effects, empty neurons (that would only characterize shape) were discarded.

Various assays indicated that the model was not highly sensitive to the D, density of surface points and MD values (typical values were D = 20 points/Å^2 and MD 0.3 Å). The best model yielded $q^2 = 0.91$, s = 0.108 with three LV. CoMFA treatment (electrostatic field only calculated over the carboxylic group) gave $q^2 = 0.92$, $s = 0.107$. If the template is the (unsubstituted) benzoic acid, neighboring results are obtained ($q^2 = 0.90$, $s = 0.121$, with nine LV in place of 0.89, 0.126 in CoMFA).

This study was pursued [399] on a data set of 41 benzoic acids, including *ortho-*substituted compounds, and 47 alkanoic acids, confirming that, for these series, CoMSA gave better results than CoMFA. For the benzoic acids (on a 4.3 log unit range) CoMFA yielded $q^2 = 0.75$, $s = 0.48$ whereas CoMSA gave $q^2 = 0.90$, $s = 0.32$ with six LV with benzoic acid as template [95]. For alkanoic acids (range of 4.6 log units) $q^2 = 0.52$ ($s = 0.77$) in CoMFA and 0.86 (0.42) in CoMSA. Slightly better results were obtained restricting the analysis to the 33 acids of the data set bearing two hydrogens at the C-2 carbon atom.

A similar treatment was carried out on the classical 31 steroid data set (see Appendix B), using validated values, corrected from some previous errors [126]. The most active molecule (**s-6**) was chosen as template. The alignment was based on carbons 5, 8, 9, 10, 13, and 14. For the training set (**s-1** through **s-21**), the best model ($q^2 = 0.88$, $s = 0.424$) outperformed CoMFA ($q^2 = 0.733$, $s = 0.657$). For the test set, after elimination of the outlier (molecule 31), CoMFA reached $r_{ext}^2 = 0.71$, $s = 0.325$ and CoMSA $r_{ext}^2 = 0.41$, $s = 0.433$.

Kubinyi [115] observed that molecules **s-1** through **s-21**, usually constituting the training set, did not cover all the structural space and proposed to substitute them with molecules **s-1** through **s-12** and **s-23** through **s-31** (structures **s-13** through **s-22** forming the test set). He obtained $q^2 = 0.713$, $s = 0.50$ and in test $r_{pred}^2 = 0.78$, $s = 0.63$, whereas, for the CoMSA model, $q^2 = 0.67$, $s = 0.59$ and in test (molecules **s-13** through **s-22**) $r^2 = 0.92$, $s = 0.21$. For the TBG affinity, similar results were obtained, either considering all neurons or only the "nonempty" ones.

From these two examples, it appears that CoMSA was at least as effective as (or outperformed) CoMFA for modeling electrostatic or steric influences.

3.7.3 MULTITEMPLATE APPROACH

In a parallel study [404], Polanski proposed a multitemplate scheme. In classical CoMSA (with single template approach [STA]), a single molecule is selected for the alignment and the training of a unique template network, which is then used to analyze the test objects (the counter-templates [CT]). In the multitemplate approach (MTA), each molecule in the series trains a separate independent network for the mapping of a unique test object (the reference molecule that is now the CT). The comparison in the series of map pairs is more difficult, since the template is always different. However, it is possible to ascribe, during training, the atoms of each template to a certain neuron. The atoms of each molecule projected on a same neuron can be identified through their Cartesian coordinates. Back projection of the SOMs on each atom of the CT molecules gives a transformed property matrix, submitted to PLS analysis.

The approach was tested on a set of 14 colchicinoid derivatives (**3.32**), inhibitors of tubulin polymerization, with IC_{50} varying from 1.4 to 15 µM. Colchicinoids and related substances have been widely studied as potential anticancer drugs, due to their capability to bind tubulin, prevent its polymerization, and stop proliferation.

A relatively satisfactory model was obtained ($r^2 = 0.991$, $q^2 = 0.333$, $s_{cv} = 4.09$ with three components) from an alignment on the most active molecule.

3.32

It was previously supposed that the inhibitory activity of colchicinoid derivatives (**3.32**) was largely dependent on the A and C rings. To test this hypothesis, three different CT molecules were selected: one covering the entire molecule, the second corresponding to a biphenyl with methoxy substituents, and the third to trimethoxy-benzene. Although cross-validated correlation coefficients q^2 were rather low, they delivered some interesting information. It might be observed that a negative q^2 was obtained with trimethoxybenzene as CT (alignment only based on ring A), and that q^2 value was lower when selecting as CT a biphenyl rather than the whole ring system. This was considered as an indication that the B ring also contributed to the interaction with tubulin.

However, for a seven-molecule test set (with one outlier discarded, since bearing a large alkyl chain, and lacking a polar group in C-10) performances were limited (SDEP = 2.8 on a range of $2.4 < IC_{50} [\mu M] < 11.0$).

3.7.4 UNINFORMATIVE VARIABLE ELIMINATION (UVE-PLS) METHOD

In another application, [110] Polanski and Gieleciak tackled the problem of variable elimination in order to improve the PLS models. The study concerned 50 steroid inhibitors of aromatase, on an activity range of 5 log units. This enzyme is capable of producing estradiol by the conversion of testosterone, and so its inhibitors appear as promising drugs in the treatment of breast cancer.

The standard CoMSA method, $q^2 = 0.77$, $s = 0.71$ (with five latent variables), is better than CoMFA ($q^2 = 0.72$, $r^2 = 0.94$, 5 PC) [405] and quite comparable to the spectroscopic approaches CoSA ($q^2 = 0.77$) and CoSASA ($q^2 = 0.67$) [406], but inferior to CoSCoSA ($q^2 = 0.86$) [407] (see Chapter 6).

A further improvement was proposed, reducing the number of variables included in the final PLS model [110,111]. UVE was carried out according to an algorithm proposed by Centner et al. [408] and based on the analysis of the PLS regression coefficients. The relation between the responses (**y**) and the predictors (**X**) can be written as

$$y = Xb + e$$

where **b** is the vector of the regression coefficients and **e** the vector of the errors. UVE analyzes a parameter t, called stability, which is the ratio

$$t = \frac{\text{mean } \mathbf{B}}{\text{std}(\mathbf{B})}$$

where
> $\mathbf{B}$ is the matrix of the b coefficients obtained during the leave-one-out cross-validation (L-O-O) process
> std($\mathbf{B}$) is the standard deviation of b

Only the variables of high stability t are maintained in the final PLS model. To define a cutoff level for this selection, a supplementary column of random noise is added to the original matrix of variables. A modified procedure, an uninformative iterative variable elimination (IVE-PLS) with L-O-O cross validation, was also introduced, leading to the best IVE-PLS-CoMSA model: $q^2 = 0.96$, $s = 0.31$, with five components. In this example, about 245 neurons survived (among 900) this selection procedure.

Basically, IVE comprises

- PLS analysis and L-O-O cross validation
- Elimination (in the matrix of coefficients) of the column with the lowest stability
- PLS analysis with the new matrix
- Iterative repetition of the preceding steps to maximize q^2 L-O-O

Although the authors stress that variable elimination is always risky, an interesting point (in addition to the performance improvement) is the indication of those areas in the molecules that are important for activity. This can be achieved by back projection of the neurons onto the molecular surfaces. In the case studied, in addition to rings A and B, a previously unsuspected region (ring D) was highlighted. This was confirmed by the good predictions also obtained in a CoMSA model visualizing only this area, thanks to the choice of 1,2-dimethylcyclopentane as template.

On the other hand, the stability of the procedure of data elimination (and the predictive ability of the technique) was confirmed on 50 runs of IVE-PLS-CoMSA on different (randomly selected) train/test sets (25 molecules each). From extensive assays, it was shown that, generally speaking, the IVE procedure did not deteriorate the SDEP performance. During the elimination procedure, a large domain existed where SDEP did not depend on q^2, indicating high stability. But the authors laid stress on the fact that, after reaching the maximal q^2 value, further variable elimination deteriorated the model, so that the external predictivity did not depend on the value of q^2, in agreement with the results of Golbraikh and Tropsha [113].

One of the major concerns in the analysis of biological effects of chemicals is the identification (and possibly the prediction) of groups and of areas within their 3D representation that are mostly responsible for activity. But, frequently, different QSAR models may provide similar statistical performances (e.g., if the variables are intercorrelated), making difficult the interpretation. This search for the important regions is usually performed in PLS analysis, by elimination and selection of relevant variables. The ability of the IVE-PLS protocol to delineate these areas was investigated by Gieleciak and Polanski [111] with the example of the ionization of benzoic acids. The data set encompassed 72 *m*- or *p*-substituted benzoic acids, previously studied by Kim and Martin [237]. The question was to examine whether the IVE-PLS procedure was able to determine contour plots indicating (as expected) the carboxylic group as the reaction center determining the respective pKs. Various options were proposed to define a robust stability criterion unaffected by the noise. This even led to propose contour plots independent of the method used for the derivation of partial atomic charges (AM1 or Gasteiger-Marsili).

3.7.5   EXTENSION OF CoMSA TO 4D-QSAR SCHEMES: SOM-4D-QSAR AND s-CoMSA MODELS

With SOM-4D-QSAR, the CoMSA method was further extended to take into account the possible coexistence of various conformers in a way similar to the 4D-QSAR model of Hopfinger et al. [63]. Another approach sector-CoMSA (or s-CoMSA) was also developed, relying only on a unique conformer [363]. Extensive comparisons were carried out for the performance of CoMSA, SOM-4D-QSAR, s-CoMSA, and 4D-QSAR (see Chapter 9 devoted to 4D approaches).

3.7.6   MULTIWAY PLS IN CoMSA

Multiway PLS was also applied to CoMSA-type analysis by Hasegawa et al. [313]. In the usual CoMSA method, the MEP values on points scattered on the molecular surface are projected onto a Kohonen map. Comparative maps (with respect to a reference compound) are drawn and the neurons encoded with the MEP values of the projected points. These maps are then unfolded into vectors (constituting the structural descriptors) submitted to PLS analysis. The originality of the treatment of Hasegawa et al. is to avoid the unfolding (that partially loses the proximity relationships conserved in the Kohonen map) and directly use the wad of maps, obtained for the different compounds, in a 3-way PLS analysis.

The method was tested on 25 antagonists of the D2 receptor, potential antipsychotics (**3.33**). Molecules were aligned on the most active compound, on the basis of benzene ring and amide group. MEP values were calculated on about 40,000 points (about 0.5 Å apart). A Kohonen neural network (KNN) was trained on the reference compound and the other molecules filtered through this KNN, giving comparative maps. The data block, constituted by 25 sheets of 50×50 matrices (from KNN), was then treated by 3-way PLS, yielding $q^2 = 0.63$ ($r^2 = 0.88$). After conversion of the model into a regression-like model, the regression coefficients were projected onto the molecular surface. This representation highlighted the importance for high activity of a positive MEP at R_3 and a negative one at R_5, in qualitative agreement with previous results. An external validation test was also performed, and for the sake of comparison, a usual 2-way PLS treatment (unfolding maps into vectors) was also carried out, giving a larger RMS error of prediction. The authors concluded that the 3-way PLS led to more stable models, with the possibility of visualizing contour maps.

3.33

This study was followed [67] by the examination of a more complex case where two different potentials (electrostatic and lipophilic) were simultaneously considered. For 36 2-phenyl indoles (**3.34**), antagonists of the estrogen receptor (on an activity range of 8 log(RBA) units) alignment was carried out on the minimum energy conformers, and two series of comparative KNN maps were drawn up. The dimensions of the **X** data block were 36 (compounds) * 50 * 50 (Kohonen map) * 2 (MEP and molecular lipophilic potential [MLP]). A 4-way PLS treatment yielded $q^2 = 0.80$ ($r^2 = 0.89$) with six components. Unfolding the maps into vectors led to $q^2 = 0.79$, $r^2 = 0.98$. The large difference between q^2 and r^2 might indicate overfitting of the 2-way model.

X, Y = OH
R_1, R_2 = H, Alkyl

3.34

The coefficient map, from the 4-way PLS, highlighted four important regions of enhanced activity (Figure 3.17): for the electrostatic contribution, a positive MEP near the hydroxyl hydrogen of Y-groups, and a negative MEP around the oxygen of X; for the MLP, a hydrophobic group in R_1 increased activity whereas a zone of negative MLP near Y was beneficial.

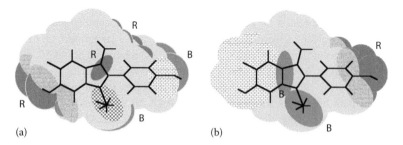

(a)   (b)

FIGURE 3.17 Coefficient map derived from 4-way PLS analysis. (a) MEP, (b) MLP. Areas encoded (B) and (R) represent plus or minus coefficient respectively. Coefficient map was drawn at 0.01 level. (Reproduced from Hasegawa, K. et al., *Comput. Biol. Chem.*, 27, 381, 2003. With permission.)

ER

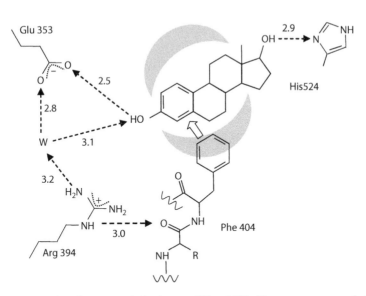

FIGURE 3.18 H-bonding network for human ER-α LBD. The gray areas symbolize the hydrophobic contacts. For these, only Phe residue is shown, on account of its anchoring role. (Adapted from Tanenbaum, D.M. et al., *Proc. Natl. Acad. Sci. U.S.A.*, 95, 5998, 1998.)

These conclusions are consistent with the X-ray structure of the complex (17β estradiol –ER) [409], indicating a charge–charge interaction of the 3-OH with Glu-353 of the receptor, an H-bond between the 17-hydroxyl and the imidazole ring of His-524, and contacts of the B and C rings with lipophilic residues (Figure 3.18).

3.8   QSAR MODELS FROM PHARMACOPHORE-ORIENTED APPROACHES

A little different from the methods presented, CATALYST [255,256] and PHASE [262] were primarily devoted to pharmacophore discovery for the search of new leads and database screening. However, they can be extended to the field of QSAR models. As already indicated (Chapter 1), the alignment of structures needed by CoMFA and related methods may take advantage of the knowledge of a pharmacophore (a group of atoms that is necessary for a molecule to be recognized by a receptor). It gives some anchoring points for superimposing the investigated structures. But, in addition, this approach may constitute a correlation tool of its own. Basically these methods start from a library of conformers of low energy and generate hypotheses according to the presence of a minimum number of features common to a subset of active molecules. Pruning may be introduced, eliminating hypotheses also present in inactive compounds. The surviving hypotheses are then refined.

3.8.1 CATALYST APPROACH

In CATALYST, a cost function (in bit units) is introduced to rank the hypotheses. It consists of three parts:

- Weight cost (increasing if a feature weight in a model deviates from an ideal value)
- Error cost (deviation between estimated and experimental activity)
- Configuration cost (penalizing the complexity of the model)

In the evaluation of the error cost, activity is estimated by

$$\log (\text{Estimated act}) = I + \text{Fit}$$

I intercept of the line logact/Fit, with

$$\text{Fit} = \sum_{\text{mapped hypothesis features}} *W\left[1 - \sum (\text{disp/tol})^2\right]$$

where mapped features correspond to the number of pharmacophore features successfully superimposed on a chemical fragment. W is the weight of the corresponding feature (default 1.0). Disp is the distance between the center of a particular pharmacophoric sphere and the center of the superimposed chemical moiety, with a tolerance tol = 1.6 Å.

CATALYST also calculates, on the one hand, the cost of a *null hypothesis* assuming that there are no relationship features/activities and that observed activities are normally distributed about their mean. On the other hand, an *ideal cost* corresponding to the simplest model fitting all data perfectly is evaluated. For good-quality models, the difference between the null and ideal costs must be large, as well as the difference between the null cost and that calculated for the hypothesis examined. Then clustering of the various pharmacophore hypotheses is carried out, according to the scores (possibly in two successive steps to reduce the number of solutions).

CATALYST was applied by Hammad and Taha [410] to neuraminidase (NA) inhibitors. These compounds are attractive targets against influenza viruses, but the problem is rather complex since NA inhibitors are slow binding/transition state analogues (TSA) particularly sensitive to steric requirements. Usual database queries to find new inhibitors are faced with the problem of too lax pharmacophore models, leading to a high rate of false positives.

Diverse inhibitors (244 chemicals) from four scaffolds (**3.35** through **3.38**) were investigated. In this study, a hybrid treatment selected, by GA and MLR, a combination of pharmacophore-based models and molecular descriptors able to explain bioactivity variations. Five pharmacophores emerged (one of them is featured in Figure 3.19), suggesting several ligand–receptor binding modes. These optimal pharmacophores merged with high ligand-shape constraints were subsequently used as 3D queries in screening the National Cancer Institute database. Several hits with good inhibitory activity were found (among them, one with a IC_{50} value of about 2 μM). Interestingly, docking supported the suggested binding modes.

3.35 **3.36** **3.37** **3.38**

For another application of CATALYST, see Tromelin and Guichard [411] and Refs. [254–261,263] quoted in Section 1.3.2.

3.8.2 PHASE Approach

PHASE, proposed by Dixon et al. [262], was primarily dedicated to pharmacophore perception and 3D database screening, but it may also be used for deriving one (or a set of) QSAR model(s). The used strategy somewhat relied on the *active analogue* approach. Each molecule is represented by a set of conformers (generated by torsional sampling or Monte Carlo method), from which a set of points corresponding

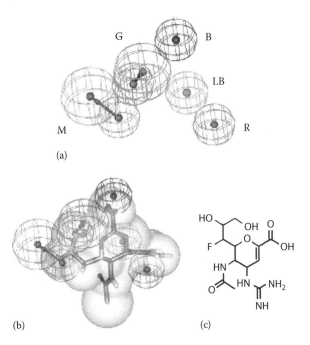

(a)

(b) (c)

FIGURE 3.19 (**See color insert following page 288.**) (a) One of the emerging pharmacophore features. Color code: (LB) Light blue spheres hydrophobic features, (R) red spheres positive ionizable features, (B) blue spheres negative ionizable features, (G) green, and (M) magenta vectored spheres H-bond acceptor and donor features, respectively. (b) inhibitor, (c) IC_{50} = 0.5 nM fitted against the pharmacophoric features and combined with the corresponding molecular shape. (Adapted from Hammad, A.M.A. and Taha M.O., *J. Chem. Inf. Model.*, 49, 978, 2009. With permission.)

to possible pharmacophoric features are extracted, such as H-bond (donor and acceptor), hydrophobic, ionisable (positive and negative), and aromatic. Sites involved in H-bonding are mapped by vector features (for privileged H-bond directions) or projected points. Hydrophobic sites in a fragment are located as weighted average of the non-hydrogen atoms (Figure 3.20).

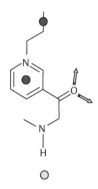

The search for a pharmacophore comes to identify common characteristic features providing hypotheses to account for ligand–receptor binding. This search for a pharmacophore of, say, k points is carried out from an exhaustive analysis of intersite distances filtered through a binary decision tree (with 1 Å bins) in order to identify all spatial arrangements of features shared by a given number of active molecules. Each hypothesis surviving this filtering is considered as a reference pharmacophore for scoring the other (nonreference) survivors aligned on it. The score is based on convergence of site point positions and direction of the vector features (H-bonds, aromatics). Conformation-independent properties (as log P), selectivity (the rarity of a hypothesis indicates that this hypothesis is more likely to be unique to the actives), or conformational energy cost may also be introduced.

FIGURE 3.20 Example of feature mapping: dark gray hydrophobic; light gray H-bond donor or acceptor (either as projected points or vector features). (Adapted from Dixon, S. et al., *J. Comput. Aided Mol. Des.*, 20, 647, 2006.)

The hypothesis ultimately retained as reference ligand corresponds to the highest sum of scores with other ligands. The ligands may be therefore aligned on at least the 3-point-pharmacophore or the entire molecule depending on flexibility or congenericity. Note that hypotheses may be augmented by consideration of the excluded volume, regions of space that cannot be occupied. Scoring is also performed with respect to inactives (in the assumption that inactivity results from a pharmacophore deficiency and not from steric clashes, desolvation penalty, or important entropy loss upon binding). Once a hypothesis is validated, that is, a pharmacophore is selected, a QSAR model may be built.

For constructing a QSAR model, a lattice encompasses the space occupied by the aligned molecules. Each cell (typically 1 Å edge) is encoded 0 (empty) or 1 (occupied if the center of the cell falls within an atom sphere). Each molecule is so represented by a bit-string, each assigned to a definite site/atom nature. These strings have a common length for all the molecules of the training set, and constitute the structural descriptors submitted to PLS (Figure 3.21). At this level, the approach is not so far from the HASL model of Doweyko [42] (see Section 3.4).

Excluded volumes may be also considered. From the PLS treatment, it is also possible to delineate which cells (if occupied) would have a beneficial or detrimental influence on reactivity, something like the CoMFA contour plots.

PHASE was applied to inhibition of human DHFR by 2,4-diamino-5-deazapteridines (**3.39**) (20 molecules in training, 57 in test) [262]. Starting from about 400 hypotheses with 5 or 6 points, the most predictive model proposed contains two H-bond donors (C) and three aromatic features (O) (Figure 3.22). This model led to good correlations

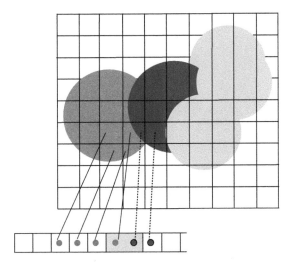

FIGURE 3.21 The mapping of a molecule to a volume bit pattern that provides the independent variables for PHASE 3D-QSAR model. For simplicity, only two dimensions are represented. (Reproduced from Dixon, S. et al., *J. Comput. Aided Mol. Des.*, 20, 647, 2006. With permission.)

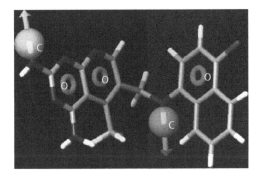

FIGURE 3.22 The PHASE hypothesis (CCOOO) that yielded the most predictive atom-based QSAR model. H-bond donors are represented by C spheres, aromatic rings by O. (Reproduced from Dixon, S. et al., *J. Comput. Aided Mol. Des.*, 20, 647, 2006. With permission.)

for both training and test sets: in training $r^2 = 0.932$, and in test $r^2 = 0.492$. It was also in rough agreement with a previous classification study (on three activity levels) carried out by Debnath [412] with CATALYST.

3.39

PHASE was also applied [413] to the inhibitors of equilibrative nucleoside transporter 1 (ENT1) that may have potential applications in case of heart disease, viral infection, or cancer chemotherapy. In the data set, 77 compounds were used for training

FIGURE 3.23   **(See color insert following page 288.)** The best generated pharmacophore model obtained from PHASE. Pharmacophore features are (R) red vectors for H-bond acceptor, (O) orange rings for aromatic groups, and (G) green balls for hydrophobic functions. Two compounds (**3.40**) and (**3.41**) were aligned to the pharmacophore. For molecules, blue indicates nitrogen, red indicates oxygen, yellow refers to sulfur, gray indicates carbon, and white indicates hydrogen. (Reproduced from Zhu, Z. and Buolamwini, J.K., *Bioorg. Med. Chem.*, 16, 3848, 2008. With permission.)

and 39 in test. Hypotheses generated from three highly active compounds led to five pharmacophore models (comprising six features), all with an *anti*-conformation (with respect to N9-C1′ glycosidic bond). The best pharmacophore was formed with three H-bond acceptors, one hydrophobic function, and two aromatics. The involved groups were the 3′ OH, 4′ oxygen, NO_2 group, and the benzyl phenyl, the imidazole and pyrimidine portion of the purine ring (Figure 3.23 where two molecules **3.40** and **3.41** are superimposed). The model led to $r^2 = 0.916$, $r^2_{pred} = 0.777$. It was further validated by a CoMFA treatment with alignment based on the selected pharmacophore: $r^2 = 0.894$ and $q^2 = 0.591$. An interesting point is that the pharmacophore hypothesis and the CoMFA contour plots complement each other, the pharmacophore treatment capturing more specific features (Figure 3.24).

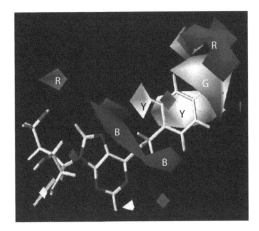

FIGURE 3.24 **(See color insert following page 288.)** Contour plots of CoMFA 3D-QSAR model. (R) Red, negative electrostatic potential enhances potency; (B) blue, positive electrostatic potential favors activity; (G) green, bulky substituents enhance potency; (Y) yellow, bulky substituents are unfavorable to activity. (Reproduced from Zhu, Z. and Buolamwini, J.K., *Bioorg. Med. Chem.*, 16, 3848, 2008. With permission.)

In CoMFA (R) red areas near the nitrogen group indicate that electronegative groups are favored, in agreement with PHASE (NO$_2$ as HB acceptor). The green area (G) near the *para* position in the 6-benzyl ring favors bulky substituent but the yellow zone (Y) on the other side of the 6-substituents limits the degree of bulkiness. The yellow area (Y), near the *ortho* position of the benzyl ring, indicates that steric bulk is disfavored. For the blue zone (B), near purine, electron-deficient groups are favored.

Part II

Around the 3D Approaches

The 3D models just presented (CoMFA, CoMSIA, and others) are based on local quantities (such as steric or electrostatic potentials) calculated on individual, discrete elements such as atom positions, or points located in the 3D space surrounding the molecules under scrutiny or on the molecular surface, or parts of the molecular surface or volume (voxels). On the other hand, "traditional" (2D-) QSARs use descriptors corresponding to scalar quantities (calculated or from experimental origin), relative to the whole molecule (such as molecular weight, element counts, log P, or K_{OW}), or characterizing substituent groups. Substituent constants (Hammett, Taft, Hansch), which express the influence that groups, introduced on a common core, have on thermodynamic (equilibrium constants), kinetic (rate constants), or physicochemical quantities (e.g., NMR chemical shifts) have been largely used as mirroring electronic, steric, or hydrophobic effects. Other common ways to characterize a molecular structure rely on fragment counts or treatments of the graph associated with the 2D structural formula, leading to numerous topological indices.

In an intermediate position some models or some descriptors, geometrical or quantum chemical, for example, are obviously sensitive to the actual 3D structure, but the spatial location of the atoms (and its influence on the calculated descriptors) is not explicit in the final model. Extending a proposal from Turner and Willett [44] this would correspond to 2.5D models or descriptors.

In the autocorrelation GRIND method [45], for example, the information collected on points in the vicinity of the molecular volume (in fact, interaction energies with an external probe, as in CoMFA) is condensed in a few numerical elements, which, however, can be traced back to the original structures, whereas in VolSurf [178] the 3D information is encoded in some numbers. In similarity-matrix treatment, each structure in the set is characterized by its resemblance to other members of the set

[47], whereas Spectroscopic Data–based methods use IR, UV, or NMR parameters as descriptors, without any in-depth examination of the relationships between these parameters and the molecular structure [52]. Finally, some approaches are quite similar to the usual 2D-QSARs, with descriptors globally evaluating geometrical elements (surface, volume) or reducing the molecular organization (including actual interatomic distances) to a few indices [53,54].

4 Autocorrelation and Derived Methods

An important constraint with comparative molecular field analysis (CoMFA) and comparative molecular similarity analysis (CoMSIA) is the need for superimposing molecules onto a common reference compound. This operation, decisive for obtaining good models, is relatively easy for a series of congeneric compounds sharing a rigid skeleton. Unfortunately, for sets of structurally varied molecules (with possibly conformational flexibility), defining an alignment is often subjective and always time consuming. This handicap led researchers to propose alignment-free systems (that is systems avoiding the superimposition of the molecules under scrutiny onto a common reference). In these methods, the values of the molecular fields are generally not directly considered but rather undergo some mathematical treatment to generate Translation, Rotation Invariant (TRI) descriptors. These approaches are closely related to 2.5D descriptors and methods but owing to the significant common features between CoMFA, CoMSIA, and GRIND, we preferred to deal with autocorrelation in a separate chapter.

4.1 AUTOCORRELATION- AND CROSS-CORRELATION-BASED METHODS

4.1.1 AUTOCORRELATION VECTOR

Autocorrelation functions are based on the concept of spatial regularity: if a property takes a given value in some spatial region, how probable is it that it takes the same value at a certain distance? Autocorrelation was first applied in 2D-QSARs by Moreau and Broto [200,414–417] with the introduction of autocorrelation coefficients: for an atom pair i, j separated by d bonds and an atomic property p, the autocorrelation coefficient A(d) is:

$$A(d) = \sum p_i p_j$$

(summation running over all pairs separated by d bonds). Considering different topological distances d, a series of coefficients A(d) is obtained, defining an autocorrelation vector (Figure 4.1). Other 2D-QSARs and QSPRs were proposed by Gupta et al. [418], Chastrette and Crétin [419], and Devillers [420,421]. Interesting characteristics of this autocorrelation vector are

- Reduced canonical description (independent of the atom numbering)
- Fixed length (independent of the molecular size)

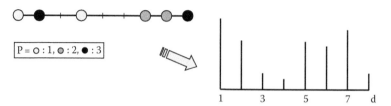

FIGURE 4.1   Autocorrelation vector. Vertical bars represent the terms of the autocorrelation vector for property p.

This concept was then adapted to 3D descriptions, replacing the topological distance d by the true interatomic distance [417], and, some years later, extended to spatial autocorrelation of molecular surface properties [146]. A mixed approach includes Euclidian distances in a topological framework [422].

Spatial autocorrelation vector: In the spatial autocorrelation method, points are randomly distributed on the molecular surface, according to a preset point density. A property p (e.g., the electrostatic potential) is mapped onto these points. The distances between surface points are ordered in preset bins ($d_{min} < d < d_{max}$), and the autocorrelation coefficient is

$$A(d_{min}, d_{max}) = \frac{1}{N} \sum p_i p_j$$

where
    summation is over all atom pairs ij for $d_{min} < d < d_{max}$
    N is the number of distances d in the interval

Experiments confirmed (as expected) that the autocorrelation vector was independent of the spatial orientation of the molecule, but sensitive to changes in the conformation (and also to the van der Waals radii chosen for defining the molecular surface). Scattering points onto the Conolly's solvent accessible surface or van der Waals' surface gave nearly the same results. It was also shown that a point density equal or greater than 5 points/Å^2 modeled the surface with enough details and that distance bins with amplitude of 1 Å or less should be used [146].

4.1.2 APPLICATIONS

A first application concerned the classical steroid benchmark: affinity for corticosteroid-binding globulin (CBG) for 31 steroids [146]. In this application, 3500 points were scattered on the van der Waals surface (corresponding to a density of about 10 points/Å^2). Coulomb-type electrostatic potential was calculated on the surface points from empirical atomic point charges. For each molecule, 12-component autocorrelation vectors were calculated for distance intervals of 1 Å, from 1 to 13 Å. Although beyond the scope of our direct concern, we can note that these autocorrelation vectors input into a Kohonen network allowed for well-separating high-, intermediate-, and low-activity molecules, in a nonlinear model, whereas the linear classification model proposed by PCA was not satisfactory.

As to QSAR, autocorrelation vectors were used as inputs in a 12/2/1 back propagation neural network. The correlation between observed vs. predicted pK_i values gave a good correlation coefficient $r = 0.99$ (standard deviation 0.18 on 2.9 log units). However, in leave-one-out (L-O-O) cross-validation $q^2 = 0.63$, with a large deviation for compound s-31 (the only one containing a fluorine atom in position -9). This compound was also an outlier in several other QSAR studies on this data set. Discarding this compound, good results were obtained: $r^2 = 0.85$, $q^2 = 0.84$, $\sigma = 0.44$. These results compared favorably with those obtained in numerous other studies on the same data set [126]. See Appendix B.

The second example concerned the binding affinity to cytosolic AhR of 78 halogenated aromatic hydrocarbons (25 polyhalogenated dibenzo-p-dioxins PCDDs, 39 polychlorinated dibenzofurans (PCDFs) and 14 polychlorinated biphenyls (PCBs) [see compounds 1.4 through 1.6]) on an activity range of 6.3 log units. The molecular property retained was in that case the hydrophobicity potential (calculated from atomic contributions to log P and a distance dependent potential). A good correlation was observed with 12 autocorrelation coefficients: q^2 reached 0.83 ($r^2 = 0.79$, $\sigma = 0.61$). The same series of compounds was previously studied by CoMFA. After alignment on 2,3,7,8-tetrachlorodibenzo-p-dioxin (TCDD) the CoMFA treatment, carried out for every family and for the whole set, yielded a similar performance. For the whole set, $r^2 = 0.878$ and $q^2 = 0.724$ with six principal components (PCs), SEE $= 0.53$ [423]. It was also noted that fitting fields would force biphenyls to an unrealistic planar geometry. In a subsequent CoMFA study [210], biphenyls were removed from the training set and used as a test set to confirm the predictive ability of the model.

A very similar approach was used by Moro et al. [424] in the analysis of a very large population of antagonists of the human A_3 adenosine receptor. These compounds intervene in inflammation processes and regulation of cell growth. A_3-receptor selective antagonists belong to two major groups: purine-type derivatives and non-purine compounds. The latter mainly encompasses flavonoids, pyridines and 1,4-dihydropyridines, quinazolines and triazolo-quinazolines, isoquinoline, and pyrazolo-triazolo-pyrimidines.

A first study [425] concerned 106 pyrazolo-triazolo-pyrimidines (4.1 and 4.2) with the aim of defining a novel pharmacophore model of the human A_3 receptor. A combined target-based and ligand-based approach was developed, where CoMFA was used in tandem with high-throughput docking. In a first step, the 106 compounds were docked in a rhodopsin-based model of the human A_3 receptor. Then CoMFA was used with the low-energy docked conformations to get a better estimate of ligand receptor interactions. A very satisfactory correlation was obtained for the pK ($r^2 = 0.922$, $q^2 = 0.840$, SEP $= 0.18$ with six components). On a test set of 17 new compounds, predictions were very close to the experimental values with errors less than 0.25 on a range of 2.0 log units. From this robust integration of a target-based and a ligand-based approach, a new "y-shape" model was proposed for the human A_3 receptor.

4.1   **4.2**

This study was pursued [424] with the aim to propose, thanks to the autocorrelation approach, a general and robust QSAR model on a data set of 358 compounds corresponding to 21 distinct chemical classes. As in Wagener et al. [146] treatment, the molecular electrostatic potential was calculated on points scattered on the molecular van der Waals surface and an autocorrelation vector of 12 components (corresponding to bins from 1 to 13 Å distances) was used as structural descriptor. But, in this study, analysis was carried out by PLS (and not by a back propagation artificial neural network). With six latent variables, the treatment led to $r^2 = 0.67$ and $q^2 = 0.65$ with RMSEC = 0.76, RMSEP = 0.79 (on 10 bootstrap groups) for a reactivity range of about 5.5 log units. The robustness of the model was checked on a new data set of 40 compounds for which the preceding model gave $r^2_{pred} = 0.67$, RMSEP = 0.7.

According to the authors, the autocorrelation vector constituted a structural fingerprint that might be efficiently used to predict activity in virtual screening of large databases, as established in a further study [426], see Chapter 8.

4.1.3   3D Topological Distance-Based Descriptors

Klein et al. [422] proposed a modified version of 3D topological autocorrelation vectors including 3D information about steric, electronic, and atom-type characteristics: the 3D topological distance-based (3D-TDB) descriptors. The component of the autocorrelation vector A(d), relating to property p, for the topological distance d is given by

$$A(d) = \frac{1}{k(d)} \sum_{i=1,n-1} \sum_{j=i+1,n} p_i D_{ij} p_j$$

where
    k is the number of atom pairs (i, j) at topological distance d
    D_{ij} is the the Euclidian distance between i and j
    For the steric component, p is the covalent radius
    For the electrostatic component, p is the sigma orbital electronegativity [427]

A third vector is computed for atom-type indicators

$$I(d) = \frac{1}{k(d)} \sum_{i=1,n-1} \sum_{j=i+1,n} \delta_{ij}$$

where $\delta = 1$ if i and j belong to the same type of atoms, 0 otherwise. With d varying from 1 to 13, the three-property autocorrelation vectors correspond to a total of 39 descriptors per molecule. The method was applied to three data sets: the steroid benchmark, 70 guest compounds for host–guest inclusion in β-cyclodextrin, and 70 TIBO-based HIV-1 reverse transcriptase inhibitors (**4.3**).

4.3

For steroids and TIBO derivatives, 3D-TDB descriptors outperformed Tsar descriptors [428, version 3.3] for both q^2 and SDEP on external test sets, but they remained inferior to the spatial autocorrelation method of Wagener et al. [146] for steroids or CoMFA for TIBO analogues [277]. This was not surprising since 3D-TDB descriptors condensed a distributed spatial information in a few descriptors. But, for host–guest complexation, Tsar and 3D-TDB descriptors showed a nearly similar predictive ability. This was attributed to the fact that complexation was much less specific a process than ligand–receptor interactions. On the same examples, 3D-TDB descriptors also appeared very efficient in cluster analyses.

4.2 SESP AND DiP MODELS

The Distance Profile (DiP) descriptor approach proposed by Baumann [429] relies on TRI–descriptors (descriptors which are Translationally and Rotationally Invariant) and so avoids the alignment phase. It is based on a previous 2D descriptor, the SE-vector (Start End-vector) [430], which was first extended to SESP (Start End for Shortest Path) in "Topologic SESP," and then completed with distance information, "Geometric SESP." Since the difference between the topologic and geometric SESP is generally rather small, except in extreme cases such as twisted systems like biphenyls, these descriptors may be considered as 2.5D rather than 3D [429].

Although such DiPs are not strictly autocorrelation models, they rely on a similar concern to evaluate proximity relationships between pairs of atoms depending upon their attributes, all combinations being considered.

4.2.1 START END FOR SHORTEST PATH

The SESP vector is a topological distance-counting descriptor in the H-depleted molecular graph:

$$[p0, p1, p2, \ldots]$$

where
  p0 is the number of atoms
  p1 is the number of bonds
  p2 is the number of two-bond paths and so on

But, at the difference of Ref. [430], only the shortest path between two atoms is considered.

Atom and bond nature are also specified. So every atom is given three attributes: The first one is topology T (in fact existence), the second specifies the atom type (such as O, C sp^2, C sp^3, N), the third one is attributed to atoms taking part in a double or triple bond. For example, for acetaldehyde O = CH—CH$_3$ (Figure 4.2)

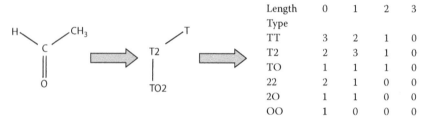

Length	0	1	2	3
Type				
TT	3	2	1	0
T2	2	3	1	0
TO	1	1	1	0
22	2	1	0	0
2O	1	1	0	0
OO	1	0	0	0

FIGURE 4.2   SESP encoding for acetaldehyde.

where
  length corresponds to the path between two atoms
  T stands for all heavy atom
  2 for sp^2 hybridized atoms
  O for oxygen
  C sp^3 and H are omitted

For length = 0, this comes to an atom count. For paths where terminal atoms are of the same type (TT, 22, OO, for example) the path is counted only once. For unsymmetrical paths (T2, 2O, etc.), the count depends on the origin atom.

To introduce stereochemical and conformational information, topological distances are replaced by the ratio Euclidian distance/topological distance, and this value is used to increment the count of distances (in place of 1, in the absence of such a weighting). This scheme was preferred to a histogram, which would largely increase the number of variables.

This encoding results in high redundancy (e.g., a carbonyl oxygen atom intervenes as T, as 2 and as O). So it is advantageous (and sometimes necessary) to reduce dimensionality: PCA may be used for classification or similarity searching. For structure-activity studies, the PLS treatment, to derive the correlation, ensures, by itself, a reduced dimensionality in its latent variables.

4.2.2 APPLICATION OF SESP VECTORS

The approach was applied to six data sets – four of them have been previously analyzed by EVA and CoMFA: binding the cytosolic Ah receptor of 25 dibenzo-*p*-dioxins, 39 dibenzofurans, and 14 biphenyls, with chloro and bromo (for dioxins) substituents [423] (see compounds **1.4** through **1.6**). The fourth data set concerns the binding affinity of 39 muscarinic (**4.4**) agonists from Ref. [431].

The other data sets refer to the prediction of octanol/water partition coefficient (log P) for 185 compounds [420], and the last one to the classical steroid benchmark (affinity to CBG). Depending upon the population, the maximum (topological) distance path varied from 4 to 7. Although the descriptors are orientation invariant, they depend on the conformation. So, in the study of flexible muscarinic agonists, the orientation independence was confirmed and a conformational search (MC simulation) was performed. Validation was carried out by 200 runs of Leave–33%-Out to define the number of latent variables. Then L-O-O was performed for the sake of comparison with previous results, since it gave a criterion not perfect but perfectly reproducible.

- For polychlorinated hydrocarbons (all 78 compounds on a 6.3 log unit range), SESP gave $q^2 = 0.77$ (topological mode) and 0.75 (geometric mode) with $r^2 = 0.84$ with seven latent variables whereas in CoMFA, q^2 was 0.724 and in EVA 0.68.
- For muscarinic agonists (39 compounds on an activity range of 2.6 log units), the various assays (using different sets of conformers) gave practically identical results of $q^2 = 0.75$, comparable to CoMFA (0.72) [431] or EVA (0.76) [51].
- For log P, the 185 objects considered, split into training (135) and test (50), exhibit a large structural diversity (from methanol to anthracene). With q^2 about 0.87, SESP outperformed EVA (0.68) but remained inferior to an artificial neural network (ANN) working on autocorrelation vectors (RMSE test = 0.46 vs. 0.30). However, the ANN was trained on 7200 molecules and not on 135 as in this study [420].

Similar results were obtained for the steroids (q^2 = 0.87 vs. 0.80, EVA; or 0.85, CoMFA) [432].

4.2.3   DISTANCE PROFILES DESCRIPTORS

The geometrical information was only a rather modest part of the SESP descriptors, whereas it is at the hearth of the DiP descriptors [116]. The basic idea in the definition of the DiP descriptors is that the occurrences of particular 3D atom pairs are different in active and inactive molecules. For p atom types, there will be

$$\frac{p(p+1)}{2} \quad \text{possible combinations}$$

(of type $N \leftrightarrow N$, $N \leftrightarrow O$, $O \leftrightarrow O$, for example, for atoms N, O).

A molecule is described as a histogram for each atom pair. For the various possible pairs, bins are characterized by a lower (d_{min}) and upper distances ($d_{max} = d_{min} + 1$ Å) and the nature of the atom pair. Each atom pair is examined in turn; its Euclidian distance is calculated and is counted in bins of a defined resolution (1 Å). This presents some analogy with autocorrelation vectors.

Atom types are encoded with some hierarchical levels: first all heavy atoms irrespective of their nature, then sp^2 and sp^3 atoms, the presence of double (or aromatic) and triple bonds. Finally, the nature of the atom: N, O, F, etc. sp^3 carbons are not specifically encoded since they are included in "heavy atoms."

The descriptors are then correlated to activity by PLS. On account of this large number of descriptors, a selection is necessary. It was carried out with a reverse elimination method with *tabu search* [433,434]. The process corresponds to a stepwise selection (steepest descent, mildest ascent): in each iteration, a variable is systematically added or removed to the model (in $\rightarrow$ out, out $\rightarrow$ in). The process avoids revisiting an already examined solution (tabu) and chooses the change that improves the objective function the most, or which is the least detrimental (avoiding to be kept in a local minimum). The search is deterministic.

4.2.4   APPLICATION OF DiP DESCRIPTORS

Two data sets were investigated. Thirty-six aryl sulfonamide antagonists (see compound **2.12**) of Endothelin receptor subtype A (ET_A) studied by CoMFA [346], EVA [44], MS-WHIM and Autocorrelation [435], and SOMFA [344]. With six descriptors retained for DiP, q^2 reached 0.79 (L-O-O), whereas WHIM, MS-WHIM, and CoMFA led to 0.65, 0.66, and 0.70, respectively. Autocorrelation and SOMFA gave q^2 inferior to 0.6.

With this kind of descriptors (counts in bins), it is often difficult to come back from the correlation (obs./pred.) to a structural interpretation. However, in this case, on account of the small number of descriptors, some insight can be gained as to the factors favoring activity. For example, the important role of NN 7.12 (two nitrogen atoms at a distance between 6.75 and 7.5 Å) highlighted the best N substitution pattern.

Interestingly, the influence of the resolution (the width of bins) was also discussed. As expected from a coarse resolution, the geometric information was vague and the performance of the models decreased. Too high a resolution complicated the model and gave many bins that were not populated. In this example, resolutions of 1 or 0.75 Å seemed a good choice. Another question is: how sensitive the model is with respect to the chosen conformation (here, the CoMFA alignment and conformations modeled on X-ray structure)? Variations on q^2 are marginal, but the descriptors retained are not the same. A suggested interpretation would be that the method searched for different descriptors to reflect the same structural features responsible for the activity [116].

A similar study was carried out for 18 HIV1-RT inhibitors, which was previously studied by Gancia et al. [435]: N-substituted 3-amino-5-ethyl-6-methylpyridin-2(1H)-one derivatives (**4.5**) differing only by the aromatic ring system. For this data set also, the DiP model outperformed MS-WHIM ($q^2 = 0.78$ vs. 0.69).

4.5

It may also be noted that the author has carefully discussed the validation of his results. For example, he preferred multiple runs of Leave-Some-Out cross-validation to avoid chance correlation that might occur in L-O-O.

4.3   GRID INDEPENDENT DESCRIPTOR APPROACH

The preceding examples led to satisfactory predictions, comparable or superior to other approaches, and the method avoided the troublesome step of alignment, but the drawback was that the original information could not be retrieved from the autocorrelation vectors. A same limitation also arose for the WHIM descriptors (see Chapter 7), but not with the GRid INdependent Descriptor (GRIND) approach proposed by Pastor et al. [45]. The GRIND approach also relies on an autocorrelation transform, but with the advantage that their descriptors, independent of the reference frame, can be traced back to the original molecules themselves and to the original information they carry. This even allows for identifying (or at least suggesting) some important ligand–receptor interactions.

The approach takes up the concept of molecular interaction fields (MIFs) introduced by Goodford [58] and transforms this information into a small number of variables characterizing the existence, at a certain distance, of couples of points representing important interactions. Originally calculated from protein active sites, the MIFs identified regions where certain groups might interact favorably. This might indicate possible locations for adapted ligand groups. But MIFs can also be calculated from ligands. They indicate regions where, similarly, a group of a potential receptor may interact favorably. With different probes, a virtual receptor site (VRS) may be inferred. Schematically, GRINDs correspond to a small set of variables representing the geometrical relationships between different important regions of the VRS.

4.3.1 GENERATING AND ENCODING THE MOLECULAR INTERACTION FIELDS

The basic steps of the approach are (Figure 4.3):

MIF computation: This is carried out with the GRID program [176]. Three probes are used: DRY (for hydrophobic interactions), carbonyl oxygen O (for H-bonding acceptor interactions), and amide nitrogen N1 (H-bonding donor). A mesh of 0.5 Å was chosen for the grid (with a 5 Å extension beyond the studied molecule).

Filtering: Among the huge number of MIF values calculated (usually more than 10^4), the most interesting regions are those of intense negative energies (favorable) but delineating these regions may mask other interactions from different parts of the ligand. Relevant regions are selected (for each MIF) on the basis of the intensity of the field at a node and the distance to another selected node, leading to the extraction of about 100 nodes to define the VRS; schematically, couples of points associated with important interactions.

Encoding: The product of interaction energies is calculated for each pair of nodes and ordered into bins corresponding to small ranges of distances. This relies on auto- and cross-correlation, but, in the maximum auto- and cross-correlation (MACC) method, only the highest product is stored. This allows for tracing back to the corresponding atoms in the molecule to make interpretation easier (which is not possible with the traditional cross-correlation transforms that consider the sum of the products in each bin). Correlogram plots display the product values vs. the distances between nodes (with only one value per bin). With the three probes indicated, the three auto-correlograms correspond to an important product for a unique type of interaction (hydrophobic, H bond-acceptor or donor) on two nodes separated by a given distance, and the three cross-correlograms to the product of two important different interactions. For example, the correlogram DRY/N1d corresponds to a hydrophobic center on one node and

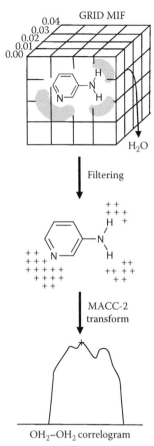

FIGURE 4.3 Computation of the GRIND. A MIF is computed with the GRID force field, and the most relevant interactions are filtered and encoded into MACC-2 correlograms. (Reproduced from Fontaine, F. et al., *J. Med. Chem.*, 47, 2805, 2004. With permission.)

H-bond donor on another node, at the distance d indicated (vide infra). Note that these nodes are outside the molecule but suggest that the corresponding sites on the molecule (a hydrophilic group and a H-bond acceptor) play an important role in the interactions of the molecule upon binding its receptor.

At the end, a matrix of descriptors is obtained, where rows represent compounds and columns, the various blocks of the correlograms. In other words, the approach focuses on pairs of important interactions (favorable or not) associated with certain node–node distances. Data are then analyzed by PLS or PCA after selection of the relevant variables, for example, with fractional factorial design (FFD) [177]. All the treatments may be carried out with the program ALMOND [55].

4.3.2 APPLICATIONS OF GRIND APPROACH

To test the efficiency of the approach, the authors [45] investigated a data set of 10 glucose analogues (**4.6**) inhibitors of glycogen phosphorylase (from 47 structures previously studied with GRID/GOLPE [189,436]). The crystal structures of the complex inhibitor/receptor have been resolved by X-ray crystallography, but to check alignment independence, two supplementary copies of each structure were built with a random orientation. With 3 probes, the 6 correlograms of 35 values each, for the 3×10 molecules, produced a matrix of 30 rows, 210 columns. A PCA treatment showed that the three copies of each compound were closely clustered in the plane PC1, PC2, stressing the alignment independence. PLS analysis (experimental vs. predicted biological activity) indicated a good correlation. Examination of correlograms shows that N1–N1 autocorrelograms were strongly correlated with activity. Using only this correlogram (and considering only one structure per compound), in PLS (10 compounds, 30 variables) one latent variable was highly significant and led to $r^2=0.83$, $q^2=0.64$ on an activity range of 3.5 log units.

R = H, OH, CH_2CN, $C(O)NH_2$, $NHC(O)NH_2$ ...

4.6

The simplicity of the model gave some insight into the important interactions. The N1 probe interacted less strongly with the hydroxyl groups present in the less active derivatives than with the carbonyl oxygens of the spiro rings in the most active ones (spiro rings formed from R_α and R_β, see Scheme **4.6**). These observations were in agreement with the published crystal structures of the complexes, and consistent with a GRID/GOLPE treatment, but the latter involves 847 variables.

For the steroid benchmark (affinity to CBG receptor), from the six correlograms of 45 variables each, two outliers appeared: compound **s-31**, already identified in some other treatments, and **s-20**, probably on account of steric effects [45]. Discarding these two compounds, and considering only O and N1 probes, the final model yielded $r^2=0.83$, $q^2=0.76$ with only two latent variables. Splitting the data into training- and test-sets (20/9 structures) led to $q^2 = 0.64$, SDEP = 0.26. These results were consistent with the data reported by Coats [126]: on 14 methods, q^2 varies from 0.23 to 0.93 (average 0.71).

Good results were also obtained on a series of 25 butyrophenones with serotoninergic 5-HT$_{2A}$ affinities and known as atypical antipsychotics [45]. These molecules are conformationally constrained (although bearing several rotatable bonds) but are of very varied structures. On an activity range of 2.9 log units, GRIND led to $q^2 = 0.83$, $r^2 = 0.93$. This model was subsequently improved by the introduction of surface descriptors (vide infra).

4.3.3   INCORPORATING MOLECULAR SHAPE: TIP FIELD

In many cases, molecular shape is crucial for biological activity. Hydrophobic interactions often play a major role in drug–receptor binding, as for instance in endocrine disruption, and GRIND did not take it well into account. In fact, the three previously proposed probes (DRY, O, N1) generally prove to be sufficient. However, the hydrophobic fields (calculated with DRY) are often very weak, so that non polar aliphatic areas may be ignored by the model.

In further developments [437], molecular shape has been incorporated in GRIND by addition of a descriptor based on the curvature of the molecular surface. Using the oxygen probe, an isosurface is generated at 1 kcal/mol interaction energy. Grid nodes at the boundary of this threshold are identified and connected to form a surface (a grid mesh of 0.5 Å gives a good approximation). Curvature is calculated as the cosine of the angle between the surface normal at the current node and the vector from this node to its neighbor. Values are averaged over each neighbor (Figure 4.4). This information, TIP "field" (since important values generally correspond to the "tips" of the molecules), is added to the MIF. In fact, it is not really a field but it is written in that form for consistency of treatment. Only negative curvatures are considered (only convex regions are interesting since they are more prone to interact with the receptor pockets), and a cubic function is used to get sharper curvature

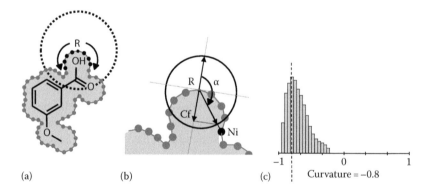

(a)   (b)   (c)

FIGURE 4.4   Surface curvature calculation. (a) Recursive nearest neighbor finding for each surface node R; (b) calculation of the partial curvature coefficient (Cf) for each nearest neighbor Ni; and (c) curvature coefficient value of node R is equal to the median of the Cf distribution. (Reproduced from Fontaine, F. et al., *J. Med. Chem.*, 47, 2805, 2004. With permission.)

variations. A distance cutoff (about 6 Å) allows for providing a global description of the molecular surface, not limited to small surface irregularities. Then PCA may be used to detect similarities between compounds and PLS to correlate the observed biological activities with the GRINDs.

4.3.4 APPLICATION OF TIP FIELD

4.3.4.1 Butyrophenones with Serotoninergic 5-HT$_{2A}$ Affinity

The study of butyrophenones with serotoninergic 5-HT$_{2A}$ affinity has been extended to 52 compounds [438], on an affinity range 4.4 log units (data from Ref. [439]) introducing surface descriptors. With two latent variables, q^2 yielded 0.74. Removing an outlier, raised q^2 up to 0.81 ($r^2=0.89$). A similar study on 43 molecules binding 5-HT$_{2C}$ yielded a more limited performance with $q^2 = 0.36$ ($r^2=0.79$). In addition to an optimal distance for the protonated amino nitrogen (H-bond donor) and a H-bond acceptor, a precise optimal distance was also found for a favorable situation involving the protonated nitrogen and the farthest extreme of the molecule (TIP) represented by the molecular shape field (Figure 4.5). According to the authors, this

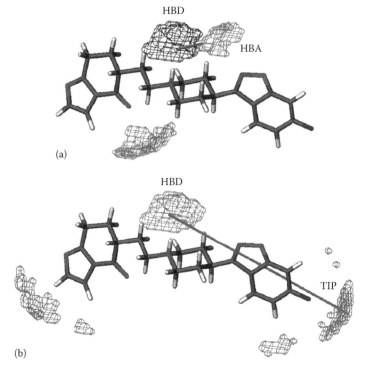

(a)

(b)

FIGURE 4.5 Regions identified as important for increasing the 5-HT2A binding affinity. (a) Shows the optimal relative location of HBD and HBA regions; (b) displays the optimal distance between the HBD region generated by the protonated amino group and the farthest extreme of the molecule (TIP) represented by the molecular shape field. (Reproduced from Brea, J. et al., *Eur. J. Med. Chem.*, 38, 433, 2003. With permission.)

distance would indicate possible interactions with hydrophobic residues of the receptor, or demarcate the depth of the binding pocket.

4.3.4.2   Xanthine Antagonists of the A$_1$ Adenosine Receptor

The importance of shape descriptors was evidenced in a study of 22 xanthine (**4.7**) antagonists of the A$_1$ Adenosine receptor [437]. Such antagonists are informative in cases of cognitive diseases, renal failure, or heart attack. The compounds mainly differ by the length of the CH$_2$ chain between the two ring systems, and the nature of the R substituent (methyl or *n*-propyl). Replacing the DRY probe by shape descriptors led to r^2=0.96, q^2 = 0.85 on an activity range of 2.7 log units. Usual GRINDs gave poor results with two LV (q^2=0.11) whereas six LV are necessary to get a comparable performance (r^2=0.98, q^2=0.83). The model revealed the importance of interactions at the two ends of the molecules and suggested, for increased activity, some cooperative effect between the length of the CH$_2$ spacer and the bulk of R: with R=*n*-Pr, the spacer can be short, if R=Me the spacer must be long enough.

4.7

4.3.4.3   Inhibitors of Plasmepsin II Proteases

The speed of GRIND treatments and the ease in carrying them out, also make them applicable to in silico screening, as shown in a study of inhibitors of plasmepsin II proteases. These proteases are needed by *Plasmodium falciparum* for metabolizing hemoglobin, its principal nutrient. Let us recall that *P. falciparum* is an agent of malaria, responsible for millions of deaths yearly. The problem was to find inhibitors (**4.8**) of high activity of plasmepsin II, but not of cathepsin D, a homologous human enzyme of importance for life.

4.8

Standard GRINDs gave no relevant QSAR, but adding shape descriptors led to q^2=0.35, and removing correlograms with little contribution raised q^2 to 0.53 (r^2=0.95). A negative influence is detected for two surface patches attributed to a butyl group in R$_1$ and a glutamine in R$_3$, whereas the less bulky isoleucine at that position induced higher activity. Influence of shape was minor with cathepsin D.

This result was in agreement with the crystal structures of the two enzymes bound to an inhibitor showing that cathepsin might accommodate bulky amino acids. Pursuing this study with activity prediction for cathepsin and plasmepsin II on 329 compounds allowed identifying of chemicals with a marked selectivity (more than 10-fold) to either enzyme. For a further study on free energy and molecular dynamics of cathepsin inhibitors, see Ref. [440].

4.3.4.4 Dopamine Transporter Affinity and Dopamine Uptake Inhibition in Cocaine Abuse

One avenue of therapy for cocaine abuse is the search for drugs binding to the Dopamine Transporter (DAT), but with a weak inhibitory activity on reuptake of dopamine. A QSAR study for both DAT affinity and dopamine uptake inhibition was carried out by Benedetti et al. [441], with GRIND descriptors, on 54 GBR compounds (N_1-benzhydryl-oxy-alkyl-N_4-phenyl-alk(en)yl-piperazines) (**4.9**) and mepyramines (**4.10**), on a range of about 3 log units for both activities.

4.9

4.10

$X = C, N$ $Y = (CH_2)_{2-5}, CO(CH_2)_{3,4}$ $Z = (CH_2)_3, CH_2{-}CH{=}CH$

Although no good correlation was obtained (q^2 about 0.4 and 0.2 for DAT affinity and dopamine uptake affinity), some insight was gained into important interactions. So compounds with a high DAT affinity exhibited hydrophobic interactions at short distances (2.4–3.6 Å), appearing in the space between the two aromatic rings of the benzhydril part whereas H-bond donors separated by 4.4–5.6 Å were detrimental as in mepyramines with an ethylene linker, but absent in the case of a propylene linker (interaction N⋯N in the central heterocyclic part). For uptake inhibition, hydrophobic interactions at long distance (about 19 Å) would be favorable; but a pair of H-bond donors at about 12 Å will correspond to a decreased uptake inhibition. These two factors intervene in compounds having an amide linker between the piperazine and phenyl ring. However, other factors may also be invoked such as change in geometry or unfavorable polar interactions.

Selectivity of DAT binding vs. SERT (serotonin) binding was also investigated, looking at the difference $\Delta = pK_{DAT\ binding} - pK_{SERT\ binding}$ as the dependent variable. The concern is to examine if compounds showing a more or less equilibrated affinity to DAT and SERT are more influential in the treatment of cocaine abuse. The correlation $\Delta_{pred.}$ vs. $\Delta_{obs.}$ yielded $q^2 = 0.73$ on 2.6 log units for 54 molecules. From the PLS coefficients, it appeared that highly selective compounds (DAT affinity $\gg$ SERT affinity) were characterized by a pair of interactions: hydrophobic (DRY–DRY) and DRY–N1 at a distance of about 18 Å. Compounds binding preferentially to SERT were characterized by strong contributions DRY–DRY at 10 Å and N1–N1 at 12 Å. They might be found when one of the piperazine nitrogen, was part of an amide group, leading to interactions involving the ether oxygen (in fragment A) and the amide oxygen (in fragment C) near the piperazine ring.

This approach gave interesting clues for rationalizing activity. But, in some cases, it might be difficult to detect the truly important intervening interactions. So, in SERT–DAT selectivity, the two interactions DD (10 Å) and NN (12 Å) both represent the effect of an amide group in fragment C.

4.3.5   Influence of Flexibility on GRINDs

GRIND descriptors are alignment-free, but, like all 3D descriptors they are inherently sensitive to molecular conformations. Caron and Ermondi [442] investigated the influence of molecular flexibility on the results of a GRIND analysis. The data set was constituted by 37 non-terpenoid oxidosqualene cyclase (OSC) inhibitors, compounds of potential interest in cholesterol-related health problems. Six series of conformers were generated, from the 2D structures, using either 2D–3D model builders or manual adjustment (including stereochemistry). The generic scaffold was constituted by an aromatic moiety and a protonated nitrogen linked by a spacer of five to eight rotatable bonds.

All models gave satisfactory r^2 values (>0.82), and correct predictions on a test set of five compounds (the same outliers being detected), but q^2 varied from 0.11 to 0.53 indicating more or less robust models. The authors concluded that, if a fast screening for good drug candidates is desired, any reasonable 3D structure may be used as an input. If the predictive accuracy is important, starting from the lowest energy conformer is better.

4.3.6   Expansion of GRINDs

4.3.6.1   GRIND and Molecular Modeling

Ragno et al. [443] carried out an extensive molecular modeling and 3D-QSAR study on 25 (aryloxopropenyl)pyrrolyl hydroxamates (**4.11**) to characterize their activity and selectivity against maize HD1-B and HD1-A. These enzymes are homologous of mammalian histone deacetylases, HDACs class I and II, respectively, which play an important role in the regulation of gene expression. GRIND models of high quality were obtained (r^2 and q^2 better than 0.96 and 0.81).

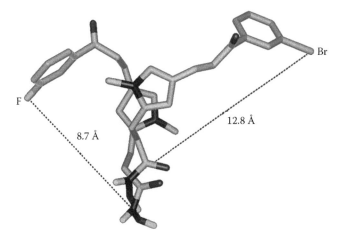

4.11

From the correlograms, common features and differences were highlighted. An important variable was the interaction DRY–N1 at 2.4 Å for HD1-B and 2.8 Å for HD1-A associated with the carbonyl group and a part of the aromatic cap. This required non coplanarity of these groups. Other important distances: O-TIP 8.4 Å (between the NH group of hydroxamic acid and the cap) for HD1-A, and TIP–TIP 10.8 Å (hydroxamic acid and cap) for HD1-B, suggested that selective HD1-A activity required a bent molecular shape, whereas a straight shape was important for HD1-B (rotation involving Φ_2). These results were confirmed by homology modeling and docking studies. Homology models of HD1-A and HD1-B were built from a known crystal structure (PDB entry 1t 64) and refined by molecular dynamics simulations. Although similar, these two structures showed striking differences at the entrance of the channel leading to the catalytic zinc ion and at the end, near the catalytic zone. Docking, carried out with AutoDOCK [444] on the two most selective molecules, with respectively a fluorine (HD1-A) and bromine (HD1-B) atom (Aro = 3-X-C_6H_4 in compound **4.11**), indicated very different best scored poses, in agreement with the minimum energy conformations, and confirmed the geometrical requisite for HD1-A and HD1-B selectivity (Figure 4.6).

FIGURE 4.6   Docked conformation of 3-F-C_6H_4 and 3-Br-C_6H_4 derivatives of compound **4.11**. (Reproduced from Ragno, R. et al., *Eur. J. Med. Chem.*, 43, 621, 2008. With permission.)

4.3.6.2 Differentiation between Transport and Specific Binding

The study of Bertosa et al. [445] seems more relevant to a classification study. But, it exemplifies a situation where a QSAR analysis allowed the evaluation of the role of specific mechanisms in the global biological response: affinity toward a receptor and its ADMET properties (absorption distribution metabolism excretion and toxicity), particularly the passive transport through cell membranes. In the case of Auxin-related derivatives (hormone of crucial importance for plant development), these two mechanisms were modeled by lipophilicity and similarity of MIFs.

In a first study [446–448], a classification was carried out on about 70 compounds, including indole derivatives (about 50%), phenoxy acetic acids (10%), benzoic acid derivatives (20%), phenoxy propionic acids, and naphthalene acids. Four categories were distinguished: actives (class 1), inactives (class 3), on a wide range of concentrations, low activity (class 2), and inhibitors (antiauxins, class 4). The model was built on the similarity of the MIFs between the molecules under scrutiny and a few representatives of the different classes. Only 10% of the set was incorrectly assigned, but the phenoxy compounds encompass a greater percentage of misclassified compounds than the other chemical families. A subsequent paper [445] extended the study to 100 compounds, (the 30 new compounds corresponding essentially to phenoxy acetic acid derivatives). The model was supplemented by characterizing the permeability (lipophilic character) represented by log P (or log D, apparent lipophilicity). Only deprotonated species on the carboxyl group were considered. As to MIFs, they were calculated with four probes H_2O, NH_2^+, CH_3, $=O$, the molecules being aligned using the SEAL (Steric Electrostatic ALignment) program [151] or comparing the MIFs at the molecular surface. For evaluating log P or log D various approaches were tried, one of them using VolSurf of Cruciani et al. [178,449] which generated 1D and 2D descriptors from the MIFs determined by GRID [58]. About 100 descriptors were calculated and selected using PLS and FFD strategy [450]. Incorporating the lipophilicity characteristic significantly improved the classification.

A second facet of this study was the combination of SAR results with docking studies, using the crystal structure of auxin-binding protein (ABP1) complexed with auxin [451] as the initial model. Selected ligands were manually inserted into the active receptor site. Monte Carlo search was performed to determine the possible binding modes, and then molecular dynamics simulations on selected conformers explored stability and possible conformational transitions. Different preferred binding modes, schematically corresponding to a monodentate or bidentate coordination with a Zn ion, are found. Comparing docking observations and SAR analysis allowed for rationalizing the experimental observations. MIF similarity was an efficient tool for predicting binding in the receptor site, but bioavailability had also to be considered. So, inactive auxin-like compounds (class 3) had significantly lower log P (or log D), but bound similarly to active compounds, suggesting that the loss of activity was due to a poor bioavailability (hindering access to the receptor); whereas class 4 compounds (inhibitors) were bound in a manner inappropriate to give the expected response.

4.3.6.3 GRINDs in Data Mining

Although not directly in the field of this book, we can also notice that GRINDs have been compared to different types of descriptors regarding efficiency in data mining, see Chapter 8, Section 2.3. UNITY fingerprints [452], ISIS keys [453], Volsurf [178,449], and log P have been considered in Ref. [454]. A somewhat similar study was also carried out by Carosati et al. [74] who also compared GRINDs to FLAP and TOPP 3D descriptors [455,456]. The use of GRINDs for diversity analysis and optimal sampling of molecular libraries has also been discussed [457].

4.4 MAPPING PROPERTY DISTRIBUTION

One of the major advantages of GRIND [45] compared to other TRI descriptors such as WHIM, MS-WHIM, is its capability of "back projection" allowing the retrieval of the original information at the level of the molecules themselves, making easier physicochemical interpretation. This interesting characteristics also exists in the MaP (Mapping Property distributions of molecular surfaces) of Stiefl and Baumann [458] who proposed the use of TRI descriptors allowing for back projection. The SESP and DiP descriptors were based on atom positions (either in a topological description or taking into account actual distances). On the contrary, the MaP descriptors are based on the molecular surface.

4.4.1 MAPPING PROPERTIES ON SURFACE POINT

MaP is based on a set of regularly spaced surface points. They are defined, from an orthogonal lattice where the molecule is embedded, as the centers of the elementary cuboids (edge about 0.8 Å) that intersect the molecular surface (Figure 4.7). To insure TR invariance, molecules are oriented along their principal axes of inertia and the origin of the grid box is at the center of mass. These surface points are encoded

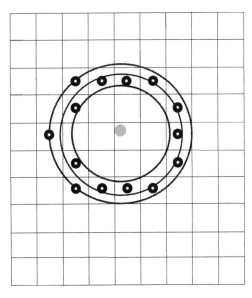

FIGURE 4.7 Extraction of surface points around an atom centered at the gray disk. Surface points are extracted as the centers of grid cells situated in a "peel" around the atom van der Waals volume (medium circle). Scheme in 2D.

depending on the nature of the nearest atom. Four properties are mapped: H-bond acceptor (A) or H-bond donor (D) propensity, hydrophilicity (H) and hydrophobicity (Ls strong, Lw weak), with the definition of a hydrophobic potential for atoms not labeled A or D. For each pair of properties (such as A–A, A–D) all relevant surface points are examined and pairwise distances are calculated. These distances are then arranged in bins (resolution 1 Å). The principle is to count the occupancy of each bin for each property combination. These counts will constitute the structural descriptors to be treated by PCR or PLS. In fact, a "fuzzy counting" is used: the two distance bins closest to the actual distance are incremented proportionally. So, a OH–NH distance of 6.75 Å is incorporated in the bins H-donor/H-donor of 6 Å (probability 0.25) and 7 Å (probability 0.75).

This method generates a large number of variables. So, a reduction of their number is desirable to make interpretation easier. The most informative variables (MIV) can be extracted in a reverse elimination technique with *tabu-search* [459,460] (vide supra). Back projection of the MIV onto the molecules may give some insight into the structural characteristics necessary for (increased) activity. MaP appears somewhat similar in spirit to GRIND [45], but one can see that MaP does not use interaction energies calculated on a lattice but considers the molecular surface. Variables here are category counts and not continuous (interaction energies). Furthermore, in GRIND, only the MACC component is retained.

4.4.2 Map Applications

This method was applied to the steroid benchmark. However, two compounds were discarded: molecule **s-31** (often appearing as an outlier as the unique molecule containing a fluorine atom) and molecule **s-1** (aldosterone, the unique structure with a cyclic hemiacetal). For the 29 remaining molecules, q^2 in L-O-O was rather low (<0.55), but a search for the MIV and a more stringent cross-validation with multiple runs of L-50%-Out led to $r_{cv}^2 = 0.84$ in PCR with four latent variables ($r^2 = 0.91$). In L-O-O $q^2 = 0.88$ (the authors stressed that L-O-O may be too optimistic).

Similarly, splitting the data set into learning and test (20/9) gave $r^2 = 0.81$ for test and 0.92 for learning ($q^2 = 0.89$ in L-O-O). In the review of Coats [126], the average r_{test}^2 on this series was 0.76. MaP treatment so appeared satisfactory. A comparable study with GRIND (on the same 29 molecules) led to $q^2 = 0.76$ in L-O-O and $r^2 = 0.83$ on the whole set. With learning and test sets $q^2 = 0.64$ in L-O-O, $r^2 = 0.82$ in learning and $r_{pred}^2 = 0.93$ in test.

Back projection of the MIV highlighted the importance for instance of the variable AL_{12} (H-bond acceptor and lipophilic atom at a distance of 12 ± 0.5 Å). In fact, this variable (chosen on statistical criteria) was highly correlated to AA_{14}, a more easily interpretable variable that corresponded to distances involving a carbonyl on position 3 and another one on a side chain bound to C17 (as in progesterone). The simultaneous presence of these groups is a well-known condition of high activity.

Eye irritation by chemicals was the subject of a second example. The irritation potential was usually evaluated in vivo, on rabbits. The actual concern for limiting animal testing makes in silico models particularly desirable. The investigated data set encompassed 38 compounds of varied chemical families, hydrocarbons, aromatics, ketones, alcohols, acetates, and acids. Reducing variables for more than one hundred to two latent variables, led to $r_{cv}^2 = 0.75$ (on 500 runs of L-50%-O) and $q^2 = 0.82$. Important variables identified in the analysis (HH_1 and HL_1) encoded the existence of hydrophilic surface areas and the vicinity of hydrophilic and lipophilic areas.

Finally, a third example showed the applicability of this approach to highly flexible structures. It relies on the modulation of the M2 muscarinic receptor. The rigid, highly active, alcuronium (compound **3.7**) molecule was used for the localization of the pharmacophore model, not for alignment but for defining the selected conformation of the studied molecules. With variable selection, the best model yielded $r^2 = 0.86$, $q^2 = 0.81$ with $r_{cv}^2 = 0.78$ in L-50%-O for the whole set. Splitting the data into training and test (29/15 compounds), results were nearly identical in training, with $r_{test}^2 = 0.68$. Among the selected MaP descriptors, HLs_6 and HLs_{14} encoded the distance between the positively charged N and the aromatic systems, and ALs_9 favored the existence of acceptor atoms in the neighborhood of the aromatic systems (e.g., the phthalimido part) (**4.12**). For another study on modulation of the muscarinc receptor see Refs. [376,378,431].

4.12

5 Similarity-Based 3D-QSAR Models

"Similarity" (and its complementary concept, "diversity") have always played an important role in Chemistry. It relies on the widely accepted "similarity-property principle," which states that structurally similar molecules are more likely to have resembling properties. At a very elementary level, this corresponds to the definition of a "chemical function," a group of atoms common within a set of molecules, and that gives them some common properties.

Relevant also to this similarity-property principle is the notion of "pharmacophore": a spatial arrangement of few atoms (or chemical groups) necessary for a drug to be recognized by a biological receptor). Once a pharmacophore is identified, this concept of similarity played, for years, a key role in drug design in the search for new leads that possess the right atoms in the right geometry. With the boom of large chemical databases, the emergence of combinatorial chemistry and high throughput virtual screening, the notions of similarity between compounds and diversity between populations, largely gained importance for organizing and retrieving chemical information. Numerous indices allowing for quantifying similarity, and for example, ranking between hits in querying a database, were defined. Our concern here being QSARs, the reader can be referred for more details on these aspects to specialized articles and textbooks [461–465] and to a recent review [46 and references therein].

We wish however to stress that similarity is, per se, a notion relative to a given problem, since it encompasses various aspects, depending on the physicochemical interactions or mechanisms involved. Numerous examples are known where a too superficial analysis of seemingly similar compounds may even lead to misleading conclusions. So, neighbor 2D structural formulae may correspond to very different shapes due to a conformational change. Neighboring behavior may come from a similarity in electrostatic potential, hidden in a simple examination of molecular shapes looking similar, but in another alignment. In other words "a similarity may hide another one." This point has already been addressed in Section 1.1.1.4 (see Figure 1.3). The pteridine parts of dihydrofolic acid and methotrexate seem very similar. However crystallographic evidence shows that they bind differently to dihydrofolate reductase [157], see also Ref. [466]. The two analgesics R4238 and R6372 have activities in a ratio 18/1 although looking very similar (they only differ by a methyl group). But a superposition of their 3D skeletons indicates a major conformational change (Figure 5.1) presumably explaining the difference in activity. Other examples may be found in Refs. [46,467].

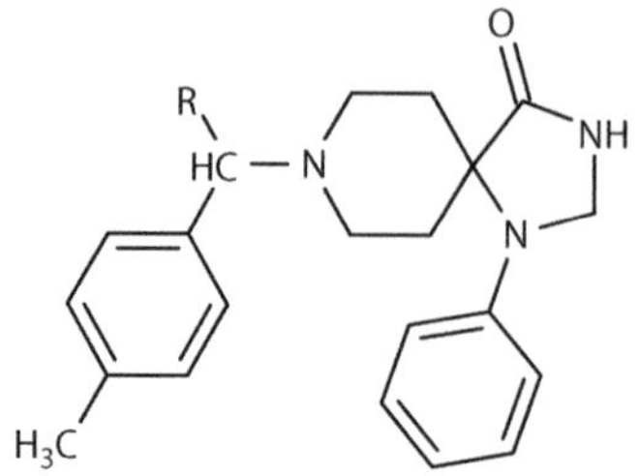

		Activity
R4238	R = H	10
R6372	R = Me	180
Reference	Morphine	1

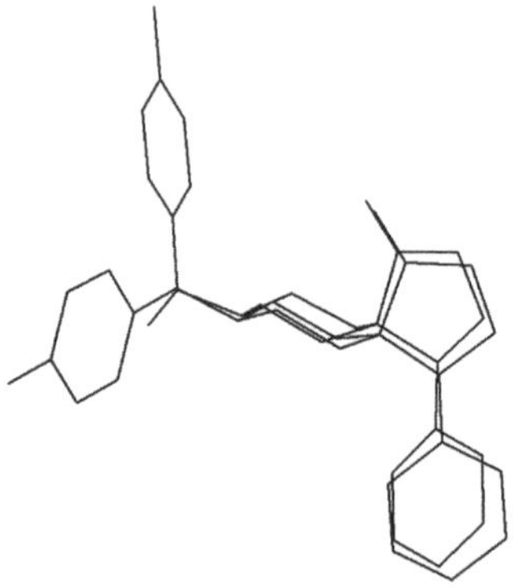

FIGURE 5.1   Analgesics R4238 and R6372 and their superimposition showing the flip of a phenyl ring. (Adapted from Doucet, J.P. and Weber, J., *Computer-Aided Molecular Design: Theory and Applications*, Academic Press, San Diego, CA, 1996.)

We will now focus interest on applications relying on the similarity-activity concept in 3D-QSAR treatments. Two main avenues may be distinguished:

- Molecules in a data set are compared to a unique reference compound. The problem is then to select this reference compound and its "active conformation" (that binding the receptor) among its possible low-energy conformations. This is, for example, at the basis of molecular shape analysis (MSA) developed by Hopfinger et al. from a comparison of fields and volumes [41,371].
- Each molecule in the set is compared to all others, leading to a (intermolecular) similarity matrix (N × N) if there are N molecules. An important aspect is that these similarity matrices can be built using pairwise superimpositions, without the need for a common alignment: a definite advantage for systems with flexible moieties or with data sets encompassing non-congeneric compounds.

CoMSIA method also relies on similarity concept, but similarity in CoMSIA is evaluated between the atoms of the sampled molecule and a probe on an external grid—a quite different situation from the intermolecular similarity, which is of concern here. At last, relevant also to the similarity concept, but not strictly devoted to 3D models, the k-nearest neighbors (k-NN) approach and counter propagation artificial neural networks (CPANNs) were used in some 3D-QSARs. We will briefly, at the end of the chapter, summarize some of them, although this chapter is mainly dedicated to similarity matrices.

5.1 SIMILARITY MATRICES

As stated by Kubinyi et al. [115], although the principle of similarity-activity was long ago accepted, the first use of similarity matrices in QSAR may be traced back to Rum and Herndon [468] who introduced in 1991, similarity matrices N × N, calculated from topological indices, to correlate steroid–protein binding constants. Distance-based similarity matrices were used by Kubinyi with his lipophilicity distance-matrix [469], or Martin et al. [470]. But in QSARs, many applications rely on the similarity index proposed by Carbó-Dorca et al. [50] for the comparison of two molecules. A SEAL-based similarity matrix was also introduced by Kubinyi [115] (vide infra).

On the other hand, Carbó-Dorca et al. [471] largely developed the theory of quantum similarity measures (QSM), starting from the idea that a quantum chemical system is completely defined by the wave function, solution of the Schrödinger equation, and in extension by the density function. This electron density function contains all information on the system and can reproduce its properties. The molecular quantum similarity concept proposed in the early 1980s was the starting point of applications in various fields of chemistry presented, for example, by Fradera et al. [472 and references therein]: QSARs, chemical reactivity, enzyme activity, solvent effect studies, ordering, and representation of molecular sets. An important development was the description of a fundamental Quantum QSPR equation (QQSPR or Q^2-SPR), which demonstrates that the empirical QSPR models can be generally founded on a well-defined relationship of quantum mechanical origin [473]. It was also established that the expected value of any given molecular property (physical, chemical, etc.) can be expressed in terms of a linear combination of QSM.

5.1.1 CARBÓ-BASED SIMILARITY MATRICES

For two molecules, A and B, the Carbó similarity index was based on the overlap of their electron density, but the definition was further extended to the electrostatic potential or polarizability. Similarity matrices were also defined for the molecular shape derived from the van der Waals atomic radii. More generally, for a property P (such as electron density, electrostatic potential, polarizability), the similarity index between two molecules A and B is:

$$R_{AB} = \frac{\int P_A P_B dv}{\left(\int P_A^2 dv \right)^{0.5} \left(\int P_B^2 dv \right)^{0.5}}$$

But it was noted that Carbó index is more sensitive to the shape of a property rather than its magnitude (if P_A becomes nP_A, P_B remaining unchanged, the index does not vary). A new definition was proposed by Hodgkin and Richards [474].

$$H_{AB} = \frac{2 \int P_A P_B dv}{\left[\left(\int P_A^2 dv \right) + \left(\int P_B^2 dv \right) \right]}$$

The similarity index obviously depends upon the relative position of the molecules to be compared. In this way, it may be used to optimize molecular superposition [475] in the search for the best alignment, that giving it the maximum value. Therefore, methods based on similarity indices are intrinsically alignment dependent (at least by pairwise comparisons).

5.1.2 SIMILARITY MATRICES AND PLS (OR GOLPE) ANALYSIS

An original application is due to Good et al. [396] who proposed, rather than a comparison between a pair of molecules, as in Refs. [41,476], to define a similarity matrix (N × N): each molecule is compared to every other in the set and so is characterized by its similarity with the other ones. These similarity indices specify the interrelations between molecules and therefore become structural descriptors.

This approach was applied, for both classification and correlation, to the steroid benchmark [38]. Two matrices were built, one for the electrostatic potential (ESP matrix) and the other (SH matrix) for molecular shape, the two matrices being possibly merged in a combined treatment. Integration was carried out analytically. For ESP, three Gaussian functions are used to reproduce the 1/r dependence over distance [380]. For estimating shape similarity, Gaussians representing the square of the STO-3G wave functions (calculated at that time with Gaussian 88 [241]) and constrained to drop to zero outside of the van der Waals volume were used [477]. These functions insure rapid analytical integral evaluation and speed up the calculation of the similarity indices.

Classification was carried out with a Reversible Nonlinear Dimensionality Reduction (ReNDeR) neural network [478]. Basically such a network encompasses five layers: input, encoding, parameter, decoding, and output (Figure 5.2). The network operates in such a way as to deliver at output values identical to those given in input. The trick is that the parameter layer has only 2 or 3 units, so that the output of the neurons of this parameter layer may be considered as the coordinates of the input patterns via a nonlinear mapping in a 2D or 3D space. This allows for a visual inspection of possible clustering. In this study, a 31/10/2/10/31 network led to a good separation of high affinity compounds. Identifying intermediate and low affinity derivatives was however less clear.

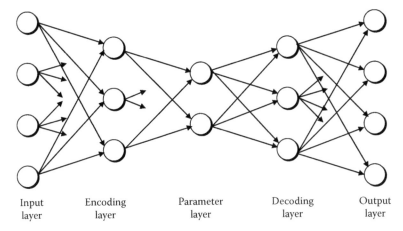

| Input layer | Encoding layer | Parameter layer | Decoding layer | Output layer |

FIGURE 5.2   Architecture of a ReNDeR neural network.

For establishing a quantitative relationship specifying activity, a PLS analysis was carried out (the number of descriptors was largely superior to the number of objects). In this example, shape matrix yielded better results than ESP or both combined. For affinity to CBG, on a training set of 21 compounds (as in Ref. [38]), the cross-validated determination coefficient q^2 equaled 0.633 (standard error 0.82 log unit on a range of 2.8). These results compared well with the CoMFA analysis ($q^2=0.761$, SE$=0.60$). For the ten compounds of the test, SE was 0.40 vs. 0.42 in CoMFA (excluding one outlier: compound **s-31**) [396].

Filtering the data from noise, with a test release of GOLPE (generating optimal linear PLS estimations) [177] improved results: $q^2=0.828$ in a three-component model, derived from 14 shape and 9 electrostatic indices.

In a subsequent publication [47], various assays were carried out to validate the technique, comparing analytical evaluation and grid point methods and considering other similarity indices. Various computational options were examined: for Carbó index, where a summation of Gaussian functions was used for an analytical integration [380], models with 1, 2, or 3 Gaussians were tried. But ESP could be also calculated by a numerical discrete calculation on grid nodes after embedding the aligned molecules in a 3D lattice (as in CoMFA). Point charges were introduced and the calculation was not performed inside the molecular volume to avoid singularities.

Shape similarity was calculated using Gaussian functions representing the square of the STO-3G wave function, but possibly modified to take into account "harder" atoms (electron density set to zero outside the van der Waals volume) or "soft" hydrogens (electron density extending largely outside the van der Waals volume, but unmodified for heavy atoms). For shape, a numerical calculation was also possible, where similarity can be evaluated by

$$S_{AB} = \frac{U_{AB}}{(T_A T_B)^{0.5}}$$

counting grid points inside the volumes of A and B (T_A, T_B) and in the union $(A \cup B)$, U_{AB} [479]. Furthermore, Hodgkin indices were considered as well as Carbó indices for both electrostatic potential and electric field. Other indices were also defined in numerical calculations for electrostatic potential [381]:

$$\text{Average similarity linear } L_{AB} = \frac{\sum (1-X)}{n}$$

$$\text{Exponential } E_{AB} = \frac{\sum \exp(-X)}{n}$$

where $X = |P_A - P_B|/\max(|P_A|, |P_B|)$ for the current node, the summations being over the number (n) of grid points involved in the calculation. The Spearman Rank Correlation coefficient was also considered:

$$R_{AB} = 1 - \frac{6 \sum\limits_{i=1,m} d_i^2}{m(m^2 - 1)}$$

d_i difference in the rank of the property value (MEP) at point i for the two structures, the summation running on m, total number of grid points.

Seven supplementary data sets were investigated (regarding affinity constants, or, for two of them, dissociation constants) [82,236,237,396,480–482]. The series under investigation concern:

- 54 anticoccidial triazines [82], on a range about 4.2 units for log (1/MEC) **(5.1)**

5.1

- 37 β carbolines, pyridodiindoles, and GCS ligands binding the benzodi-azepine receptor inverse agonist site (6.4 units of $\log(1/IC_{50})$ **(5.2** through **5.6)** [480]

5.2   **5.3**   **5.4**

5.5   **5.6**

- 14 3β-(*p*-substituted phenyl) tropane-2β-carboxylic acid methyl esters binding the cocaine binding site (2.5 units of $\log(1/IC_{50})$ (**5.7**) [481]

5.7

- MPTP (1-methyl-4-phenyl-1,2,3,6-tetrahydropyridine) and 40 analogues binding to monoamine oxidase (only qualitative data on three relative activity levels) [482] (**5.8** through **5.12**)

5.8   **5.9**   **5.10**   **5.11**   **5.12**

- pK$_a$ of 28 clonidine-like imidazoline analogues (4 unit range) (**5.13**) [236]
- pK$_a$ of 16 2-substituted imidazoles (9.3 unit range) (**5.14**) [236]
- 49 Hammett substituent constants (σ) determined from the dissociation of substituted benzoic acids (range of variations about 1.5 in σ values) (**5.15**) [237]

5.13   **5.14** **5.15**

Generally speaking, for ESP, 1-Gaussian evaluation and grid-based calculations were outperformed by 2- and 3-Gaussian ESP; whereas, for shape, the grid-based model would be slightly better. As to the grid mesh, 2 Å for ESP and 0.5 Å for shape seemed correct, although the authors recommended analytical evaluation. Hodgkin and Carbó indices, for both potential and field, worked similarly but linear and exponential indices (L$_{AB}$ and E$_{AB}$) seemed superior.

It may be noticed that using an N × N matrix (or N × 2N if both types of interaction are treated together) involved many variables, some of them, not really significant, and possibly introducing noise. GOLPE [177] allowing for selecting the most relevant variables, significantly improved cross-validated q^2. It drastically reduced (about 50%) the original variable set, a definite advantage over CoMFA (Table 5.1).

5.1.3 Other Applications

Several other applications emphasized the interest of the SM/PLS approach for deriving reliable QSARs [47,483,484]. Relation between properties of 28 N-terminus fragments of tachykinin NK1 receptor antagonists (**5.16**) and their overall affinities (pIC$_{50}$ range of 4.1 units) was investigated by Horwell et al. [483] in a comparative study with similarity matrices and with classical QSAR. Such antagonists are potential drugs for pain, inflammation, and asthma. Only the fragments bound to the N terminus are considered. This simplification is justified by the fact that previous data indicated a single mode of binding. Furthermore, since the central and C-terminal parts of the molecules remained the same, it was likely that changes in reactivity were associated with structural variations in the N-terminus moiety.

R = CH$_2$ – substituted aromatic group
5.16

TABLE 5.1
Statistical Data for Combined Electronic and Shape Similarity Data Matrices

	q² Values/LV		
	Original[a]	GOLPE[b]	CoMFA[c]
Steroids (TBG data)	0.73/3	0.77/2	0.44/4
Steroids (CBG data)	0.79/2	0.82/2	0.69/2
Triazines	0.74/5	0.73/3	0.47/2
β-Carbolines + ⋯	0.69/4	0.72/3	0.59/4
Cocaine analogues	0.54/2	0.64/2	0.57/4
MPTP + ⋯[d]	0.50/2	0.56/2	0.57/4
Imidazolines[e]		0.32	0.27
Imidazoles[e]		0.90	0.69
Benzoic acids[e]		0.12	0.05

Source: Reproduced from Good, A.C. et al., *J. Med. Chem.*, 36, 2929, 1993. With permission.

[a] Unmodified similarity matrices.
[b] After variable selection by GOLPE.
[c] CoMFA results on the same data sets.
[d] Relative values (three levels).
[e] Only electronic matrix used for these series. Since q² were not reported, only predictive standard errors are given.

Charges were calculated according to the Abraham et al. approach [485] relying on electronegativity and polarizability effects in saturated molecules and Hückel MO calculations for π systems. For classical QSAR, the structural descriptors were computed with Tsar [428 (version 2.20)]. A total of 37 descriptors were retained: topological, constitutional, geometrical, plus log P, and its square, dipole moment, and the molecular surface area. For the evaluation of similarity matrices, the Gaussian function approximation was chosen for Carbó potential indices and a grid-based method for Carbó atomic field indices.

In a first phase, a classical QSAR treatment was performed. For PLS analyses, three criteria to select the more significant latent variables were examined: stopping when the statistical significance of the current vector went above a fixed value, stopping at the first decrease in PRESS, or at its lowest value. However, the models obtained were not satisfactory since not robust enough ($r^2 = 0.540$) for the first criterion, or overfitted for the other two, with r^2 about 0.83 and q^2 about 0.55.

This mediocre performance prompted to use PCA in order to reduce the number of descriptors, and particularly to select only few representative elements within each class of highly correlated descriptors. Then MLR analyses were performed, with or without stepping procedures. On an activity range of 4.13 log units, the selected model yielded $r^2 = 0.671$, $q^2 = 0.576$, $s = 0.72$ with five classical variables (surface area, log P and its square, total dipole moment, and number of methyl groups), whereas similarity-PLS analysis (with one latent variable) gave only $r^2 = 0.540$, $q^2 = 0.366$.

A second treatment was performed with Carbó indices regarding electrostatic and lipophilic potentials associated with size- and shape-related properties of the N-terminus groups. Various combinations of these indices were analyzed by PLS method yielding about 20 models of satisfactory quality, with, for each, neighbor values of r^2 and q^2, which indicated that no overfitting occurred. Models using field descriptors (giving information about location and strength of ionic interactions) were better than those using potential descriptors. The best one, obtained with electrostatic and size properties (refractivity) yielded $r^2 = 0.846$, $q^2 = 0.737$, PRESS = 7.10 (with two latent variables).

The robustness of the models derived from similarity indices was internally validated by a mixed treatment with first a PCA analysis of the similarity matrices, to extract within each matrix of field and shape similarity indices, a smaller number of components (combination of the original independent descriptors). Then a stepwise MLR treatment was performed to get the best model ($r^2 = 0.829$, $q^2 = 0.756$, $s = 0.50$ with three of these latent variables). In this study, 3D similarity-based models (with PLS or MLR) outperformed the conventional QSARs ($r^2 = 0.671$, $q^2 = 0.576$). This observation on the efficiency of 3D models contrasted with the conclusions of Agarwal et al. [486] who indicated, on another system, that CoMFA methodology offers "little or no advantage over the classical Hansch treatment."

These best models (PLS and MLR) obtained from the similarity indices were used to predict the activity of 11 new molecules. Among them, three "interesting" structures were synthesized and experimentally tested. The predicted activities agreed with the experimentally determined values, within the range of average experimental error of the biological tests.

5.1.4 SIMILARITY MATRICES AND GENETIC NEURAL NETWORK

Whereas similarity matrices were analyzed in several cases [47,483] with PLS, possibly with the help of GOLPE, So and Karplus [487,488] proposed to use a genetic neural network (GNN) coupling a genetic algorithm, for the selection of relevant descriptors, with a BNN for their treatment. "GA selects the most relevant descriptors; NN constitutes a model-free nonlinear mapping able to optimize their use." This approach gave good results for descriptor selection and subsequent processing in cases where the dimensionality of the data was large and the interrelations between variables were intricate [487,488]. It was then applied to similarity matrix treatment [48,49].

Shape similarity matrix (SSM) was calculated on the basis of grid points inside or outside the van der Waals volume. Alternatively a van der Waals similarity matrix was based on van der Waals potentials with the Carbó formula. A numerical calculation (on a grid) was carried out for the electrostatic similarity matrix (ESM), according to Hodgkin index. For GA, 200 individuals evolved in 50 generations, in evolutionary programming, to select six variables (to maintain a correct ratio between the number of compounds and that of descriptors). These variables were transmitted to a 6/2/1 BNN to calculate the affinity. The correlation coefficient for the training set was used for the fitness function, except for the final cycle where a classical L-O-O validation was performed to determine the quality of the model.

A first paper [48] concerned the "steroid benchmark." Suffice it to stress here that on the whole set (31 compounds, with corrected values) with six descriptors, q^2 yielded 0.94, with no outlier. Best results were obtained with electrostatic similarity matrix (ESM) $r^2 = 0.951$; $q^2 = 0.903$, in place of, respectively, 0.885 and 0.825 with shape (SSM). These results corresponded to the mean on 50 simulations (since GA usually gave several nearly equivalent solutions with different descriptors). van der Waals potential gave no advantage compared to shape matrix. Using the two matrices simultaneously gave better results ($q^2 = 0.941$) contrary to Good's conclusion [47] ($q^2 = 0.76$ with the shape similarity matrix [SSM] and PLS).

In order to validate the approach, various options were examined, as to changes in the parameters of the model:

- Type of charges: AM1-Mulliken charges appeared to be better.
- Type of indices: no clear advantage was seen.
- Grid parameters (size, location, mesh): extension and location around the molecules did not induce significant variations. A mesh of $2\,\text{Å}$ for MEP, $0.5\,\text{Å}$ for shape was convenient.
- Number of input descriptors to BNN: From a systematic variation of this number, the best result ($q^2 = 0.843$) corresponded to six descriptors for the vdW similarity matrix. Beyond eight inputs, the decrease of q^2 may indicate overfitting. Note that for the electrostatic similarity matrix (ESM) where a similar behavior is observed (best $q^2 = 0.903$, for six inputs), q^2 was even 0.677 for only one descriptor. This fairly good result may be compared to that of Montanari et al. [484] with bisamidines, where a comparison with only the least active analogue yielded a "reasonable" model. Combining similarity matrix with PLS, the best model yielded $q^2 = 0.707$ with five components, but with descriptors different from the GNN selection.

To get some insight about the factors improving the results in the GNN procedure (selection by genetic algorithm or use of a nonlinear correlation model), the authors combined GA with MLR ("genetic regression model"). This led to $q^2 = 0.819$ with six descriptors, highlighting the efficiency of GA for descriptor selection, further advantage being gained with NN, which raised q^2 up to 0.941. The results obtained by SM/GNN competed well with other approaches. On the first 21 steroids, MSA led to 0.56.

It was interesting to observe that the six descriptors retained (rows of the similarity matrix) correspond to similarity indices with *identified* compounds of the set that may be considered as representative for that problem: two of them are among the more active, and three exhibit low binding affinity. This contrasted with PCA or PLS analyses, where the composite descriptors selected were combinations of the original individual elements of the similarity matrix. This GNN approach also substantially differed from substituent parameter-based methods, in as much as the indices represented a global measure of the resemblance between a pair of molecules. So a global characteristic was associated to the activity variations among a series.

One may also remark that the similarity matrix approach significantly reduced dimensionality compared to CoMFA since it depended only on the number of compounds. In a companion paper on applications [49], this approach was extended to diverse properties previously investigated by various laboratories. Eight data sets were examined. Some examples are given below.

5.1.5 VALIDATION OF GNN MODEL

Among these examples, the affinity to AhR receptor of 73 polyhalogenated aromatics (see compounds **1.4** through **1.6**) (a subset of that from Ref. [423] after elimination of redundancies and incorrect values) was investigated. After alignment of the molecules (same criteria as those of Waller and McKinney [423]), every molecule was characterized by its similarity to the others in the set, both at size and electrostatic levels, on a grid evaluation. A genetic neural network (GNN) was then fed with these descriptors. To fix the best number of descriptors, various n/2/1 architectures were tested, with the number of inputs n varying from 1 to N/5, where N is the number of training compounds. The optimal model was that giving the smallest standard error in cross-validation prediction. It should be noted that this criterion penalized models with a large number of descriptors (see the definition of SE).

For an activity range of 6.4 log units in pEC_{50}, the model led to $q^2 = 0.72$ with the electrostatic index, and 0.85 with the shape index (with seven descriptors), consistent with the limited importance of electrostatic effects in that series. Considering both indices at a time gave no improvement. These results favorably competed with those of Waller and McKinney [423] (CoMFA, $q^2 = 0.72$) and Wagener et al. [146] (autocorrelation descriptors, $q^2 = 0.83$).

The other investigated series encompassed:

- Inhibitory activity on dopamine β hydroxylase (DβH) of 47 1-(substituted benzyl)-imidazole-2(3H)-thiones (see compound **3.2**), with data extracted from Refs. [390,368] (activity range of 4.1 log units for pIC_{50}). A genetic neural network with four electrostatic descriptors led to $q^2 = 0.64$, whereas with six shape descriptors $q^2 = 0.76$. Using a combined matrix gave only marginally better results ($q^2 = 0.77$ with one electrostatic and three shape descriptors). These results are quite comparable to previous studies on the same data set. So, Burke and Hopfinger [368] proposed, in MSA, a six descriptor regression (with quadratic variables) involving the partial charges on the 6- and their sum on 3-, 4-, 5-positions, the hydrophobic character (molecular lipophilicity) π_0, the fragment constant π on 4-position and the common overlap steric volume V_0 (and its square) against the most active compound. This model yielded $q^2 = 0.76$.

  On the other hand, Hahn and Rogers [389,390] obtained $q^2 = 0.79$ with their receptor shape model (RSM) using spline functions of two variables: interaction energy E_{inter} (van der Waals and electrostatic) between the inhibitor and a pseudoreceptor, and E_{inside} (intramolecular energy of the ligand within the receptor environment).

- 43 β-carbolines, pyridodiindole and CGS (see above **5.2** through **5.6**) acting as inverse agonists and antagonists of the benzodiazepine receptor (BzR), a series already largely investigated [47,238,480,489,490]. SM/GNN with a q^2 of 0.73, on an activity range of 6.4 log units (with four descriptors), appeared as efficient as the SM-GOLPE/PLS model of Good et al. [47] (0.72), but was outperformed by the best CoMFA/GOLPE model with $q^2 = 0.825$ (five components) [490], CoMMA giving the worst results (0.39) [489]. This comparison highlighted the interest of GNN and GOLPE to select the most relevant descriptors.

- Acetylcholinesterase (AChE) inhibitors. For 60 AChE inhibitors (**5.17** through **5.23**), the CoMFA treatment of Cho et al. [249] with region focusing [184,491] yielded $q^2 = 0.73$ with seven components. SM/GNN gave better results on a 6.2 log unit range for pIC_{50} ($q^2 = 0.80$).

- For the activity of 37 bisamidine analogues (**5.24**) against *Leishmania*, Montanari et al. [484] proposed a three-variable MLR treatment involving a Carbó similarity index (calculated with respect to the least active compound) plus log P and an electrotopological state index, with $q^2 = 0.51$. But, a previous MLR model, with six indicators variables and two physicochemical properties, yielded $q^2 = 0.63$. These results are surpassed by SM/ GNN with $q^2 = 0.80$ with a shape-only model (range 2.1 log units for pIC_{50}).

5.24

- Another series concerned 30 α-D-glucose derivatives (**5.25** through **5.28**), inhibitors of glycogen phosphorylase [492]. Regulation of glycogen metabolism is an important concern for diabete treatment and search for more potent inhibitors than α-glucose as physiological regulators is of potential interest. The study was made easier by the knowledge of the co-crystallized structure of glycogen phosphorylase with an inhibitor [492–494]. Using a combined matrix (shape plus electrostatic) led to $q^2 = 0.82$ with five components, for a variation of pK_i about 4.1 log units. This result compared with that of GRID/GOLPE (q^2 about 0.8) [436].

5.25   **5.26**

5.27   **5.28**

- The last two series related to the prediction of ionization constants of imidazoles (see **5.14**) on one hand and, on the other hand, to that of substituted benzoic acids (see **5.15**) (benzoic acids' pK is the property of definition of the widely used Hammett's σ constants). These studies allowed for estimating the importance (as expected) of electronic effects compared with shape influence.

For 49 substituted benzoic acids (with σ varying between 0.93 and -0.84, from p-SO$_2$CF$_3$ to p-NHMe$_2$), GNN with a q^2 of 0.83 is better than two similarity-matrix treatments (Hodgkin HSM/PLS and Carbó CSM/PLS) that both gave $q^2 = 0.75$ [470]

or CoMMA (0.69) [489]; but it was outdone by CoMFA (0.89), and a distance matrix DM/PLS $q^2 = 0.90$ [470]. However, it may be noted that a linear relationship using only the sum of the partial atomic charges of the two oxygen atoms and the acidic hydrogen yielded an excellent q^2 of 0.91 [236]. For 16 imidazoles (8.7 variation in pK), SM (electrostatic, shape)/GNN led to $q^2 = 0.92$, better than CoMMA 0.69 [489] or SM GOLPE/PLS 0.77 [47].

As a general result, SM/GNN appeared as an efficient method (q^2 generally greater than 0.73), well competing with other 3D-QSAR studies. Substituting GNN to PLS for variable selection was only slightly more intensive. Not only for quantitative predictions, but also on a qualitative point of view, models are consistent with previous conclusions regarding steric and electronic influences. However reducing description to similarity indices (a global description) and destroying spatial references induce some loss of information. No visual display is available, at the difference of the CoMFA contour plots. The (sometimes) tricky alignment is still necessary. So, it was suggested [49] that evaluating similarity from molecular invariants (autocorrelation vectors, for example) would eliminate this critical phase; and that considering subgrids as in Ref. [184] might indicate the most important regions. The results of this study also emphasize the interest of variable reduction by GOLPE.

5.2   SEAL-BASED SIMILARITY MATRICES

Kubinyi et al. [115], nearly at the same time as So and Karplus, proposed a quantitative similarity-activity relationship based on SEAL similarity matrices, a different approach from that of Carbó-Dorca or Hodgkin. We already presented the SEAL program of Kearsley and Smith [151] (see Chapter 1). SEAL proposed an automated alignment without subjective hypotheses, thanks to a similarity score: the best alignment of two molecules is that leading to the largest mutual similarity score. SEAL was also adapted to multiple alignments of flexible systems [341]. On the other hand, SEAL is also at the basis of the CoMSIA approach [40] (see also Refs. [152,153]).

The similarity score (A_F) is based on a weighted sum of electrostatic and steric similarity components, but other influences may be added, such as hydrophobic effects, H-bond donor or acceptor properties.

$$A_F = -\sum_{i=1,m}\sum_{j=1,n} w_{ij}\exp(-\alpha r_{ij})^2 \quad \text{with } w_{ij} = w_E q_i q_j + w_S v_i v_j + \cdots$$

summation on the i atoms of molecule A and the j atoms of molecule B. q_i, q_j charges on atoms i and j, v_i, v_j powers (usually three) of the van der Waals radii of i and j. The α coefficient introduces a distance dependence. Default parameters have been calibrated on large set of ligands bound to a same protein.

Similarity scores were calculated in four options: electrostatic, steric, hydrophobic fields alone, and all components (including H donor and acceptor) in a weighted sum. It was checked that the SEAL (objective) pairwise superposition gave results better than (or as good as) the (subjective) atom-by-atom rigid fit.

The method was applied to the steroid benchmark (using duly corrected data). SEAL similarity fields, calculated on nodes of a $24\,\text{Å}$ edge box, were submitted (after normalization) to PLS analysis or alternatively a relationship was searched for with one or few columns of the matrix. In other words, the activity of the population is then compared to that of one or few molecules, somewhat considered as references. Although a strict comparison was difficult (because of errors in some data previously used), it appeared that the various treatments with similarity matrices led to better results than the preceding CoMFA or CoMSIA studies, with r^2 about 0.90, 0.95, and q^2 about 0.6–0.7 but in prediction r^2_{pred} was significantly inferior. For example, with SEAL alignment (all fields considered), and the original splitting (train/test), a regression model yielded $r^2 = 0.918$, $q^2 = 0.880$ and $r^2_{pred} = 0.598$.

This study also evidenced that the training set must include as varied structural features as possible in order to allow reliable predictions. Statistical parameters obtained in L-O-O on the training set could not prejudge the performance in prediction. A strict limitation of the number of PLS latent variables was more able to produce stable models and reliable predictions, at the expense of internal performance. For a better spanning of the structural space, a modified splitting of the data set was proposed: in training compounds **s-1** through **s-12**, plus **s-23** through **s-31** and in test **s-13** through **s-22** (see Section 1.1). The predictive ability of the model largely increased with $r^2 = 0.774$, $q^2 = 0.713$, and $r^2_{pred} = 0.784$. A more anecdotal point concerned a simple Free-Wilson correlation. Taking into account the presence (or not) of a C4–C5 double bond in the steroid A ring yielded $r^2 = 0.778$, $q^2 = 0.726$ and $r^2_{pred} = 0.477$. Although inferior in training, it was better than CoMFA, CoMSIA in prediction.

Some methodological remarks:

- Similarity matrices have also been successful in a classification problem [495]: induction of aneuploidy (a genetic mutation) in *Aspergillus nidulans*, by halogenated aliphatic hydrocarbons. In this study, similarity was evaluated with electrostatic potential, lipophilicity and refractivity, using a three-term Gaussian approximation to fit the 1/r dependence. However, some methodological points were addressed: for example, it was shown that different results may be obtained depending on the initial orientation of molecules submitted to the alignment program ASP [496]. So, a user-chosen initial orientation was recommended. Furthermore several local minima might exist (a common problem in minimization procedures).

- It was indicated that in certain cases, overall structural similarity might correspond to similar biological activity, whereas in other cases it was only the similarity of certain active moieties that intervened. The situation was made still more complex since there are various definitions of similarity indices [46].

- The authors stressed that classical descriptors and similarity matrices carry out a largely overlapping chemical information, so that classical descriptors and similarity matrices might lead to satisfactory QSAR. However it was concluded that similarity matrices "neither complemented the classical descriptors nor improved on their performance." Two hypotheses were put forward. Similarity matrices could not quantify the spatial modulation of molecular properties or this factor was not essential on that specific example. This might correspond to the fact that the cellular system involved in xenobiotics metabolization might not require a strict spatial arrangement compared to a pharmacological receptor.

5.3 MOLECULAR QUANTUM SIMILARITY MEASURES

Since the proposal of a molecular quantum similarity index, based on the overlap of electron density distributions [50], various QSAR studies directly relied on the molecular quantum similarity measures (MQSM) approach that allowed for quantifying the similarity between two molecules. As previously quoted, a system can be completely characterized by its wave function (solution of the Schrödinger equation), which is not an observable, but, in extension, by the density function (an observable). The problem is to get a way to avoid too heavy calculations. The method was developed along two ways: on one hand evaluation of similarity matrix analogous to those just seen, and on the other hand extension of topological and topographical indices with inclusion of quantum quantities. This last point will be developed with 2.5D indices in Chapter 7.

5.3.1 DEFINITION OF MQSM

Generalizing the Carbó index [50], MQSM are based on the overlap between the first-order density distributions, possibly weighted by a definite positive operator. For two molecules A, B:

$$Z_{AB}(\Omega) = \int\int \rho_A(\mathbf{r}_1)\Omega(\mathbf{r}_1,\mathbf{r}_2)\rho_B(\mathbf{r}_2)d\mathbf{r}_1\, d\mathbf{r}_2$$

where
  ρ is the density function
  Ω a positive-definite operator

The most usual is the Dirac operator $\delta(\mathbf{r}_1 - \mathbf{r}_2)$ leading to the overlap-like MQSM:

$$Z_{AB} = \int \rho_A(\mathbf{r})\rho_B(\mathbf{r})d\mathbf{r}$$

But other operators may be introduced [497]:

$$\text{Coulomb like}\ \ \Omega(\mathbf{r}_1,\mathbf{r}_2) = |\mathbf{r}_1 - \mathbf{r}_2|^{-1}$$

$$\text{Gravitational}\ \ \Omega(\mathbf{r}_1,\mathbf{r}_2) = |\mathbf{r}_1 - \mathbf{r}_2|^{-2}$$

It was indicated that the overlap operator encompassed steric feature whereas Coulomb operator reflected electrostatics characteristics [498]. A Triple MQSM (TMQSM) where Ω is substituted by another molecular density function was even introduced (vide infra). These transformations of the MQSM produce new discrete descriptors called "quantum similarity indices." Coming back to the first one proposed in the literature, the "Carbó index"

$$C_{AB} = \frac{Z_{AB}}{\left(Z_{AA}Z_{BB}\right)^{0.5}}$$

may be generalized, after normalization, for TMQSM

$$C_{AB:C} = \frac{Z_{AB:C}}{\left(Z_{AA:C}Z_{BB:C}\right)^{0.5}}$$

C varies from 1 (identical molecules) to 0 (totally dissimilar molecules). The connection between QSAR and MQSM was explicated by Amat et al. [499] who stressed that any measured molecular property y_I can be expressed as the expectation value of an operator w

$$y_I = \int w(r)\rho_I(r) \quad \text{which can be written} \quad y_I = \Sigma_K c_K z_{KI}$$

where c_K are the components of the vector representing the operator w on the basis of the density functions $\rho_I(r)$.

MQSM depends on the relative positions of molecules. But, it may be intuitively assumed that the maximum value of the similarity index corresponds to the best superimposition of the two molecules. So, to evaluate MQSM, we have to find this best alignment. And, vice versa, a way to find the best match is to maximize MQSM [498,500]. The process relies not on an alignment on a common template but on an evaluation of the similarities in the fields created by the different structures. Here it corresponds to the maximization of indices in pairwise comparisons to build the similarity matrix for the molecular data set (all molecules are supposed frozen). If molecule A is fixed and B moving, MQSM is defined by a six-variable function involving three translation components and three Euler angles (for example) for rotations, the two molecules being supposed rigid. But, the hypersurface to explore is generally complex, with (possibly) many local energy minima. Two methods have been proposed to "align" molecules for the calculation of the similarity matrix.

- The topo-geometrical superposition algorithm (TGSA) approach [501,502] is based in a first step on similarity between atomic diads or triads, in order to obtain optimal superposition on atomic positions (the score being calculated on the squared difference between the coordinates of the atoms in the two molecules after superimposition). Then a restricted search is performed starting from the best alignment found to refine the solution. If efficient for congeneric structures, the algorithm does not work very well for structurally diverse molecules where the number of common fragments may be low.

- A new approach was proposed in TMQSM [500] for the measure of $Z_{AB:C}$ The molecule operator C being fixed, overlap MQSM Z_{CA} and Z_{CB} are separately optimized (A with respect to C, and B with respect to C), and then a simplex method is used to maximize the $Z_{AB:C}$ measure (depending on 12 parameters: 3 translations and 3 rotations for both A and B molecules). When the three molecules share a common substructure, a good superposition is obtained. The quantum similarity superposition algorithm (QSSA) approach [500], more adapted to non-congeneric structures, is based on electron density functions, and uses a Lamarckian genetic algorithm. Lamarckian GA is characterized, in each generation, by an intermediate local optimization of children (by simplex algorithm) before they become parents in turn. Here 100 individuals are randomly created (and optimized) for the first generation and in every following generation, 10 new random sets are introduced, mimicking migration of news species. In general, 20 generations are sufficient.
- Alternatively, the lowest energy conformer may be used, but not without risk. Several conformers of comparable low energy may be found, and there is no guarantee that the minimal energy conformation at the receptor site is that found for the free molecule in the gas phase.

Once MQSM are computed between all molecular pairs, a similarity matrix is constructed (as already seen). In this $N \times N$ matrix, each line (or column) corresponds to a discrete representation of a molecule. The infinite-dimensional density functions are so represented in an N-dimensional subspace. Such a discrete representation of a molecular density function was called by Carbó a "point molecule" [472].

5.3.2   EMPIRICAL ATOMIC SHELL APPROXIMATION: AVOIDING HEAVY AB INITIO CALCULATIONS

Whereas approximate values of the electrostatic potential may be obtained without too heavy calculations, overlap integrals involved in MQSM are computationally intensive, and some approximations had to be introduced. For the construction of accurate density functions, in the empirical atomic shell approximation (EASA), the molecular electronic density is fitted to a sum of atomic densities $\rho_a(\mathbf{r})$ of the atoms a constituting the molecule. One advantage of the EASA approach is that it needs only atom coordinates and charges that may be easily obtained by semiempirical methods. Simple overlap integrals over s functions replace (and avoid) the intensive determination of ab initio density function.

$$\rho_A(\mathbf{r}) = \sum_{a \in A} \rho_a(\mathbf{r})$$

summation over all atoms a of molecule A, and

$$\rho_a(\mathbf{r}) = Q_a |S_a(\mathbf{R_a} - \mathbf{r})|^2$$

where
  Q_a is the electronic population on atom a
  $S_a(\mathbf{R_a} - \mathbf{r})$ an atom dependent exponential function centered at a-atom

Many shell atomic densities can be used, the atomic densities being written as a sum of contributions from the diverse shells of the atoms:

$$\rho_a(\mathbf{r}) = \sum q_{ia} |S_{ai}(\mathbf{R_a} - \mathbf{r})|^2$$

where
  summation over the number of shells i for atom a (chosen according to the location of a in the periodic table)
  q_i electronic population on the ith shell

Various choices are possible with either Gaussian- or Slater-type functions: from one $1s$-GTO function per atom to one ns-STO function per atomic shell. These functions are optimized to reproduce atomic ab initio self similarities. A new algorithm was proposed with a fast promolecular ASA on the basis of $1s$-GTO functions. The necessary parameters (coefficients and exponents) were calculated by Amat and Carbó-Dorca [503,504] and are available on the web [505]. In the work of Amat et al. [497] one function was used for Hydrogen, three for C, N, and O and four for Chlorine.
  Rather than a single similarity matrix, one can use a linear combination of similarity matrices associated with various operators (Ω).

$$Z = \sum c_a Z(\Omega_a)$$

where the coefficients c_a fulfill convex conditions $c_a > 0$, $\Sigma c_a = 1$. $Z(\Omega_a)$ represent the different kinds of MQSM, corresponding to the different weight operators that can be used. Each molecule is represented in a vector space spanned by the density function weighted by operator Ω_a. These coefficients c_a are determined so as to give the best MLR model for experimental data. This corresponds to a "tuned QSAR" [506].
  So, EASA gives an easy way to compute reliable MQS indices that can constitute molecular descriptors as previously seen. Comparing every molecule to all others in the set leads to a similarity matrix where every column (or line) can be viewed as a point-molecule [472]. Once the similarity matrix for the data under scrutiny is built, PLS or PCA may be used to reduce the number of variables and build a regression model, as in the approaches just presented. Generally the eigenvectors associated to the highest eigenvalues are selected in decreasing order. Alternately, another method was proposed: the "most predictive variables" method. It selects the columns of the data matrix in descending order of absolute correlation with the data (a parameter filter rejects eigenvectors with too small eigenvalues that might indicate noise).

Fradera et al. [472] presented two QSAR studies, regarding the inhibitory activity on DHFR of 33 Baker's triazines, previously studied by Hopfinger [41] (the activity corresponded to the concentration necessary to produce a 50% inhibition of the DHFR enzyme) and 23 indoles binding the bovine brain membrane [507]. These compounds were known to bind the benzodiazepine receptor; hence their interest as potential sedatives or anticonvulsants.

For triazines (see compound **3.1**), using the same frozen torsional angle (Θ = 270°) as in Hopfinger's study, slightly better results were obtained using PLS rather than PCA. The best model, obtained with 1s-STO-EASA (and three PLS factors) led to $r^2 = 0.602$ and $r^2_{pred} = 0.485$ (on a range of 5.1 log units). No strict comparison could be made with the results of Hopfinger (due to the lack of a separate test set in the present study). For 27 compounds, Hopfinger's treatment yielded $r^2 = 0.908$.

Similarly, on the population of 23 indole derivatives (**5.29**), with three PLS latent variables, 1s-STO and 1s-GTO functions gave nearly identical results $r^2_{pred} = 0.600$, $r^2 = 0.731$ (on a range of 2.3 log units). Interestingly, in this family, the molecular skeletons were well superimposed except for two chloro-derivatives (in R_1 or R_3, respectively) for which the chlorine–chlorine superposition has more weight than that of the carbon chain.

5.29

Another study [497] concerned antitumor agents: camptothecin (CPT) (**5.30**) a natural product isolated from the fruit of a Chinese tree, *Camptotheca acuminate* and 11 of its analogues, acting as cytotoxic drugs, inhibitors of topoisomerase-1. Comparison of the known X-ray structure of one member of the series with results of geometry optimization at different levels of sophistication (from MM+ to 3-21G) prompted the authors to use AM1 to fix the 3D-geometries (MM+ giving the worst results, and 3-21G being too computer-time intensive). For this process, 15 similarity matrices were used (overlap-like, Coulomb-like, gravitational-like, and 12 TMQSM where the Ω operator is replaced by the density function of each molecule).

5.30

The resulting q^2 are rather mediocre with three or four PCs. With four PCs and TMQSM, using as operator another molecular density function, led similarly to many negative q^2, indicating a poor prediction capacity. A tuned QSAR (TQSAR) approach provided a better predictive capability than a simple QSAR model, yielding $r^2 = 0.899$ and $q^2 = 0.858$ with three matrices and four PCs. Using four matrices gave a marginal improvement but with a supplementary cost.

MQSM was also used by Gironés and Carbó-Dorca for an evaluation of the toxicity of chemicals, measured on three populations of fishes. This is an important field on account on the international concern about environmental problems [498]. Toxicity was evaluated as the minimal necessary concentration of a compound to cause death of 50% population within a given period of time (96 h, for example, for *Poecilia reticulata*), and expressed as $-\log LC_{50}$.

Molecules were aligned using the TGSA [502], and electron density functions were built in the PASA approximation (promolecular ASA) from linear combinations of $1s$ Gaussians previously fitted on the ab initio ones. For *Poecilia reticulata* (92 aromatics on a toxicity range of about 3.3 log units) $r^2 = 0.754$, $r^2_{CV} = 0.673$ and $s = 0.37$ with four latent variables. For the second example (toxicity of 69 benzene derivatives to *Pimephales promelas*) $r^2 = 0.691$, $r^2_{CV} = 0.540$ with $s = 0.51$ on a range of 3.4 log units, with four LV. Extending the data set to 114 compounds gave slightly inferior results (which was not unexpected due to an increased heterogeneity of the structures): The best model corresponded then to 6 LV and s was only 0.70, for a range of 5.4 log units ($r^2 = 0.739$, $r^2_{CV} = 0.517$). For this application, the overlap operator was preferred to the Coulomb one.

5.3.3   3-Way PLS Treatment

Similarity matrix–based treatments can be also improved by 3-way PLS [66]. Indeed Hasegawa et al. introduced 3-way PLS on the example of eight neonicotinoid compounds (with insecticidal activity) binding the nicotinic acetylcholine Ach receptor. Compounds are defined by two torsion angles. Four conformers were generated, by 180° rotations of one or two of these angles, from the X-ray structure of template imidacloprid ($\alpha = \beta = 0°$) (Figure 5.3). With four conformers and two alignment

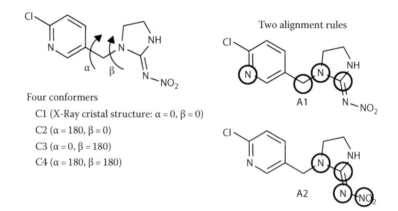

FIGURE 5.3   Neonicotinoid compounds: conformers and alignment rules. (Adapted from Hasegawa, K. et al., *Chemom. Intell. Lab. Syst.*, 47, 33, 1999. With permission.)

rules (focusing on the pyridine or imidazolidine-like moieties), eight combinations are built. The Carbó index calculated from a Coulombic molecular electrostatic potential was chosen as structural descriptor (numerical integration from MEP values on a node lattice). This led to a similarity matrix 8×8 (each of the eight molecules characterized by its similarity indices by respect to the others, and that, for each combination conformer/alignment, leading to a 3-way array).

A 3-way PLS analysis led to $q^2 = 0.855$ with three components. Once conformer and alignment were chosen, a standard PLS method might be applied. The resulting scores could be converted back to the original Carbó similarity indices, leading to a MLR-like correlation. Interestingly, from the sign of the coefficients, it appeared that, to be active a molecule must be similar to the most active molecule and dissimilar to the less active one.

5.4 FRAGMENT QUANTUM SELF-SIMILARITY MEASURES

When looking at a series of molecules with a common scaffold and varied substituents on a few positions, the free energy relationship (FER) formalism links up changes in activity to the substituent-induced changes in electronic (and/or steric or hydrophobic) effects. A similar approach can be applied for characterizing the changes in electronic distribution thanks to MQSM and particularly molecular quantum self-similarity measures (MQS-SM) defined by comparison of a molecule with itself. If it is possible to separate density functions into fragments, QS-SM can be defined separately for the (common) molecular scaffold, and for the "substituents." These QS-SM constitute molecular descriptors that can be used in multilinear regression [333,508,509]. The fragment electron densities may be easily calculated from the total electron density of the whole molecule. An advantage of the self-similarity measures is to avoid the need for alignment.

Three examples of application were studied [333]: affinity of serine proteinase inhibitors (benzamidine-like compounds) (see compound **2.1**) to three receptors: thrombin, trypsin, and factor Xα. The results of MQSM approach were compared to those of the largely used CoMFA and CoMSIA methods (Table 5.2). The data set encompassed 88 compounds [331] split in training and test sets (72/16) on a reactivity range of about 4.5 log units in pK$_i$ (only three units for Xα inhibitors). Structural variations were restricted to positions R$_1$ and R$_2$. A pre-alignment was carried out on the basis of the resolved X-ray structure of a trypsin-inhibitor complex [331].

From the training set, a MQS matrix (72×72) was built. The dimension of this matrix was then reduced retaining only the most important PCs and reducing the columns of the original matrix to 20 PCs. Then a variable selection process was performed generating all possible combinations of them, and selecting the subset the most representative in MLR, according to the usual criteria (highest q^2 in L-O-O). The performance on q^2 in L-O-O depending only slightly on the number of PC retained, the authors (to avoid the risk of overfitting the training set) preferred a decision based on both q^2 and the standard error in prediction SDEP (on the test set).

The other approach was the fragment-based quantum self-similarity measure (Fragment QS-SM). QS-SM was constructed from fragments of one, two, or three

TABLE 5.2

Comparison of MQS, Fragment QS-SM, COMFA, and COMSIA Approaches

		PC	q^2	r^2	SDEP
MQS[a]	Thrombin	6	0.551	0.627	0.75
	Trypsin	8	0.628	0.721	0.47
	Xa	9	0.545	0.644	0.30
Fragments	Thrombin	7	0.513	0.616	0.67
	Trypsin	8	0.690	0.753	0.51
	Xa	7	0.565	0.637	0.34
CoMFA	Thrombin	4	0.687	0.881	0.66
	Trypsin	5	0.629	0.916	0.52
	Xa	3	0.374	0.680	0.28
CoMSIA	Thrombin	6	0.757	0.950	0.75
	Trypsin	9	0.752	0.972	0.35
	Xa	6	0.594	0.915	0.32

[a] Training on 69 compounds only. The last right column gives the SDEP standard error of prediction for the test set of 16 compounds.

atoms leading to 95 fragment contributions. PLS reduced the number of variables to 15–25. Then, all possible k-dimensional combinations of variables were input as parameters in a MLR. One advantage of this method was to give some insight about the regions responsible for activity. The more important fragments in MLR, after a systematic search, could be displayed to indicate regions where differences in electron density influenced the binding affinity of the ligand. Comparison of these regions might give information regarding selectivity toward a specific receptor (Figure 5.4). For thrombin, they involved the α- and β-carbon atoms of the aliphatic chain, the *m*- and *p*- carbons of the aromatic ring and one amino group. The situation was globally similar for trypsin but differences arose in the removal of the amino group, the presence of a carboxyl group and the substituent R_2. Whereas for Xα, optimal fragments were R_1 and R_2, the α-C on the aliphatic chain, and the *o*-carbon of the aromatic ring. However some caution was advised since correlations were rather mediocre for trypsin and Xα factor.

The largely used CoMFA and CoMSIA methods [331], with many descriptors, allowed for a very good adjustment on training (better than that of SM), but led to significantly lower results in validation, which suggested overfitting. It was also noteworthy that with only its few dozens of initial descriptors, SM are not so much inferior as to performance, and are not affected by the mesh size or by the position of the molecules in a lattice. Spatial information is intrinsically contained in the fragments of the QS-SM treatment and can be recovered in contour plots.

In another publication [499] relying on the steroid benchmark the fragment QS-SM approach gave better results ($q^2 = 0.886$, $r^2 = 0.917$) than MQSM ($q^2 = 0.759$),

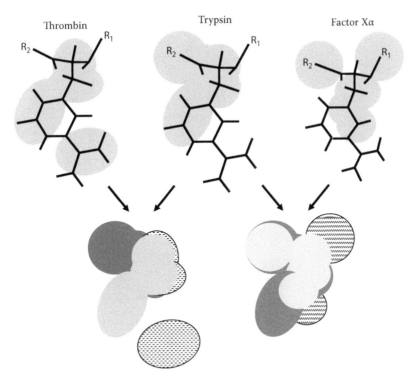

FIGURE 5.4 Fragment QS-SM patterns for elucidation of the optimal molecular regions with respect to thrombin, trypsin, and factor Xα. In the lower part, light gray regions correspond to common area between trypsin (dark gray) and either thrombin or factor Xα. (Adapted from Robert, D. et al., *Int. J. Quantum Chem.*, 80, 265, 2000.)

with scaling for variable reduction and selection of PCs (otherwise 0.705, 0.781) [616], whereas tuned MQSM ($q^2 = 0.842$) and TQSI ($q^2 = 0.775$, vide infra). The standard error of prediction SDEP on test compounds **s-20** to **s-30** (SDEP = 0.49, excluding steroid **s-31**) competed well with previous approaches such as CoMFA (0.36), COMPASS (0.34), or SOMFA (0.37). The treatment also suggested, as important for activity, the role of a carbonyl group on the steroid A-ring.

The fragment-QS-SM approach was also used in a study of antimycobacterial benzoxazines [510]. Antituberculotic activity is an important concern since tuberculosis becomes a major emerging opportunistic affection with particularly a rising occurrence among HIV-infected patients. A total of 39 molecules with minimum inhibitory concentration (MIC) between 0.6 and 2.7 log units were investigated. MIC (in μmol/L) is the lowest concentration of a drug at which the inhibition of the growth occurs. The QS-SM formalism was applied from seven fragments (from two to six atoms) including exocyclic atoms (Figure 5.5), possibly combined to QSM for the whole molecule.

All possible one, two, or three parameter models were examined. Whereas traditional QSARs are only valid for limited subseries, a single QS-SM model accounted for the whole series with only two fragments, f1 and f5 that involved the oxo (or thio) groups known for playing an important role in antimycobacterial activity.

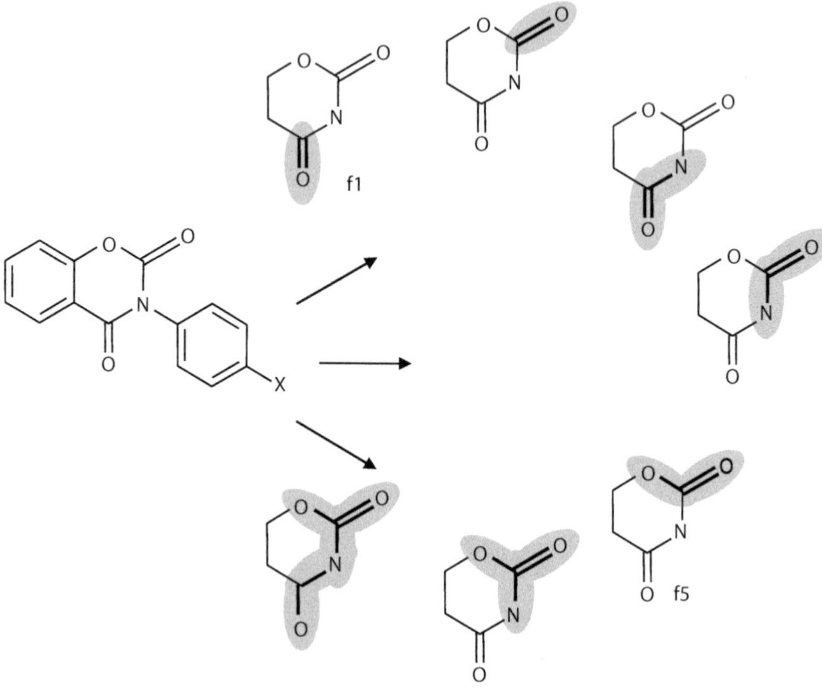

FIGURE 5.5   Fragment f1 and f5 are involved in all models and particularly in the model valid for the whole series. (Adapted from Gallegos, A. et al., *Int. J. Pharm.*, 269, 51, 2004. With permission.)

5.5   OTHER SIMILARITY-BASED METHODS: CoMASA, k-NN, OR CPANNS

Similarity-based methods largely rely on similarity matrices where a molecule is characterized by its resemblance to all other molecules in the data set (except applications derived from the self-similarity indices). But, similarity may be also evaluated by pairwise comparisons with pseudoatomic positions extracted from the aligned molecules, as in CoMASA [43]. On the other hand, other methods relying on the similarity-activity concept but initially dedicated to classification problems have been extended to correlations of biological activity with structural characteristics.

5.5.1   CoMASA: COMPARATIVE MOLECULAR ACTIVE SITE ANALYSIS

Comparative molecular active site analysis (CoMASA), proposed by Kotani and Higashiura [43] relies on similarity evaluation (somewhat neighbor to that of CoMSIA). It is based on a previous proposal of similarity indices derived from pairwise calculation of the nearest atomic distances [511]. According to the authors, this technique may appear as "coarse and rough but is extremely rapid and accurate" (two or three times faster than usual grid-based evaluations).

5.5.1.1   Generating the Molecular Represented Points

Basically, the CoMASA approach may be characterized by two major components. First, interactions are calculated using these simplified similarity indices. Second, instead of working on nodes in a lattice, some representative points are extracted from the superimposed molecules. Similarity is calculated between the atoms of the molecule under scrutiny and these "molecular represented points." Three advantages are claimed: The results are simple to interpret and would easily transform to pharmacophore or queries in 3D database screening; there is no problem of orientation or translation in a lattice; calculation is easy and fast.

The first step consists of superimposing the molecules under scrutiny and extracting a minimum collection of occupied atomic positions MRP (molecular represented points) by means of a cluster analysis. In the superimposed molecules, atoms at a distance inferior to a threshold (arbitrarily 0.75 Å) are iteratively eliminated and replaced by the molecular represented points (black dots in the last but one step in Figure 5.6). These points will be used in replacement of the nodes of CoMFA or CoMSIA lattices, but with an important reduction of their number: in the quoted examples, about 90 in place of 7200. These molecular represented points may be seen as a hypermolecule not too far from that created in the MTD method [512]. Centroids of aromatic rings have been considered but they did not lead to any improvement.

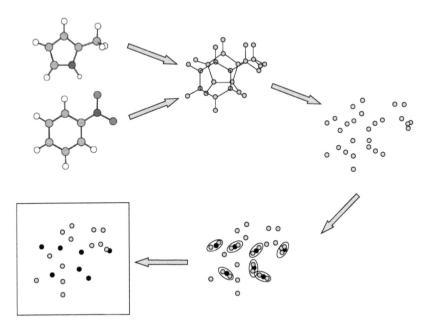

FIGURE 5.6   Determination of the molecular represented points. (Adapted from Kotani, T. and Higashiura, K., *J. Med. Chem.*, 47, 2732, 2004. With permission.)

5.5.1.2 Evaluating Similarity

Then, interactions with each molecule are evaluated at these "Molecular Represented Points" (MRP) and constitute the descriptors submitted to a PLS treatment. Diverse evaluation functions are available for steric, electrostatic, or hydrophobic contributions.

A *rapid similarity index* between a compound and a probe atom systematically located on the MRP (k) was proposed. This index was previously defined in pairwise molecular shape comparisons and proved to be very efficient in optimization of molecular superimposition processes [511]. For the molecular represented point k of molecule j

$$S^k(j) = I_k$$

I_k is evaluated from the distance r_k of point k to the nearest atom i of molecule j

$$I_k = 1.0 \quad \text{if } r_k \leq rlim1 \quad \text{and}$$

$$I_k = 0.5 \quad \text{if } rlim1 < r_k \leq rlim2$$

$$\text{with } rlim1 = \frac{\alpha_1(vdw_i + 1.0)}{2} \quad \text{and} \quad rlim2 = \frac{\alpha_2(vdw_i + 1.0)}{2}$$

where vdw_i is the van der Waals radius of the nearest atom i of molecule j (the van der Waals radius of the probe being arbitrarily fixed at 1 Å as in CoMSIA). Optimal values of α_1 and α_2 are set to 1.3 and 0.6, respectively. One advantage is that the calculation time may be reduced: for example, if $r_k < rlim1$, further distances based on atom i are not required, since the maximum value of I_k is reached.

SEAL [151] constitutes another possible similarity index. At the MRP k for molecule j

$$A_{F,q}^k(j) = -\sum_{i=1,n} (W_{probe,q} \ldots W_{i,q}) \exp\left(-\alpha r_{ik}^2\right)$$

summation over all atoms i of molecule j

r_{ik} is the distance between the probe (at MRP k) and atom i of the tested molecule j
$W_{i,q}$ is the actual value of property q at atom i

For the probe $W_{probe,q}$, charge is set to 1 and the radius to 1 Å. A Gaussian-type attenuation factor is used ($\alpha = 0.3$) as in CoMSIA. At 1 Å distance, the property value is 26% reduced and 93% reduced at 3 Å.

Good's Gaussian similarity index [380,477] may also be used for the evaluation of shape and electrostatic similarity. Electrostatic and shape similarity indices P_{el}, P_{st} between compound of interest j and a probe atom (at point k) are expressed as a summation of two or three Gaussian functions:

$$P^k(j) = P_{el}^k(j) + P_{st}^k(j)$$

$$P_{el}^k(j) = -\sum qi\ (G_{el1} + G_{el2})$$

summation over atoms i of molecule j (i = 1 to n)

$$P_{st}^k(j) = -\sum (G_{st1} + G_{st2} + G_{st3})$$

where

$$G_z = \gamma_z \exp\left(-\alpha_z r_{ik}^2\right)$$

qi is the charge on atom i

γ_z is a proportionality constant

α_z is the exponential constant for Gaussian function approximation, defined from
  the atom type

Simple indicator variables are a variant of the topological index as in MTD where atomic positions are ordered in bins (beneficial, detrimental, or irrelevant). They were introduced to reduce calculation time. They define three categories depending on the interatomic distance r_k between the molecular represented point k and the nearest atom of molecule j.

$$I^k(j) = I_{el}^k(j) + I_{st}^k(j)$$

$$I_{el}^k = q(k) \quad \text{if } r_k \le \text{th}$$

$$\text{or} \quad = 0.5\ q(k) \quad \text{if th} < r_k \le 2\ \text{th}$$

And, similarly

$$I_{st}^k = 1 \quad \text{if } r_k \le \text{th}$$

$$\text{or} \quad = 0.5 \quad \text{if th} < r_k \le 2\ \text{th}$$

th is the threshold for cluster analysis (0.75 Å); and q the charge at the atom of molecule j closest to the MRP k. As for the rapid similarity index, note that if the distance between the atom j and the MRP k is less than th, other distances need not be calculated.

Hydrophobic interactions may be evaluated by various options. The AlogP hydrophobic parameter of atom i can be introduced in the SEAL formula (hydrophobic parameter of the probe being set to 1). The simple indicator variable methodology can also be used.

$$I_{lipo,k} = a_i(k) \quad \text{if } r_k \le \text{th}$$

$$= 0.5\ a_i(k) \quad \text{if th} < r_k \le 2\ \text{th}$$

Alternately, FlexS hydrophilic parameters are substituted to AlogP values and used with SEAL-type function or indicator variable type function. In FlexS method

TABLE 5.3

Hydrophobicity Classification (FlexS)

Type	th1	th2
H	0.1	0.06
C, N, O, F, B	0.2	0.1
P, Cl, Br, I, S	∞	0.1

Source: Reproduced from Lemmen, C. et al., *J. Med. Chem.*, 41, 4502, 1998.

(Fast Flexible Ligand Superposition, see Chapter 11) [513], atoms are classified as hydrophobic, hydrophilic, or ambiguous depending on the absolute value of their partial charge: chr(i) for atom i, according to:

$$\text{If } |\text{chr}(i)| > \text{th1}(\text{type}(i)) \quad \rightarrow \quad \text{hydrophilic}$$

$$\text{Else if } |\text{chr}(i)| < \text{th2}(\text{type}(i)) \quad \rightarrow \quad \text{hydrophobic}$$

$$\text{Else} \quad \rightarrow \quad \text{ambiguous}$$

The thresholds th1 and th2 according to the atom types are indicated in Table 5.3. Hydrophobic, hydrophilic, and ambiguous characters are encoded 1, −1, and 0, respectively.

Once the descriptor matrix is built, a classical PLS treatment is carried out. Results may be visualized drawing isocontours of the type "CoMASA coefficients × standard deviation": (STDEV × COEFF) as in CoMFA or CoMSIA. These contours delineate volumes where differences in molecular active sites are related to variations of the dependent variables. An interesting point is that the model is not ruled by the choice of a template, since extraction of MRP considers the whole population.

5.5.1.3 Applications

The first example treated concerned the classical 31 steroid benchmark. Cluster analysis restricted the number of MRP to about 90 points in CoMASA, to be compared with the 7000 (and more) lattice nodes considered in CoMFA or CoMSIA. With the alignment proposed by Cramer et al. [38], the various possible options (using, for evaluation function, similarity index calculation, SEAL, Gaussian similarity indices, or simple indicator variables) were systematically explored.

Compared to CoMFA ($r^2 = 0.719$, $q^2 = 0.662$) or CoMSIA ($r^2 = 0.763$, $q^2 = 0.662$), CoMASA gave very promising results with generally r^2 about 0.93 and q^2 about 0.750, although some options were less satisfactory. The authors favored the use of simple indicator variables which yielded $r^2 = 0.982$, $q^2 = 0.798$ with four LV and an equal importance of steric and electrostatic fields. As to hydrophobicity, FlexS parameters (considered alone) seemed more convenient (r^2 about 0.930, q^2 about 0.750).

Including simultaneously steric, electrostatic, and hydrophobic contributions (with simple indicator variables) and selecting, as usual, the training set of 21 molecules led to 0.929 and 0.779 with two LV. The steric, electrostatic, and hydrophobic fields contributed for 0.33, 0.26, and 0.40, respectively. Standard deviations were 0.31 (training) and 0.58 log units (test set of 10 molecules, including the outlier molecule **s-31**). SEAL-type function gave similar results.

A second example concerned 40 1,5-diarylpyrazole analogues COX-2 inhibitors (**5.31**). Binding conformations and alignment were derived from a previous docking study [351]. Best results were obtained with simple indicator variables for steric, electrostatic, and FlexS hydrophobic parameters ($r^2 = 0.796$, $q^2 = 0.411$ with two LV). Fields contributed for 24% (steric), 38% (electrostatic), and 38% (hydrophobic).

5.31

Although some differences regarding sterically favorable regions, contour plots were in agreement with CoMFA and CoMSIA results for sterically unfavorable zones around the five-membered ring. As to electronic effect, plots also suggested that a CF_3 group on this ring, or substituents on the second benzene ring (that devoid of sulfone/sulfonamide group) enhance inhibitory activity (Figure 5.7). Hydrophobic

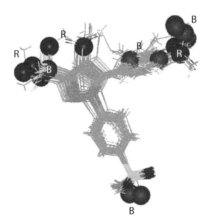

FIGURE 5.7   (**See color insert following page 288.**) Map obtained for the set of COX-2 inhibitors (indicator variables for steric and electrostatic contributions, FlexS parameter for hydrophobicity). (R) Red balls indicate regions where an increase of negative charge will enhance affinity. (B) Blue balls indicate regions where more positively charged groups will improve the binding properties. (Reproduced from Kotani, T. and Higashiura, K., *J. Med. Chem.*, 47, 2732, 2004. With permission.)

favorable regions were detected around the first benzene ring, suggesting interactions with the receptor hydrophobic residues. For an extensive study of COX-2 inhibitors (227 compounds from five chemical families) involving CoMFA, CoMSIA, and HQSAR, see Chen et al. [187].

It is noteworthy that the CoMASA contour plots concerned points (pseudoatoms) generated from the real atom locations and so directly showing interactions of the molecules, whereas CoMFA contour plots delineated regions outside the molecular volume.

5.5.2   k-Nearest Neighbor QSAR Models

The principle of similarity-activity is also at the basis of the k-nearest neighbor approach proposed by Zheng and Tropsha [514]. The method was first used with 2D descriptors [113,514] and [358] where a consensus model was built. But, the approach can be applied, with suitable descriptors, to elaborate 3D-QSARs [171] as for example the k-NN-MFA method (k-nearest neighbors-molecular field analysis). In this interpolative method, the activity of a compound is not predicted by fitting a regression equation, but rather by evaluating an average weighted activity over its k most similar compounds. Insofar as the method mainly relies on distances between compounds (in the descriptor space), it may be used, with suitable tools, to select the structural descriptors that best characterize the resemblances or differences between compounds, and so can become an efficient way for variable selection.

5.5.2.1   k-NN Molecular Field Analysis

Basically, building a k-NN model comes to optimize (from a training set characterized by a pool of descriptors) both the number of neighbors to consider, and the subset of descriptors to select; the objective function being the L-O-O cross-validated correlation coefficient q^2.

In k-NN-QSAR, for a compound i, predicted activity y_{i-pred} is given by:

$$y_{i-pred} = \sum_k w_i y_j$$

The weighting coefficient w_i is:

$$w_i = \frac{\exp(-d_{ij})}{\sum \exp(-d_{ij})}$$

summation over the k-nearest neighbors of i. d_{ij} is the Euclidian distance between molecules i and j (calculated on the subset of descriptors in the current step).

$$d_{ij} = \left[\sum_n (X_{i,n} - X_{j,n})^2 \right]^{0.5}$$

In the current step, a subset of descriptors is selected. From the calculated y_i for each compound that is in turn left out, q^2 may be calculated. The process is repeated for $k = 1, 2, 3$, etc. in order to find the best model (highest q^2). Generally speaking, the optimal model is found for a small number (<5) of neighbors. Various methods, relying on the general concern of variable selection, can be used to find the best sub-set of descriptors [171]. They will be developed below.

5.5.2.2 Variable Selection with k-NN

- *Stepwise k-NN (k-NN-MFA-SW)*, a trial model is generated starting from a single descriptor and adding other descriptors one by one while examining the fit of the model.
- *Simulated annealing (k-NN-MFA-SA)*, already used in 2D k-NN. Schematically, the procedure here works as follows:
  1. Choice of an initial "pseudo" temperature (1000°K, in this example)
  2. Random selection of a subset of descriptors
  3. Prediction of activity, calculation of q^2 in L-O-O, and choice of the best number of neighbors
  4. Change of a (user-defined) proportion of the descriptors by other descriptors randomly chosen
  5. Evaluation of q^2, as in step (3)
      If $q^2_{new} > q^2_{current}$ the change is accepted. Otherwise the Metropolis crite-rion is examined: The change is accepted if

$$ rnd < \exp\left[\frac{-\left(q^2_{current} - q^2_{new} \right)}{T} \right] $$

  where
      rnd is a random number between 0 and 1
      T is the "pseudo" temperature [515]

  If the change is accepted, go to (4), otherwise, if q^2 does not change after many of these loops, decrease the temperature (by 10%, for exam-ple) and go to step (4), or stop if $T = T_{minimum}$ ($=10^{-6}°$K here).
- *Genetic Algorithm (k-NN-MFA-GA)*, already presented as optimization tool. Here the procedure is
  1. Generate an initial population of chromosomes by random selection of genes (descriptors).
  2. Predict activity for each chromosome, calculate q^2 in L-O-O, and choose the best k.
  3. Select chromosomes for mating pool from roulette wheel selection.
  4. Create a new population of offspring (by crossover and mutation) and calculate the fitness (q^2) for each offspring.
  5. Replace the least fit chromosomes by the best offspring.
  6. Repeat the steps until convergence (or maximum number of steps) is reached.

5.5.2.3 Applications of k-NN 3D-QSAR Models

3D k-NN method was applied in a CoMFA treatment [171] as a substitute of the PLS step. After energy minimization and alignment, the pool of descriptors was constituted by the field (electrostatic and steric) values on the nodes of the lattice. Three data sets were examined:

- Usual steroid benchmark
- Selective COX-2 inhibitors (1,5-diarylpyrazole analogues) [516]
- Anticancer 1-N-substituted imidazoquinoline-4,9-dione derivatives) [517]

For the sake of comparison, alignment and choice of training and test sets were those proposed in the original publications. For the steroid set, with the usual 10 test molecules, q^2 yielded 0.95 (SA-k-NN) and 0.93 (GA-k-NN) with k = 2 and 6 or 8 descriptors, whereas SW-k-NN was slightly inferior ($q^2 = 089$) but with only 2 descriptors and 3 neighbors. Things were similar in prediction; r^2_{pred} were, respectively, 0.78 (SA), 0.85 (GA), and 0.58 (SW). These results compared favorably with CoMFA ($q^2 = 0.80$, $r^2_{pred} = 0.56$). It was often noted that the choice of Cramer for the training and test set was not the best. Here the authors used the sphere exclusion method for choosing an optimal splitting [518]. The statistical parameters were improved (for SA and GA, $r^2_{pred} = 0.87$ and 0.86, whereas for SW, $r^2_{pred} = 0.72$, and $q^2 = 0.94$).

For COX-2 inhibitors [516] (with 25 1,5-diarylpyrazoles in training and 5 in test) the approach, with r^2_{pred} between 0.81 and 0.90, and q^2 between 0.82 and 0.85, (2 neighbors and 6–10 descriptors) clearly outperformed CoMFA (0.68 for both r^2 and q^2). Results were nearly similar with the optimal test set of eight molecules.

A similar conclusion could be drawn for the anticancer drugs [517] where predictions of CoMFA were quite bad (r^2_{pred} negative) whereas the three options of k-NN-MFA led to r^2_{pred} better than (or equal to) 0.65. In training (16 compounds) CoMFA ($q^2 = 0.625$), CoMSIA (0.52), HQSAR (0.501) were inferior to k-NN where q^2 was between 0.82 and 0.96.

In a comparison with the reference method CoMFA, the authors emphasized that their approach was alignment dependent (since it works from field values to evaluate distances), but also largely more time consuming. One advantage, however, was that stochastic methods generated several models of comparable quality giving a wider choice for model selection. On the other hand, the method highlighted a limited number of nodes (the descriptors retained in the model) playing an important role in the evaluation of activity.

5.5.3 Counter-Propagation Artificial Neural Networks

A CPANN is a generalization of the Kohonen network and its Self Organizing Maps (SOM). But, although a Kohonen network is dedicated to classification, a CPANN allows—inter alia—for property prediction. In our concern, this involves prediction of a biological activity value from the knowledge of descriptors characterizing a chemical structure.

The architecture of CPANN is shown in Figure 5.8. It consists of two active layers: a Kohonen layer (the same as in SOM) and an output layer associated with output values (in our case, the output layer is of dimension one, but it may be multidimensional,

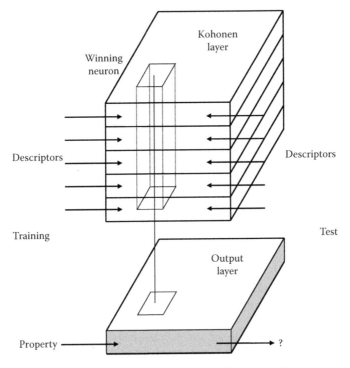

FIGURE 5.8   Architecture of a counter-propagation artificial neural network.

depending on the nature of the properties treated). The first part works as a usual Kohonen network. Structural descriptors are given as inputs and a SOM is obtained for a given input vector. A winning neuron is determined by competitive (unsupervised) learning. Compounds are organized on the basis of the similarity of their structural descriptors [401,519]. In the second step, the winning neuron of the SOM, for each object (here a chemical compound), is projected to the output layer (in the same position) and the weights between the Kohonen layer and the output layer are adjusted to the corresponding property value.

In the prediction phase, the new object presented is first located on its best-fitting neuron in the SOM. This winning neuron is projected onto the output layer. The corresponding weights between these two layers deliver the property value. This step corresponds to a supervised process: weights between the two layers (Kohonen and output) are adjusted on known examples.

5.5.3.1   Weight Adjustment

Remember that for the current iteration cycle (epoch) of the Kohonen part, for each compound, in turn:

- All inputs are connected to the neurons of the Kohonen map, and a winner neuron sc (the "central" neuron for sample s) is selected, choosing the neuron j with weight vector $\mathbf{w}_j$ (w_{j1}, w_{j2}, ..., w_{jm}) most similar to the input $\mathbf{x}_s$ (x_{s1}, x_{s2}, ..., x_{sm}).

$$\text{out}_{sc} \leftarrow \min\left[\sum_{i=1,m}\left(x_{si} - w_{ji}\right)^2\right]$$

$j = 1,n$ n is the number of neurons of the SOM

At the beginning weights are randomized (but other options exist) and are iteratively modified during the training process.

- Once the winner neuron (sc) is selected, the corresponding weights $\mathbf{w}_c$ (w_{c1}, w_{c2}, ..., w_{cm}) are adjusted to improve the response (for the next cycle). The weights w_{ji} of the neighboring units on the map are also corrected to make their response w_{ji} closer to the input x_{si} (smoothing the map). As for a Kohonen network,

$$w_{ji}^{new} = w_{ji}^{old} + \eta(t)a(d_c - d_j)(x_i - w_{ji}^{old})$$

where a $(d_c - d_j)$ is a function decreasing with the topological distance on the map between central neuron c and the current neuron j. Triangular or "Mexican hat" functions, for example, have been proposed. These functions allow for enhancing the contrast between the winner and its surroundings. $\eta(t)$, the learning rate, monotonously decreasing as the network goes on, scales the weight correction. The learning rate $\eta(t)$ is calculated from

$$\eta(t) = (a_{max} - a_{min})\frac{(t_{max} - t)}{(t_{max} - 1)} + a_{min}$$

a_{max} and a_{min} (in the interval $[0,1]$) determine the maximum and minimum corrections of the central atom (c) at the beginning of the training ($t = 0$) and at the end ($t = t_{max}$).

- Simultaneously, once the winning neuron is selected, the output neuron at the same position is selected. The weight between the Kohonen and the output layers is fixed by the property value y_s, and the neighbor units are corrected with the same process as in the Kohonen layer.
- The process is repeated for all compounds (s) and a next epoch begins.

Some particular aspects deserve attention:

- A CPANN does not provide a function to calculate a property from structural descriptors (as do MLR or BNN, for example). Predicted properties are already stored as weights. CPANN may be viewed as a look-up table and only delivers a fixed number of responses, depending on the numbers of cells in the maps.
- The response is not obtained from all neurons of a preceding (hidden) layer as in BNN, nor from a single unit as in a Kohonen network, but is stored as the

weight between a unique neuron (the winner) of the Kohonen layer and the corresponding unit of the output layer (units for a multidimensional target).

- The name "counter-propagation" came from the fact that the data flow runs in opposite directions: from the input vectors to the Kohonen layer and from the target value to the output layer.

5.5.3.2 Application of CPANN in QSARs

Applications of CPANN in QSARs are relatively scarce. In a study of the toxicity of 41 benzene derivatives [520], structural description was achieved by atom positions, supplemented in a second model by atomic charges (Mulliken partition scheme, at the 6–31G level). Atom positions were characterized by gnomonic projection leading to a "spectrum like" description. Satisfactory models were obtained, particularly with consideration of charges. However their quality varied according to the constitution of the selected training set (which is not quite surprising on account of the rather limited population and the heterogeneity of the compounds).

Performance of CPANN (as nonlinear method) [521] was compared to that of a MLR linear approach [522] on the mutagenicity of 95 aromatic amines, on a range of 7 log units. An initial pool of 378 descriptors, was reduced to 21 topological descriptors, 10 geometrical (including Wiener indices and van der Waals volume), and 6 quantum-chemical and log P. Various combinations of subsets of descriptors were tried. The best results, comparable to those obtained by the linear approach of Basak et al. [522], were obtained with topological descriptors plus 3D descriptors; log P or the quantum chemistry descriptors slightly disturbing the model. CPANN was also used in the modeling of endocrine disruptors [523].

CPANN and SVR (support vector regression) were compared by Ghafourian and Cronin [524] on estrogen activity of 131 compounds, in conjunction with variable selection method. Stepwise regression, partial least squares, and recursive partitioning (formal inference-based recursive modeling [FIRM]) were applied to an initial pool of 151 descriptors. Some overall similarity appeared as to the selected descriptors. But the predictive power varied with the method. With selection via stepwise regression, SVR outperformed CPANN, whereas with some FIRM choices, CPANN was the best.

See also Kuzmanovski and Novic [525] for the recent development of a new program and its application to the prediction of the boiling point of 185 saturated acyclic compounds.

An interesting application of CPANN, although not yet directly applied to 3D-QSAR, was published by Arakawa et al. [526] in a study of 77 anti-HIV HEPT analogues. The originality of the treatment comes from the coupling of a counter-propagation neural network with multi-objective genetic programming (GP), and the recourse to the Pareto multi-objective optimization concept.

GP is a branch of GA. In classical GA, a possible solution is represented by a bit string. In GP a possible solution is represented by a parse tree (Figure 5.9). In this tree, each leaf is a variable, a constant or a function without parameters (such as +, −, ×, /, if, sin). A solution is said to be "Pareto optimal" if it is impossible to improve it regarding an objective value without decreasing at least one other objective value. A situation that can be improved is said "dominated" (Figure 5.10).

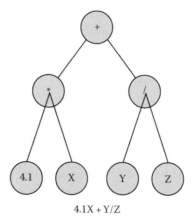

4.1X + Y/Z

FIGURE 5.9 Example of genetic programming for the expression 4.1X + Y/Z.

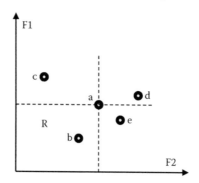

FIGURE 5.10 Example of Pareto optimization for a system with two cost functions F1 and F2, and five solutions (a–e). c and b are Pareto optima. a would be dominated by any solution in rectangle R. (Adapted from Arakawa, M. et al., *Chemom. Intell. Lab. Syst.*, 83, 91, 2006.)

The GP was used as optimization method for finding a non-dominated Pareto optimum. Each individual of the GP was constituted of the structural descriptors, and the number of dominated solutions (according to the results of a PLS analysis) defined the fitness function.

With 34 2D descriptors, a classical PLS treatment yielded unsatisfactory results. Using GP-PLS led to a reduced set of 11 descriptors. A counter propagation neural network fed with these descriptors gave a good model ($r^2 = 0.875$) that outperformed a direct PLS analysis with the same 11 descriptors. Furthermore, a comparison of the output map with the maps corresponding to the individual descriptors gave some indication about the influence on activity of the substituents of the common core.

A neighboring set of 67 HEPT derivatives and 13 TIBO compounds was studied by CoMFA and CoMSIA (Chen et al. [527]). Docking showed that these two types of inhibitors presented similar mechanisms of interaction with HIV-1 RT. For the whole set (80 compounds) $r^2 = 0.940$, $q^2 = 0.720$ in CoMFA (0.920, 0.675 in CoMSIA). These approaches were validated on a test set of 27 compounds. They significantly outperformed the classical 2D treatment of Hansch et al. [528].

6 Spectroscopic QSARs: Quantitative Spectroscopic Data Activity Relationships

Many approaches in 3D analysis methods rely on the so-called spectroscopic QSAR methods, sometimes also called quantitative spectroscopic data–activity relationships (QSDARs) [406]. The basic idea is that spectroscopic characteristics constitute, from an empirical point of view, a very sharp structural fingerprint of the molecular structure. But from a more theoretical point of view, they are also closely related to the eigenvalues of the corresponding Hamiltonian (e.g., normal mode frequencies in the IR, orbital energies in the UV–visible spectra), and so reflect intrinsic electronic and physicochemical properties of molecules (implicitly related to their 3D structure). Consequently, they may be considered as molecular descriptors. These descriptors are "physical observables" and not some artificially calculated descriptions of the molecular structure. Such descriptors are also orientation-invariant and no alignment is necessary when using them in QSAR models [52] unlike the comparative molecular field analysis (CoMFA)-type methods. Strictly speaking, these methods come within the 2.5D approaches since the geometrical information is not explicit, except for some interatomic distances involved in the nuclear Overhauser enhancement (NOE) effects. Finally, with these methods, no contour plots are available to specify regions where appropriate structural modifications may increase biological activity.

The first method EigenVAlue (EVA), based on the wave numbers of the absorption spectrum in the infrared or Raman range, was proposed by Turner et al. [51,529,530]. The approach was then extended to the orbital energies in the UV–visible in Electronic EigenVAlue (EEVA) [531] and to NMR chemical shifts (^{1}H or ^{13}C) in comparative spectra analysis (CoSA) [52] and its derived techniques (Figure 6.1). Basically, these approaches rely on experimentally determined or simulated spectra. These simulated spectra may be calculated, after energy minimization, from quanto-chemical methods. The semiempirical level (AM1, for example) was claimed to be sufficient for evaluating molecular orbital energies and vibrational frequencies (at least for such applications), but high level ab initio wave functions are required for NMR shift calculations (as, for example, the DFT/GIAO method) [532]. Alternatively, spectra may be reconstructed from substructure/subspectrum pairs (as in ACD/LabsCNMR software [533]) using HOSE [534], a substructure similarity technique from the concentric environment method or, in the IR, via a counter-propagation neural network [535]. Generally

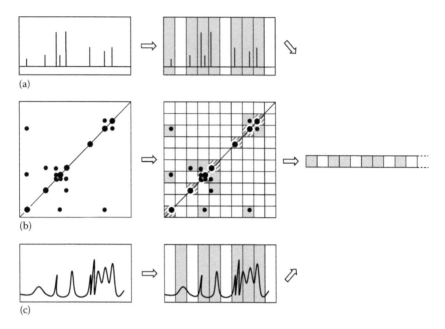

FIGURE 6.1 From spectral information to linear numerical descriptors. (a) CoSA (1D-NMR), (b) CoSCoSA (COSY 2D-NMR), and (c) EVA (IR).

speaking, only the peak position is considered (with normalized intensity), without, for example, a specific assignment of an NMR peak to a given atom. The common process (that we will exemplify in the case of EVA) consists in determining (usually calculated or simulated) the spectroscopic quantities (MO energies, vibrational frequencies, NMR shifts), and converting them into a "theoretical" spectrum.

6.1 IR EIGENVALUES

In EVA, for example, after calculation of the normal modes of vibration, the set of the wave numbers (for each molecule) is projected onto a linear-bounded scale (typically, in the IR 0–4000 cm^{-1} range). A Gaussian-shape factor is placed over each "peak" of normalized intensity. A summation of these profiles over the 3N—6 normal modes (3N—5 for linear molecules), if N is the number of atoms, leads to the EVA component at the wave number x.

$$EVA(x) = \sum_{i=1,3N-6} \frac{1}{\sigma(2\pi)^{0.5}} \exp\left[-\frac{(x-f_i)^2}{2\sigma^2}\right]$$

where
   f_i is the wave number of the normal mode i
   σ is the user-defined Gaussian standard deviation, related to the half-width at the half-height, depending on the technique (IR, UV, NMR) but constant for the whole spectrum (however, vide infra, EVA-GA)

The "spectrum" is then built up adding to the "intensity" of the peaks, i.e., the EVA(x) at regular sampling intervals (L), for something like a "resolution." For each molecule, the spectroscopic EVA descriptors are constituted by these components in the different sampling intervals. The data matrix of these descriptors (with the number of columns = the number of sampling intervals, and the number of rows = the number of compounds) is then submitted to the usual statistical treatments (mainly PLS, but BNN was also used). It must be remarked that this kind of "spectrum" is a conceptual representation and not really a simulated spectrum as that wanted by spectroscopists to mimic experimental data. The combination band and the overtones are ignored, as well as the intensity variations due to transition moment values. For such an EVA analysis, two parameters must be fixed: the band width (σ) and the sampling interval ("resolution" L). L determines the precision achieved in describing the spectrum: if L is too large, details are blotted out. Too small a value increases the number of data to treat and slows down the treatment. Typically, L must not be greater than twice the band width σ, and is often set at $\sigma/2$ [347]. The σ value determines to what extent proximal peaks overlap or not. This smearing procedure populates more intervals around each peak, leading to a more complete description. It also helps in coping with small variations in the peak position (e.g., the lack of precision of the wave number calculation, the inaccuracy of spectrum simulation with empirical models, or the solvent effects if an experimental spectrum is used). An extensive study [44] on various data sets indicate that a default value of $10\,cm^{-1}$ for σ is a "good starting point" that may be refined depending on the PLS performance.

As to correlation, the effectiveness of EVA descriptors was checked on a wide range of data sets and the robustness of the models was validated either with test sets or by using data-scrambling techniques. Some results on largely studied chemical families will be given in the course of the chapter and compared with those from other approaches. EVA was also tested for similarity by searching structure databases. EVA provided a performance similar to that obtained with conventional 2D bitscreens descriptors, but it was remarked that EVA tended to return different sets of nearest-neighbors and might suggest new ideas [44]. However, it requires beforehand the calculation of the normal modes of vibration for each compound on an energy-minimized structure.

The information content carried by IR spectra was examined by Benigni et al. [536,537]. The first study concerned 112 noncongeneric compounds. Descriptors extracted from the fingerprint region of IR spectra were compared to classical physicochemical or quantum-chemical properties (such as log P, molecular refraction, MR, and HOMO energy), molecular connectivity indices, and molecular distances. The distance matrices associated with these parameters indicated much redundancy and overlap, except for IR spectra that carried markedly different information.

In a second study [537], different classes of congeneric compounds were investigated. The 13 data sets represented different biological activities and interaction mechanisms. Three approaches were compared: the classical QSARs ("mechanistically oriented" models, according to the authors), experimental IR spectra, and eigenvalues of the modified adjacency matrix (BCUT descriptors) [538]. Gas phase spectra were sampled on 91 values in the range $600-1500\,cm^{-1}$. In adjacency matrices, the off-diagonal elements represented bonds (encoded 1–3 for simple to triple bond) and diagonal elements were roughly proportional to atom electronegativity. From the corresponding descriptors, the principal component analysis extracted 16 PCs to build correlation models (using 1–4 of them). With very few exceptions, on the 13 series investigated, the mechanistic models were superior, although, for this study on congeneric compounds, IR information provided a basis for generating QSAR models. From these results, the authors emphasized the importance of sound scientific reasoning, and mentioned that this (obvious) remark was not unimportant when QSAR research goes in for pure chemoinformatics approaches.

In the preceding correlations established with EVA, all peaks were equally weighted (shape, width, and height). A further refinement EVA_GA consisted in introducing different widths ("local σ") for different parts of the spectrum (typically for bins of about 100 sampling intervals). A genetic algorithm was used to optimize the solution [539]. For example, for the oral bacteria *Porphyromonas gingivalis* inhibition by phenols ($-\log$ MIC), classical EVA yielded $q^2 = 0.81$ or 0.69 for two training sets of 62 compounds, whereas EVA_GA reached 0.89. Similarly, for 53 melatonin-receptor ligands, the pK_i for chicken brain melatonin receptors was modeled with $q^2 = 0.46$ (training set of 44 compounds) and $r^2_{pred} = 0.81$ (on a test set of seven compounds, after elimination of two outliers), whereas EVA_GA yielded $q^2 = 0.65$ and $r^2_{pred} = 0.89$. So, this procedure might significantly improve the results. However, a caveat was put on the risk of overfitting when looking at the GA fitness score.

6.2   USING NMR SHIFTS

NMR shifts sharply reflect the environment of the resonating atom. Particularly, ^{13}C shifts that cover a wide range of more than 200 ppm (for neutral molecules) are very sensitive to structural influences within formal fragments of four to about eight atoms. These shifts are strongly dependent on both the geometry of the molecule and its electronic features, such as electron density and bond order, elements that are important in determining the activity in biological systems. So, despite a weak sensitivity, when compared to ^1H NMR, ^{13}C shifts convey potentially useful information for the development of QSDARs.

6.2.1   COMPARATIVE SPECTRA ANALYSIS

Bursi et al. [52], in the CoSA method, extended the QSDAR approach to RMN spectra (^1H and ^{13}C) and mass spectra. The electron ionization mass spectra also give information on the molecular structure by identification of real fragments created by bond breaking (but possibly with rearrangements!). For a set of 45 steroidal progestagens, covering a range of 2.8 log units in the relative binding affinities (RBA) (data from Ref. [540]), the CoSA results were compared to the CoMFA approach [52]. In addition to the simulated spectra, Bursi et al. also considered (depending on the technique) the use and the effectiveness of experimental, digitalized, spectra. According to the authors, spectral descriptors led to better performance than CoMFA, either with rigid alignment or a field-fit option, both for training- and test-sets. In training, the best CoMFA led to $r^2 = 0.871$, $q^2 = 0.550$, whereas EVA (simulated IR spectra) yielded $r^2 = 0.986$, $q^2 = 0.638$, and nearly similar results were obtained from the mass spectra. However, predictions were mediocre for ^{13}C (simulated) $q^2 = 0.397$ or IR (experimental) $q^2 = 0.323$. Interestingly, combining spectral descriptors with other spectral descriptors or molecular fields generally provided better models than using individual descriptors: (CoMFA + $^1H_{exp}$), ($IR_{sim} + {}^1H_{exp}$), or ($^1H_{exp} + {}^{13}C_{sim}$) gave nearly similar results: q^2 about 0.62 or better, s about 0.46, with r^2, and SEE better than 0.980 and 0.10. For a test set of seven compounds, RMSE was about 0.49. For other 3D studies on progestagens, see Ref. [328].

The CoSA method was also applied to the binding affinity of polychlorinated aromatics (biphenyls (**1.4**), dibenzodioxins (**1.5**), and dibenzofurans (**1.6**)) to the aryl hydrocarbon receptor (AhR). [541]. Simulated ^{13}C spectra were sampled at a resolution of 1 (or 2 ppm) in the 106–160 ppm range (the domain of aromatic carbon shifts). Intensities were normalized: 100 if there is one peak in the bin, 200 for two peaks, etc. For the whole population of 52 compounds (12 PCBs + 14 PCDDs + 26 PCDFs), a PLS treatment led to $r^2 = 0.85$, $q^2 = 0.71$, considering the 12 more correlated bins. These results seem quite satisfactory compared to the EVA approach ($q^2 = 0.68$ but with only 3 PCs on a more extended data set of 77 chemicals) [44] as well to a quantum mechanics–based QSAR model using Mekenyan et al. parameters (HOMO, LUMO, log P, Lmax) $q^2 = 0.60$ ($r^2 = 0.72$) for 51 compounds [541].

^{13}C QSDARs have been also developed in a study of 50 steroids binding the aromatase enzyme on an activity range of 5.1 log units [406]. The simulated ^{13}C spectra were sampled at a 1 ppm resolution on a 0–256 ppm range. Two CoSA models were proposed from the PLS analysis: one selecting the better (more correlated) 5 bins ($r^2 = 0.82$, $q^2 = 0.77$), the other one considering the 87 occupied bins and selecting the five principal components ($r^2 = 0.78$, $q^2 = 0.71$). This choice of five components corresponds to the optimal value in a previous CoMFA study [405] for the sake of a comparison of the methods (CoMFA $r^2 = 0.94$, $q^2 = 0.72$). The fact that the correlation based on the best PC was of slightly inferior quality was interpreted by considering that the whole spectrum would include some noise (irrelevant information). Although the ^{13}C shift assignment is not used in CoSA, it is easy in this family, to identify the "important" bins, and the corresponding carbons positions. Shifts of positions—3, 9, 6, and 7 of the steroidal framework are roughly correlated to activity. If the role of positions 3 and 7 had been previously identified, that of position 9 was not quoted before.

^{13}C and mass spectra were also used similarly in classification problems for the estrogen receptor binding affinity of 108 chemicals [542], and for selectivity toward α or β ER subtypes for 30 estrogens [543].

6.2.2 COMPARATIVE STRUCTURALLY ASSIGNED SPECTRA ANALYSIS

For the same family, a slightly different procedure, comparative structurally assigned spectra analysis (CoSASA) used the assigned ^{13}C shifts at the 17 positions of the steroid framework (17 bins in which the corresponding intensity is the shift of each carbon). Here, the two models also considered either the five more correlated atoms or the PCs built from all carbons. The results were equivalent ($r^2 = 0.75$, $q^2 = 0.67$) [406]. These QSDAR treatments compared well to a previous CoMFA analysis [405]

($r^2 = 0.94$, $q^2 = 0.72$), and a classical QSAR–PLS analysis built from 256 descriptors, Cerius2 [56], including constitutional, topological, electrostatic, thermodynamic, and E-state descriptors. ($r^2 = 0.73$, $q^2 = 0.66$); all models comprising five latent variables (the optimal number in CoMFA). A BNN (87/29/1) was also used for CoSA analysis ($r^2 = 1.0$, $q^2 = 0.75$); but the very high r^2 value as well as the number of epochs (45,000) may suggest overtraining.

6.2.3 COMPARATIVE STRUCTURAL CONNECTIVITY SPECTRA ANALYSIS

In a further development, additional structural information was introduced with the indication of carbon–carbon connections. This was conveyed in comparative structural connectivity spectra analysis (CoSCoSA) by ideal (simulated) 2D-COSY spectra [407,544]. Roughly said, in a 2D-NMR-COSY map (see Figure 6.1b), the ^{13}C shifts are indicated on the two axes. For two carbons i and j, nondiagonal spots (δ_i, δ_j)—respectively (δ_j, δ_i)—appear if atoms i and j are directly bonded (expressing a through-bond $^1J(^{13}C-^{13}C)$ coupling mechanism between i and j. Diagonal spots correspond to the usual 1D spectrum. In CoSCoSA [544], structural descriptors are constituted by the $^{13}C-^{13}C$ COSY spots: the occupied bins in an array (δ_i, δ_j). They may be considered as characterizing one-bond C–C fragments augmented by the chemical shift values (an indirect information on their connectivity, hybridization, and environment) but without any peak assignment.

This method was applied, for example, on a large set of 130 chemicals binding the estrogen receptor (on 7.1 log RBA units) and including steroids, phytoestrogens, DESs, DDTs, PCBs, alkylphenols, and parabens. Experimental data from Refs. [545,546] corresponded to rat uterine cytosol ER competitive assays (a good standard for in vitro assays, well correlated with yeast-based reporter gene assay or MCF-7 cell proliferation assay and also with results on hER-α). This data set was previously analyzed by Shi et al. [188] who revisited a CoMFA/QSAR comparison performed on a more limited data set [547]. Two test sets (27 compounds each), came from the data of Kuiper et al. [548] and Waller et al. [173].

In CoSCoSA, nondiagonal spots were distributed into bins of 2 ppm in each dimension (a rather wide interval but which allowed many bins to be populated and which got rid of possible small errors in the spectra simulation). The map was symmetrical: so, on a 240 ppm range, 7381 bins had to be considered. In this application, 605 bins were occupied but only 337 corresponded to more than one spot and were retained for the analysis by forward multilinear regression. With 16 bins, the model yielded $r^2 = 0.827$ and $q^2 = 0.78$. To cope with distance influences (through-space interactions), the authors introduced a compacity indicator (1 if the maximum distance between two heavy atoms is less than 7.5 Å, 0 otherwise). This information was, relatively speaking (with a different distance threshold), somewhat comparable to that provided by NOESY spectra expressing short-range, through-space interactions. With 15 bins and this indicator, the analysis yielded $r^2 = 0.833$, $q^2 = 0.79$. However, nearly 50 compounds (mostly weak binders) in the training set did not have a hit in any of the 16 selected bins in the COSY map and, for them, a constant activity was predicted, whereas experimental values spanned a range of about 4 log RBA units. As to the external test sets of Waller [173] and Kuiper [548], for some compounds, the occupied COSY bins unfortunately did not fall in the original 605 bins retained for the training set, but in neighbor bins. Merging them with these 605 bins (with reduced intensity) was considered, but that did not really improve the performance. The best CoSCoSA model yielded $q_{pred}^2 = 0.74$ (after elimination of four compounds that presented one suspect ^{13}C shift). With Kuiper's test, $q_{pred}^2 = 0.53$.

These results were comparable to those obtained in the CoMFA analysis ($r^2 = 0.903$, $q^2 = 0.707$). In this study, alignments were based on the crystal structure of four ligands, representative of the chemical family. Standard CoMFA gave $r^2 = 0.908$, $q^2 = 0.665$. The electrostatic field played the major role (57%) consistent with results on smaller sets [359,547].

We already saw in Section 1.2.1.3 that, in this series, adding a phenol indictor increased the performance to $r^2 = 0.903$, $q^2 = 0.707$ (whereas "region focusing" [184], the "all orientation, all placement search" of Wang et al. [186], or the introduction of log P, did not significantly improve the results). The robustness of the model was established in extensive runs (100) of leave-N-out cross-validation. For the two external test sets, q_{pred}^2 reached 0.71 and 0.62 (with phenolic indicator). On the other hand, the fragment-based 2D-HQSAR treatment led to $q^2 = 0.585$ ($r^2 = 0.756$) in training but only to 0.15 and 0.22 in testing.

However, a parallel study [549] to identify significant descriptors associated with ER binding affinity was carried out on 131 compounds (same origin as the compounds studied by Shi et al. [188]). Starting from 151 descriptors (such as quantum mechanical, graph theoretical indicators, log P), a rigorous selection by stepwise regression had a predictive power $r_{pred}^2 = 0.627$ (in L-33%-O), a result comparable to that of CoMFA, 0.623 (in a leave-25%-out cross-validation) [188].

CoSA and CoSCoSA methods were also applied to the classical benchmark of steroids binding the corticosteroid binding globulin [550,551]. A first ^{13}C CoSA treatment led to a robust model with $r^2 = 0.80$ and $q^2 = 0.78$ (with three bins). Considering assigned chemical shifts in CoSASA did not really improve the results (0.80, 0.73, with three LVs). With the help of simulated 2D-COSY spectra (CoSCoSA approach), which provided information about one-bond connections, the performance was improved ($r^2 = 0.84$, $q^2 = 0.74$ with three PCs, and even reached 0.93, 0.88, if eight PCs are retained). Additional structural information might be gained from other components of the interatomic distance matrix (and not just only one-bond connections). Following the presentation of the authors, the distance matrix might be considered as a wad of 2D maps. The first one corresponds to the COSY map just considered (one-bond connection, or in other words, distances inferior to 2 Å), the other ones, to more distant atoms with off-diagonal spots indicating carbons separated by a distance between 2.0 and 3.6 Å (short range); then between 3.6 and 6.9 Å (medium range), and greater than 6.9 Å (long range). From these excerpts of the distance matrix, distance spectra were generated that may be used as a COSY map to develop a QSDAR model. Owing to the importance of the structural environment of carbons 3 and 17 in the steroid binding affinity, the study considered only long-range distance spectra.

The treatment was then similar to that of Beger and Wilkes [407] performing a forward MLR on the most correlated PCs from the COSY- and the distance-spectra.

Various options were examined: COSY or Distance only; COSY + Distance; or combining COSY and Distance before PC extraction. The best model (COSY + Distance) yielded $r^2 = 0.84$, $q^2 = 0.74$ with three PCs, but with eight PCs, better results were obtained (0.96, 0.92). Various cross-validation runs testified to the robustness of the model.

An artificial back propagation neural network was also used with a parallel distribution scheme (Parallel Distributed Artificial Neural Network [PD-ANN]) to optimize the neural network configuration. A BNN model with a (593 [number of bins]/198/1) architecture led to $r^2 = 0.96$, q^2 about 0.75. But the huge number of connections compared to the number of patterns (30 molecules) suggests that the system was strongly overfitted. These results compared well with other studies (the molecular similarity matrices [396], the HE-state/E-state fields [221], the electrotopological index [552], the SOMFA [403], and other 3D methods reviewed by Coats [126]).

However, two remarks can be made:

- This study concerned only 30 molecules (aldosterone was discarded because of its two conformations and not **s-31** as in the other approaches).
- Similar (or better) performance is reached compared with other methods, but with a higher number of principal components (eight in place of three or four). The authors argued that CoSA and related methods treat information of a "digital" form, which needs more components than the other approaches working in the "analogue" format [551].

For 50 steroids binding the aromatase enzyme, combining the COSY one-bond connectivity with the 2D long-range distance spectra [407], also significantly improved the CoSA and CoSASA models (provided that more PCs are retained) [406]. Compared to the previous CoMFA study [405], the r^2 values were comparable (CoMFA 0.94, CoSCoSA 0.92), but q^2 was slightly better (CoMFA 0.72 vs. CoSCoSA 0.86).

6.3   ELECTRONIC EIGENVALUES

The spectrometric QSAR technique EEVA [531], a modification of the EVA approach [51], uses molecular orbital energies in place of vibrational wave numbers with appropriate scale and bandwidth values (typically: scale from -45 to -10 eV, and 0.1 eV for σ, but larger values, $\sigma = 0.5$ eV were also used in some families [553]). EEVA was applied to model the binding affinity to the aryl hydrocarbon receptor AhR for a set of halogenated aromatic hydrocarbons [553], a data set also largely studied with very varied methods: see Ref. [328]. On 73 compounds (the same initial set as that of Waller and McKinney [423], but with five benzofurans) $r^2 = 0.912$, $q^2 = 0.818$ (eight PCs), SPRESS (in prediction) = 0.69, SEE = 0.48 [553]. These results were quite comparable to those of So and Karplus [49] from the similarity matrices. The EEVA model was then used to predict all possibly existent PCDDs and PCDFs (210 compounds in all).

The authors suggested that EEVA (alignment free) was well adapted to reflect electronic substituent effects, but a loss of efficiency existed for nonplanar structures where these effects were partially hindered (as in PCBs). They also stressed that although very efficient for prediction, this approach might "not be the best for understanding the mechanism of the electronic effect." Even if a clear explanation was lacking for the success of EEVA, the authors emphasized that MO energies (and corresponding squared orbital coefficients) were key parameters from which other physicochemical properties might be (in principle) calculated.

The effectiveness of EEVA was then validated using the steroid benchmark set [554]. The influence of the bandwidth (σ) on the performance was carefully investigated (from this example, 0.09 eV was the best, although the choice was not very critical around this value). The robustness of the method was checked with multiple runs on different randomized training sets, and with scrambling. At last, it was established that the method was not very sensitive to conformational changes (however, with the rather rigid steroidal framework, conformational freedom was mainly limited to lateral chains). With the usual training and test sets (respectively compounds s-1 through s-21 and s-22 through s-31), $r^2 = 0.97$, $q^2 = 0.84$, SEE = 0.24, SPRESS = 0.52, and in prediction $r^2_{pred} = 0.64$, SDEP = 0.58 (and respectively 0.85 and 0.40 without the outlier compound s-31). From an extensive comparison with a dozen of previous approaches, it appears that EEVA compared favorably with other QSAR models although the best results were obtained with Molecular Field Topology Analysis (MFTA) [555].

A comparative study was carried out by Asikainen et al. [347] on 36 estrogens previously examined [59,185,274], and the quality of the models was carefully checked by external validation on a large number of randomized test sets. σ parameters were set at 10 cm^{-1} (EVA), 0.075 eV (EEVA), 1.0 ppm (CoSA ^{13}C), and 0.05 ppm (CoSA ^{1}H). According to the authors, CoSA ^{13}C had good predictive ability ($q^2 = 0.69$), whereas EEVA ($q^2 = 0.42$) was a borderline case, and EVA ($q^2 = 0.22$) or CoSA ^{1}H with $q^2 = 0.35$ only gave semiquantitative information. But CoSA remained inferior to standard CoMFA ($q^2 = 0.796$) [185], and significantly inferior to the more refined CoMFA treatment of Sippl [59,274] with a receptor-based alignment and a smart region definition for variable selection ($q^2 = 0.921$). A SOMFA model was also proposed, giving, in a refined version using statistically inspired modification of the PLS method (SIMPLS) or multicomponent self-organizing regression (MCSOR), comparable results ($q^2 = 0.698$) [348].

In Ref. [556], for 30 estradiol derivatives (from Napolitano et al. [557]), EEVA outperformed EVA ($q^2 = 0.77$ vs. 0.68 for a mean over 500 randomized runs with one-third compounds in the test). These performances of EVA and EEVA were also compared to a classical Hansch-type relationship using Molecular Refraction (MR), the hydrophobic parameter π and an indicator (for the presence of a 16-α OH group) yielding $r^2 = 0.821$, but without any cross-validation [558], and to the Molecular Electronegativity Distance Vector (MEDV-4) approach of Sun et al. [559]. In MEDV-4, nonhydrogen atoms are split into four categories according to their connectivity. Ten descriptors are calculated corresponding to the different pairs i and j (with $0 < i, j \leq 4$)

$$M_{ij} = \sum \frac{q_i\, q_j}{d_{ij}^2}$$

where

q_i, q_j are the relative electronegativities

d_{ij} is the (topological) interatomic distance

Then, an MLR is performed. An optimized model (with only six terms) gives $r^2 = 0.852$, $q^2 = 0.747$. But no external validation was performed.

In the same paper, the activity of 28 coumarin (**6.1** through **6.3**) inhibitors for a cytochrome P450 mouse (CYP2A5) and a human (CYP2A6) were investigated (data from Poso et al. [560]) for activity ranges of about four units of log pIC_{50}). EVA and EEVA were compared to Hansch-type relationships with McGowan's characteristic volume (MgVol) [561], and the authors also examined models combining MgVol (and its square) plus the first principal components of EVA-PLS or EEVA-PLS. With 23 compounds in training and five in test, as in the work of Poso et al., the internal performance of modified-EVA and -EEVA (including MgVol parameters) were comparable to CoMFA or GRID/GOLPE (q^2 about 0.8) for the CYP2A5 data set, EVA being the poorer ($q^2 = 0.57$). All methods were nearly equally good in external prevision with r^2_{pred} about 0.9 (mean over 500 randomized runs, one-third compounds in the test). But one could note that a simple MLR with MgVol (and its square) was nearly as efficient. With the CYP2A6 for internal performance, CoMFA and GRID/GOLPE were the best, and EVA, EEVA, and the Hansch-type QSAR the weakest. As for external prediction, EEVA with MgVol ranked first, and the classical QSAR the worst (presumably because of a different sensitivity to MgVol parameters).

6.1   **6.2**   **6.3**

In conclusion, spectroscopic QSARs appear as a very appealing alternative to usual 3D-QSARs. One of their major advantages is, without doubt, that they are alignment free. As previously said, they use descriptors that are intrinsic characteristics of the molecules. They may be calculated "once and for all" to be used for various processes, without the trouble of selecting the "good" descriptors in the huge number now available from specialized packages. When spectroscopic data (such as ^{13}C shifts) are directly extracted from experimental records or from an empirical simulation system (built on a database of experimental spectra), they correspond, in usual conditions, to the conformational mixture, and their use in QSDAR avoids the need to do an energy-weighted conformational analysis, as may be required in 4D-QSARs.

However, QSDARs present some limitations. Chirality is not considered (in an inactive solvent enantiomers give the same spectra). Symmetry (reducing the number of signals) may also cause problems. From a practical point of view, recording experimental spectra that obviously require availability of the product is time consuming and may suffer from several constraints: solvent, temperature (in some cases), and shimming (in NMR) may modify the position or the shape of the signal. In view of general applications, spectroscopic QSARs (QSDARs) are therefore largely dependent on reliable spectra simulation systems, covering a wide structural range (as those already available in ^{13}C RMN or infra red [535,562–565]. This is particularly true for CoSCoSA due to the low natural abundance of the ^{13}C nucleus and its intrinsic low sensitivity, although ^{13}C–^{13}C COSY spectra can be partially reconstructed from more sensitive ^1H–^1H COSY and HETCOR ^{13}C–^1H sequences [566].

Interesting advances may be expected from further developments mimicking, from simulated spectra, other RMN pulse-sequences, such as Relayed COSY in ^1H 2D-NMR or 3D-, 4D-NMR involving ^{15}N, ^{13}C, ^1H shifts in isotope-edited spectra. See, for example, Ref. [461, pp. 103 and 425].

7 2.5D Descriptors and Related Approaches

Faced with a huge number of descriptions that allow the characterization of a molecular structure, it may be useful to classify the molecular descriptors according to the nature of the encoded information and their dimensionality. Remember that such descriptors correspond to scalar quantities (or small vectors) condensing the information (structural and/or chemical) associated with the molecular formula. A fruitful advance in the field was to consider the structural formula of a molecule as a graph, where vertices and edges correspond to atoms and bonds. This opened the way to the definition of a large number of indices, later extended to 3D structures. On the other hand, diverse encoding schemes were proposed to express the chemical information in a numerical form, which was to be the input in a statistical package.

7.1 USUAL STRUCTURAL DESCRIPTORS

Some descriptors are more currently used either because of the time elapsed since their introduction or due to their "intuitive" interpretative power.

7.1.1 Descriptor Dimensionality

0D descriptors correspond to molecular weight and atom counts (absolute or relative). 1D descriptors, based on local connectivity, include the count of cycles and of fragments of a predefined length (although a molecular graph is invoked for extracting the fragments). On the other hand, considering the 2D structural formula as a molecular graph leads to the generation of 2D descriptors often referred to as topological indices or "graph invariants" [34,36,567]. Such holistic descriptors condense in a number or in a brief numerical vector the chemical information contained in a structural formula. It will suffice to recall that representing the chemical objects (such as molecules, reactions) by graphs constituted a very fruitful concept leading to countless successful applications [568] in similarity/diversity analysis, database mining, combinatorial library design [569,570], virtual screening [571,572], and QSAR/QSPRs [36, pp. 307–360].

In an intermediate position between 2D and 3D models, 2.5D descriptors also correspond to holistic descriptors, but tend to incorporate some aspects of the geometrical information contained in a 3D structure that were ignored by a 2D description from the structural formula [435,489]. In other words, these 2.5D descriptors depend on 3D molecular geometry, but spatial positions of atoms are not explicitly considered. A large variety of these 2.5D descriptors have been proposed. Some constitute stand-alone sets, whereas others are often used in conjunction with "classical" descriptors and have been integrated in several commercial packages. Among their advantages, one can note their translation, rotation independence (TRI), which alleviates orientation or alignment problems.

Another classification of descriptors, which partly conceal the preceding one, relies on the nature of the conveyed information: constitutional (in fact mainly 0D and 1D descriptors), topological (which will be discussed later), geometrical, quantum-chemical (electronic), thermodynamic, or mechanistic (with substituent constants characterizing, for example, polar, steric, or hydrophobic effects). At last, partition coefficients log P or log K_{OW} correspond to physicochemical quantities (experimental or calculated) as well as spectroscopic data (^{1}H or ^{13}C NMR shifts, IR wave-numbers) that may also be measured or simulated.

7.1.2 GEOMETRICAL DESCRIPTORS

Geometrical information may be directly included in the structural description with quantities, such as principal moments of inertia, gyration ratio, volume, or shadow areas delineated by the contour of the projection of the molecule on the plane defined by two of its principal axes. An important aspect is also the account of quantities related to the molecular surface (which undoubtedly plays an important role for interactions with the surroundings). Molecular surfaces were considered long ago in molecular modeling with the van der Waals surface, the solvent accessible surface (a contour pushed out from the van der Waals surface by a distance equal to the radius of a rolling water probe) or the molecular surface of Connolly, (contact surface plus re-entrant surface). For details, see Ref. [461, pp. 239–265].

An innovative notion was that of *charged partial surface areas* proposed by Stanton and Jurs [573] to encode polar interactions. Such descriptors were initially used in quantitative structure–property relationship (QSPR) studies (boiling point, chromatographic retention, surface tension) before their introduction in QSARs. These parameters combined atomic contributions to the solvent accessible surface area (SASA) with partial-charge information derived from the empirical model of Abraham and Smith [574], which evaluates σ and π contributions with a parameterization reproducing experimental dipole moments. Twenty-five charged partial surface area (CPSA) parameters were proposed, including partial positive and negative surface areas, the associated quantities weighted by total charge or atomic charge and their differences, fractional charged partial surface areas and their surface-weighted counterparts, plus the relative charges and their corresponding charged surface areas.

Directly derived from the consideration of the molecular surface area, 11 supplementary descriptors were proposed to encode H-bonding features [575]. They include the count of H-donor or acceptor groups, the solvent accessible surface area of the acceptor groups and of the hydrogens that can be donated, as well as their percentage with respect to the total molecular surface area.

Conformational dependence of CPSA descriptors has been investigated by Stanton et al. [576]. CPSA weighted by partial atomic charges appear sensitive to conformational changes, making them interesting for use in dynamic (conformationally related) QSARs. CPSA have also been found capable of replacing lowest unoccupied molecular orbital (LUMO) energies for describing electrophilicity in case of noncovalent interactions. For 216 narcotic chemicals, a CPSA parameter associated with log K_{OW} in a response surface model, led to a good representation of toxicity ($r^2 = 0.892$). It was proposed [576] that CPSA may be considered as an *intrinsic electrophilicity* parameter, whereas LUMO characterizes a *potential electrophilicity* (ability to accept endogenous electrons). CPSA were also important parameters in the reactivity pattern separating agonists and antagonists of the estrogen receptor [577].

7.1.3   QUANTUM-CHEMICAL DESCRIPTORS

The molecular wave-function can, in principle, express all electronic properties and molecular interactions. These quantities are now available from semiempirical methods with good accuracy and limited computational efforts. Thus, they become an attractive source for conveying molecular information. Descriptors generated from quantum-chemical calculations are now largely used, alone or in addition to conventional descriptors, for building QSAR models. As stated by Karelson et al. [578], the approximations inherent to the calculation method, and the (usual) neglect of solvation or entropic terms (although they may be taken into account by some programs) are not too detrimental. Indeed, these terms may be considered as either constant or regularly evolving in a series of homogeneous compounds, so that it makes sense to use quantum-chemical descriptors to analyze structure-induced variations in biological activity (or physicochemical properties). The most frequently used quantum-chemical descriptors have been reviewed by Karelson et al. with extensive literature (Ref. [578]).

Atomic charges (although dependent on the calculation method and the partition scheme for electronic populations) are widely used, since they give an easy representation of electrostatic-type effects, with some variants, such as considering the maximum or minimum net atomic charges. Atomic charges, from a rigorous point of view, are in fact true 3D descriptors since they are localized, but as they are available with the same programs and at the same time as the other quantum parameters, we group them with 2.5D descriptors. Related to the charge distribution, the *dipole moment* encodes the global polarity of a molecule. *Orbital energies* (HOMO, LUMO, and their gap) as well as their localization (frontier electron densities) were frequently called for in chemical reactivity studies in relation to susceptibility to electrophilic or nucleophilic reagents. The derived concept of *superdelocalizability* corresponds to an evaluation of the contribution of an atom in charge-transfer processes. *Atom–atom polarizability* representing the effect on an atom of a charge variation on another atom, and *molecular polarizability* (response to perturbations due to an external field) are related to molecular volume and hydrophobicity. The total energy or the *heat of formation*, was also considered as a measure of global stability or of nonspecific interactions.

Such descriptors have a clear physical meaning and are rather closely related to common concepts about molecular interactions or chemical reactivity. However, more recently, several other quantities (less commonly interpretable) have been proposed in the framework of QSAR program packages, such as the atomic-orbital population, the partition of overlap populations into one-center and two-center terms, the free valence of atoms or the diverse interactions (attraction, repulsion) involving electrons and the nucleus of given atom species [579].

Some early applications of quantum-chemical descriptors were reviewed by Franke [580]. Numerous, more recent results can be found in Karelson et al. [578]. For example, a good linear correlation was proposed (in 1987) between anhydrase inhibition by heterocyclic sulfonamides and the CNDO/2 calculated total net charges of the $-SO_2NH_2$ group and its oxygen atoms More recently, toxicity of aromatic compounds to *Tetrahymena pyriformis* was correlated in a "response surface model" involving the octanol/water partition coefficient and the LUMO energy or the average acceptor superdelocalizability characterizing the electrophilicity [99,103,581,582].

At the Hartree–Fock level, the ab initio methods give a good representation of the electron-nucleus interactions in a molecule. However, the prohibitive calculation-time often limits their use (the more extended the basis set, the better the results). Fortunately semiempirical methods (at varied levels of simplification) have been proven to give satisfactory results. Actually, AM1 (Austin Model-1) [162] and PM3 (Parametric Model 3) [163] are the more frequently used in QSAR models on account of the acceptable computational time they need and the reliability of the results they give. The effect of the precision of Molecular Orbital (MO) descriptors was investigated by Seward et al. [583]. Although some variability appeared (more for HOMO than for LUMO), depending on the selected package and the Hamiltonian, it did not affect the significance of QSARs derived from these calculations.

On another hand, the empirical charges of Gasteiger et al. [158,159] are based on a Partial Equalization of Orbital Electronegativity (the PEOE method). The method was extended to conjugated systems with the concept of π orbital electronegativity. This approach deserves particular attention since, in many applications, it compared well with charges derived from semiempirical quantum methods, despite a great simplicity of calculation. Empirical charges are also included in the Merck force field MMFF94 [242]. This concept of electronegativity equalization also underlay several other calculation methods: see, for example, Abraham et al. [485,574] or Mullay [584,585].

At last, it has been also proposed to carry out quantum-mechanics calculations on fragments and then patch them together, the transferability being assumed, a solution adopted for polypeptides in the earlier releases of AMBER [586]. This idea can be found again [510] with fragment-based MS-SM (vide infra).

We now recall some of the basics about topological indices before going on to a generalization of "topographic" indices.

7.2 FROM TOPOLOGICAL TO TOPOGRAPHICAL INDICES

From the perspective of QSAR models, we indicate that a molecule (in 2D) may be represented by a planar graph where vertices represent atoms and edges bonds. Generally, only "heavy atoms" are considered ("hydrogen depleted graph"). At the so-called *topostructural* level, only the existence of atoms and links is taken into account, which obviously corresponds to some loss of information, whereas for the derivation of *topochemical* descriptors, additional chemical information, such as type of atoms and bonds is included. Then, one talks about a "colored graph." We will only give a few examples of the most widely used indices.

7.2.1 TOPOLOGICAL INDICES

These approaches opened the way to the introduction of *topological matrices* expressing the graph in a numerical form. From these matrices, mathematical manipulations define univocally calculated *topological indices*, simple numbers that encode the topological information embedded in the molecular structure. These indices have the advantage of being molecular invariants (independent, inter alia, on the numbering of the atoms, and of course of the molecular geometry). Their successes also prove that they convey significant molecular information.

According to Balaban [567] the topological indices (TI) can be classified into three classes or generations. TIs of the first generation (Wiener, Hosoya) were integer numbers based on integer local graph-vertex invariants (LOVI). Their high degeneracy limited their use. The second-generation TIs (Randić, Balaban, Kier-Hall) were real numbers obtained from the manipulation of local graph properties that are integers. Katritzky and Gordeeva [587] showed that for QSARs, these indices must be supplemented by additional parameters (hydrophobicity, log P) although they are often sufficient for QSPRs. The third generation corresponded to real numbers derived from graph invariants expressed by real numbers. Further types of indices originated from the information theory.

Two types of topological matrices took a preeminent role: the adjacency (connectivity) matrix and the distance matrix (Figure 7.1).

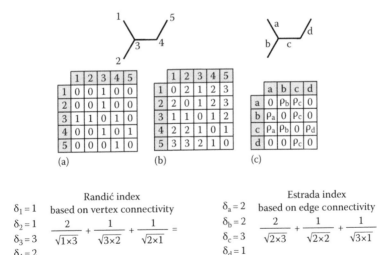

FIGURE 7.1 Adjacency (a) and distance (b) vertex matrices and adjacency (c) bond matrix. Randić index is calculated from the atom connectivities, which correspond to the sum of the rows. The Estrada index is similar from the bond connectivities. Here, the calculation is carried out on an unweighted matrix ($\rho \equiv 1$).

In the *adjacency* matrix, often symbolized by [**T**], off-diagonal elements (i,j) are set to 1 if atoms i and j are directly connected, 0 otherwise. Diagonal elements are null, but they can be replaced by some information, such as an atomic number.

In the *distance* matrix [**D**], off-diagonal elements encode the number of bonds (topological distance) in the shortest path to the atoms i,j. Diagonal elements are zero.

But molecular graphs may be weighted on vertices or edges giving rise to a large variety of topological indices [588].

Topological indices that mainly derive from these matrices can be traced back to the pioneering works of Wiener, Hosoya, and Randić. Wiener [35] introduced (in 1947) a description of the molecules by path-counts from the distance matrix in the molecular graph. The Wiener path number, i.e., the total number of bonds among all the atom pairs in the graph, was applied to characterize paraffin properties. The Hosoya index [589], proposed in 1971, encodes the number of ways to choose k bonds in the graph so that two of them are not connected. On the other hand, Randić (1975) used the vertices of the adjacency matrix to characterize connectivity and molecular branching [590]. This approach was taken up by Kier et al. [591] with the connectivity index, later generalized to higher degree indices. Randić also considered various path numbers [592,593].

In line with the development of these approaches, some desirable requirements that a newly proposed index should meet have been stressed by Randić [594]. According to Gallegos and Gironés [595,596], about 400 different indices have been proposed, most of them derived from the distance or adjacency matrices (Table 7.1). Extensive reviews of the principal indices are provided in the books of Devillers and Balaban [36] or Todeschini and Consoni [37].

Anyway, ignoring the actual 3D structure may be an important limitation, hence the introduction of topographical indices that took into account the molecular geometry [567]. As to these 2.5D descriptors, often referred to as "3D-"TIs, some are directly derived from the corresponding 2D quantities by just introducing weights in the matrices currently used to generate 2D topological indices or by using true Euclidian distances. Others result from rather different approaches.

7.2.2 TOPOGRAPHICAL INDICES

The improvement that might be expected by the introduction of geometrical information into topological indices aroused interest as to the definition of topographic indices. Spatial characteristics were first introduced by Randić who proposed the embedding of molecules in a honeycomb or a diamond lattice to get some distance information [597–599], and by Bogdanov et al. [600], who used real distances to compute a 3D Wiener index.

With the definition of topographic indices of centricity and centricomplexity, Dudea et al. [601] were able to reproduce physicochemical properties, such as the van der Waals areas and toxicity of 25 aliphatic ethers. Randić's definition of molecular shape or bonding profiles can also be referred to [602,603]. Other indices were later proposed on the model of 2D indices, where the topological distances (count of bonds) on the nondiagonal elements of the distance matrix were replaced by the actual interatomic Euclidian distances. For example, the 3D Wiener [600]

TABLE 7.1
Some Currently Used Topological Indices

Wiener path number

$$\text{WPN} = \sum_{i=1,n-1} \sum_{j=i+1,n} D_{ij}$$

D_{ij} elements of distance matrix

Wiener index

$W = \text{WPN} + p_3$

p_3 polarity number: number of paths of length 3

Platt index

$$F = 2N_2 = \sum_{i,j=1,n} (v_i + v_j) - 2$$

v: vertex degree

Hosoya index

$$Z^A = \sum_{i=0,n/2} p^A(k)^A$$

p^A (i) number of ways in which k bonds are chosen so that no two of them are adjacent

$p^A (0) = 1; p^A (1) =$ number of bonds

Randić index

$$\chi = \sum_{i=1,n-1} \sum_{j=i+1,n} \frac{T_{ij}}{\left[\left(v_i * v_j \right)_k \right]^{0.5}}$$

Generalized connectivity $\ ^m\chi_t^Z = \sum_{i=1,nt} \Pi_{i=1,m+1} \left[\left(v_i \right)_Z \right]^{-0.5}$

$\kappa = \Sigma_{\text{all edges}} (v_i v_j \ldots v_m)^{-0.5}$

order m (number vertices), type t (path, cluster, chain, etc.), nt number of connected subgraphs of type t

Balaban index

$$B = \left(\frac{n_e}{\mu+1} \right) \sum_{i=1,n-1} \sum_{j=i+1,n} \left[(D_i)(D_j) \right]^{-0.5}$$

D_i sum of topological distances from vertex i to all other vertices, n_e total number of bonds, μ number of cycles

Schultz index

$$\text{MTI} = \sum_{i=1,n} \left[v(T+D) \right]_i I$$

Sum of elements E_i in the row matrix $v(T + D)$ where v is the vertex degree

Harary number

$$H = \sum_{i=1,n-1} \sum_{j=i+1,n} D_{ij}^{-2}$$

D^{-2} is the squared inverse distance matrix

Zagreb indices

$$M_1 = \sum_{i=1,n} v_i^2$$

$$M_2 = \sum_{\text{all edges}} v_i * v_j$$

(continued)

TABLE 7.1 (continued)
Some Currently Used Topological Indices

Largest eigenvalue $\quad x_i = \max [\text{Eigenvalue (T)}]$

Xu index $\qquad\qquad Xu = n^{0.5} \log \dfrac{\sum_i v_i (D_i)^2}{\sum_i v_i (D_i)}$

n number of atoms, B, i^{th} row of matrix B, μ number of cycles.

n_e number of edges in the graph. $(D)_i$ sum of distances from vertex i to all other vertices, n_t number of connected subgraphs of type t.

Source: Adapted from Gallegos Saliner, A., Molecular quantum similarity in QSAR; Applications in computer-aided molecular design, PhD thesis, Girona University, Girona, Spain, 2004.

path numbers were defined with the exclusion or not of hydrogen atoms. Similarly, the 3D-Schultz index or the 3D-Harary number has been proposed. For reviews, see Refs. [567, pp. 1–24] and [604].

However, according to Basak et al. [605] these 3D indices were not largely used. For example, for the mutagenicity of 95 aromatic and heteroaromatic amines, ${}^{3D}W$ and ${}^{3D}W_H$ only slightly improved the results of the usual 2D-TIs. For 107 benzamidines, inhibitors of the complement system, a good correlation was obtained between the activity and ${}^{3D}W$ as the only descriptor, but a nearly as good relationship was also observed with degree complexity (I^D) alone [606]. In the same paper, ${}^{3D}W_H$ indices were also invoked in multilinear relationships (four or six parameters) established for the acute toxicity of 69 benzene derivatives to *fathead minnow*.

7.2.3 INDICES: AVAILABILITY AND USE

From the first free-energy relationships (Hammett, 1937 [19]), and the proposal of the first topological index (Wiener, 1947 [35]), we have witnessed the introduction of a flood of descriptors (currently more than 3000), in particular, topological indices (aimed at characterizing various aspects of the molecular structure: shape, branching, cyclization), or (characterizing electronic features) quantum-chemical quantities. These descriptors are now easily available (nearly "on the fly") with powerful packages, such as (among others) CODESSA [53,578], DRAGON [37,54], Cerius -2 [56], and ADAPT [57]. Some appropriate tools for descriptor selection and correlation programs are also included.

However, this deluge is not without problems. Multiplying the indices has led to intricate schemes. It is now not uncommon for authors to only specify the acronym of the descriptors they use and refer the reader to a user manual! The physical meaning is easily lost. Faced with a huge number of descriptors, an automated variable selection is an essential step, often needed to replace the intuitive guess of the good descriptors, which was common at the beginnings. The collinearity of descriptors, within a (limited) data set, may obscure the analysis. Preferring one descriptor to another, on account of a minimal benefit on statistical criteria, may hide the real mechanism, a criticism that is partially relieved in some approaches, such as those developed by Estrada et al. [607] or Carbó-Dorca et al. [510] where the role of molecular fragments can be identified (vide infra). It is quite out of the question to detail all the structural descriptors now available (well over thousands!). We have just mentioned those more commonly used in 2.5D-QSAR models, as a direct extension of the traditional models and prefer to put more stress on specific 3D preprocessing modules to encode 3D information and condense it into 2.5D descriptors.

Let us also remark that whatever the descriptors may be (1D, 2D, 2.5D), the same statistical tools may be used, taking into account the fact that the descriptors are generally quite numerous (much more than the number of experimental data) and thus variable selection becomes an inevitable step to address. Some details about the mathematical or statistical methods will be given in the appendix, joined to some examples that are not in fact formally too different from "classical" QSARs. It would suffice to say that in the beginnings, linear methods were nearly exclusively used. A major advance seems to be the introduction of nonlinear methods, particularly Neural Networks (and their different architectures) or Evolutionary Algorithms and Genetic Algorithms (GAs).

These methods mainly intervene in two important steps in the elaboration of QSAR models: in variable selection and in building up the relationship (descriptor/activity). For some applications of neural networks and genetic algorithms see the reviews of Livingstone and Manallack [89] and Niculescu [90]. Genetic algorithms and evolutionary algorithms have been used, for instance, by So and Karplus [487,488] for variable selection and have been coupled with a back-propagation neural network for data treatment. One advantage of GAs is that they generally lead not to a unique solution but rather to an ensemble of models of similar quality, which may be interpreted as different visions of the same phenomenon.

An example may be found in the comparative study of a set of 56 steroids known to bind the progesterone receptor (PR) in vitro, on a 2.8 log units in relative binding affinity (RBA). This data set was the subject of a comparative study involving varied approaches [608–611]. The 52 structural descriptors encoded 43 characteristics of substituents (or individual atoms): charge, volume, surface, plus nine whole-molecule properties, such as the dipole moment, HOMO or LUMO energies, and the heat of formation (from AM1 calculations). In a preliminary study, comparative molecular field analysis (CoMFA) outperformed a 10-10-1 back propagation neural network (that was clearly overfitted) [608]. In subsequent publications, diverse methods for descriptor selection were compared on the same data set. Combining genetic functional approximation (GFA) regression for descriptor selection and artificial neural network (ANN) for model building, led to better results ($r_{train}^2 = 0.64$ on 43 compounds, $r_{test}^2 = 0.49$ for 11 compounds). Using the 10 most frequently used descriptors in a 10/5/1 ANN yielded slightly improved results (respectively 0.880 and 0.570) [609]. Comparison was pursued [610] on the same data set, but with only eight descriptors to maintain a 6/1 ratio between the number of patterns and descriptors. A subjective selection from known models, such as forward stepping regression and GFA, where a genetic algorithm selected the descriptors input in a nonlinear regression (with inter alia splines) were examined. In two other models, data were

mapped from a neural network and descriptors selected either by simulating annealing in generalized simulated annealing (GSA), or by GA in the genetic neural
network (GNN). It was then established [610] that a 8-2-1 BNN combined to GSA
or GA gave the best results, $q^2 = 0.635, 0.717$, and $r^2 = 0.860, 0.880$, respectively.
The more limited performance of GFA may result from its difficulty to treat nonlinearity. However, GFA ($q^2 = 0.626, r^2 = 0.722$) or stepwise regression ($q^2 = 0.535$,
$r^2 = 0.721$) may be useful in preliminary studies since it is easier to implement. On
another hand, genetic algorithms gave several models of comparable quality making the interpretation less easy. It may be noticed that in this study, GSA and GFA
selected five common descriptors, and that half of the descriptors retained concerned
the "key" positions 11, 13, 17 of the steroid scaffold. Niculescu and Kaiser [611]
compared these results to the performance of a probabilistic neural network (PNN)
with the same descriptors. In training PNN ($r^2 = 0.863$) was comparable to GSA
($r^2 = 0.860$) and slightly lower than GNN ($r^2 = 0.880$) but performed a little better in
the test (0.769 vs. 0.488 and 0.610 on 10 compounds).

Other nonlinear regression models, in 2.5D treatments (such as Support Vector
Regression, Projection Pursuit Regression, Radial Basis Functions Neural Network)
have been used by Ren [103], Panaye et al. [99], Yao et al. [97], Sun et al. [612], and
Tang et al. [613].

In a different approach but in a formalism close to the classical indices, Estrada
et al. [614,615], Lobato et al. [616], and Gallegos et al. [595] introduced quantum-
chemical quantities to encode some electronic features. On account of the treatments
involved, these applications lie a little apart from the most widely used descriptors
incorporated in "classical QSARs" and will be developed in a separate paragraph.

7.2.4   3D Quantum-Connectivity Indices

In 1993 [614], Estrada and Montero proposed an adjacency matrix based on graphs
weighted with bond order or valency indices between bonded atoms in order to
account for heteroatom differentiation. By introducing the charge density as a diagonal entry in the topographic adjacency matrix, they defined a new topographic index
Ω (weighted vertex connectivity index).

$$\Omega(q) = \sum_{\text{edges } i,j} \left[\delta_i(q) * \delta_j(q) \right]^{-0.5}$$

With $\delta_i(q) = q_i - h_i$; electron charge density connectivity = electron charge on atom i,
corrected by charge carried by attached hydrogens. Ω correlated well with the physical properties of alkanes and was used in a study of the chromatographic retention
of alkenes [615].

A new concept was also introduced: the *edge-adjacency matrix* (**E**-matrix) proposed by Estrada [617,618]. Distance or adjacency matrices usually refer to graph vertices (atom positions). In edge-adjacency matrices, rows and columns refer to bonds
rather than atoms. Non-diagonal elements are set to 0 or 1 depending if the corresponding bonds are adjacent or not. This led to the definition of an edge-connectivity

index independent of the other TI [619]. The edge adjacency matrix was later extended to a 3D **E**-matrix [620], where nondiagonal elements were weighted by the corresponding bond orders ("valence indices"), calculated, for example, at the PM3 level, which implies that the matrix is no longer symmetrical (Figure 7.1).

The edge degrees $\delta(e_i)$ are defined as the sum of the bond orders of the adjacent bonds (sum along a row i), and the bond order weighted edge connectivity index $\varepsilon(\rho)$ is calculated according to Randić's expression:

$$\varepsilon(\rho) = \sum_s \left[\delta(e_i) * \delta(e_j) \right]_s^{-0.5}$$

summation over all adjacent bonds

The good correlation obtained with the molar refractivity of 69 alkanes suggested that these edge connectivity indices can convey information on the bulkiness and the polarizability of molecules. This edge connectivity index appears better than the topographic Ω index of Estrada and Montero [614] that may be considered as a topographic Randić's branching index, and is also better than the ε index derived from the edge topological matrix (**E**-matrix) of Estrada [617]. The methodology was applied to a study of urinary excretion of unchanged drugs (Percent Excreted Unchanged, PEU) for amphetamine derivatives. Interestingly, according to experimental conditions (influence of pH on ionization), the best results were obtained with a topographic index (when electronic effects intervene) or a topological one (if they are absent). Higher degree indices were also proposed [568].

In a parallel development, Estrada [607,620,621] proposed the use of the *spectral moments* of the edge-weighted adjacency matrix, with diagonal terms weighted by bond distances (extracted from a table of standard values). The kth spectral moment is defined by the trace of the corresponding matrix.

$$\mu_k = tr\ (\mathbf{E}^k)$$

An interesting point is that such spectral moments can be evaluated by algebraic expressions in terms of the structural fragments present (Figure 7.2).

For example, for the first moments

$$\mu_0^d = |F_1| = m$$

$$\mu_1^d = \sum_i d_i$$

$$\mu_0^d = 2|F_2| + \sum_i (d_i)^2$$

$$\mu_3^d = 6|F_3| + \sum_i (d_i)^3 + 3 \sum_i F_2^i d_i$$

where

    d_i is the distance corresponding to edge i

    F_k is the number of fragments of type k in the graph, with F_k^i among them containing edge i

    summation over all adjacent edges in the graph

For the sake of simplification, the figure corresponds to a nonweighted matrix (null diagonal elements) so that the d_i terms vanish.

    These various moments may be used to derive, with Multiple Linear Regression (MLR), a QSAR model. The approach was validated on an estimation of various physicochemical properties of alkanes [621] and later extended to the boiling point of 58 alkyl halides, and the antifungal activity of benzyl alcohol derivatives (although with this example, a good correlation was also obtained with the simpler Hansch Fujita method [28]).

    In Vilar et al. [622], the TOPological Substructural MOlecular DEsign (TOPS-MODE) based on the spectral moments of the edge adjacency matrix was used to screen 206 nucleosides (7.1) for their anti-HIV1 activity. In addition to good predictive models (built with bond dipole moments as the weighting scheme), the approach allowed the specification of the contribution of identified structural fragments, which may be a useful tool for the design of new, possibly active, chemicals.

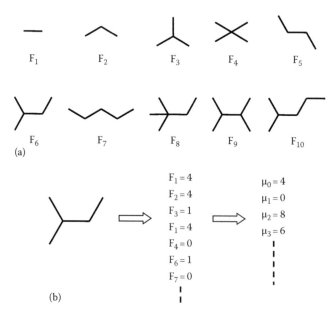

FIGURE 7.2   (a) Some structural fragments F_k of simple molecular graphs and (b) the calculation of spectral moments in the hypothesis of an unweighted matrix (diagonal elements = 0). (Adapted from Estrada, E. and Ramírez, A., *J. Chem. Inf.*, 36, 837, 1996; Estrada, E., *J. Chem. Inf. Comput. Sci.*, 37, 320, 1997.)

7.1

Estrada and Molina [568] extended the definition of topographic indices and introduced quantum-mechanical properties (bond order or partial charge) in the calculation of vertex or edge molecular connectivity indices of different types and orders. A comparison of the performance of 2D, 3D, indices and quantum-chemical parameters–based QSARs was performed by Estrada and Molina [568] on 34 2-furylethylenes derivatives (**7.2**) (antimicrobial, antitumoral, and cytotoxic activity). The models were built from 41 topological descriptors, 55 topographic, and 51 derived from quantum-chemical calculations (such as HOMO, LUMO, and superdelocalizability indices). For both calculations of log P and the classification of biologically actives (or not), topographic indices outperformed the other two methods. Similarly, these quantum connectivity descriptors outperformed the classical QSAR in a study of the solubility of environmentally important organic compounds [623].

7.2

On the other hand, an edge distance-based topological index was proposed by Estrada and Gutman [624] in analogy with the graph invariant of Schultz [625]. For a review, see also Ref. [626].

7.2.5 3D "Topological" Quantum-Similarity Indices

These indices are based on an extension of the molecular quantum similarity measure introduced in Chapter 5. According to our scheme, such indices rely on topography, but we have retained the original name from the authors.

7.2.5.1 Definition of TQSI

Lobato et al. [616] introduced the Topological Quantum Similarity Index (TQSI) taking into account the distance matrix but using, rather than the information of bonds, the scaled interatomic quantum similarity values. Let us note that this approach is different from the Quantum Topological Molecular Similarity (QTMS) method proposed by Popelier et al. [627], in the framework of the theory of quantum-chemical topology, i.e., "Atoms in Molecules" [628] (vide infra).

In the classical topological distance matrix, integer off-diagonal elements indicate the shortest topological path between atoms, and the valence vector (the sum of row or column entries) specifies the connectivity. In the TQSI approach, the integer elements of the topological matrix are replaced by the interatomic Coulomb quantum similarity measure, calculated within a molecule, between each pair of atoms (diagonal elements are set to zero). Atomic densities are scaled by the square root of the atom connectivity [595]. The valence vector is similarly derived from the entries of this matrix. The topological distances are substituted by the true 3D Euclidian distances.

These indices now involve pairs of atoms instead of full molecules. Atomic electron densities are represented by Atomic Shell Approximation (ASA), which expresses the density function as a sum of atomic density contributions (taken as a linear combination of $1s$ Gaussian functions fitted to the ab initio 3-21G basis set). In TQSI applications, the alignment phase (e.g., by the Quantum Similarity Superposition Algorithm [500] or the Topo Geometrical Superposition Algorithm [502]) is not necessary, since the similarity matrix only concerns the atoms within a unique molecule. However, energy minimization is preferable to get consistent results.

So, any topological index definition that is expressed as contributions of connectivity, such as the Randić or Zagreb indices, can be translated from the classical- to the QS-formulation by including the interatomic QSM and the 3D distances, which ever is appropriate [595]. Whereas classical indices only depend on connectivity, TQSI are affected by the geometrical dependence of both interatomic distances and quantum similarity indices (QSI). Conformation, tautomerism, stereoisomerism, (except for enantiomers) differentiate TQSI. In various applications, models were built using MLR after generating all combinations of descriptors and selecting the model with the highest q^2 for each dimension.

In the inhibitory activity of benzenesulfonamides against Carbonic Anhydrase II (therapeutic agents against glaucoma and neurological alterations) for 29 compounds, q^2 reached 0.86 with three descriptors, but 0.89 with four ($r^2 = 0.922$). These results might be compared with the 2D topological correlation obtained by Thakur et al. [629] on the same data set: $q^2 = 0.974$ with four parameters. See also [342]. In the same paper, the heat of formation of 60 hydrocarbons was very precisely evaluated ($r^2 = 0.990$).

Another example [10] concerned the inhibitory growth concentration of 30 aliphatic alcohols and amines expressed as $-\log (IGC_{50})$ on a range of about 5 log units, for *Tetrahymena pyriformis*. With four descriptors from 13 calculated indices, $r^2 = 0.887$ and $q^2 = 0.859$. The study was pursued on 48 anilines with similar results. On a range of 2.9 log units $r^2 = 0.823$ and $q^2 = 0.790$ with seven descriptors (the greater number of descriptors being presumably required on account of the larger size of the set). In a previous study, Schultz et al. published slightly better results ($r^2 = 0.952$ or 0.872) in a correlation of $-\log (IGC_{50})$ with $\log K_{ow}$, the log of the 1-octanol/water partition coefficient. But no validation was performed [630].

This study was extended to 92 aromatics [595]. With four descriptors, r^2 yielded 0.631 and $q^2 = 0.553$, a result somewhat inferior to those of Roy and Gosh [631] who obtained for r^2 and q^2, respectively, 0.885 and 0.865 with extended topochemical atom (ETA) indices and 0.738, 0.718 with non-ETA descriptors. This rather modest performance of quantum indices was attributed to the difficulty of modeling the various interaction mechanisms intervening in toxicity studies.

The TQSI methodology was also tested on two other examples [10]. The first application concerned the dermal penetration of 60 polycyclic aromatic hydrocarbons bearing three to seven fused rings. Activity is expressed as the percentage of the applied dose penetrating the skin 1 day after application of the dosing solution, and varies in the range 0.7%–50% [632]. After calculation of the indices, Partial Least Squares (PLSs) regression was performed and the best model (the optimal prediction ability, taking into account the complexity of the model) led to $r^2 = 0.694$ and $q^2 = 0.652$ with five LV derived from 13 indices. These results were marginally better than those previously obtained [633] on the same set by the Molecular Quantum Similarity Measures (MQSM) theory, leading to $r^2 = 0.684$ and $q^2 = 0.634$ with three PCs. These qualified results may be attributed, according to the authors, to a crude description of observed activities (expressed as integers, which masked small variations). For example, 16 molecules with a same experimental activity (20%) were attributed-predicted affinities from 10% to 40%.

7.2.5.2 MQSM and TQSI Approaches of Antimalarial Activity

Antimalarial activity was the subject of an extensive investigation of various chemical families carried out in several papers, using either the MQSM formalism or the TQSI approach. Malaria is an endemic disease that affects hundreds of million people, mainly in tropical and subtropical areas. Due to the increasing immunity of protozoan *P falciparum* clones to traditional therapies, there is an urgent need for new drugs against these pervasive strains.

QSM approach: The first study (at QSM level) [634] concerned two series (20 and 7 compounds) of synthetic 1,2,4-trioxanes (**7.3**) with different responses of the parasite *P. falciparum*. Two strains were scrutinized: W2 from Indochina and D6 from Sierra Leone with activities expressed as the concentration (ng/mL) of the drug able to inhibit 90% of synthesis and reduction of hydrofolate in the parasite in vitro (IC_{90}). In addition, seven other compounds were tested in vivo on *Plasmodium berghei* (activity expressed with ED_{90} values in mg/kg). An interesting point was that geometrical optimization was performed from the SYBYL MM force field [125] after verification that this method gave results similar to more sophisticated semi-empirical AM1 or ab initio HF/3-21G$^+$ methods, but at a much lower cost. The density function was calculated by Promolecular ASA and the overlap operator used for the QSM calculation (after alignment by TGSA). After data reduction and selection of the variables by PCA, and the "most predictive variable" method, the best model was chosen in L-O-O. Results were also validated by a randomization test confirming that the proposed models represented real QSARs and not fortuitous correlations (Table 7.2).

7.3

TABLE 7.2

Statistical Results for Antimalarial Activity

Strain	r^2	q^2	σ	nb PC	Range (log units)
W2	0.757	0.589	0.55	4	3.4
D6	0.789	0.662	0.44	4	3.0
P. berghei[a]	0.929	0.708	1.23	3	6.8 to 22.5

Source: Gironés, X. et al., *J. Comput.-Aided Mol. Des.*, 15, 1053, 2001. With permission.

[a] Correlation on ED_{90} and not on log!

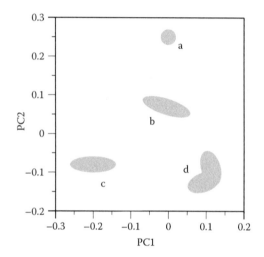

FIGURE 7.3   Plot of the first vs. the second PC for the molecular set of 20 1,2,4-trioxanes. (a) and (c) clusters presented low activity, (b) and (d) clusters corresponded to high activity.

On a qualitative point of view, a plot of the molecular set in the plane of the first two principal components (PC1 vs. PC2), highlighted the existence of four clusters (Figure 7.3). In particular, compounds devoid of aromatic substitution (b) or possessing phenyl groups on both sides of the fusion bond (d) exhibited high activity. On the other hand, molecules with a phenyl ring fused to the main backbone (a), or molecules with an aromatic substitution and a phenyl group in the region where the two main rings fused (c), presented low activity.

In a second QSM study [635], inhibition of the metabolism to hydrofolate by 20 cyclic peroxy ketals (**7.4**) was investigated. The reported IC_{50} (in nM) span was about 3 log units. Using the Coulomb operator for the MQSM, the best model leads to $r^2 = 0.778$, $q^2 = 0.691$, and $\sigma = 0.16$ with four PCs. Gironés et al. [634] investigated the inhibition of synthesis and the reduction of hydrofolate (expressed as log IC_{50}) in the NF54 strain, with 18 artemisinin (**7.5**) derivatives. A second series encompassed 15 other artemisinin derivatives studied in vitro over the strains W2 (Indochina) and D6 (Sierra Leone). These two clones presented different sensitivity to chloroquine and mefloquine (the W2 strain being chloroquine-resistant and mefloquine-sensitive, whereas the reverse was true for the D6 strain). For these three strains, good models were established with $r^2 > 0.75$ and $q^2 > 0.52$ (with four PCs). It is interesting to see that the selected PCs were not those for maximal variance. Furthermore, they only explained less than 40% of the whole variance. Nevertheless, they provided an acceptable description. In this application, the Kinetic Energy Operator was preferred for the MQSM calculation.

7.4   **7.5**

Remark: Kinetic energy operator [634]:

The expectation value of the kinetic energy operator $\langle K \rangle$ can be written as

$$2\langle K \rangle = -\int \psi * \nabla^2 \psi \, dr = \int (\nabla \psi)^*(\nabla \psi) dr = \int \kappa(\mathbf{r}) \, dr$$

with

$$\kappa(\mathbf{r}) = \sum_i w_i \left| \nabla \varphi_i(\mathbf{r}) \right|^2$$

φ_i and w_i corresponding respectively to the MOs of the system and their occupation number

Z_{AB} provides a measure of resemblance between two molecules based on their kinetic energy density function.

$$Z_{AB} = \int \kappa_A(\mathbf{r}) \kappa_B(\mathbf{r}) \, dr$$

A graphical comparison of the electronic and kinetic energy density functions respectively, on the example of artemisinin may be found in Ref. [634].

TQSI approach: This study was pursued using the topological quantum similarity indices. Atomic densities were described using a simple set of $1s$-GTO basis-functions. In addition to the classical topological matrices (distances and adjacency), matrices resulting from the Atomic Quantum Similarity Measure with Coulomb and Ciolowski [636] operators extended the classical formulation for the common indices. Forty-six indices were calculated, such as the Wiener path number, and the Wiener-, Randić-, Schultz-, Balaban-, Hosoya-indices (distances were the true 3D Euclidian distances). In fact using classical indices and topological quantum similarity indices with both Coulomb and Ciolowski operators amounted to 46 indices*3 operators = 138 descriptors per molecule. All combinations of two, three, and four descriptors were generated, and after elimination of highly intercorrelated descriptors, the best models were looked for using either MLR, PCA, or PLS.

Two data sets of respectively 15 derivatives of 3-alkyl substituted artemisinins (**7.5**) and 17 analogues of 10-deoxoartemisinin were investigated. For the two series, r^2 is about 0.84 and q^2 about 0.73, with three descriptors. Four descriptors would lead to a better performance but might correspond to overfitting on account of the limited number of samples. This result contrasted with an earlier paper by Avery et al. [637] who did not obtain a good QSAR from topological and connectivity/shape indices. Gironés et al. [635] concluded that their combination of several kinds of TQSI activated a synergic effect leading to acceptable linear models.

Other studies concerned 21 β-methoxyacrylates (**7.6**), 17 3-methyl-10-(substituted-phenyl) flavins (**7.7**), and 27 phenothiazine derivatives (**7.8**) [638]. Molecular quantum similarity measures (MQSM) have also been successfully used in classification problems, as for example, in estrogen activity [639] or carcinogenic activity of polycyclic aromatic hydrocarbons (PAH) [633].

7.6   **7.7**

7.8

7.3 3D FRAGMENTS

3D fragment-based descriptors have been frequently used as alignment independent descriptors in drug research. Structural keys from a predefined fragment dictionary are mainly dedicated to structural search or classification tasks, whereas fingerprints are currently used in correlation. A molecular fingerprint is a bit string encoding the presence or the absence of specific fragments, as for example, Hologram QSAR treatment [125]. The hashing function allows the breaking of the fingerprint into smaller parts which is easier to handle but which also has some drawbacks (difficulty of interpretation, collision).

7.3.1 FINGERPRINTING ALGORITHM

A new approach, the Fingerprinting Algorithm (FINGAL) program, proposed by Brown et al. [640], provides geometrical fingerprints encapsulating 3D geometric information. The drawback is that (as for 2D fingerprints) reversing mapping to the original structure is impossible. In the generation of 2D fingerprints, all atom-bond paths between a minimum and a maximum bond path-length are considered. In a 3D encoding, for each pair of atoms in the molecular graph, the Euclidian distance is calculated. During the enumeration of a path in the graph, the distances of the newly added atoms to all of the previous atoms in the path are added. These distances are classified into variable bin ranges allowing sufficient but not too much discriminatory ability: for example, in this study, the bins were 2.3–2.71 Å and 9.32–12.44 Å. The individual paths are then hashed into the fingerprints characterizing the geometric fragment.

These 3D fingerprints may be applied to substructure screening, conformation sampling, or similarity search. In the field of 3D-QSARs, several series were investigated.

- 38 dopaminergic D2 receptor antagonists partitioned into training (32 structures) and test (6), previously examined by Böstrom et al. [332]
- 58 estrogen receptor agonists from Waller et al. [641]

For D2 antagonists, Fingal and Dragon models (2D or 3D) competed well with CoMFA or CoMSIA, Dragon 2D yielding the best performance in prediction. This suggested that approximate geometries (from a model builder or semiempirical methods) did not convey, in this example, useful additional information.

With ER ligands, the FINGAL model performed well on the entire data set. As in Waller's work, to test the predictive activity, a chemical class was in turn extracted from the data set and used in the test. Results showed a rather important variability depending on the structural series, and the rationalization was somewhat difficult. This variability prompted the authors to repeat the treatments on five partitions: training (40)/test (18). Models performed well, but still, a significant degree of variability was observed between the runs.

As a conclusion, it was stressed that FINGAL yielded competitive models. It was also observed that 3D fingerprints did not significantly improve the corresponding 2D models. The study also confirmed (once more) that for reduced populations, results were strongly dependent on the partition (train/test).

7.3.2 TRIPLETS OF PHARMACOPHORIC POINTS DESCRIPTORS

Sciabola et al. [642] examined the applicability in QSARs of the recently defined Triplets Of Pharmacophoric Points (TOPPs) descriptors and compared their performance to that of various other 2D or 3D descriptors. The investigated dataset encompassed 80 apoptosis-inducing 4-aryl-4H-chromenes. Apoptosis is the normal process of cellular suicide. It enables organisms to control their cell number and to eliminate unneeded cells. Disorders in this process underlie various pathological problems.

The TOPP descriptors rely on the 3D molecular structure and the GRID force field parameterization, but spatial information remains limited to distance, so that these descriptors may be considered as 2.5D in agreement with the speed of their calculation. In TOPP descriptors, the atoms are classified into four types (as in GRID): DRY (hydrophobic), HBD (hydrogen bond donor), HBA (hydrogen bond acceptor), and

HBD/A (hydrogen bond donor/acceptor). Given a 3D molecular structure, all 3-point fragments are extracted. The descriptor denotes the atom type and their distances.

In 3D, TOPP descriptors have been compared to the GRid INdependent Descriptors (GRIND) (specifying geometrical relationship between pairs of nodes of relevant molecular interaction fields [MIF]) and to the GRID descriptors (on the aligned molecules). These descriptors represent the sum on the grid nodes, of the interaction potentials (van der Waals, H-bonding, electrostatic plus an entropic term) for various probes (DRY, methyl, carbonyl oxygen, N1 amide, OH phenolic). After the elimination of uninformative variables, about 2900 GRID descriptors survived. About 1500 DRAGON descriptors were also considered. As to 2D descriptors, the study included Minimum Description Length (MDL) keys (166 for small topological substructure fragments, atoms, rings or more complex patterns), in tandem with 960 keys for atom pairs algorithmically generated [643] and then selected by FREDs (fast random elimination of descriptors), an evolutionary algorithm [644,645]. 4016 BCI fingerprints (2D-features in a predefined list) [646], ECFPs (extended connectivity fingerprints), and FCFPs (functional connectivity fingerprints) [647] were also considered.

In the correlation with chromene activity, the best results are obtained with TOPPs and GRIND, whereas GRID, DRAGON, and (in 2D) MDL are largely outperformed. However, the authors indicated that these results might depend on the substructure composition of the data set. They also stressed the advantage of a consensus analysis. Another application of the TOPP descriptors was the identification of pharmacophoric elements favorable or detrimental to activity.

7.4 ENCODING QUANTUM-CHEMICAL 3D INFORMATION

We will now examine tools more removed from the preceding 2D- or 3D-descriptor generations and which may constitute stand-alone modules. We first begin with two models directly derived from quantum mechanical calculations.

7.4.1 QUANTUM TOPOLOGICAL MOLECULAR SIMILARITY

The QTMS method proposed by Popelier et al. [627,648], involved the theory of quantum chemical topology "Atoms in Molecules" [628] in which an atom system is partitioned into atomic constituents on the basis of electron density (ρ). The method uses a new type of quantum descriptor calculated at the bond critical points (BCPs), which are points where the gradient of the electron density vanishes between two bounded atoms ($\nabla\rho = 0$). For each bond, four descriptors (that will be further indicated) are calculated and submitted to a PLS treatment, with a prior GA analysis to select relevant BCP properties submitted to PLS.

This approach rests on (rather) high level ab initio wave functions, but examples exist where the optimized bond-lengths (at the semiempirical AM1 level) yield efficient QSARs. As indicated by the authors, like classical QSARs, the method addresses congeneric series of compounds (a common core with variable substituents) and essentially captures the electronic effects. But the descriptors generated may be used in conjunction with substituent constants describing steric or hydrophobic effects in classical PLS correlations.

We now give a brief description of the BCP parameters, adapted from the original paper [627]. In this treatment, four functions are evaluated at the BCP appearing between two bonded atoms.

- Electron density
- Ellipticity (null for a cylindrically symmetrical bond)
- Laplacian of the electron density
- Kinetic energy density

Some examples also involve the equilibrium bond length.
The components of the Hessian matrix of the electron density (ρ) are

$$\partial^2\rho/(\partial u_i \partial u_j), \quad \text{where } u_i, u_j \text{ are the coordinates x, y, or z.}$$

Among the three eigenvalues of the Hessian, one (λ_3) is positive and measures the curvature along the bond. The other two are negative and characterize the curvature perpendicular to the bond. Ellipticity is defined by

$$(\lambda_1/\lambda_2) - 1$$

It was shown that the bond order is related to these functions at BCP via a relationship

$$a + b\rho + c\lambda_3 + d\,(\lambda_1 + \lambda_2)$$

where a, b, c, and d are fitted parameters. $(\lambda_1 + \lambda_2)$ measures the degree of the π character, and ρ (at BCP) and λ_3, the σ character. The Laplacian $\nabla^2\rho$ is

$$\nabla^2\rho = \lambda_1 + \lambda_2 + \lambda_3 \text{(a negative value indicates a locally concentrated charge)}$$

The last terms are the kinetic energy density terms K(r) and G(r), already seen.

$$K(r) = (-1/2)\,N \int d\tau'(\overset{*}{\psi}\,\nabla^2\psi)$$

and

$$G(r) = (1/2)\,N \int d\tau'(\nabla\psi)*(\nabla\psi)$$

(the integral running over the spin coordinates of all N electrons minus one), a quantity not readily interpretable in chemical terms.

This leads, for each bond, to an eight-dimensional vector of property components evaluated at the BCP. The equilibrium bond length (r_e) may be added. This set (for a common ensemble of bonds in the investigated molecules) constitutes the structural descriptors to be used. Others may be added, such as atomic charges and dipole moments. After energy minimization, these quantities are evaluated at the (HF/6-31G*//HF/6-31G*) level, or for larger molecules (HF/3-21G*//HF/3-21G*).

Besides a correlation model, PLS analysis also provides the variable importance in the projection (VIP), i.e., the importance of the individual independent variables in explaining the activity. If the highest VIP values correspond to neighboring bonds, this may indicate the active part of the ligand, although "contamination" may occur (vide infra). The method was also presented as less resource demanding than the MQSM of Carbó-Dorca et al. [471,509].

Several data sets were analyzed, corresponding to series where electronic effects are largely predominant on steric effects or hydrophobicity. For example, ionization of carboxylic acids was approached in Ref. [648]. Data sets, more closely related to biological activity, were scrutinized in Ref. [627]. Generally speaking, the models competed favorably with those previously developed, including similarity matrices, CoMFA, or classical QSARs. Furthermore, VIP may give new insights into the involved mechanisms.

The first series concerns the acidity constants of imidazolines (**5.13**) (compounds neighbor to the antihypertensive drug clonidine) and imidazole derivatives (**5.14**) (imidazole is a constituent of histidine, an important catalyst of biological processes). Good correlations are derived from the BCP analysis (Table 7.3). The models developed outperformed those of Good et al. for the similarity matrices, that of Kim and Martin with CoMFA [236] and Silverman and Platt with CoMMA [489]. If the VIPs confirmed the role in protonation of the exocyclic nitrogen of imidazolines, curiously the protonable nitrogen did not appear in the VIP of imidazoles.

The inverse agonist activity of indole derivatives (**5.29**), measured by the displacement of [^{3}H] flunitrazepan from the binding to the bovine brain membrane, was also investigated. These compounds, that also bind the benzodiazepine receptor, may possess potential interest as sedatives and anticonvulsants. The VIPs emphasized the role of the phenyl ring, which may be compared to the hypothesis implying an initial recognition of the phenyl part by a receptor [649]. Inhibition constants of the influenza virus by alkylated benzimidazoles were also studied. The treatment outperformed a previous E-states-derived model [650].

In the following series, where the electronic effects were not largely predominant, the CBP descriptors were used in conjunction with classical substituent parameters. Amides may inhibit the action of alcohol dehydrogenase that catalyzes the oxidation-reduction processes between alcohols and aldehydes, a process studied in connection with alcohol intoxication problems. Adding log P to the BCP descriptors significantly improved the correlation performance compared to the LFER (σ^* and log P) proposed by Hansch et al. [651]. From the sign of the σ^* coefficient in that LFER, Hansch concluded that the carbonyl is the important moiety in binding to the enzyme, whereas the VIPs in the QTMS treatment attached more importance to the adjacent C–N bond.

Although the QTMS approach is clearly dedicated to congeneric series, it may be possible to limit the analysis to only a common fragment from series with different common cores. Inhibitors of carbonic anhydrase (that catalyzes the hydration of CO_2 to CO_3H_2), such as sulfonamides, deserve much interest as diuretics or for the treatment of glaucoma. For two data sets of such sulfonamides, previously investigated by Kakeya et al. [652], good results were obtained with the analysis limited to the common –SO_2NH_2 fragment. Inclusion of the lipophilicity parameter π marginally improved the results.

TABLE 7.3
Comparaison between Bong Critical Point Treatment and Some Other QSAR Models

Ref	Method	N	r^2	q^2	LV
PK imidazolines (3.5)					
[627]	BCP	23	0.995	0.993	2
[47]	SM			0.96	8
[236]	CoMFA	28	0.99		
PK imidazoles (9.3)					
[627]	BCP	15	0.989	0.956	3
[489]	CoMMA		0.788(a)	0.698	
[47]	SM		0.910	0.77	2
[49]	GNN/SM			0.92	3
[236]	CoMFA	16	0.995		
IC_{50} Indoles (2.3)					
[627]	BCP	23	0.821	0.623	3
		20	0.855	0.743	
[509]	Carbó index	20	0.761	0.631	
[655]	MQS-SM	23	0.603	0.507	
[472]	Carbó SM	23	0.855	0.776	3
[507]	$\sigma + 2$ indic.	20	0.81		
Inhibition constants benzimidazoles (3.8)					
[627]	BCP	15	0.913	0.797	4
[650]	E-states		0.91		
Inhibition constants amides (2.0)					
[627]	BCP	17	0.761	0.596	2
	BCP+log P		0.923	0.734	3
[651]	$\sigma + \log P$	15	0.738		
Inhibition constants sulphonamides (2.5)					
[627]	BCP	19	0.910	0.891	2
	BCP + π		0.943	0.914	
[652]			0.887		
EC_{50} Chlorophenols (2.5)					
[627]	BCP	14	0.928	0.840	2
	BCP+log P		0.958	0.905	
[653]	log P alone		0.884	0.875	
	$\Sigma\sigma$ alone		0.829	0.765	

For each series, the activity range (log units) is indicated between brackets. (a) Gasteiger's charges; SM = similarity matrix

In the toxicity of chlorophenols, Argese et al. [653] showed that the most important factors were log P and $\Sigma\sigma$ (sum of Hammett constants), which, taken separately, yield good correlations with the observed EC_{50} (for log P, $r^2 = 0.884$ and $q^2 = 0.875$, whereas for $\Sigma\sigma$, $r^2 = 0.829$ and $q^2 = 0.765$). Including log P with the other BCP parameters selected by GA leads to a better model ($r^2 = 0.958$, $q^2 = 0.905$).

Finally [654], BCP descriptors were used in a PLS treatment for correlating the pIC_{50} values of 13 PCDDs, binding the AhR receptor. At the HF/3-21G(d)//HF/3-21G(d) level, bond critical points descriptors led to $q^2 = 0.57$ ($r^2 = 0.74$). But surprisingly, using only bond lengths (calculated at AM1 level), q^2 reached 0.89 ($r^2 = 0.84$). Localizing "active centers" also highlighted the importance of halogen substitution on lateral positions. However, it was mentioned in Ref. [627], that generally speaking, the AM1 level is not able to generate accurate BCP descriptors (due to the absence of atomic cores in the molecular electron density), even if, in some cases (unfortunately not all), bond lengths only are sufficient for good correlations.

These examples clearly evidence that the QTMS method accurately captures the electronic effects but not hydrophobicity or steric effects. However, QTMS may be efficiently combined to the corresponding descriptors (such as log P, Es), if necessary. They so appeared as a low-cost alternative to the Carbó-index since no molecular superposition was involved. The influence of conformational changes was not addressed, but it was indicated that the variations on the BCP induced by changing a substituent are much more important than those originating from conformational changes.

More recently [656], Roy and Popelier applied the QTMS descriptors in a study of the toxicity of nitroaromatics to *Saccharomyces cerevisiae*. For this data set, the usual response surface model (with log K_{OW} and LUMO energy as the electrophilicity index) gave acceptable results in training but poorer predictive ability in the test. Adding QTMS descriptors largely improved the results.

7.4.2   COMPARATIVE MOLECULAR MOMENT ANALYSIS

Comparative molecular moment analysis (CoMMA) [489,657], like other methods, aims to correlate variations of activity in terms of variations of shape and charge distribution. But its originality is to define as structural descriptors, molecular moments of mass and charge distribution, calculated in a multipole expansion up to the second order. So, no alignment is required to establish QSARs.

Different from several other 3D-QSAR approaches (such as CoMFA, CoMSIA, COMPASS), CoMMA does not involve a detailed local description of steric or electrostatic features, but works on a more global description. However, the information about charge, shape, and quantities characterizing their relationship is more or less implicitly present. Shape is characterized by the distribution of atomic masses. The moment of mass at the zeroth order, is the total molecular mass. The first-order moment is null if the origin of the reference frame is chosen at the center of mass (it is the definition of the center of mass). The second-order moments are the principal moments of inertia expressed along the three principal axes of inertia. Parallel expressions may be given for the charge distribution. We will give some more details, since things are perhaps less familiar.

Given a distribution of charges $\rho(\mathbf{x}')$, we wish to calculate the potential at position $\mathbf{r}$, external to the distribution (bold letters correspond to vectors):

$$V(r) = \int \frac{d^3x\ \rho(\mathbf{x})}{|\mathbf{r} - \mathbf{x}|}$$

In a Taylor expansion, around the origin, $\mathbf{x} = 0$

$$V(r) = q_T \frac{1}{r} + \frac{1}{r^3} \int d^3x \rho(\mathbf{x}) \mathbf{x} \mathbf{r} + \frac{1}{6r^5} \left[\int d^3x \left(3\mathbf{x}\mathbf{x} - |\mathbf{x}^2|\mathbf{I} \right) \rho(\mathbf{x}) \right] \left(3\mathbf{r}\mathbf{r} - |\mathbf{r}^2|\mathbf{I} \right)$$

where $\mathbf{I}$ is the identity matrix. In Cartesian coordinates, for discrete charges

$$V(r) = q_T \frac{1}{|\mathbf{r}|} + \frac{1}{|\mathbf{r}|^3} \sum_{\alpha=x,y,z} P_\alpha r_\alpha + \frac{1}{6|\mathbf{r}|^5} \sum_{\alpha,\beta=x,y,z} Q_{\alpha\beta} \left(3r_\alpha r_\beta - \delta_{\alpha\beta}|\mathbf{r}|^2 \right) + \cdots$$

with
$q_T = \Sigma q_i$
$P_\alpha = \Sigma_{i=1,N}\ q_i x_{i\alpha}$
$Q_{\alpha\beta} = \Sigma_{i=1,N}\ q_i\ [(3x_{i\alpha}\ x_{i\beta} - \delta_{\alpha\beta}|x_i|^2)]$
$\alpha, \beta = x, y, z$
Sums running from 1 to N

The zeroth-order moment is the total charge. The first moment is the dipole moment, the third moment (quadrupolar contribution) defines a three-dimensional (traceless) tensor that (written in diagonal form) gives the principal quadrupolar components plus the three principal quadrupolar axes (defined with respect to the molecular structure). Things look alike for mass contributions. But the trouble is that (for a neutral molecule) only the dipole moment is invariant for translations. The components of the quadrupolar tensor depend on the origin of the multipolar expansion. The ability to calculate the second-order moments of charge distribution (for neutral molecules) depends on the ability to define an appropriate center of electrostatic multipolar expansion. The trick is that one can find a "magic center" (the "center-of-dipole") so that in a multipolar expansion starting from that point, the so-calculated dipolar potential most closely approximates the actual far-field potential (at positions far from the charge distribution).

How is the center-of-dipole to be found? This comes by minimizing

$$\frac{1}{\Omega} \int d\Omega(\mathbf{x}) |V(\mathbf{x}) - V(\mathbf{x}, \mathbf{x}_0)|^2$$

$V(\mathbf{x})$ is the actual potential at point $\mathbf{x}$ and $V(\mathbf{x}, \mathbf{x}_0)$ the calculated potential at $\mathbf{x}$ with the center of expansion in $\mathbf{x}_0$, in a solid angle average (Ω) on a sphere at a given distance of x_0. For a dipole moment **p** and a quadrupolar tensor Q, calculated at an arbitrary origin, the displacement to the center of a dipole is given by

$$\mathbf{x}_0 = \frac{2}{3\mathbf{p}^2}\left[Q\mathbf{p} - \left\{ \frac{\mathbf{p} \cdot Q \cdot \mathbf{p}}{4\mathbf{p}^2} \right\}\mathbf{p} \right]$$

When the moments are calculated at the center-of-dipole, in the trirectangular trihedron of the principal quadrupolar components, one component is null and the dipole moment is oriented along this axis. On the other two axes, the principal quadrupolar components are equal in magnitude and opposite in sign.

A molecule is so defined by 13 descriptors, among them, the three moments of inertia for the mass distribution, and for charge distribution, the magnitude of the dipole moment, and the magnitude of the principal quadrupolar moment. Other descriptors involve the relationship between shape and charge, i.e., the components of the dipole moment along the principal inertial axes, plus the displacements between the center of mass and the center-of-dipole. Two additional descriptors are the diagonal components of the (traceless) tensor (calculated at the magic center-of-dipole in the inertial-axe referential). This description of the molecular structure only implies a limited number of parameters and is independent of rotations or translations (TRI description). However, although no preliminary alignment is needed, the descriptors depend on the molecular conformation chosen.

So, after energy minimization, the center-of-dipole is calculated from an arbitrary Cartesian frame, and the principal quadrupolar moments and axes are obtained. Then a PLS treatment is carried out with the 13 descriptors. (Molecular weight may be also added amounting to 14 descriptors [657].) A calculation of these descriptors is strongly dependent on the charge distribution, and in the absence of experimental measurements, it was important to examine (as a first criterion) the consistency of the results with respect to the method used for the charge calculation. Charges were therefore calculated by either the Gasteiger–Marsili method (G-M) [158,159], or AM1 (Mulliken analysis), and ab initio STO-3G* and 6–31G*. For the studied examples, STO charges are generally better but not always.

Five data sets were investigated, which we will examine in sequence.

- First, for the 21 steroids used in training in the classical 31 steroids benchmark, $q^2 = 0.674$ with two components for the Gasteiger–Marsili charges. But for TBG affinity, q^2 was only 0.412. However, reorienting the hydroxyl group of estriol (an outlier) led to 0.715 (CBG) and 0.574 (TBG). AM1, and ab initio charges gave better results ($r^2 = 0.828$ in ab initio for the CBG affinity of the 21 steroids).
- The second set corresponded to 15 imidazoles (**5.14**) (pKa measurements), one compound from the data set in Refs. [47,236] being eliminated since ionized. A model with two components gave $q^2 = 0.788$ (G-M charges) and 0.698 (STO-3G*) on a range of 9.3 pK units. The Hammett σ constants [47,237] were also investigated leading to $q^2 = 0.398$ (G-M charges), but in STO-3G*, $q^2 = 0.690$, for a range of variations of 1.44 σ units.

- For 37 β-carbolines, pyridodiindoles, and CGS derivatives (**5.2** through **5.6**), with affinity for the benzodiazepine receptor inverse agonist site [480], results were rather mediocre (q^2 between 0.493 and 0.394) probably (according to the authors) because the structural variations in the series mainly concerned the nature and position of alkyl groups, with little influence on the electronic properties.
- The last study concerned the inhibitory activity 33 derivatives of the TIBO series (**4.3**) (non-nucleoside HIV-1 reverse transcriptase inhibitors) on a range of 4 log units for IC_{50} [658] q^2 varied from 0.468 to 0.358 (STO-6-31G*). But, since determinations of the molecular structure were lacking, a common scaffold and a unique conformation were supposed, although the seven-membered ring might suffer conformational flexibility.

Silverman et al. [657] also reported an interesting counter example on Phosphodiesterase PDE type III inhibitors, as for example, milrinone (**7.9**) (30 compounds on a 3 log unit activity range). q^2 reached 0.58 with only the mass contribution, whereas considering the electrostatic terms led to a q^2 nearly null ("no model"). Such observations were considered as consistent with the fact that, in that series, steric effects were likely to be important, and that the main interactions might intervene in a region distant from the polar cyclic amide group.

7.9

7.5 ENCODING FROM ATOM POSITIONS

7.5.1 MoRSE AND RDF CODES

The codes developed by Gasteiger et al. (MoRSE [564] and RDF [563]) meet the requirement from various statistical or pattern recognition tools (as for example, neural networks) to propose descriptors of fixed length, a length that can be fixed independently of the size of the molecule. Furthermore, they offer the advantage to be translation–rotation invariant (and so avoid any alignment phase).

The 3D-MoRSE code (molecular representation of structures based on electron diffraction) [564,565] is derived from an adaptation of the molecular transform intervening in electron diffraction. The intensity of the scattered radiation in direction s by an ensemble of N spherical scatters (atoms) located at points r_i is

$$I(s) = \sum_{j=1,N} f_j \exp(2\pi i r_j s)$$

where r and s are vectors. The f_j form factor takes into account the directional dependence of scattering from a spherical body. The reciprocal distance s is a function of both the scattering angle θ and the wavelength λ

$$s = \frac{4\pi\left[\sin(\theta/2)\right]}{\lambda}$$

These formulae were adapted to generate molecular descriptors. On replacing the form factor by an atomic property p on atoms i and j

$$I(s) = \sum_{i=2,N} \sum_{j=1,i-1} p_i p_j \left[\frac{\sin(sr_{ij})}{sr_{ij}}\right]$$

$I(s)$ is calculated for a fixed number of discrete values of the reciprocal distance s (e.g., 32 values evenly distributed in the range 0–31.0 Å^{-1}). These values constitute the 3D-MoRSE code of the molecule. Depending on the application, the property may be, for example, the atomic mass, volume, electronegativity, partial charge, e-donor or -acceptor capability.

The radial distribution function (RDF) code [563] is closely related to the MoRSE code, and is based on the radial distribution function intervening in X-ray diffraction. This function may be viewed as the probability distribution to find an atom in a spherical volume of radius r, or in other words, the probability to find atom pairs (i,j) with atom j in a "peel" at the mean distance r from atom i (Figure 7.4). For an atomic property p, the RDF code is

$$g(r) = f\left\{\sum_{i=2,N}\sum_{j=1,i-1} p_i p_j \exp\left[-B(r-r_{ij})^2\right]\right\}$$

where
   f is the scaling factor
   N is the number of atoms in the molecule
   p_i, p_j are the property values on atoms i and j, distant to r_{ij}
   B is a smoothing factor

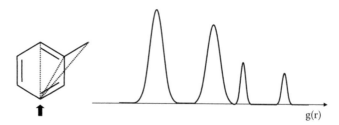

FIGURE 7.4   Contribution of the specified atom in the RBF code.

Similar to the MoRSE code, the chosen property depends on the application, and g(r) is calculated at a number of discrete points within defined intervals.

As from diffraction spots that allow the retrieval of the 3D structure of a molecule, it would be possible to reconstruct the 3D formula from MoRSE or RDF codes.

Applications: The MoRSE code was used for the classification of 31 steroids binding the CBG receptor [564] and for the simulation of IR spectra [535,564]. More recently, the MoRSE and RDF codes were used by Saiz-Urra et al. [659] in a study of benzophenazines (**7.10**) acting as topoisomerase inhibitors. DNA topoisomerases are involved in various cellular processes (replication, transcription, recombination, and chromosome segregation) and the design of new inhibitors is of great interest to overcome multidrug resistance in anticancer treatment.

For the cytotoxicity of 64 compounds (expressed by IC_{50} values), the best model ($r^2 = 0.725$, $q^2 = 0.619$) was obtained with 9 MoRSE descriptors weighted by mass (3 terms), volume (2 terms), polarizability (2 terms), and Sanderson electronegativity (2 terms). RDF codes (with 12 descriptors) gave similar results ($r^2 = 0.700$, $q^2 = 0.538$). These results were compared to analyses using varied descriptors collected in the DRAGON package [37,54] (topological, geometrical, BCUT, 2D autocorrelation, Galvez topological charge index, Randić molecular profiles, and WHIM). Relevant descriptors were selected by a genetic algorithm before MLR analysis. Except for topological descriptors ($q^2 = 0.527$), all the other methods led to q^2 being inferior to 0.5, indicating relatively poor predictive ability. The collinearity of descriptors may result in instable regression coefficients and make interpretation difficult. Orthogonalization (to get rid of these problems) and the elimination of outliers finally yielded a model involving seven MoRSE terms with $r^2 = 0.822$ and $q^2 = 0.761$.

MoRSE codes were also used by Gupta et al. [660] in a study of oxazolyl thiosemicarbazone analogues (**7.11**) in a search for new antibiotics for tuberculosis (TB), an important concern given the persistent pandemic of TB and the damages due to the synergy between HIV and TB. Seventeen compounds with inhibitory concentration MIC on a range of 3 log units were investigated. Correlations with good predictive ability were obtained combining a MoRSE descriptor with an empirical aromatic index (ARR) function of the percentage of aromatic bonds in the total number of bonds in the H-depleted molecule.

7.10   **7.11**

In a QSAR study of the agonist effects of adenosine analogues (**7.12**) on the A1 receptor subtype [661], the descriptors generated with the MoRSE and RBF codes were compared to BCUT, GETAWAY (GEometry, Topology and Atom Weight AssemblY) (vide infra), and 2D autocorrelation descriptors also available in DRAGON [37,54]. Such molecules have potential therapeutic interest as analgesics, antiepileptics, and neuroprotective agents. The data set was constituted by 32 compounds, the affinity of which was measured by the displacement of specific [^{3}H] R-PIA binding in rat-brain membrane, on a range of 3.4 log units.

7.12

From an initial pool of 470 descriptor, after elimination of constant or strongly intercorrelated descriptors, a genetic algorithm led to a 6-descriptor model with $r^2 = 0.797$ and $q^2 = 0.699$. After orthogonalization, a satisfactory model was obtained with only five descriptors ($r^2 = 0.784$, $q^2 = 0.677$). The RBF descriptors outperformed the others descriptors, although the results obtained from GETAWAY are fairly close. The descriptors retained in the model are associated to mass, volume, and polarizabilities. This point was judged in agreement with experimental observations that small polar substituents (such as C=O, Cl, I) in the 2-position are tolerated and may increase activity and that substitution on N, with bulky residues enhances activity, high atomic volumes increasing the hydrophobic character and permeability through biological membranes.

7.5.2   WEIGHTED HOLISTIC INVARIANT MOLECULAR DESCRIPTORS

The weighted holistic invariant molecular (WHIM) strategy condenses molecular information about size, shape, symmetry, and atomic distribution in a few global descriptors [662,663] invariant to translation–rotation. This compactness makes statistical analysis faster, but at the expense of easy interpretability. Due to the holistic nature of the description, it is impossible to trace backward from the WHIM descriptors to the original molecular features. Furthermore, no graphical display of local influences on the molecular surface, or in its neighborhood, is available as in field analysis approaches.

7.5.2.1 Calculation of WHIM Descriptors

In the first implementation, WHIM [662,663] referred to atomic positions. The approach was then modified with MS-WHIM [664–666] where the starting information is provided by points scattered on the molecular surface. This would allow the consideration of contributions arising from molecular surface recognition in specific ligand-receptor interactions. In WHIM, after geometry optimization, descriptors are calculated from the atomic coordinates (x,y,z) within different weighting schemes (i.e., different property types) according to the following scheme:

- Atomic coordinates are gathered in a matrix $\mathbf{M}(n*3)$ and the weights (the atomic property considered) in a diagonal matrix $\mathbf{N}(n*n)$, n being the number of atoms in the molecule at hand. A principal component analysis is carried out on the centered data by using, for each weighting scheme, a weighted covariance matrix.

$$s_{jk} = \frac{\sum_{i=1,n} w_i * (q_{ij} - \bar{q}_j) * (q_{ik} - \bar{q}_k)}{\sum_{i=1,n} w_i}$$

where
w_i is the weight of the ith atom
q_{ij} represents the jth coordinate (1, 2, or 3) of the ith atom
$\bar{q}_j$ is the average of the jth coordinates

This leads, for the molecule and for each weighting scheme ($\mathbf{N}$), to a score matrix $\mathbf{T}(n*3)$ indicating the coordinates of the atoms in the three principal axes. According to the kind of weights selected, different sets of descriptors are obtained. If the weight for each atom is its atomic mass, the principal axes are the *principal axes of inertia*, and the elements of $\mathbf{T}$ are the coordinates of the atoms on these axes. Other weighting schemes correspond to

- Unitary value for each atom. The information is only geometrical, without distinguishing the atom type.
- van der Waals volume.
- Mulliken atomic electronegativity.

Subsequently, weighting schemes also included atomic polarizability and electrotopological indices of Kier and Hall (with an offset to get only positive values), and so amounted to six properties [667]. For each weighting scheme (property), all weights are scaled on the carbon atom weight (only "heavy atoms" are considered). Then, the WHIM descriptors are calculated along the corresponding three principal axes.

- WHIM *size* property is defined by the variance of the atom positions along the (new) axes: the principal components. These are the three eigenvalues

$$\lambda_m = \frac{\sum_i w_i t_{im}^2}{\sum_i w_i}$$

where i denotes the atom and m, the axis.

- WHIM *shape* property is defined by the variance proportions along each principal component

$$\theta_m = \frac{\lambda_m}{\sum_m \lambda_m}$$

Due to the closure condition, only two of these are independent, and rather than θ_3, the acentric factor $\omega = \theta_1 - \theta_3$ may be used.

- WHIM *symmetry* property is defined by the skewness along each principal axis (absolute value)

$$\gamma_m = \left| \frac{\sum_i (w_i t_{im}^3)}{\left(\sum_i w_i\right)} \right| \lambda_m^{3/2}$$

Skewness represents the molecular asymmetry along each component. A different expression (from an information content index) was proposed in [663].

$$\gamma_m' = -\left\{ \frac{n_s}{n} \log_2\left(\frac{n_s}{n}\right) + n_a \left(\frac{1}{n}\right) \log_2\left(\frac{1}{n}\right) \right\}$$

$$\text{and} \quad \gamma_m = \frac{1}{1 + \gamma_m'} \qquad 0 < \gamma_m < 1$$

with n_s, as the number of central symmetric atoms along the mth component, n_a, the number of nonsymmetric atoms, and n the total number of atoms in the molecule.

- WHIM *distribution* property is defined by the kurtosis along each axis

$$\kappa_m = \frac{\left(\sum_i w_i t_{im}^4\right)}{\left(\sum_i w_i\right)} \lambda_m^{-2} \qquad \text{for } m = 1, 2, 3$$

The reciprocal $\eta_m = 1/\kappa_m$ is the emptiness and represents the surface property density.

For each weighting scheme, 12 descriptors are computed corresponding to the values along the 3 principal axes: λ, θ (in fact, θ_3 or ω can be eliminated), γ, and η. The data matrix formed with these descriptors is then submitted to PLS analysis, with autoscaled variables (mean = 0, SD = 1). A column (property) may be discarded in case of poor representation, a high intercorrelation with another property, or if its removal considerably improves the results. Alternately, ordinary least squares regression may be applied after variable selection by a GA [668]. In a further publication [669], nondirectional WHIM descriptors, related only to a global view of the molecule, have been added.

$$T = \lambda_1 + \lambda_2 + \lambda_3$$

$$A = \lambda_1 \lambda_2 + \lambda_1 \lambda_3 + \lambda_2 \lambda_3$$

$$V = \prod_{m=1,3} (1 + \lambda_m) - 1 = T + A + \lambda_1 \lambda_2 \lambda_3$$

The shape is represented by

$$K = \frac{\sum_m \left| \frac{\lambda_m}{\Sigma_m\ \lambda_m} - \frac{1}{3} \right|}{\left(\frac{4}{3} \right)} \qquad 0 \le K \le 1$$

which can be substituted to the acentric factor ω. The global molecular symmetry is defined by

$$G = (\gamma_1 * \gamma_2 * \gamma_3)^{(1/3)}$$

And the total density of atoms is characterized by

$$D = \eta_1 + \eta_2 + \eta_3$$

Nondirectional descriptors are five for each of the weighting schemes T, A, V, K, D, plus Gu, Gm, and Gs (corresponding to unitary, mass and electrotopological weights) giving a total number of 33 descriptors.

7.5.2.2 Applications of WHIM Descriptors

WHIM descriptors, either alone or joined to other molecular descriptors (from inter alia the Dragon package [54]) were applied to some physicochemical properties (total surface area, molar refraction, log K_{OW} [663], and to the toxicity evaluation of various chemicals [670–673]). For example, the model was applied to the aqueous toxicity on *Daphnia magna* [671] of a highly heterogeneous set of 49 compounds, on an activity range of 7 log units. After a variable selection by GA and PCA treatments, some insight may be gained about the important features, showing for example, that accessibility and size increase toxicity. Toxicity to *Pimephales promelas* (Fathead Minnow) was also studied on an extended set of more than 400 chemicals in a direct toxicity prediction model (irrespective of the diverse possible MOA) [673].

Other studies concerned the atmospheric half-life of persistent organic pollutants [674], or the chromatographic relative retention time of polychlorinated biphenyls (PCB) [675].

7.5.2.3   Calculation of MS-WHIM Descriptors

In MS-WHIM [664–666], descriptors are related not on atomic positions but on points scattered on the molecular surface (MS) with the MS program of Connolly [676] at a density of 10 points/Å^2 and selected properties are calculated for every surface point. Weighting schemes (the selected properties) are unitary and MEP (in fact positive and negative MEP values are separated, since weights must be semipositive definite values). But in further applications, the H-bonding capacity (donor or acceptor) and the hydrophobicity were taken into account, amounting to six weighting schemes. For the MEP, a Coulombic potential is chosen with a distance-dependent dielectric constant.

$$P_i = \sum_h \frac{q_h}{\varepsilon r_{i,h}}$$

with q_h as the charge on the atom h, and $r_{i,h}$, the distance of the surface point i to the atom h. In a further publication [435], it was established that results were slightly better with semiempirical ESP-fitted charges. H-bonding propensity (HBA, HBD) and hydrophobicity (HYD) were expressed as binary values (1,0) depending on the nature of the nearest atom (vide infra).

The treatment is then identical to that of WHIM: a weighted PCA delivers the score matrix that specifies the coordinates of the surface points in the trihedron of the three principal axes. As in WHIM, each molecule is thus defined by 12*6 descriptors.

7.5.2.4   First Application of MS-WHIM Descriptors

The MS-WHIM methodology [664] was applied to the steroid benchmark. With a training set of 16 compounds (determined by the fractional factorial design from the 21 usually considered), SDEP = 0.67 (in prediction on the 10 test compounds) was quite comparable to that obtained by COMPASS [367] 0.71 or CoMFA 0.72 (in this study), and clearly outperformed WHIM (1.6).

WHIM and MS-WHIM were also compared to the autocorrelation approach (see Chapter 4) since they both condense information onto holistic descriptors [435]. Two series were investigated, 18 N-substituted 3-aminopyridin-2(1H)-one derivatives (see compound **4.5**), inhibitors of HIV-1 RT, and 36 arylsulfonamides endothelin inhibitors (data from [346]), a series previously largely investigated by SOMFA [344] (see structure **2.12**), CoMFA [346], EVA [44], autocorrelation. DiP [116] with six descriptors gave $q^2 = 0.79$ (L-O-O), and outperformed WHIM, MS-WHIM, and CoMFA; autocorrelation and SOMFA being the worst (see Section 4.2.4).

The autocorrelation method was applied on both atomic properties (as in WHIM) and on the surface point MEP (as in MS-WHIM). The two methods referred to the same property and thus the observed differences (if any) would only be in the treatment used to condense information. At the atomic level, the WHIM protocol could provide models equal or better than autocorrelation. With surface points, MS-WHIM could take into account information on shape and size that were not considered in autocorrelation (looking only at the electrostatic contribution) that led to poorer performance when steric effects may be important (as for endothelin inhibitors). Comparison of pairs of compounds also showed that MS-WHIM is highly sensitive to details in structure description since molecules differing only by one position could be distinguished. In this study, various models were successively tried; 1 for the 6 properties together, 6 models excluding in turn 1 of these 6 properties, then excluding 2 and so on, leading to 63 different PLS models.

Property contribution: Drawing contour plots similar to those of CoMFA or CoMSIA is not possible. However, the most important contributing properties can be globally identified with the evaluation of

$$\text{Contribution}_q = \sum_p \left[\frac{SD(X_{pq}) * ABS\_COEFF(X_{pq})}{SD(y)} \right]$$

(then scaled to a sum of 100)

$SD(X_{pq})$ is the standard deviation associated with the descriptor $p(1-12)$ of the property $q(1-6)$, $ABS\_COEFF(X_{pq})$ is the absolute value of the coefficient for that descriptor, and $SD(y)$ is the standard deviation for the response y.

7.5.2.5   Further Applications of MS-WHIM Descriptors

Expended MS-WHIM descriptors have been applied to the binding affinity of 16 coumarin-type substrates (see compounds **6.1** through **6.3**) and inhibitors of Cytochrome P4502A5 (CYP2A5), with pK varying on three pK_i units. This series was already studied by Poso et al. with CoMFA [677]. The best model included (with nearly equal contributions) UNI, PMEP, HBA, and HYD and yielded $r^2_{cv} = 0.706$; SDEP $= 0.40$, with $r^2 = 0.986$; SDEC $= 0.09$ with 5LV. This model would give q^2 (L-O-O) $= 0.683$, and so compared well with the CoMFA treatment ($q^2 = 0.562$ with 5LV, but it reached 0.723, adding the LUMO energy as the supplementary descriptor and discarding an outlier, that was however correctly calculated with MS-WHIM). The authors highlighted the convergence of the important MS-WHIM contributions (UNI, PMEP, HBA characterizing size, positive MEP, and H-bonding acceptor) and the CoMFA contour plots (suggesting a H-bond interaction with CYP2A5 near the lactone moiety, and a positive MEP area required for binding affinity).

Let us remark that a more extended data set (28 coumarins from Poso et al. [560]) was later studied by Asikainen et al. [556]. EVA (or EEVA) plus MgVol gave results comparable to CoMFA or GRID/GOLPE. On another hand, Ekins et al. [678] used MS-WHIM after a multiconformer generation in a 4D analysis of cytochrome P450 (CYP)2C9.

Another study concerned the binding affinity of polyhalogenated aromatic compounds to the cytosolic Ah receptor (pEC_{50}). This population of polyhaloge-nated aromatics has been largely investigated by numerous approaches (for a mini review, see Ref. [328]). We only compare here, the MS-WHIM analysis to other methods on the same data set of 78 halogenated aromatic hydrocarbons, with a few exceptions (25 polyhalogenated dibenzo-p-dioxins PCDDs, 39 dibenzo-furans PCDFs, and 14 chloro-biphenyls PCBs, see compounds **1.4** through **1.6**) on a range of 6.3 log units [44,49,51,146,210,423,531,541,553]. The best selected MS-WHIM models retained four properties (48 descriptors), including UNI, PMEP, HBA, and HYD. It led to $q^2 = 0.732$, rmse 0.76, $r^2 = 0.794$, rmse = 0.66, with two components. Let us remark that here, q^2 was calculated not in L-O-O but in a less optimistic process: the mean of 100 leave-20%-out trials.

The CoMFA treatment led for the whole set $q^2 = 0.724$, $r^2 = 0.878$, and SEE = 0.53 [423] with 6 PCs. So and Karplus [49] using similarity matrices and a genetic neural network obtained $q^2 = 0.85$ and $r^2 = 0.89$ (with a shape index) on a population of 73 derivatives. These results compete well with those of Waller et al. [423] ($r^2 = 0.72$) and Wagener et al. [146] $q^2 = 0.83$ and $r = 0.89$. The spectroscopy-based approach EEVA [531,553], using MO energies, was applied to the same initial set (but with five benzofurans) with $q^2 = 0.818$ (8 PC), SPRESS = 0.69, $r^2 = 0.912$, and SEE = 0.48 [553].

MS-WHIM was also applied to the prediction of various physicochemical prop-erties in QSPR models [666]. Log P (corresponding to the octanol/water partition coefficient), plays an essential role in many models of biological activity. For log P values, the data set consisted of 268 small organic molecules (data from Schaper et al. [679]). The best model yielded $r_{cv}^2 = 0.709$ and $r^2 = 0.771$ (with RMSEC = 0.58). The model was validated on a test set (50 compounds) with RMSEP = 0.66. These results compared well to those of Schaper et al. An ANN with binary descriptors encoding the presence (or not) of atoms and bond types led to s = 0.25 in training, but in prediction, s was only 0.66.

Caco-2 cell permeability of 17 structurally diverse compounds (log P_{erm} vary-ing on 2.7 log units) was also investigated. The data set, from van der Waterbeemd et al. [680], encompassed varied structures, such as testosterone, salycilic acid, and mannitol. The best MS-WHIM model led to $r_{cv}^2 = 0.797$ ($r^2 = 0.981$) with 3LV (SDEP = 0.46; SDEC = 0.13).This model emphasized the importance of H-bonding, in agreement with a previous PLS model, with H-bonding and size descriptors yield-ing $q^2 = 0.852$ [680].

MS-WHIM was also compared to the CoMMA approach of Silverman et al. [489] on the dissociation constant prediction pK_a of 15 imidazoles (**5.14**) (on a range 9.3 log units). The best model, $r^2 = 0.987$, $r_{cv}^2 = 0.728$ (with SDEP = 1.50 and SDEC = 0.34) would correspond to $q^2 = 0.833$, which compared well with the Silverman and Platt results ($q^2 = 0.788$).

7.6 ENCODING INTERACTION ENERGY

7.6.1 VOLSURF

VolSurf [178,449,681] aims to characterize in a molecule shape, polarity, hydrophobicity, and the balance between them, by means of a set of about 100 descriptors. These numerical, alignment-free descriptors are typically of a 2.5D nature since they are generated by compressing the detailed 3D information provided by the molecular interaction field (MIF) maps produced by the GRID program.

As indicated by Crivori et al. [179], in molecular modeling, molecular surfaces are often partitioned into separate polygons, spreading information into small contiguous pieces. In contrast, VolSurf builds a single surface or volume for a specific property thanks to an approach derived from image analysis. 3D maps may be considered as composed of voxels (volume elements) comparable to the pixels constituting a 2D image. These voxels correspond to the elementary cells defined by the lattice in GRID (or in CoMFA or CoMSIA). Contouring the voxels at various definite energy levels allows the delineation and the computation of surface areas and volumes.

The interaction fields used by VolSurf can be calculated with the varied probes introduced in GRID: H2O, DRY (for hydrophobic effects), BOTH for amphipatic interactions, carboxy sp^2 oxygen, N1 neutral planar nitrogen (from amide), N sp^3 (with lone pair), N3$^+$ (amine cation), and O$^-$ (phenolate oxygen). However, most interest was put on polar and hydrophobic interaction sites around the target and on hydrophobic/hydrophilic interactions calculated with the H_2O and DRY probes.

In addition to the usual descriptors (molecular weight, surface area, volume, molecular polarizability), VolSurf proposes nonstandard descriptors: For example, hydrophilic regions are characterized by a *capacity factor* (hydrophilic surface per surface unit). The volumes of hydrophilic and hydrophobic regions, at different energy levels, define INTEGY moments (INTEraction EnerGY), i.e., the distance between the molecular center of the mass and the barycentre of polar interaction sites. Amphiphilic moments are calculated from the centers of hydrophilic and hydrophobic regions.

Other descriptors involve hydrophilic/lipophilic balance, critical packing (ratio between hydrophilic and lipophilic parts as to shape, a notion different from the preceding one), local interaction energy minima for a water probe (or for the MEP), and their distances, etc. H-bonding capacity (water probe) and polarizability (based on an additive model and not on interaction fields) are also considered.

Applications of VolSurf. The relevance of VolSurf to evaluate physicochemical quantities was presented with the example of the hydration of 25 brassinosteroids [178,449], where an excellent correlation was established with the corresponding free energy variations ΔG calculated at the AM1acqSM2 level [682].

In the domain of pharmacokinetics, VolSurf was also successfully applied to skin permeation, an important phenomenon in drug administration, cosmetics, and toxicology. For 46 varied compounds (mostly drugs), on a scale of 4.4 log units for the skin permeability coefficient, experimental and VolSurf calculated values correlated well with $q^2 = 0.78$ (with two latent variables) [178].

Another application concerned the binding affinity of 26 basic drugs with the A-variant of the human serum α1-acid glycoprotein. Interestingly, in this application, where polar areas played a critical role, appropriate VolSurf descriptors (determined on a single conformer) correlated well with the Boltzmann-averaged polar area determined by high temperature molecular dynamics [178]. VolSurf was also applied, as a classification tool, to the prediction of blood–brain barrier permeation of very varied chemicals and drugs [179].

In the study of Lo Piparo et al. [683], the affinity of 93 aromatics (from the 95 ones of Waller and McKinney [210]) to the AhR receptor was approached using CoMFA, VolSurf, and the 2D approach HQSAR. With five probes, (water, hydrophobic, carbonyl oxygen, carboxy oxygen, and amphipatic), 118 descriptors were generated; log P was added as a crude measure of desolvation energy and bioavailability (in HQSAR). A binary indicator for torsionally constrained compounds (intervening about 5%) was introduced in the CoMFA and HQSAR treatments. CoMFA and HQSAR gave similar results ($q^2 = 0.62$), whereas VolSurf was slightly inferior. But a hybrid model, combining VolSurf and HQSAR gave the best q^2 (0.70), although r^2 was not improved. On a test set of nine compounds (on the upper half of the affinity range), r^2 (CoMFA) was only 0.42 instead of 0.69–0.72 as in the other methods. In this example, the authors stressed on the good results obtained with VolSurf, which is alignment-free, and HQSAR, which works on 2D descriptors only, two methods less intensive than CoMFA. Consistent with CoMFA, VolSurf indicated an increased activity with the presence of high hydrophilic regions, and with the delocalization of the hydrophobicity in a few areas of the molecular surface [683].

VolSurf was also used in a study of an extended set of 101 antivirial quinolones (a new interesting class of anti-HIV derivatives). PCA and PLS extracted the key structural features responsible for the antivirial activity [684].

7.6.2 3D HOLOGRAPHIC VECTOR OF ATOMIC INTERACTION FIELD

With the 3D holographic vector of atomic interaction field (3D-HoVAIF), Ren et al. [352] proposed an extension of the 2D molecular edge distance vector (MEDV) and the molecular electronegativity-interaction vector (MEIV), a sort of electrotopological approach introduced by Liu et al. in 2D-QSPRs and QSARs [685–688].

These new translation–rotation invariant descriptors were proposed for an easy, but efficient extraction of structural information directly related to bioactivities. Schematically, atoms are separated in categories according to their chemical family (the column of the periodic table) and their hybridization state, leading to 10 types: H, three classes for C (sp^3, sp^2, sp), three for N (or P), two for O (or S), one for the four halogens. The model is based on the evaluation of interactions between pairs of nonbonded atoms. Three types of interactions were considered: electrostatic, van der Waals, and hydrophobic. A summation of these various types led to (to the maximum) $3*[(10*9/2) + 10] = 155$ values (if all atom classes are present). These values constitute the 3D-HoVAIF descriptors.

Electrostatic interactions are expressed by a Coulombic expression from the partial atomic charges. Van der Waals terms are evaluated with a Lennard Jones potential with the usual combination rules for ε and r*, and the hydrophobic contribution is derived from the solvent accessible surface area (SASA), according to the formalism of HINT [192].

The approach was evaluated in a study of 32 analogues of Artemisinin (**7.5**), a compound of high antimalarial activity, isolated from Qinghao (*Artemisia annua*) and used for more than 2000 years in Chinese traditional medicine. Activities measured on a *P. falciparum* clone span a range of 2.6 log units in relative values. To determine their active conformation, compounds were docked in the crystal structure of heme (extracted from the complex heme-artemisinin). The statistical analysis was carried out on 25 compounds in training and seven in the test (from a careful selection) with a descriptor pool of 81 descriptors (owing to missing values). A first treatment with PLS led to a model seemingly good but a little instable ($r^2 = 0.874$, $q^2 = 0.552$) leading the authors to limit the number of variables with a genetic algorithm before the PLS analysis. This treatment led to a model with better internal stability and external predictive ability. After a selection of 26 variables (9 electrostatic, 10 for van der Waals terms, and 7 hydrophobic), PLS analysis yielded (with 2 components) $r^2 = 0.852$, $q^2 = 0.778$ with rmse = 0.30 log units (training) and 0.37 (in the test).

The treatment was compared to the WHIM [663] and SOMFA [348] approaches (in 3D) and MEDV (in 2D). This choice was guided by the fact that WHIM was another system of TRI descriptors, and that SOMFA took into account steric and electrostatic features, while MEDV considered atom classification and pairwise interactions. The analyses were carried out with GA/PLS for WHIM and MEDV, and by MLR for SOMFA. On this data set, MEDV was the most inferior (presumably due to difficulty in modeling steric effects), whereas WHIM, with an insufficient grasping of the nonbonding potential, did not show good stability and predictability. SOMFA was comparable to 3D-HoVAIF, but it was stressed that contrary to SOMFA, 3D-HoVAIF needed no alignment. However, it may be remarked that the choice for the active conformations resulted from docking the studied compounds into the receptor.

7.7 MATRIX TREATMENTS

7.7.1 BCUT METRICS

The BCUT metrics (Burden–CAS-University of Texas eigenvalues) of Pearlman [689–691] constituted an extension of the original Burden parameters [538,692]. These parameters were first designed for similarity/diversity studies on large databases and they proposed a highly compact molecular identification number with minimal redundancy (number of digits needed to avoid degeneracy). But the approach was also applied to the working out of QSAR models.

Burden's identification number: Given a molecule and its symmetrical connectivity matrix **B** (vide infra), the method starts from the eigenvectors **V** and the (diagonal) eigenvalues **E** matrices:

$$V' * B * V = E$$

The identification number consists of the smallest eigenvalues taken to m significant figures (user defined according to the size of the database to be indexed). The connectivity matrix concerns only heavy atoms (numbered arbitrarily). The diagonal elements b_{ii} are the atomic numbers of the atoms. Nondiagonal elements, for a bond ij are set to 0.1 (single bond), 0.2 (double bond), 0.3 (triple bond), and 0.15 for an aromatic delocalized bond; 0.01 is added for terminal atoms. All other terms are set to 0.001 (this scale on powers of 10 preserves details from each type of element without mixing with other categories).

BCUT metrics expand the number and types of atomic features and introduce a great variety of proximity measures and weighting schemes. Four classes of matrices are considered. Their diagonal elements incorporate information about atomic properties (e.g., atomic charge, polarizability, H-bond donor, or acceptor ability). The off-diagonal matrix elements are composed of topological information including connectivity and/or interatomic distances possibly weighted by quantities, such as fractional surface area.

BCUT descriptors are the highest or lowest eigenvalues of these matrices. For example, [693] a descriptor (among others) may be the highest eigenvalue of a matrix formed by products of Gasteiger–Hückel charges and the fractional surface area on the diagonal, and for off-diagonal terms, the inverse interatomic distances to the sixth power scaled by 0.60 (calculation on the Hydrogen-depleted skeleton of the most stable conformation).

Note, however, that BCUT may be calculated at the 2D level since a 2D connection table is sufficient to calculate Gasteiger–Marsili charges, while on the other hand, the polarizability or the H-bond ability is tabulated. Off diagonal elements then encode topological distances, whereas in 3D, geometrical interatomic distances are taken into account.

Applications of BCUT metrics: The performance of BCUT descriptors has been examined in various publications. Stanton [694] investigated the efficiency of BCUT descriptors in QSARs on the activity of 74 substituted benzyl-pyrimidines (see compound **3.3**). As example of quantitative Structure/Property Relationship, the study also encompassed the boiling point of 179 chemicals. To 85 BCUT descriptors were added 105 ADAPT descriptors (topological, geometric, electronic, including charged partial surface areas, and H-bonding descriptors) [57]. Variable selection was performed by generalized simulated annealing. From the initial pool, 64 descriptors survived (and among them, 25 BCUTs). A comparison between ADAPT and BCUT showed that the information provided by BCUT was more or less similar to that provided by the polar surface areas.

The best model, outdoing ADAPT, yielded $r^2 = 0.878$ in training and the prediction for a test set of 10 compounds was excellent. Interestingly, the 10 best 6-descriptor models included three or more BCUT descriptors. It was concluded that BCUT descriptors performed better than CPSA descriptors in capturing structural information important for understanding polar intermolecular interactions. It was also remarked that the metrics appearing the most useful were related to SASA (in line with the idea that intermolecular interactions were related to atom exposure on the surface of the molecule).

BCUT descriptors were more recently applied to the study of 32 adenosine analogues (7.12) agonists of the A_3 adenosine receptor [695]. After the usual elimination of descriptors that are constant or highly intercorrelated, a genetic algorithm (typically, 300 populations, 10,000 generations) selected relevant variables. The best 6-descriptor model, led to $q^2 = 0.724$ ($r^2 = 0.808$) and outperformed the Galvez topological charge indices, the Randić molecular profiles and the geometrical descriptors.

BCUT encodes not only topological and topographic information but also atomic information relevant to the strength of ligand-receptor interactions. With the concept of the receptor-relevant subspace [690], BCUT was also the basis of an efficient approach for reducing the dimensionality of the chemistry-space, discarding metrics (axes) that are irrelevant to a given receptor while retaining metrics that convey receptor–relevant information. This task is important in comparing libraries, searching for neighbors of lead compounds, or for other diversity-related applications. BCUT descriptors have also been used in library design, see Mason and Beno [693].

7.7.2 GEOMETRY, TOPOLOGY, AND ATOM WEIGHT ASSEMBLY

The GETAWAY methodology is related to the notion of leverage and autocorrelation treatment [696]. It encodes both topological information (molecular graph) and geometrical information (from the influence molecular matrix) weighted by chemical information encoded in selected atomic weighting. Different weighting schemes are provided (atomic, polarizability, van der Waals volumes electronegativity, mass).

A first set of descriptors, H-GETAWAY, are derived from the molecular influence matrix, whereas R-GETAWAY combines this information with geometric interatomic distances in the molecule. For a molecule of A atoms, the molecular influence matrix (MIM) is defined from the molecular matrix $\mathbf{M}$ (A,3) of centered atomic coordinates:

$$\mathbf{H} = \mathbf{M} * (\mathbf{M}' * \mathbf{M})^{-1} * \mathbf{M}'$$

$\mathbf{H}$ is a symmetrical (A $*$ A) matrix of elements h_{ij}.

with
$$0 \le h_{ii} \le 1 \qquad \sum_{i=1,A} h_{ii} = D$$

h is the D/A average value of diagonal terms
D is the rank of M: 1 (linear molecule), 2, (planar), 3 (3D molecule)
$\sum_{j=1,A} h_{ij} = 0$

This formalism recalls the leverage matrix used in statistics.

Hat matrix: In statistics, if a linear least squares model is used to relate the observed values y to the matrix $\mathbf{X}$ of the descriptors (independent variables)

$$\mathbf{y} = \mathbf{X}\mathbf{w} + \mathbf{\varepsilon}$$

where
   $\mathbf{w}$ is the vector of the coefficients (to be determined)
   ε is the error vector

As for the "hat matrix" $\mathbf{H}$,

$$\mathbf{H} = \mathbf{X}\,(\mathbf{X}'*\mathbf{X})^{-1}\,\mathbf{X}'$$

which relates the fitted values $\hat{y}$ to the observed values y (it puts a hat on y to get $\hat{y}$).

$$\hat{\mathbf{y}} = \mathbf{H}\mathbf{y}$$

Its diagonal elements are the leverages which describe the influence that each observed value has on the fitted value for the same observation.

In the GETAWAY H matrix, the diagonal elements h_{ii}, called leverages, encode the influence of each atom in determining the whole shape of the molecule. Nondiagonal elements h_{ij} represent the degree of accessibility of atom j to interactions with atom i (a negative sign indicates that the two atoms occupy opposite regions with respect to the center, and, hence, have less chance to interact).

H-GETAWAY descriptors are calculated from this $\mathbf{H}$ matrix. They encompass

- The geometric mean on the leverage magnitude

$$H_{GM} = 100*(\textstyle\prod_{i=1,A} h_{ii})^{1/A}$$

with $0 < H_{GM} < 100$. High values correspond to branched molecules (almost spherical).
   Other descriptors may be conceptually divided into three main groups.
- Information indices, which are the total and standardized information content on leverage equality. These indices rely on the concept of molecular complexity, considering for example symmetry, branching, and cyclicity

$$I_{TH} = A_0 * \log_2 A_0 - \sum_{g=1,G} N_g * \log_2 N_g$$

$$I_{SH} = I_{TH}/A_0 * \log_2 A_0.$$

where
   N_g is the number of atoms with the same leverage value
   G is the number of equivalence classes into which atoms are partitioned according to the leverage quantity

Only heavy atoms (number A_0) are considered. These indices encode some information about molecular entropy.

Another index is the mean information content on the leverage magnitude (HIC)

$$\text{HIC} = \underline{I}_{\text{H}} = -\sum_{i=1,A} \frac{h_{ii}}{D} \log_2\left(\frac{h_{ii}}{D}\right)$$

where
D is the matrix rank (sum of all leverages)
A is the total number of atoms (H included)

- Autocorrelation descriptors, within the framework of which property values will be considered to explicitly include chemical information.
  For a property defined by an atomic property vector **w**

$$\text{ATS} = \mathbf{w}'\mathbf{U}*\mathbf{w} = \sum_{i=1,A} w_i^2 + 2 \sum_{i=1,A-1} \sum_{j>i} w_i * w_j$$

$$= \text{ATS}_0 + 2 \sum_{k=1,d} \text{ATS}_k$$

with

$$\text{ATS}_0 = \sum_{i=1,A} w_i^2$$

$$\text{ATS}_k = \sum_{i=1,A-1} \sum_{j>i} w_i\, w_j \delta(k, d_{ij})$$

where
U is the unity matrix
ATS are the autocorrelation descriptors, up to the order k
d is the maximum topological distance in the molecule

In fact, the distance bins are defined from topological distances (and not Euclidian distances), and the property **w** is weighted by the h_{ii} term.

$$\text{HATS}_0(\mathbf{w}) = \sum_{i=1,A} (w_i h_{ii})^2$$

$$\text{HATS}_k(\mathbf{w}) = \sum_{i=1,A-1} \sum_{j>i} (w_i * h_{ii}) * (w_j * h_{jj}) \delta(k; d_{ij}) \quad k = 1, 2\ldots, d$$

And a HATS total index is calculated by

$$\text{HATS}(\mathbf{w}) = (w_i h_{ii})' * \mathbf{U} * (w_i h_{ii}) = \text{HATS}_0(\mathbf{w}) + 2 \sum_{k=1,d} \text{HATS}_k(\mathbf{w})$$

For each weighting scheme w, a sequence of $HATS_k(w)$ and a total autocorrelation index $HATS(w)$ are generated.

Considered properties are those selected for WHIM methodology (except the electrotopological weight). The components of the autocorrelation vector (up to a limit distance) define the HATs indices. Positive off-diagonal elements of the molecular influence matrix (for atoms that have a chance to mutually interact) similarly lead to the H indices (H_0 plus a sequence of $H_k(w)$).

This leads to a series of H descriptors

$$H_0(w) = \sum_{i=1, A} w_i^2 \, h_{ii}$$

$$H_k(w) = \sum_{i=1, A-1} \sum_{j>i} h_{ij} w_i * w_j \, \delta(k; d_{ij}; h_{ij})$$

$k = 1, 2, \dots d \quad \text{and} \quad \delta(k; d_{ij}; h_{ij}) = 1 \quad \text{if, and only if, } d_{ij} = k \quad \text{and} \quad h_{ij} > 0$

The H-GETAWAY descriptors, for each **w** weight vector, consist of $\{HATS; HATS_0, HATS_1, \dots HATS_L\}_w$; $\{HT, H_0, H_1, \dots H_L\}_w$, where L (user defined) fixes the common length of the descriptor.

- *Local vertex invariants*: An A-dimensional vector derived from **H*w**; **H** is the molecular influence matrix and **w** an atomic property vector.

R-Getaway descriptors: Other descriptors may be defined following the same approach, replacing the influence matrix **H** by other molecular matrices. For example, the **G** matrix of the actual 3D distances (r_{ij}) leads to the influence-distance matrix **R** with elements

$$[R]_{ij} = \left[\frac{\left(h_{ii} * h_{jj}\right)^{0.5}}{r_{ij}} \right]_{ij} \quad \text{for } i \neq j$$

Diagonal elements are nul, and the R-GETAWAY descriptors are similar to the H-descriptors, see Ref. [696] for more details.

Schematically, for a molecule of N_w weights and the upper autocorrelation distance L, the total number of GETAWAY descriptors is $7 + 2N_w*(3 + 2L)$. For example, if $N_w = 5$ and $L = 8$, $N = 197$. In QSARs, GETAWAY descriptors (197 in this study) have been tested, alone or combined with other descriptors, with WHIM (99 descriptors), and topological descriptors (69 terms) on varied physicochemical properties, such as melting points, heat of vaporization, heat of formation, and the motor octane number of the 18 octane isomers. In these diverse applications, they led to satisfactory results.

An extensive validation of the GETAWAY descriptors was carried out by Consoni et al. [697] on various data sets: 82 polycyclic aromatic hydrocarbons (PAHs),

22 N,N-dimethyl-2-halo-phenethylamines (adrenergic blocking activity), 47 nitro-benzenes (toxicity toward *Tetrahymena pyriformis*), 14 polychlorinated biphenyls PCBs, 25 polychlorinated and polybrominated dibenzo-*p*-dioxins, PDDs, and 34 polychlorinated dibenzofurans PCDFs (*Ah* receptor binding affinity). For some series, several properties (biological activity, $logK_{OW}$, boiling point) were investigated, amounting to a total of 13 studies. Various sets of molecular descriptors were used: GETAWAY alone, or GETAWAY plus WHIM, WHIM, BCUT, Moreau-Broto aurocorrelation, and topological or constitutional descriptors amounting to a maximum number of 517 descriptors.

Correlations were carried out by MLR, looking at models with one to four descriptors, with a variable selection performed by genetic algorithm. We present as example, the results obtained on the 73 polyhalogenated derivatives binding the *Ah* receptor (data from Tuppurainen and Ruuskanen [553]). With four variables, the best model is obtained with all descriptors together ($q^2 = 0.856$), followed by GETAWAY plus WHIM (0.850), whereas CoMFA yielded 0.724 (with six latent variables) and EVA 0.818 (with eight latent variables). If the various sets of descriptors are used alone, WHIM ($q^2 = 0.940$) outperforms GETAWAY (0.795), BCUT (0.790), and the topological descriptors (0.775). A more global analysis was performed by PCA on the matrix formed by (on rows) the best models (one to four variables for each set of descriptors) and (on columns) their prediction power for the various properties. This synthetic presentation confirmed the superiority of the 4-variable models using the all-descriptor set, but highlighted also the predictive ability of WHIM and GETAWAY models alone.

Considering mixed models (all descriptors together) with four variables, GETAWAY, WHIM, and topological descriptors were the more frequently included in the best solution. But selected descriptors were not strictly the same when mixing two types of representation. It was remarked that topological and WHIM descriptors were the most frequently selected in one-dimensional models, whereas GETAWAY descriptors were preferred in higher dimensional models. This probably corresponded to the fact that topological and WHIM descriptors gave a holistic representation, whereas GETAWAY descriptors better characterized portions of the molecular structure.

For a comparison with other 2D or 2.5D approaches, see also Saiz-Urra et al. [659] for a study of benzophenazines (**7.10**) acting as topoisomerase inhibitors, where the best results were obtained with 3D-MoRSE and RDF codes.

7.8 TWO-DIMENSIONAL vs. THREE-DIMENSIONAL MODELS

As we have seen in the preceding chapters, practitioners now have at their disposal a wide panel of 3D methods for building QSAR models, and an obvious question arises as to which QSAR method should be followed, a problem still complicated by the choice between 2D and 3D (or 2.5D) methods. When this question was stressed on some years ago, it sparked off contradictory opinions and aroused impassioned debates in QSAR meetings. Schematically, two lines of arguments conflicted: 3D models are more calculation intensive; 2D models forget the spatial structure of molecules. To the authors it now seems clear, however, that the answer essentially depends on the problem under scrutiny in terms of resources and objective.

Strictly speaking of performance, an extensive and objective study of various approaches regarding the most efficient modeling QSAR method was carried out by Sutherland et al. [72]. An extended data set was investigated, encompassing eight series of chemicals (altogether 1245 compounds). Details on this data set have been given in Chapter 1 (because they were reused in the study of Peterson [141,182] that discussed the best CoMFA settings). About 1/3 of the compounds constituted the test sets (they were chosen in each series so as to represent a maximum dissimilarity in order to widely span the structural space). An interesting point is that the data compiled for that study are available to the chemical community and so constitutes an extended benchmark to carry out objective comparisons for new QSAR methods.

Sutherland's study compared CoMFA, CoMSIA, and EVA to HQSAR and classical 2D- and 2.5D-QSARs (the latter method corresponding to the addition of descriptors, such as volume and charged partial surface areas information, i.e., about 40 and 70 descriptors for these methods after variable selection).

Comparison of EVA and field-methods CoMFA, CoMSIA with HQSAR: In the hologram QSAR (HQSAR), [125] all possible fragments of varied lengths (typically four to seven atoms) are systematically generated. These fragments are encoded according to various characteristics (atom and bond type, hybridization, connectivity, chirality, H-bond donor, acceptor propensity can be considered, and H atoms may also be included). Thus, a molecule is represented by the count of each kind of fragment that it contains. These numbers are stored in a string of integers (hologram), each position in the string representing the occurrence of a given fragment.

Since different molecules generate different fragment, strings of different lengths would be obtained. To get an array of descriptors of fixed length, a hashing procedure is used to form a molecular hologram. Its length will be selected on the basis of its performance.

In HQSAR, looking at the weights of the different fragments made it possible to obtain some insight about the individual atomic contributions, from one molecule to another one. However, HQSAR, with regard to fragments, gives only a global contribution to reactivity, without identifying the various components (steric, hydrophobic, electrostatic), whereas CoMFA or CoMSIA can analyze them.

In the derivation of the models, Sutherland et al. examined a large number of options, such as a combination of fields for CoMSIA, a bandwidth for EVA, and the characteristics of the HQSAR fragments. Similarly, for the data treatment of 2D/2.5D models, settings of neural network, of GA, nonlinear GA, or GA-PLS were widely investigated. A detailed presentation of these results is beyond the scope of this book. We will now recall some elements the authors have stressed on in their extensive work.

The predictive accuracy was examined at two levels: in L-O-O (which corresponds to accuracy in interpolation) and on external test sets (which is more important for prioritizing the synthesis of novel derivatives, or proposing structural alerts). The results somewhat depended on the series under scrutiny, but some definite trends might be distinguished.

In cross validation of PLS models, all methods performed reasonably well in most cases. CoMFA and CoMSIA (with only electrostatic and steric fields) are comparable. An addition of other fields in CoMSIA was beneficial for four sets. HQSAR, giving results similar to those of EVA, performed as well or better than field-based methods in six series and was generally better than the 2.5D method, which itself outperformed the 2D approach.

As to external predictivity, field-based methods were the most predictive. CoMSIA with additional fields appeared preferable to CoMFA (simpler contour plots, less sensitivity to orientation, and grid mesh). HQSAR compared well with field-based methods, except for three sets among eight. EVA was comparable to HQSAR but with the most costly calculation of normal modes. Incorporation of 2.5D indices (CPSA, volume) to classical 2D models is desirable.

For these 2D or 2.5D methods, only neural network ensembles are interesting alternatives to PLS. The authors also stressed that other points to be considered, in addition to the predictive ability, are the interpretability of the model and its capacity to suggest promising structural modifications (for optimizing biological activity or conversely decreasing toxicity). We already evoked this problem, for example, with autocorrelation methods (except GRIND), spectroscopic methods, topological or topographic indices, etc.

Miscellaneous: Various other, more isolated, similar comparisons between 2D and 3D approaches appeared. We summarize here some examples in the field of endocrine disruptors. For more details see Ref. [328]. In a study of binding affinity to the estrogen receptor, Tong et al. [359] compared CoMFA to CODESSA [578], on a set of 53 2-phenylindoles plus three steroids, two triphenylethylenes and Hexestrol. CoMFA ($r^2 = 0.97$, $q^2 = 0.61$) outperformed CODESSA ($r^2 = 0.68$, $q^2 = 0.54$). It was suggested that CODESSA seemed unable to capture some structural features (e.g., the position of the hydroxyl group on the indole phenyl ring). However, in prediction, results are more similar.

CoMFA and CODESSA results were further compared to HQSAR on three datasets: the first two sets encompassed 31 compounds (among them 19 steroids) binding human ER-α and rat ER-β, respectively, the third one 47 compounds mainly congeners of 2-phenylindole studied with calf ER. As the same PLS analysis was used, differences in results originated from the structural description. HQSAR was slightly inferior to CoMFA, CODESSA exhibiting the lowest performance (presumably because it condenses structural information in numerical indices). For example, for human ER-α binding $q^2 = 0.70$ in CoMFA, 0.67 for HQSAR, and 0.46 with CODESSA.

Similarly, on an extended data set of 130 estrogen receptor binding compounds (steroids, phytoestrogens, DESs, DDTs, alkylphenols, and parabens) on seven log RBA units. HQSAR only led to $q^2 = 0.585$ in place of 0.665 with standard CoMFA (and 0.707 if adding a phenol indicator) [188]. However, with 151 2D descriptors, a q^2 value of 0.627 was reached in classical QSAR [549]. On the other hand, on Bisphenol A derivatives, CoMFA, CoMSIA, and HQSAR gave nearly identical results [196].

In another study on 58 compounds binding the estrogen receptor, [641] FRED/SKEYS (already quoted in Section 7.3.2) appeared to be the best with $q^2 = 0.700$ vs. about 0.58 for CoMFA and HQSAR.

A comparison between 3D-QSAR models COMFA and CoMSIA with the 2D approach HQSAR was presented by Zhang et al. [698] on novel neuronal nAChRs ligands-open ring analogues of 3-pyridyl ether (**7.13**).

7.13

Neuronal nicotinic acetylcholine receptors (nAChRs) are closely involved in important functions of the central nervous system, and ligands with high affinity and selectivity may have great potential therapeutic interest in case of CNS disorders. 3-pyridyl ether analogues are excellent candidates but the development of QSAR models is faced with the great flexibility of these compounds. The aim of this study was to get some information that would help in the synthesis of new ligands with therapeutic potency. The data set (64 compounds) was split into training and test sets of respectively 50 and 14 compounds, on an affinity range of 5 log units. For each compound, after a preliminary energy minimization, a conformational Monte Carlo search was performed and the 25 lowest energy conformations were submitted to AM1 optimization to obtain the global energy minimum. Molecules were considered in the unprotonated form, in agreement with the conclusions of a previous study [699].

HQSAR was slightly superior to CoMFA and CoMSIA. A same outlier appeared in the three methods, but discarding it was beneficial mainly for HQSAR. In fact, the outlier was the unique molecule bearing a CH_2-CH_2-Cl group. This suggested that HQSAR is less efficient for molecules possessing a group not present in the training set, whereas field-based methods better "extrapolate" steric and electrostatic features. In this example, the conclusions are consistent with CoMFA or CoMSIA contour plots.

The inhibitor potency of 73 factor Xα inhibitors (**7.14**) and oral bioavailability (for 15 of them) have been investigated, after docking with FlexX, by CoMFA, CoMSIA, classical QSAR (with CPSA and VolSurf parameters and analysis by a support vector machine), and on the other hand, the HQSAR method. [612]. These different methods led to consistent good results (although HQSAR was slightly inferior) and provided complementary information.

7.14

CoMFA and CoMSIA were the subject of most comparisons. Horwell et al. [483] compared a similarity matrix treatment and classical QSAR (with Tsar parameters [428]) on the activity of 28 N-terminus fragments of tachykinin NK1 receptor antagonists (**5.16**), see Section 5.1.3. We also recall the study of Sciabola et al. [642], see Section 7.3.2, comparing HQSAR, TOPPs, GRID, GRIND, and DRAGON on apoptosis-inducing chromenes.

8 QSARs in Data Mining

Although seeming a little apart from the strict domain of QSARs, and primarily developed for activity prediction, an interesting application is the development of virtual high-throughput screens (VHTS) [700]. Virtual compound libraries and VHTS are now promising tools to prioritize library synthesis in both lead optimization and identification studies. Indeed, as stated by Tropsha et al. [701], there is a close parallel between the selection of those pharmacophoric features responsible for specific biological activity and the selection in QSAR models of relevant descriptors, those associated with increased activity. 3D-QSARs provide two important components for a VHTS:

- The definition of the spatial pharmacophore, necessary for binding
- The identification of sites nonavailable, due to the receptor

This prompted various authors to propose an extension of QSAR models to database mining. It was also emphasized [76] that such applications, proposing potentially promising chemical structures rather than delivering good statistical models, best come up to the expectations of medicinal chemists. However, as remarked by Hillebrecht and Klebe [702], all models (except the k-nearest neighbors [514]) must be set up on a training set, and their predictive ability is to be limited to a structural space with no more than a reasonable structural extrapolation (see, e.g., Ref. [703]). So, QSAR models cannot be applied to data mining in huge databases of wide structural diversity. They seem more suited to the screening of focused databases or to reduced sets of compounds after initial filtering. But it may be noted that for such uses, the prediction accuracy is not crucial since the problem is only to categorize chemicals into a few classes (highly, medium, weakly active, inactive). Rather than the classical correlation coefficient (generally used to examine the quality of the QSAR), Spearman's rank correlation coefficient better characterizes the ability of a model to rank compounds according to their activity.

From the number of false responses, false positives (FPs) (negatives predicted as positives) and false negatives (FNs) (positives predicted as negatives) vs. the number of true diagnoses, true positives (TPs), and true negatives (TN), various measures may be defined [702]. The most frequently used are

Sensitivity (or "recall") $Se = [TP/(TP + FN)]\ 100\%$
Specificity $Sp = [TN/(TN + FP)]\ 100\%$
Hit rate (precision) $H = [TP/(TP + FP)]\ 100\%$

Another useful indicator is the enrichment curve, a plot of the fraction of good selection when the screening of the base moves forward. The enrichment factor is the fraction of binders found in a subset with respect to the random selection:

$$\text{Enrichment factor} = \frac{\left[\text{nb binders/nb compounds}\right]_{\text{subset}}}{\left[\text{nb binders/nb compounds}\right]_{\text{total}}}$$

Alternately, the receiver operating curve (ROC), a plot of the sensitivity (signal) vs. $1 - $ specificity (noise), specifies the performance for a given level of "noise."

8.1 CoMFA, CoMSIA IN DATABASE SCREENING

8.1.1 APPLICABILITY OF CoMFA, CoMSIA ON LARGE-SCALE SCREENING

The paper of Hillebrecht and Klebe [702] examined the applicability of CoMFA and CoMSIA approaches for database screening. One critical step of these methods was the alignment phase, and it was crucial to check whether a fully automated method could give reliable results. Additionally, the performance of two 2D models was examined in order to determine if such simpler and rapid methods might be applied. The selected example concerned the binding constants (pK$_i$ values) of sulfonamide-type inhibitors of the human carbonic anhydrase, an enzyme involved in pH and CO_2 regulation (see Section 2.1.4.2). Activity was characterized by the affinity constants pK$_i$(II) to the hCA II receptor, and selectivity by the difference between the affinities for hCA I and hCA II: ΔpK$_i$(I–II). After training on a set of 138 ligands, the models were applied on a (small) database of 663 molecules. For CoMFA and CoMSIA, an automated alignment was carried out with FlexS [513]. Flex uses Gaussian functions to represent molecules (consistent with the CoMSIA approach) and is claimed to be fast and "easily extensible to combinatorial libraries." To initiate Flex, nine scaffolds were selected as anchoring fragments (e.g., HN$^-$–SO$_2$–Ph–, HO–N$^-$–SO$_2$–, HO–N–(CO)–, SO$_3^-$–NH–). A reference ligand must also be specified.

For the sake of comparison, 2D models were also investigated: one using the MACCS keys from [643] based on the presence of 166 distinct molecular fragments. A structure was here defined by 166 integers counting the occurrence of the fragments in the molecule [644]. A second 2D model used the 32 VSA descriptors. These descriptors were based on atomic contributions to Van der Waals' surface area, log P, molar refractivity, and partial charge. Each VSA-type descriptor is characterized as the amount of surface area with the selected-property value within a given range [704].

In terms of correlation, for a pK$_i$(II) prediction on the learning set, q^2 (about 0.8) values showed minor variations with the models used. An important point was the satisfactory alignment proposed by Flex, so that this method might be planned for large-scale screening. For the selectivity ΔpK$_i$(I–II), not unexpectedly, results were slightly inferior since the quantity to evaluate was a difference of two single variables (and errors added up). 2D methods appeared inferior to CoMFA and CoMSIA, the difference being greater for VSA descriptors that nevertheless remained usable (q^2 = 0.584). However, the authors stressed that the two MFA methods used several thousand values (values of the fields on the lattice nodes), whereas there were only 166 MACCS keys and 32 VSA descriptors. In the external prediction, for 663 sulfonamide-type inhibitors, CoMSIA gave the best result with $r_{pred}^2 = 0.482$ for $pK_i(II)$, marginally superior to CoMFA, while MACCS keys were still acceptable and VSA clearly failed. The authors saw in these result an illustration of the famous "beware of q^2" (a model with a good q^2 in L-O-O is not automatically good for external prediction) [113]. Spearman's correlation coefficients r^2 varied between 0.443 (CoMSIA) and 0.288 (VSA) for $pK_i(II)$, but much lower values were obtained for selectivity (<0.12).

Despite their qualified results, these approaches may still be useful in screening, where the membership of a class (active, inactive) is by far more important than a numerical activity prediction. To categorize compounds as actives vs. inactives, a threshold must be fixed. Selecting the lowest 5% of $pK_i(II)$ or $\Delta pK_i(I-II)$ is aimed at retrieving compounds with very high activity, or those that are highly selective toward the target hCA II. Selecting the highest 5% favored affinity to hCA I (in other situations, this would correspond to "antitarget modeling," where a given receptor must not be inhibited). Compared to 3D approaches, 2D methods exhibited a higher sensitivity (fewer actives omitted) but a lower specificity (more false positives). Results were better in the identification of those compounds that were the most active (or selective) toward hCA II. This corresponded to the fact that compounds of low activity for hCA II were often overpredicted.

The study also gave for the four approaches, the ROC that specified the performance for a given level of noise. On the other hand, in the enrichment curve, the percentage of actives retrieved was plotted vs. the size of the subset screened. The example given concerned the search for the 5% more active (the lowest $pK_i(II)$). The published curves [702] indicated that, when the first percentage of the base was screened, the four methods separated nearly 50% of the actives (Figure 8.1). Then, exploring the base in a larger manner, the performance was judged "satisfying" [702] although "not perfect." At 50% screening, CoMSIA (the best) and CoMFA selected about 90% of the actives, whereas the VSA model only reached 80%. As to the retrieval of the 5% highest selective toward hCa II $\Delta pK_i((I)-(II))$, in the early stages, discrimination was good for CoMFA, CoMSIA, and MACCS (VSA not being better than the random selection). When the screening went on, all models were said to be "disappointing" with a slight advantage for the 3D approaches, but VSA still remained close to the "no model."

A definite, general conclusion would be difficult to draw since, as stated, relevance and applicability depend on the specific problem dealt with. In this example, 3D methods significantly outperformed 2D approaches as to numerical prediction. In classification, they were more specific (but with a lower sensitivity) that the MACCS approach, easier and less time consuming. But, according to the authors, using 3D methods in screening would not be considered as "superfluous."

8.1.2 Joint Application of CoMFA and Docking

CoMFA also intervened as a component of the strategy developed by Zhang et al. [75] in the virtual screening of a rather extended database. In this study, the authors emphasized that conjointly using CoMFA analysis and structure-based virtual screening constituted an attractive approach in the search for new drugs. The aim was to find

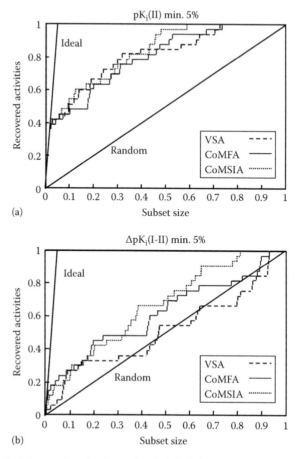

(a)

(b)

FIGURE 8.1 Enrichment plots for the retrieval of the 5% compounds with the lowest $pK_i(II)$ (a) and $\Delta pK_i(I–II)$ (b), respectively, applying three QSAR approaches. The main diagonal corresponds to a random selection, the steep line to the left to an ideal retrieval. (Reproduced from Hillebrecht, A. and Klebe, G., *J. Chem. Inf. Model.*, 48, 384, 2008. With permission.)

drugs with the ability to lower the concentration of low-density lipoprotein cholesterol (LDL-c) in humans, starting from known statin derivatives. Statins (see compound **3.13**) present a HMG-like moiety (**8.1**) that is able to bind the active site of the HMG-CoA reductase (HMGR), inhibiting its catalytic role. In addition, statins possess a hydrophobic anchor. The structure of six statins, complexed with the catalytic portion of human HMGR, had already been resolved by X-ray crystallography [705].

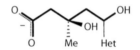

HMG-like moiety of the statins
Het = heteroaromatic

8.1

Thirty-five compounds (with pIC_{50} spanning 3.8 log units) constituted the training set and 12 the test set. For the CoMFA treatment, two alignments were tried. One with the standard field fit method (with the potent compound rosuvastatin (**8.2**) as a template); the other one derived by docking into the crystal structure with FlexE [283,706] that included not only the flexibility of the ligand (as in FlexX [280]) but also the protein structural variability. These two approaches led to rather similar results with a slight advantage to docking. With region focusing (docking model) $r^2 = 0.947$ and $q^2 = 0.731$ ($s = 0.21$). In the test, the mean error was 0.24. The CoMFA model was further consolidated by the good complementarity observed between CoMFA contour plots and the solvent accessible molecular surfaces.

*Indicate alignment points

8.2

Virtual screening was then started by the examination of a database of 41,393 structures. Lipinski rules (a maximum of five H-bond donors, of 10 H-bond acceptors, log P lower than 5, and a molecular weight inferior to 500 Da) constituted a first screen, followed by filtering with a flexible pharmacophore derived from known binding compounds with the tuplets approach [707]. In all, 4138 entries survived. These candidates were then evaluated by both CoMFA analysis and the FlexE docking score (the latter approach attaching greater importance to H-bonds, whereas CoMFA was more sensitive to steric fields). In the final step, the 2291 best-ranked compounds were docked into the active site using FlexX-Pharm [708] with constraints (H-bonds with five residues of the binding site selected from the known crystal structure of the HMGR complex) (Figure 8.2). Eight structures of predicted high activity (in both CoMFA and Flex) were so proposed. It is noteworthy that they were not statin-like. Furthermore, the ninth hit found was α-asarone (**8.3**) (a constituent of a bark already used against hypercholesterolemia).

8.3

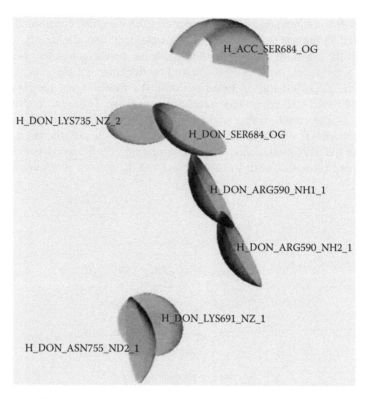

FIGURE 8.2   Pharmacophore constraints from FleX-Pharm. H_ACC and H_DON denote the H-bond acceptor and the donor, respectively. (Reproduced from Zhang, Q.Y. et al., *J. Comb. Chem.*, 9, 131, 2007. With permission.)

The H-bond donor of Lys-735 and Asn-755 were essential. Interactions with Arg-590, Ser-684, Lys-691 (H-Donor), and Ser-684 (H acceptor) were set as optional. Three constraints at least must be satisfied.

8.1.3   Topomer CoMFA

It is also worth mentioning that topomer methodology, with its ability to generate 3D models capable of being superposed, is of great interest not only in a combinatorial library design but also for 3D-based QSAR analyses downstream of the huge number of data generated by high throughput screening. The approach was carried out in a direction opposite to that just presented, where models were first developed for 3D-QSAR and then applied to data mining. By contrast, the topomer methodology was first introduced for database screening and was only later extended to structure-activity relationships. For more details, see Chapter 1.

8.2   AUTOCORRELATION DESCRIPTORS IN DATA MINING

From among those approaches relying on autocorrelation descriptors applied to virtual screening, GRIND was used on various examples.

8.2.1   SCREENING WITH GRIND DESCRIPTORS

A first application may be found in Ref. [74] in the quest for new openers of pancreatic K_{ATP} channels. ATP-sensitive potassium channels couple changes in blood glucose concentration to insulin secretion. They constitute a promising target for the treatment of metabolic disorders. However, in this study, the use of QSAR descriptors only intervened after the a priori screening of pharmacokinetic criteria. First, a subset of roughly 65,000 commercially available chemicals was extracted from the ZINC database [709], and a first (very drastic) filtering was carried out with the VolSurf software, on pharmacokinetic criteria [178,681], see Chapter 6.

From the 94 descriptors obtained with GRID probes (OH_2, DRY, and O), predefined library models were used to calculate solubility, Blood Brain Barrier (BBB) penetration (the most stringent criteria), absorption plus molecular weight, and log P. These filters allowed the selection of compounds noncrossing the BBB, which exhibited good solubility, adequate absorption, and lipophilicity. Only 1913 compounds passed these filters. Then, a pharmacodynamic similarity search was performed. Six molecules (thienothiadiazine 1,1-dioxides and benzothiadiazine 1,1-dioxides) were chosen as templates (**8.4**) (with IC_{50} between 0.01 and 1 µM) and assigned weights reflecting their activity. This stage involved GRIND descriptors [45]. Let us just recall the procedure:

8.4

- Calculation of the Molecular Interaction Fields (MIFs) with DRY, O, N1 probes plus TIP characterizing the molecular shape.
- Filtration to select the most important positions that might be viewed as defining a virtual receptor site.
- Determination of the correlograms, retaining only, for each type of interaction, the maximum value. They characterized interactions between important node pairs and the corresponding distance.

Ten conformers were generated for each of the 1913 selected compounds. For each molecule (including the templates), the descriptors were reduced to their projections on the plane of the first two principal components in a PCA treatment. This data reduction retained only 40% of the variance of the system. For a molecule I, the GRIND score was defined from the distances d_{it} to the six template molecules t (taking into account the weight w_t reflecting activity).

$$\text{GRIND score}_i = \sum_{t=1,6} w_t d_{it} \quad (\text{summation from } t = 1 \text{ to } 6)$$

From the 200 top ranked compounds, after elimination of unstable or toxic chemicals, 16 finally survived.

Two other methods were used with Fingerprint for Ligand And Proteins (FLAPs) [455] and Triplets Of Pharmacophoric Points (TOPPs) [456] 3D descriptors. Schematically, in FLAP, a molecule was described by the exhaustive set of all atom quadruplets, encoded with interatomic distances (in bins of 1 or 1.5 Å) and the nature of the atoms (hydrophobic, H-bond donor or acceptor, positively or negatively charged) [455]. Other properties might be considered, plus an additional flag for the chirality of the tetrahedron. For the FLAP descriptors, by encoding only the presence of quadruplets, a supermolecule was built from the six templates. After alignment, a score was defined by the number of common quadruplets, in a pairwise comparison. TOPP used triplets of atoms, similarly encoded (see Section 7.7.3.2). For the exhaustive set of triplets, these descriptors were organized into a unique vector. The fingerprint (presence or absence of each triplet) could be stored for each molecule and the template. Usual chemometric treatments, such as GOLPE were performed, whereas with FLAP, only pairwise comparisons were used. With the TOPP descriptors, a PCA treatment (3 PCs for 31% of the variance) permitted the calculation of the similarity score (as for the GRIND score).

Eight hits were found with each of the FLAP and TOPP descriptor sets. But, they were not the same as those found by GRIND. So, the three methods identified leads, but none was able to find all of them. The solutions they proposed were completely different and differed also from the structure of the templates. Experimental tests showed that three compounds, selected from GRIND (**8.5**), FLAP (**8.6**), and TOPP (**8.7**), and the reference compound BPDZ (**8.8**), efficiently inhibited insulin release.

8.5

8.6

8.7

8.8

8.2.2 IMPORTANCE OF SIMILARITY (TRAINING/TEST) FOR A GOOD PREDICTIVE POWER

This point is addressed in the work of Benedetti et al. [703] to antagonists of the CysLT1 receptor, acting as antiasthmatics. This study with GRIND descriptors affords a good illustration of the importance of the similarity between the training and test sets for a good predictive power. The training set comprised 54 compounds of the quinoline (bridged) aryl type where the bridging rings were a benzene or naphthalene moiety (**8.9**). GRIND analysis proposed, as important variables, a hydrophobic interaction (DRY–DRY) at 25 Å and an H-acceptor interaction (N1–N1) at 27.5 Å as favoring activity, whereas an interaction DRY–DRY at 31 Å was detrimental. This model led to $r^2 = 0.67$ and SDEC = 0.47 (with two components on 3 pK_i units). These results compared well with the previous CoMFA results, and the derived pharmacophoric features were also in good agreement with the proposal of Palomer [710] (except the central naphthyl moiety not detected with GRIND).

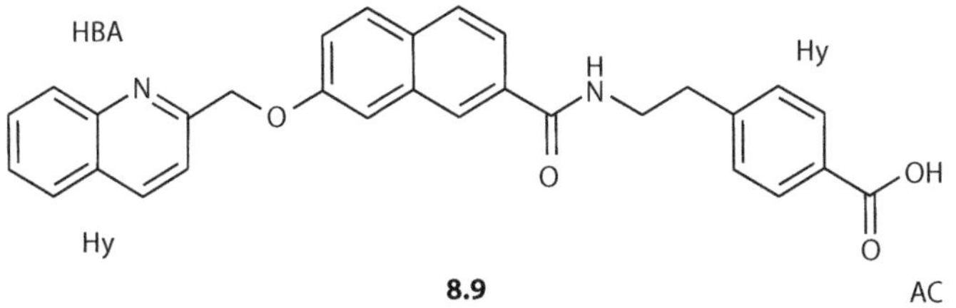

8.9

AC

Hy: hydrophobic, HBA: H-bond acceptor, AC: acidic or negative ionizable function

The model was then applied to a test set of 69 compounds. It established the importance of a high activity of the variable 22–55, corresponding to the H-bond acceptor interaction (nodes N1–N1 encoded 2–2 at 27.5 Å), associated with the carboxyl group and the quinolinic nitrogen present in the highly active compound (**8.10**), but missing in the weakly active compound (**8.11**) (Figure 8.3). On the other hand, whereas about 50% of the compounds (closely similar to the structures of the training set) were correctly predicted, more dissimilar compounds (with a bridging ring constituted by a nitrogen heterocycle, such as piperazine, homopiperazine, and piperidine) largely deviated. A discriminant analysis allowed the identification of GRIND variables associated with this breaking behavior. For example, two DRY–DRY interactions at 13 and 29 Å were absent in the outliers that, in addition, presented a DRY–DRY interaction at 18 Å.

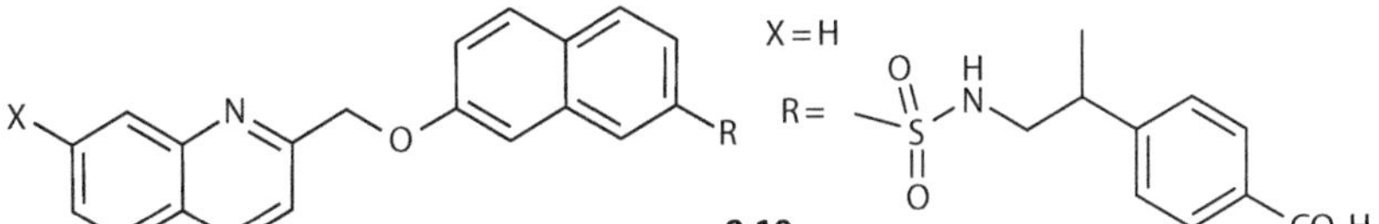

8.10

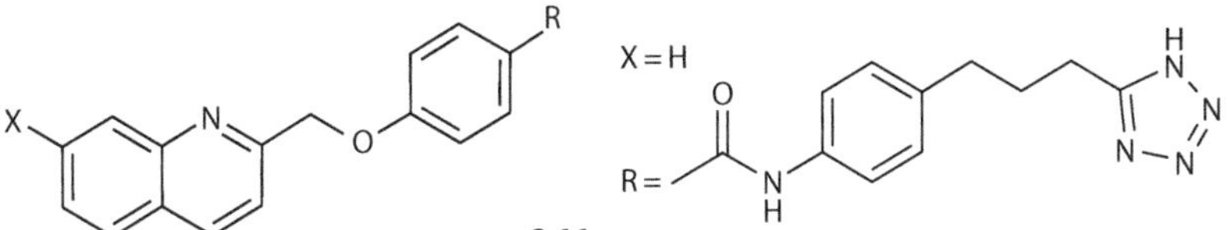

8.11

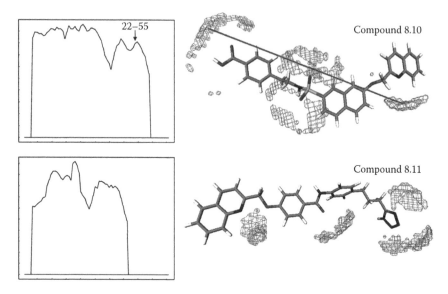

FIGURE 8.3   Graphical display of GRIND 22–55 variables with high impact on CysLT1 receptor affinity. The correlograms (left part) indicate that the variable 22–55, corresponding to H-bond acceptor interaction (nodes N1–N1 at 27.5 Å) is present in the highly active compound (**8.10**), but missing in the weakly active compound (**8.11**). (Reproduced from Benedetti, P. et al., *Bioorg. Med. Chem.* 12, 3607, 2004. With permission.)

8.2.3   2D vs. 3D Descriptors

A comparison of the information given by different 2D or 3D descriptors, including GRINDs, and descriptors derived from physicochemical properties was tackled by Cruciani et al. [454] who also examined their suitability for practical applications. Rather than data screening for the search of targeted new leads, the aim was mainly to evaluate the ability to detect pharmacodynamic similarity. The descriptors used in this study were

- The calculated log P with either substructure-based or whole-molecule approaches
- 2D UNITY fingerprints [452]
- 2D ISIS keys [453]
- GRIND Molecular Interaction Fields [45], where an autocorrelation function transforms the structural information into a small set of alignment-independent descriptors
- VolSurf [178,449] that analyzes the MIFs and computes the volume and surface delineating the regions enclosing values of interaction energies, and some other variables expressing their distribution in space

Three databases were analyzed. The first one, for pharmacodynamic studies, encompassed 1007 compounds, composed of four classes of pharmaceuticals of varied chemical structures (penicillins, β-blockers, benzodiazepines, or class I antiarrhythmics). For pharmacokinetic studies, two key properties were investigated: the solubility data of 100 molecules, and the blood/brain barrier permeation for 229 compounds.

Data were analyzed with consensus principal component analysis (CPCA) [711] where the descriptors were not variables but blocks of variables. Results were analyzed at two levels: at a block level and at a superlevel traducing the consensus of all blocks. For the pharmacodynamic database, CPCA was applied using 1771 variables corresponding to the different descriptor-types investigated. A two component model (20% variance explained) fairly separated the four categories. UNITY fingerprints gave the best clustering, followed by GRIND. ISIS keys were less successful (β-blockers and benzodiazepines were less separated and interfered with antiarrhythmics), VolSurf and log P being the worst. Loading plots (which represent the original variables in the space of the principal components) evidenced some proximity between UNITY fingerprints and ISIS keys as well as between log P and VolSurf descriptors, the GRIND descriptors that favored 3D potential interactions rather than the presence of similar patterns, being apart.

As to the pharmacokinetic aspects, for the two studied examples, VolSurf descriptors gave a better clustering while GRIND showed intermediate behavior. Of these examples, the parameter log P was of limited applicability.

8.2.4 *auto*MEP/PLS Approach in Data Mining

The combination of PLS with MEP surface properties, in an autocorrelation model *auto*MEP/PLS proposed by Moro et al. [424] gave robust ligand-based QSAR for prediction of human adenosine A3 receptor antagonists (see compounds **4.1**, **4.2**). So, for 358 antagonists, $r^2 = 0.67$ and $q^2 = 0.65$. This approach was extended [426] as an efficient pharmacodynamic-driven filtering method in a small-size combinatorial library. A total of 841 compounds were generated from the scaffold of the known human A_3R antagonist pyrazolo-triazolo-pyrimidine together with 29 fragments. The method took advantage of the characteristics of the autocorrelation vectors (fixed length, invariance to translation-rotation avoiding tedious alignment processes) and used a 2D → 3D model builder to generate, as a prerequisite, a "good" conformation.

Among the 100 structures predicted as active, nine were prioritized for synthesis (according to stability, solubility, and ease of synthesis) and their activity experimentally determined. All were found active and two of them at a subnanomolar level.

8.3 VIRTUAL SCREENING WITH COMBINE

Ortiz et al. [246] examined the influence of various settings (data scaling, variable selection) in a CoMFA treatment using the GRID energy function. The data set was constituted of 26 inhibitors of Human Synovial Fluid Phospholipase A_2 inhibitors HSF-PLA$_2$. The structure of HSF-PLA$_2$ was solved by X-ray crystallography both in its native form and in a complex. CoMFA was also compared with COMparative BINding Energy (COMBINE) [247,248]. COMBINE analysis started with the generation of the 3D structure of the complexes inhibitor-HSF-PLA2 (using the cocrystallized structure of an analogue) and of the free ligand. The total free energy of binding was then evaluated as a sum of contributions representing the different interaction energy terms involving all ligand residues with all receptor residues. The **X**-matrix of the individual contributions was similar to a CoMFA matrix with the energy terms in place of the field values. In this application, it encompassed 3310 columns and 26 rows (number of inhibitors).

The authors examined the influence of various pretreatments before the variable selection with GOLPE or q^2-guided region selection. They involved variance scaling (autoscaling or block scaling) and dielectric scaling (constant dielectric, distance dependent $\varepsilon = r$, and the Wharshel model

$$\varepsilon = 1 + 60 \, [1 - \exp(-0.1r)]$$

or the Hingerty model

$$\varepsilon = 78 - 77 \left(\frac{r}{2.5} \right)^2 \left[\frac{\exp(r/2.5)}{\exp(r/2.5) - 1} \right]$$

that had given results comparable to those of the Poisson–Boltzmann model).

Variable selection was carried out with the q^2-guided region selection or with GOLPE, relying on the validation of a number of reduced models. In GOLPE, variable combinations were selected according to a fractional factorial design (FFD) strategy, where successive models were generated by elimination of 10% of the variables.

In conclusion to this study, it appeared that the dielectric constant, energy cutoff, grid spacing, and variable scaling showed a strong influence on the results. With the GRID energy function, a constant dielectric ($\varepsilon = 4$) and unscaled data, seemed the best choice yielding with GOLPE $q^2 = 0.754$. It was also indicated that a region-guided selection involved a smaller number of variables than GOLPE, and might include variables only because they belonged to the same box as important variables. GRID-CoMFA models were consistent with COMBINE in that they identified approximately the same intermolecular interactions.

The authors also stressed that a validation of the CoMFA models was always difficult. "Any index of predictive ability should be interpreted in relation to the underlying structure of the data to avoid misleading conclusions."

Lozano et al. [712] confirmed the reliability of COMBINE analysis in a comparison with a GRID/GOLPE approach studying the genotoxicity of 12 heterocyclic amines N-oxidated by the human cytochrome P450 1A2 (CYP1A2h). In this application, a model of the enzyme was first built by homology with cytochrome P450BM3 and the complexes HCA-CYP1A2h were generated with AutoDOCK.

To improve efficiency in virtual screening approaches, Murcia and Ortiz [248] proposed to associate the receptor-based QSAR model COMBINE and a flexible docking program in a fully automated procedure. The COMBINE methodology [247] estimated binding energy differences from computed interaction energies

u_i (van der Waals and electrostatic) between the ligand and each protein residue in the energy-minimized complex.

$$pK_i = \sum w_i^{vdw} u_i^{vdw} + \sum w_i^{ele} u_i^{ele} + C$$

The evaluation of these interaction energies was based on the Amber MM program. The weights w_i were then determined by PLS regression. Nonbonded interaction energies were calculated using an all-atom model for van der Waals and electrostatic terms. It was stated that splitting the global ΔG into separate contributions allowed the highlighting of interactions that might be important in the design of new active compounds (and also that errors in the model might be filtered by the PLS analysis). Before this regression analysis, a ligand conformational search was performed. It worked by the systematic traversal of the torsional-angle space of the ligand, using only predefined values for dihedral angles. All possible dihedral angle combinations were generated and the intramolecular energy calculated. No minimization was performed, but all conformers within 30 kcal/mol were saved for docking. A flexible docking program was developed. It used an underlying grid in the receptor active site with precalculated contributions of every receptor atom at every grid point. All possible orientations of each ligand conformer in the active site of the rigid protein were examined. After this initial search, a rigid-body (off lattice) energy minimization was performed. The lowest energy pose was chosen as the binding mode.

The approach was developed on a series of 3-amidino-1H-indole-2-carboxamides analogues (**8.12**) (133 compounds on an activity range of 4 log units), acting as inhibitors of the factor Xα (fXα) that played a crucial role in the formation of blood clots. Inhibitors were docked on three different fXα structures (1fjs, 1f0r, 1xkα) known by X-ray crystallography. After docking, each complex was defined by 286 van der Waals and 286 electrostatic interaction energy terms with the receptor. A PLS regression analysis (after elimination of some outliers) led to $r^2 = 0.752$ and $q^2 = 0.628$ (with eight latent variables) for 114 compounds bound to 1fjs. A similar performance was obtained for 1f0r, whereas 1xkα led to bad results, presumably because of wrong docking on a fXα X-ray structure less precisely defined.

X = –NH, –O

R_1 = –H, –CH$_2$Ar, –Ar, –Alk

R_2, R_3, R_4, R_5 = –H, –OH, OBn, –OCH$_3$
      –F, –Cl, –Br, –I, –NH$_2$, –NO$_2$
      –CH$_3$, –CF$_3$, –CO$_2$CH$_3$

R_6 = -3-amidinino, -4-amidino, -3-CSNHBn

8.12

This two-step procedure was then applied to the virtual screening of a small in-house library of 112 ligands containing both active and inactive compounds. According to the authors, docking energies alone were adequate for mining general databases and selecting hits, but the inclusion of COMBINE provided improved rankings. So, COMBINE would be well suited for screening focused libraries where a scaffold was already identified.

8.4   OTHER APPROACHES

For the calculation of free energies of binding, several approximate methods have been developed, that were able to deliver rapidly accurate estimates and to correctly express small activity variations in a series of molecules, largely outperforming usual scoring functions. Although their (limited) throughput precludes using them in screening huge databases, they may give interesting results as a second diagnosis tool on sets of about 1000 compounds surviving a first rough filtering. This point will be developed in Part IV dedicated to receptor-related models.

Let us also remember that the EVA spectroscopic descriptors [51] have been used in structural similarity searching. Simulated property-prediction experiments established that this approach outperformed 2D fingerprints, and suggested that data fusion might significantly improve the performance for similarity searching in chemical databases. However, a drawback of EVA is the need for the calculation of the normal modes of vibration, even if this can be done once and for all [529].

We also mention that similar approaches were carried out using MOLCONN-Z 2D descriptors [713] by Tropsha's group, with firmly validated QSAR models, and by screening huge data sets (up to 750,000 compounds from three publicly available databases in Ref. [76]). Novel anticonvulsant molecules were so proposed after a k-NN analysis [701], whereas the support vector machine (SVM) approach was also used (parallel to k-NN) in the search for D1 dopaminergic antagonists [76]. In this study, CoMFA (as GA-PLS) proposed good models on a set of 48 antagonists, but it was not used in screening, due to the need for conformational analysis and alignment.

As previously stated, identification of pharmacophoric sites, and conversely of positions nonavailable due to the receptor, are important components in VHTS. 4D-QSARs resting on the grid cell occupancy descriptors are particularly able to supply such information, directly and not via the bias of some abstract descriptors. Furthermore, inclusion of external parameters, such as log P, may give information about in vivo transport. Another important requirement is the necessity of exploring the largest possible (real) space. The 4D-QSAR models that sample the conformer states and explores the thermodynamic probability of atom-type site occupancy, are particularly well suited for the development of such applications [700], see Chapter 9.

Part III

Beyond 3D

3D-QSARs take into account the spatial arrangement of atoms in a molecule, but consider only a unique bioactive conformation in a frozen geometry. This bioactive conformation and its location within the receptor active site may be guessed by examination of rigid active compounds, by analogy with known drug–receptor complexes solved in X-ray crystallography, or determined among a set of plausible candidates in the search for the best relationship between predicted and observed activity. For example, N-way PLS treatments [66,67,311–313] allow the consideration, at once, of different types of conformations (associated to a reference compound) for the molecules under scrutiny, but in the final model, only one type of alignment and one conformation are retained and not a conformational mixture.

In another approach, Lukakova and Balaz [62] examined the case of several competitive binding modes of a rigid molecule on the same receptor (on the example of polychlorinated benzodioxins binding the ArH receptor). The problem comes then from possibilities of flip-flop motions (up–down, right–left). In a more general concern, the need to take into account simultaneous multiple conformations, alignments, and substructures prompted the development of 4D-QSARs. The 4D-QSAR formalism, introduced by Hopfinger [63], addressed some problems posed by classical 3D-QSARs, like CoMFA:

- The identification of the active conformation
- The selection of an alignment for comparing molecules
- The identification of the interactions between the receptor and various parts of the ligands

However, with its new developments (docking, molecular dynamics simulations) CoMFA and CoMSIA now afford some reply to the first two problems. A neighboring concern also prompted the development of the Quasar methodology by Vedani et al. [714]. Some flexibility of the receptor on ligand binding (induced fit) will be considered in the so-called 5D approaches. The possibility of various solvation schemes added a sixth dimension.

9 4D-QSARs

As stated by Hopfinger et al. [63], the fourth dimension of 4D-QSAR analysis is the dimension of the "conformational ensemble sampling." The differences in activity of the ligands correspond to the differences in their interaction with the receptor, but now, for each compound, a Boltzmann conformational average is considered and not a unique conformer [715]. It was also suggested that 4D-QSAR, allowing the prediction of active conformations and preferred alignments, might also be used as a preprocessor for a subsequent CoMFA treatment.

We consider here the receptor-independent (RI) approach (as in CoMFA). But the analysis can also be used, with minor modifications, for receptor-dependent (RD) problems (when the structure of the receptor is known). Let us note, however, that as in the previous ligand-based 3D-methods, the receptor is here kept rigid.

9.1 4D-QSAR: OCCUPANCY ANALYSIS

We now briefly indicate the main steps in building a 3D-QSAR using the 4D-QSAR paradigm of Hopfinger et al. [63,715], before presenting some applications.

9.1.1 FORMALISM AND MODEL BUILDING

Schematically, the method is based on embedding the molecules in a 3D lattice. The descriptors are constituted by the occupancy of the elementary cells of this lattice (possibly specifying some types of atoms). But the method considers for each compound several conformers (sampled by molecular dynamics [MD]), and envisages various possible alignments. Partial least squares (PLSs) and genetic algorithms (GAs) are then used to perform a very important data reduction and select, from these different assays, the top model(s) that lead to the best correlation(s). A visual display of the "interesting" regions is also available, giving some insights about the intervening interactions. We now go into more details:

- First, a grid is generated and the 3D-(initial) structures of all compounds of the training set are specified. The grid mesh is one of the methodology parameters that may be further optimized for deriving the best QSAR. Each of the 3D structures is the starting point for a conformational ensemble sampling. In practice, a common low-energy conformer is chosen (it is not necessarily the energy minimum).
- In step 2, the atoms of the molecules are partitioned into five classes or IPE (Interaction Pharmacophore Elements), later extended to seven: polar (positively charged or negatively charged) or nonpolar, H-bond donor or acceptor, aromatic atom, and "all atoms" (no differentiation). A user-defined type may be added [716].

- The third step carries out a conformational ensemble sampling. Rather than a systematic conformational search (that is time consuming and requires a Boltzmann scaling), MD simulations were used, the delicate point being to determine sampling convergence (e.g., a resulting Boltzmann distribution independent of the sampling size or a same distribution reached from different starting points). This leads to the Conformational Ensemble Profile (CEP) that provides information about molecular flexibility. For other approaches of conformational sampling, see Ref. [717].
- In step 4, a trial alignment is chosen (this trial alignment is defined with three atoms for each molecule). Every conformation from the CEP of every compound is placed in the reference grid cell space according to the trial alignment under scrutiny.
- In step 5, the grid cell occupancy profiles are computed and used as the basis set of trial 3D-QSAR descriptors: the grid cell occupancy descriptors (GCODs). The basic assumption is that the observed differences in reactivity are related to the differences in cell occupancy. Three types of cell occupancy may be considered.

Absolute occupancy that indicates the number of atoms of a given compound in the cell under scrutiny (in fact, the center of the atom can be anywhere in the cell). The calculation takes into account the fraction of time the cell is occupied during the MD simulation. For the grid cell (i, j, k), and molecule c:

$$A_o(c, i, j, k, N) = \sum_{0-\tau} O_t(c, i, j, k)$$

where
  τ is the time-length of the MD simulation
  $\tau = N$ (number of sampling steps) $*$ Δt (time step)
  $O_t = m$, if there are m IPE atoms in the cell at time t

Joint occupancy with some reference compound

Self-occupancy: absolute occupancy minus joint occupancy. The cell is occupied by just the given compound.

In fact, values are normalized (divided by N) and deliver a fractional occupancy.

- In step 6, data reduction is performed with PLS, since, as for CoMFA or CoMSIA, a huge number of descriptors are generated (owing to the number of grid cells and the various IPE types). Generally speaking, examples showed that, usually, a small number (about 15) GCODs were significant (among more than thousands, initially). The optimal number of components in the PLS treatment was determined (as usual) by one of the following criteria:
  - The highest value of q^2
  - The smallest value of the cross-validated standard error of estimate

- The highest amount of explained variance in y (the gain, using in the model one more component, must be greater than some tolerance value)

PLS without cross-validation is subsequently used to derive the optimal equations by extracting the established best number of components.

- 3D models are generated in step 7, which includes also evaluation and optimization via a genetic algorithm, starting from the most highly weighted PLS descriptors (about 200) and possibly other user-selected descriptors, such as log P, and molecular refractivity. PLS or MLR may be chosen as the objective function of the GA. Two types of GA may be used
  - The genetic function approximation (GFA) [108]
  - The GERM genetic algorithm [393,394]
- The preceding process was for a fixed alignment. The operations are repeated again with other alignments and with other conformers for all molecules in step 8.
- After this loop is performed, for all alignments and all conformers, a set of a few top models, those giving the best 3D-QSAR (since GA often does not give a unique solution but some solutions of comparable performance), is selected for step 9. Rogers [716] also suggested that the cross-correlation matrix of the residuals, from the selected top models, gives supplementary information. Two models with a high correlation coefficient (similar residual errors) express the same information. But, two models with different residual errors may provide different information. An extreme situation is that of two good models poorly intercorrelated. This may be viewed as corresponding to different binding modes or different views of the same binding mode.

  The cross-correlation matrix of the GCODs within a model indicates which pairs of GCODs (if any) are highly correlated. This may indicate redundant information but may also detect an "allosteric" effect in connection with the conformational behavior of the molecule [63,716].
- In the tenth and final step, *active conformation* is determined. The low-energy conformers, for example, less than 2 kcal/mol with respect to the global minimum of the CEPs, are individually evaluated with the best QSAR. Since only one conformer is now used, cell occupancy is either 0 or 1, as if this conformation was the unique one sampled at 0 K. The conformation giving the best result is selected as the active conformation.

It may be noted that 4D-QSAR does not aim at finding the **best** alignment, but rather allows for a rapid evaluation of the quality of an alignment in terms of QSAR. As to the choice of a starting trial alignment, it was suggested that molecular accessibility [718] might be a useful rough guide. A bond ij is considered as spatially accessible if, $A_{ij} = (\delta_i - \delta_j)^{-0.5}$ is high (where δ_i represents the hydrogen-suppressed valence of atom i).

9.1.2   APPLICATIONS OF THE 4D-QSAR FORMALISM

Various application examples were presented.

9.1.2.1   Example 1: 2,4-Diamino-5-Benzylpyrimidines, Inhibitors of *Escherichia coli* Dihydrofolate Reductase

For a data set of 20 inhibitors, four trial alignments were investigated, based on the pyrimidine ring, the three-atom methylene bridge, the benzyl ring, or a combination of the pyrimidine ring, the benzyl ring, and the central bridge. The lattice was a cube of 20 Å edge, with a mesh of 1 Å. The reference compound was trimethoprim (**9.1**). In all, 5000 conformations were sampled from each of the five starting configurations (low-energy conformers of trimethoprim). Only "all-atom" IPEs were considered, leading to 205 descriptors to be used in the GA.

9.1

All top-10 GFA models implied only eight cells (from the 1736 occupied cells), and the best model corresponded to four GCODs descriptors, giving $r^2 = 0.957$, $q^2 = 0.885$, and SD = 0.34 on a range of 5.7 log units.

$$\log (1/IC_{50}) = 0.0205 \ GC1(J_0) - 0.0324 \ GC2(J_0) + 0.1662 \ GC3(J_0)$$
$$+ 0.1794 GC4(J_0) + 5.85$$

J_0 denotes joint occupancy on the specified GCODs (From GC1 to GC4, Figure 9.1). The important GCODs (appearing in the equation of the preceding correlation) may be visualized in space. One cell is near the benzyl side of the CH_2 bridge, the other near the 2-NH_2 group of the pyrimidine ring. The other two cells are near the 4- and 5-substituents on the benzyl ring. It was a little unexpected to see that a constant portion of the inhibitors structure intervened in the expression of the QSAR. The authors suggested that this might be caused by slight differences between compounds due to freedom in conformation and alignment.

Since the structure of a complex (inhibitor-DHFR) has been solved, more information might be gained on the mechanisms by placing the model into the active site of *E. coli* DHFR (note that the structure of the complex was not used in the construction of the model). As to the "interesting" GCODs, GC1 would be located in a hydrophobic region between Ile-94 and Ile-50; CG2 might correspond to a steric interaction with Ile-50 and GC4 would be in a hydrophobic space in the vicinity of Ala-7. GC3 presumably indicated a favorable H-bonding between 2-NH_2 and the carboxyl group of Asp-27. It was also shown that the active conformations of the active compounds were located in a conformational map (Figure 9.2), near the global energy-minimum and near the trimethoprim-bound state.

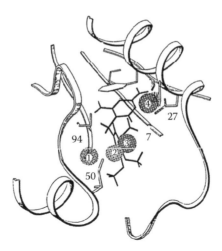

FIGURE 9.1 Schematic representation of trimethoprim bound to *E. coli* and the GCODs of the best QSAR model (grid cells are represented as spheres with 1Å diameter). (Adapted from Hopfinger, A.J. et al., *J. Am. Chem. Soc.,* 119, 10509, 1997. With permission.)

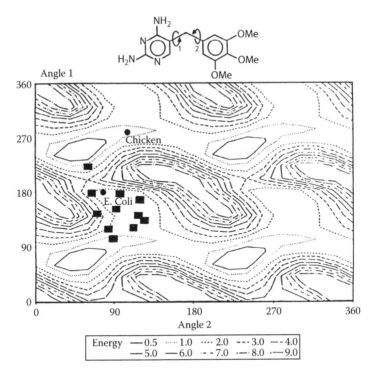

FIGURE 9.2 Conformational map of trimethoprim relative to torsion angles 1 and 2. Energy isocontours in kcal/mol relative to the global minimum. Solid rectangles indicate the postulated active conformation of the 11 most active inhibitors. The enzyme bound conformations of trimethoprim to *E. coli* and chicken are indicated by disks. (Reproduced from Hopfinger, A.J. et al., *J. Am. Chem. Soc.,* 119, 10509, 1997. With permission.)

Such results are clearly better than those obtained in a previous tensor-based 3D-QSAR ($r^2 = 0.913$, $q^2 = 0.816$ with five terms) using 3-way PLS [310].

9.1.2.2   Example 2: Ecdysteroids

The 4D-QSAR formalism was also applied to 71 ecdysteroids [719]. These structures were previously studied by CoMFA [720], but on this series, the need to take into account several conformations of side chains and various possible alignments prompted the development of 4D-QSAR. Ecdysteroids are hormones responsible for the regulation of insect metamorphosis. They are structurally different from vertebrate steroids, and so they may be good starting structures for crop-protection agents, such as insecticides, while at the same time being devoid of detrimental effects on vegetables. Moreover, such studies may orient strategies for generating plants with enhanced levels of potent phytoecdysteroids.

Activity values were measured from absorbance values at 450 nm on the *Drosophila melanogaster* B11 cell line screen on 5.5 log units. They were expressed as $-\log EC_{50}$ (half-maximal effective concentration). According to the methodology specified above, after generation of initial 3D structures and the definition of a 3D-grid, CEPs were determined from MD simulations (about 1000 frames among 10,000 time steps). For building the 4D-QSAR models, five alignments were tested from three-ordered-atom alignment rules. But only two of them gave good results. They largely involved placing the steroid ring system of the compounds upon one another. Inclusion of MlogP as an additional descriptor improved the results, leading to four very similar models. For the best ones, $r^2 = 0.872$, $q^2 = 0.799$ for one alignment; $r^2 = 0.884$, $q^2 = 0.800$ (with MlogP) for the other one (four outliers excluded). The top models identified for both alignments only retained 14–15 GCODs. The cross-correlation matrix indicated that the GCODs of the best models for the two alignments were not correlated. In other words, the two models were distinct and each GCOD grasped different information.

An extensive comparison with CoMFA results [720] was then carried out. For 67 compounds, CoMFA yielded (with steric and electrostatic fields) $r^2 = 0.923$, $q^2 = 0.593$, SEE = 0.37, and SPRESS = 0.85 with five components. For the best models, r^2 values for the two approaches were quite similar, but q^2 in CoMFA was appreciably lower than those of 4D-QSARs. A test set of 20 compounds was also studied with three outliers discarded, which showed that the average residual (0.7 log unit) was twice as large as the corresponding residuals for the training set. But the CoMFA results were not so good.

As to 4D-QSAR analysis, GCODs near C-2 (site numbering is the same as for steroids [**9.2**]) indicated that the presence of a polar acceptor or H-bond acceptor increased activity whereas occupancy by the "any atom" type decreased activity. This suggested that C-2 might act as an H-bond acceptor from the receptor and there were space requirements for that. C-22 was also identified as an H-bond acceptor, enhancing activity. The C-12 region might favorably fit into a hydrophobic area of the receptor. In one of the alignments considered, the C-20 hydroxyl group was involved in H-bond donation, whereas in the other alignment, O-6, C-20, and C-14 were predicted as not involved in binding or equally participating in all the compounds.

The best models suggested for the chain bound to C-20 a largely extended conformation directed away the steroid ring system. For the most active compound, Ponasterone (**9.2**), there might be a receptor cavity, cylinder-shaped and largely hydrophobic, but with an H-bond donor and acceptor at the top of this cylinder able to H-bond with the hydroxyls of C-20 or C-22. Conversely, for the "relatively inactive" 20-*iso*-22-*epi*-ecdysone (**9.3**), the interactions near C-2 and C-3 were maintained, but the carbon C-22 hydroxyl was not near the important GCODs in three of the four predicted active conformations that corresponded to a quasi-folded chain maintaining the substituent distance from the GCODs favoring activity. The fourth conformation corresponded to some groups occupying GCODs associated with diminished activity.

The low values of cross-correlation coefficients of the residuals led the authors to the conclusion that 4D-QSAR and CoMFA differently "explained" the structure–activity relationship. This also suggested that a consensus model of all the best QSAR models would be more likely to give the most robust prediction. For example, the low activity of 20-*iso*-22-*epi*-ecdysone was explained in CoMFA by the lack of occupation of an important negative charge region. In 4D-QSAR, GCODs favorable to the H-bond acceptor/polar negative group were unoccupied and detrimental steric regions were not avoided.

However, it is also interesting to see that CoMFA and 4D-QSAR analyses, although nonidentical, highlighted some common structural features. For example, in CoMFA, regions near C-20 and C-22 were negative charge-favored regions. In 4D-QSAR, occupation by polar atoms and the H-bond donor or acceptor was also considered as favorable. Similarly, according to CoMFA, oxygen substitution at C-24, C-26, and C-27 lowered activity, and in agreement with this observation, 4D-QSAR favored occupation of GCODs in this region by nonpolar atoms. However, due to the 1/r dependence of the electrostatic field, polar and/or H-bond donor groups were associated to rather large regions in CoMFA, whereas in 4D-QSAR, only some GCODs were concerned.

9.1.2.3 Example 3: 4-Hydroxy-5,6-Dihydropyrones: A Novel Class of Nonpeptidic HIV Protease Inhibitors

The study was about inhibition potency $pIC_{50}(M)$ and antiviral activity in infected cells $pEC_{50}(M)$ of 32 derivatives (**9.4**) on a range of 3 log units for both activities [715]. Seven three-atom alignments were considered: points selected on one of the two rings, the central sulfur atom plus the attached carbons, or plus the carbons in *meta* or *para* with respect to the carbons of attachment. The last one was, at the end, determined as the best one (note that it allowed to cope with some twists of the rings). For each of the seven alignments, the top-10 models were evaluated. In all cases, no more than six GCODs were retained. According to the alignment, q^2 varied from 0.6 to 0.9, but for a given alignment, the variation between the models was no more than 0.06.

Het : N-heterocycle
9.4

For the best model (inhibition potency) $r^2 = 0.96$, $q^2 = 0.91$, the mean error on test (five compounds) = 0.8. It is noteworthy that the 10 top models retained the same IPE, and that all but one of the selected GCODs suggested the importance of hydrophobic interactions with receptor pockets. No IPE characterized the hydrogen bonds involving the enolic hydroxyl and the carbonyl group of the common core as suggested by Hagen et al. [721]. In the same study, the antiviral response for cultivated cells was also investigated.

9.1.2.4   Additional Examples

Other examples concern prostaglandin PGF2α antinidatory analogues with high conformational flexibility and dipyridodiazepinone inhibitors of HIV-1 reverse transcriptase. For the latter example, it was found that the same set of inhibitors bind differently to the wild-type WT-RT enzyme and the cysteine 181 mutant enzyme (Y181C-RT) [63]. Interphenylene 7-oxabicycloheptane oxazole thromboxane A$_2$ receptor antagonists [722] and *Plasmodium falciparum* dihydrofolate reductase inhibitors [723] were also studied.

9.1.3   Application of 4D-QSARs to Virtual High-Throughput Screening

Although a little apart from the strict domain of QSARs, an interesting application of 4D-QSAR is the development of virtual high-throughput screens (VHTS). The 4D-QSAR models that explore the conformer states and evaluate site-occupancy are particularly well suited to develop such applications in both lead optimization and identification studies [700].

Basically, the analysis was similar to the prediction of activity for a compound of a test set, but it must be automated for the treatment of extended databases. Each new compound was submitted to the selected alignment. From MD simulation, its GCODs [63,715] were determined and its activity predicted from the 4D-QSAR model. However, an important point was that, such applications required maximum information and not only a unique "best QSAR" for a given training set. Cells that might be important for activity prediction on a large scale may have been ignored when constructing a model on a structurally limited data set. This could be performed using the *manifold of 4D-QSAR models*. It used more GCODs than necessary to get the best correlation, but this appeared useful for those more largely spanning the space (overfitted model). An *additive enhancement model* (see Ref. [700] for details) constituted another possibility.

An application was developed on a focused combinatorial library of 225 virtual analogues of glucose (see compound **4.6**) inhibitors of Glycogen Phosphorylase *b* (Gp*b*) [700]. The basis of the treatment was a 4D-QSAR model established on 47 glucose analogues inhibitors [64]. The best model for the free energy of binding to Gp*b*, on a range of 4 kcal/mol, was given by a six-term linear relationship:

$$\Delta G = 5.04 \ GC1(hbd) - 2.68 \ GC2(np) + 11.22 \ GC3(p-) + 4.87 \ GC4(any)$$
$$+ 2.67 \ GC5(p+) - 1.35 \ GC6(any) + 2.89$$

$$N = 47, \quad r^2 = 0.87, \quad q^2 = 0.83$$

The terms GCX(yz) denoted the GCODs selected by PLS and their nature, such as H-bond donor (hbd), nonpolar (np), etc. In the cited application, starting from the glucose core structure, 15 common substituent groups, such as H, Me, NHMe, and NO_2, for α and β sites were combined leading to 225 virtual analogues. The most important GCODs could be determined looking at their normalized significance. The normalized significance of a model is the square of the correlation coefficient of this model weighted against the sum of squares of the correlation coefficients of all the models in the manifold. Here, three GCODs dominated the 4D-QSAR models and were used as a basis for models with four, five, or six GCOD-based models. These models predicted compounds enlarging the known activity range; from a ΔG range of 1.77–6.65 kcal/mol in the training set to predicted values from 1.05 to 8.65 kcal/mol.

For example, the most active ligand in the training set corresponded to

$$\alpha \, C(=O)NH_2 \quad \beta \, NHC(=O)OCH_3 \quad \Delta G = 6.65 \text{ kcal/mol}$$

The most active virtual ligand was predicted to be

$$\alpha \, SO_2NH_2 \quad \beta \, NHC(=O)CH_3 \quad \Delta G = 8.65 \text{ kcal/mol}$$

Similarly, for the least active,

$$\text{In training } \alpha \, CH_3 \quad \beta \, H \quad \Delta G = 1.77 \text{ kcal/mol}$$

$$\text{Predicted } \alpha \, CH_2C(=O)CH_3 \quad \beta \, SH \quad \Delta G = 1.05 \text{ kcal/mol}$$

9.2   UNIVERSAL 4D FINGERPRINTS

Hopfinger et al. [65,724,725] extended 4D formalism to an estimation of molecular similarity thanks to the introduction of universal 4D fingerprints. As this fourth dimension plays a predominant role, this approach is presented here rather than in Chapter 5 dedicated to molecular similarity-based methods.

9.2.1   BUILDING UP THE UNIVERSAL 4D FINGERPRINTS

Schematically, these descriptors correspond to the eigenvalues of molecular similarity eigenvectors determined for a molecule from its absolute molecular similarity main distance-dependent matrix (MDDM) [65,724]. These MDDMs are, however, noticeably different from the usual interatomic-distance matrices since their elements are a conformational average of the distance between pairs of defined atom types. So they convey information somewhat similar to that of the grid-independent descriptors (GRINDs) that encode selected pairs of atoms of given types at a given distance [45]. But a new dimension is added with the Boltzmann average.

The derivation of the universal 4D fingerprints is developed in Ref. [65] (see also Ref. [725]). The constitutive atoms of molecule (α) are characterized as eight IPEs. These IPEs correspond to the categories already seen (any, np, p^+, p^-, hba, hbd, aro) plus the hs, nonhydrogen atom. For each of the IPE pairs, a unique MDDM is constructed, the elements of which are

$$E(v, d_{ij}) = \exp\left(-v\langle d_{ij}\rangle\right)$$

Constant v(=0.25) increases the difference in the sum of the eigenvalues of any two arbitrary compounds with the same number of IPEs of a given type. $\langle d_{ij}\rangle$ is the Boltzmann conformational average distance between atoms i, j for IPE-type (u,v).

$$\langle d_{ij}\rangle = \sum_k d_{ij}(k)p(k)$$

where
   p(k) is the thermodynamic probability of the conformer state k (computed from the ensemble of conformational energies)
   $d_{ij}(k)$ is the distance between atoms i and j in the IPE pair of type (u,v) for the conformer k

The diagonalization of MDDM provides its eigenvectors and eigenvalues. If the two IPEs are the same (u = v), MDDM is a square triangular matrix. Eigenvalues are normalized and ranked in numerically descending order. The nth normalized eigenvalues for the IPE-type (u,v) of a molecule (α), $\varepsilon_{uv}(\alpha)$ are obtained by scaling the non-normalized eigenvalue $\varepsilon_{uv}'(\alpha)$ relative to the rank of the MDDM. If the members of the IPE pairs are not the same ($u \neq v$), the numbers of IPE elements (n_u, n_v) may be different. MDDM is rectangular, but two square matrices may be built:

$$MDDM(u,u) = MDDM(n_u, n_v) * MDDM(n_u, n_v)'$$

and

$$MDDM(v,v) = MDDM(n_v, n_u) * MDDM(n_v, n_u)'$$

These two matrices have the same set of eigenvalues (same rank, same trace). For each IPE pair with $u \neq v$

$$\varepsilon(\alpha)_{uv} = \{[\varepsilon(\alpha)]_{MDDM(u,u)}\}^{0.5}$$

Each IPE pair corresponds to one MDDM from which one similarity eigenvector can be formed. With eight IPEs, there are 36 pairs (u, v), so there are 36 eigenvectors for each molecule α. Dissimilarity between molecules α and β is given by

$$D_{\alpha\beta} = \sum_i |\varepsilon(\alpha)_i - \varepsilon(\beta)_i|, \quad \text{summation on i (i}^{th}\text{eigenvalue in the specific IPE pair (u,v))}$$

whereas, similarity is defined by

$$S_{\alpha\beta} = (1 - D_{\alpha\beta})(1 - \varphi)$$

where

$$\varphi = \frac{|rank(\alpha) - rank(\beta)|}{[rank(\alpha) + rank(\beta)]}$$

Since the rank of MDDM is the number of atoms of the specific IPE-type, φ incorporates information of the molecular size. S and D vary between 0 and 1. Values close to 1 indicate a high degree of similarity for S and a high degree of dissimilarity for D.

The Universal 4D Fingerprint Descriptor set for a molecule is composed of all the eigenvalues of all the eigenvectors derived for all the MDDMs for molecule α. A threshold (0.002) discards very low eigenvalues. For a given training set, all compounds are assigned the same number of eigenvalues (the maximum number found in the set) and missing values are set to 0. The total number of universal descriptors for each compound is the sum of the $n_{max}(u, v)$ values for the 36 eigenvectors. Each $\varepsilon_i(u, v)$ represents the ith eigenvalue of the eigenvectors for the pair (u, v). So, notation $\varepsilon(5)12(p-,any)$ would represent the 12th eigenvalue (in descending order) for the pair (polar negative, any atom) in molecule 5.

To build a QSAR model, these descriptors (as other "classical" descriptors), generally in high number (about a few hundred) are submitted to PLS or Genetic Function Approximation. A comparison of these two methodologies is developed in Ref. [65]. For each descriptor, the value and its squares may be considered in the correlation.

9.2.2 APPLICATION OF UNIVERSAL 4D FINGERPRINTS

In the seminal paper [65], examples of application of the universal 4D fingerprints were presented on diverse data sets, previously studied by traditional 3D approaches (CoMFA, CoMSIA, HQSAR) and by the 4D-QSAR model of Hopfinger et al.:

- Limited series of rigid phenols analogues of Propofol (general anesthetics) on a restricted reactivity range
- Semiflexible glucose analogue inhibitors of glycogen phosphorylase b [64,700,726] (see compound **4.6**)
- Flavonoids (**9.5**) with affinity for the benzodiazepine binding site on the $GABA_A$ receptor (compounds with potential interest for anxiety-related disorders) as example of large molecules [727,728]

9.5

- Two series of HIV-1 protease inhibitors: 3(S)-amino-2(S)-hydroxy-4phenylbutanoic acids (AHPBAs) (**9.6**) and THP (tetrahydropyridine-2-one) (**9.7**) derivatives, typical of large systems with extended conformational freedom [729–731]

9.6   **9.7**

Generally speaking, universal 4D descriptors led to models of quality comparable to that obtained from 3D models or 4D-QSARs, both in fitting the training set or predicting test compounds. The authors stressed that universal 4D descriptors generated independently of any receptor structure or alignment information "exhibit excellent statistics of fit."

Recently (2009), the same methodology (4D-QSAR and in parallel 4D fingerprint QSAR) was applied to a study of the cytotoxicity of 25 lamellarins (**9.8**) (extracted from a marine mollusc) against human hormone-dependent T47D breast cancer cells [732]. Owing to the small size of the data set (for a family of complex, hard to synthesize, structures), the concern was to extract the maximal binding information in order to make possible virtual screening for driving analogue synthesis or to select alternate scaffolds. This strategy was carried out choosing as the training set only six of the more potent compounds and subsets with constraints on molecular weights or lipophilicity.

9.8

The pseudo consensus obtained between the 4D-QSAR model and the corresponding 4D fingerprint model suggested that the latter model, which did not require alignment, might be particularly useful in virtual screening to guide future drug development.

9.2.2.1 Application to Skin Permeability Study

Skin permeability to chemicals is an important subject, particularly for both drug administration and for toxicological alerts in the cosmetic industry. Penetration enhancers that facilitate the transport of compounds of low percutaneous absorption are therefore of great interest. A molecular modeling of the involved process(es) is, however, very complex since models derived for one drug and a given class of skin penetration enhancers may not apply to a second drug. This problem was recently tackled by Iyer et al. [725]. The aims of the paper were, tentatively, to establish significant QSAR models across a large diversity of penetration enhancers and drugs, and secondly to examine if different models may suggest different mechanisms. Four different data sets were examined. Three of them considered skin enhancers for the nonpolar and highly planar hydrocortisone (HC) and hydrocortisone acetate (HCA) molecules. The fourth series was relative to the enhancement of the small, polar, fluorouracil molecule with a view for use in cancer chemotherapy.

Set 1 (61 compounds) and set 3 (42 molecules) corresponded to surfactant-like molecules with a long alkyl chain bearing a functionalized moiety (lactam, amide, dioxolanes, or ureas for set 1; acyclic azones, amides, or amines for set 3); the reference penetrating drug being HC (set 1) or HCA (set 3). Set 2 encompassed 44 small and relatively polar enhancers (pyrrolidine, amide, urea, DMSO derivatives, and terpenes), also measured using HC as reference. Set 4 gathered 17 terpenic enhancement compounds (reference: 5-fluorouracil). Skin penetration enhancement (mg/cm^{2}h^1) was measured from the permeation curve as the ratio:

$$ER(J) = \left[\frac{\text{Flux for skin treated with enhancer}}{\text{Flux for untreated skin}} \right]$$

In addition to classical QSAR descriptors, Universal 4D Fingerprints: [65,724] were also used, leading to three types of QSAR models: models built from only one class of descriptors or from both classes. Classical QSAR descriptors involved, as usual, the highest occupied molecular orbital (HOMO) and the lowest unoccupied molecular orbital (LUMO) energies, dipole moment, molecular volume and surface area, Kier and Hall topological descriptors, Jurs surface area descriptors, plus descriptors implicitly related to intermolecular interactions (solvation free energy, hypothetical glass transition temperature) amounting to about more than 300 descriptors values.

In the analysis, for each data set, 15% compounds were randomly set apart to constitute a test set. Ten runs were performed with such a splitting and the mean values of r^2 and q^2 were reported. The complexity of the model (number of components to retain) was determined by an examination of these values. Scrambling further validated the results. As to the results, the authors indicated that it was not possible to use a unique model for both nonpolar (surfactant-like) and small polar enhancers, suggesting two different mechanisms (sets 1 and 2). Similarly, comparing sets 1 and 3 (where the enhancers were generally similar, but the reference penetrant different, HC vs. HCA) yielded no satisfactory model, also indicating possible differences in mechanism or transport.

For set 1 (61 compounds), with 79 classic QSAR descriptors and 441 4D fingerprints

	r^2	q^2	N_{des}
Classical descriptors:	0.732	0.661	6
4D descriptors	0.743	0.673	5
Mixed	0.767	0.724	6

Note: N_{des} was the number of descriptors retained in the model.

Although variations in the statistical parameters were not very important, the authors remarked that with the 4D descriptors, only linear terms intervened, whereas classical formalism mainly involved quadratic terms. More interestingly, for training set 2, no good model was obtained with the classical QSAR (76 descriptors), whereas 4D descriptors (with 357 starting values) yielded $r^2 = 0.772$, $q^2 = 0.710$ with six descriptors. For training set 3, the classical QSAR was significantly better than 4D ($q^2 = 0.473$ vs. 0.621), but the combination of the two descriptor sets still improved the statistical criteria ($q^2 = 0.753$, six descriptors). Results were less spectacular in training set 4 (17 compounds only) with 76 classical and 334 4D descriptors, even with the use of the activity value rather than its logarithm (as usual).

A full discussion is out of the scope of this book, but we just indicate that according to the authors, some insight may be gained about the interaction mechanisms, by examination of terms intervening in the selected models: For example, for training set 1, the models highlighted the importance of aqueous solubility (involved by the classical descriptor solvation energy) and suggested an optimal size, indicated by the simultaneous occurrence of a positive contribution 18(hs, hs) and a negative one 22(any, any). Similarly, the intervention of the short-range distribution of polar positive atoms, in a term $\varepsilon 1(p+, hs)$, might be an indication of specific binding interactions. For set 2, the inability to derive a significant model from classical descriptors characterizing the whole molecule might be indicative of mechanisms involving local regions and be best described by specific IPE pairs, as the joint distribution of H-bonding groups and nonpolar ones, $\varepsilon 1(np, hbd)$.

The authors also concluded that the clearly different QSARs established for the four data sets suggested different mechanisms dependent on the enhancer and the penetrant.

9.2.2.2 Blood–Brain Barrier Penetration: A Consensus Approach via 4D Molecular Similarity Measures and Cluster Analysis

The similarity analysis from the MDDMs is also at the basis of a composite approach to model blood–brain barrier (BBB) penetration. This is a fundamental property since drugs acting on the central nervous system (CNS) must go over this barrier, whereas for peripherally acting drugs, penetration may cause undesired side effects [733]. The study was carried out on 150 compounds of largely diverse chemical structures (from methane to indinavir, $C_{36}H_{47}N_4O_4$) with partition coefficients (blood/brain) varying over 3.7 log units.

A correlation with ClogP and topological polar surface area (TPSA), including the whole population, was not very satisfactory ($r^2 = 0.69$, $q^2 = 0.60$). The authors preferred to split the complete data set into subsets based on 4D molecular similarity measures from MDDMs. Using the H-bond acceptor propensity gave the best separation into three clusters. Then, within each of them, training and test sets were defined, and a Membrane Interaction QSAR (MI-QSAR) analysis was carried out, with a model membrane monolayer and MD simulations [734]. MI-QSAR analysis was previously proposed to better predict the transport of organic compounds. It included a set of membrane-solute intermolecular properties that were added to the usual intramolecular solute descriptors.

Using models elaborated with "classical" intramolecular solute descriptors (2.5 and 3D), intermolecular descriptors (dissolution, solvation), and intermolecular-interaction descriptors (membrane-solute), distinct QSARs of better performance were built. It is noteworthy that the best prediction was achieved with models trained within the same subset, highlighting the interest of such a consensus approach compared to a global treatment. These results also suggested, according to the authors, that specific properties acting on BBB permeability might vary, depending on the chemical classes.

9.3   EXTENSION OF CoMSA TO 4D-QSAR SCHEMES

The CoMSA method, developed by Polanski et al. [95,399] was further extended, adapting the 4D-QSAR model of Hopfinger et al. [63] and incorporating the fourth dimension in the SOM-4D models [735]. A systematic comparative study of the performance of CoMFA, 4D-QSAR, and SOM-4D was also carried out.

9.3.1   SELF-ORGANIZING MAP 4D-QSAR MODEL

In fact, the three approaches, CoMFA, CoMSA, and 4D-QSAR, differ in the way they encode the molecular structures. CoMFA uses a lattice of discrete external nodes for the calculation of the steric and electrostatic fields. In CoMSA, as previously indicated, the compared properties correspond to averaged values calculated on certain areas of the molecular surface. On another hand, in 4D-QSAR, the important variable is the occupancy of the grid cells (not so different from the polycube structure representing the molecule as a set of cubic cells in the molecular shape analysis method [41]). SOM-4D is based on an atomic representation and uses GCODs (grid cell occupancy descriptors) but the treatment involves a Kohonen map (Figure 9.3). These processes rely on both a fuzzy molecular-representation depending on the cube resolution and a fuzzy neighbor representation in a Kohonen mapping.

In 4D-QSAR, as in the original model, after an energy minimization (by AM1), an MD simulation samples the conformational ensemble and provides a CEP for each molecule. Then, each conformer is placed in the reference grid (alignment) and GCODs (grid cell occupancy descriptors) are calculated with formulae similar to that indicated for 4D-QSAR. But, apart from the usual GCODs defined by

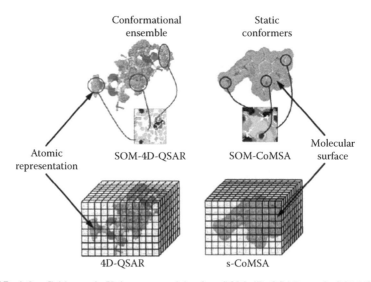

FIGURE 9.3 Grids and Kohonen models for SOM-4D-QSAR and SOM-CoMSA approaches. (Reproduced from Polanski, J. et al., *Molecules*, 9, 1148, 2004. With permission.)

Hopfinger, partial atomic charges may be taken into account. The absolute charge occupancy is now defined by

$$A_q(c, i, j, k, N) = \sum_{t=0;\tau} O_t(c, i, j, k) \frac{q}{m}$$

where
    m is the number of atoms of c present at time t in the cell (i, j, k)
    q is the sum of their partial atomic charges

Similar definitions hold for the joint- and the self-charge occupancy with the reference compound.

In the SOM-4D model, coordinates of the individual conformers sampled during the MD simulation (about 1000 conformations) are given as inputs to a Kohonen NN, trained on a reference compound. Molecules are superimposed, as in the 4D-QSAR procedure, before processing. The neurons, where they are projected, are encoded with the corresponding GCODs, for example, occupancy (output neuron encoded 0 or 1) or partial atomic charges. Each comparative map characterizes the conformational space of the compound analyzed. If several alignments are considered for a molecule with its own CEP, each alignment gives a unique cell-occupancy assessment. Data generated by these comparative maps are then analyzed by PLS.

The method was presented on two data sets: 41 benzoic acids with particular interest in the role of electronic and steric influences ("*ortho* effect") and the 31 CBG binding steroids. In the latter example, the best model on the training set (compounds **s-1** to **s-21**) obtained by grid-4D-QSAR ($q^2 = 0.84$, s = 050, with 2 LVs) slightly outperformed CoMFA ($q^2 = 0.73$, s = 0.64 [126] but was inferior to CoMSA (0.88, 0.42) [95]. In SOM-4D, considering only occupancy yielded $q^2 = 0.86$, s = 0.47 (3 LVs) and with charges $q^2 = 0.83$, s = 0.54. For external prediction (test set), SOM-4D reached stable results (provided that the Kohonen map was larger than 10*10 neurons) with SDEP = 0.75, whereas CoMFA yielded SDEP = 0.837.

These (fairly similar) results with SOM-4D and 4D-QSAR are not really surprising in as much as there is some analogy between the neuron "winning distance" (the attraction basin of a neuron) and the grid resolution. It is also interesting to note that CoMFA and 4D-QSAR, although based on very different approaches to generate molecular data, led to comparable results. Another remark is also that, at least for these series, the influence of the different modes of alignments on the quality of the model was much less pronounced with SOM-4D than with COMFA or 4D-QSAR.

9.3.2 DERIVATING THE s-CoMSA MODEL (SECTOR CoMSA)

To be free of the intrinsic indeterministic character of neural approaches, a possible source of difficulties, Polanski et al. [363] proposed a modified nonneural method **s-CoMSA** based on a grid formalism similar to that adopted by Hopfinger et al. for the 4D-QSAR [63]. This approach allows for both quantitative modeling and for finding possible pharmacophores. In the 4D-QSAR approach, after conformational sampling and alignment, GCODs, including partial charges are calculated.

For the s-CoMSA method, cells of the grid are encoded with the mean value of the electrostatic potential of the surface points found in each cell. Grid cells are then "unfolded" into vectors, ordered in a matrix for all the molecules constituting the series under scrutiny. After the elimination of empty columns (cells empty for all molecules), this matrix is submitted to PLS analysis. However, an important difference with the 4D-QSAR is that this new method (s-CoMSA or 4D-PLS) now compares single conformers, and so relies on the 3D approach.

A comparison of SOM-4D-QSAR with CoMSA or s-CoMSA (3D methods) was performed in Ref. [736] on the example of a series of dihydrofolate reductase (DHFR) inhibitors previously studied by Hopfinfer et al. [63] with 4D-QSAR. In the paper by Polanski et al., special attention was devoted to the interest of variable elimination in 4D-QSAR-PLS treatment, although PLS analysis, basically, implies some reduction in the number of variables to be taken into account.

The data set encompassed 20 2,4-diamino-benzylpyrimidine compounds (see compound **3.3**) inhibitors of *E. coli* DHFR with $\log(1/IC50)$ spanning 5.5 log units. The basis of the comparison was the 4D-QSAR model of Hopfinger et al., optimized by GA, in which a final regression equation was proposed with $r^2 = 0.957$, $q^2 = 0.885$, and $s = 0.34$. The 4D-QSAR-PLS model only yielded a modest performance ($q^2 = 0.43$), but using either UVE or IVE (vide infra) for elimination of uninformative variables largely improved the results. Globally, 4D-PLS or SOM-4D showed a nearly similar performance. With IVE, for example, charge occupancy descriptors led to $q^2 = 0.98$ for both the 4D-QSAR and SOM-4D models ($s = 0.14$ and 0.17).

These results looked better than those of Hopfinger et al. However, a strict comparison cannot be made since one compound was eliminated. As q^2 alone could not ascertain the predictive ability of a model, Polanski et al. split the data into learning (14 compounds) and test (five compounds). Limiting the complexity of the model (four latent variables in PLS) led to $q^2 = 0.96$, $s = 0.12$, and $SDEP = 0.61$ with joint occupancy descriptors.

As expected, 3D models (usual CoMSA or s-CoMSA) yielded poorer results: with CoMSA, $q^2 = 0.87$, $s = 0.59$, $SDEP = 0.72$ and for s-CoMSA $q^2 = 0.73$, $s = 0.83$, $SDEP = 0.93$.

9.3.3 SOM-4D AND (ITERATIVE) UNINFORMATIVE VARIABLE ELIMINATION

The performance of s-CoMSA, compared to SOM-4D, was also examined on other series corresponding to significantly different conditions [363,737]. At the same time, the applicability of iterative variable elimination (IVE) [110], an adaptation of the method of uninformative variable elimination (UVE), was examined, see Chapter 3, Section 7.4. The investigated series encompass:

(a) Three processes typically relevant of QSAR models.
   • The CoMFA steroid benchmark (CBG binding), a drug-receptor interaction problem
   • 50 steroids-inhibiting aromatase enzymes
   • The anti-HIV activity of 107 HEPT derivatives (**9.9**)

(b) An intermediate situation: interaction between azo dyes (**9.10 and 9.11**) and cellulose fibers. This is, a complex and still poorly understood phenomenon for which it was suggested that binding sites may exist in the cellulose crystalline region, forming cavities able to incorporate dye molecules. An efficient working hypothesis is to treat it as drug-receptor interaction (perhaps with less specificity).

9.9

X = NH, O, S
R = Substituted naphtyl
fragment

9.10

9.11

(c) A situation where there is no evidence of any biological receptor-like environment—the ionization of substituted benzoic acids (the definition process of Hammett σ constants). However, various 3D-QSPRs have already been successfully proposed for modeling chemical reactions.

Some of these series had been previously investigated [110,363], but, in this paper, the authors devoted interest to the pharmacophore mapping ability of the approach, and used a novel stochastic procedure to verify the predictive ability of the method. In these studies, comparative Kohonen maps (typically 20×20 or 30×30 units) providing the sum occupancy or the mean charge were established, and the PLS models analyzed with the UVE or IVE protocols [110,408]. IVE tended to limit the complexity of the model (with a small number of latent variables). But, it was better to choose a number of components still inferior to that minimum to get a better stability at the expense of a slightly inferior q^2. Data validation was also carefully carried out by Stochastic Model Validation (SMV) [738].

On the CBG-steroid benchmark, when using uninformative variable elimination (UVE or IVE) in the PLS analysis, it appeared that s-CoMSA only slightly depended on the grid resolution (for a mesh in the range 1–1.5 Å) and led to $q^2 = 0.88$–0.90 with SDEP = 0.78 to 0.73 (for the 21/10 sets). For the same data sets, CoMFA only yielded $q^2 = 0.73$ and s = 0.66 with SDEP = 0.84, on the test set [126,344], whereas SOM-CoMSA gave $q^2 = 0.88$ (for all compounds) with SDEP test = 0.69 [95] and Quasar $q^2 = 0.90$, for the set (1–21) [739]. Points on the molecular surface, surviving the variable elimination process and so efficiently contributing to activity, showed clear similarity with the "interesting" regions visualized by Quasar and might suggest possible pharmacophoric sites (near C-2, C-3, C-12, and the chain on C-17).

However, the authors indicated that adding or subtracting a compound in the set might change these areas, and stressed the stochastic character of the process. The authors also examined whether variable elimination (very efficient in CoMSA) was also beneficial to 4D-QSAR. Results ($q^2 = 0.94$, $s = 0.38$, SDEP = 0.77) became comparable to those of s-CoMSA, and outperformed the classical 4D-QSAR without variable elimination ($q^2 = 0.84$, $s = 0.50$, SDEP = 0.83) [735]. The authors also suggested that for the relatively rigid steroid scaffold, the time-consuming exploration of the conformational space was not very beneficial.

For **steroids inhibiting aromatase** [110], the 50 molecules were split into training (25) and test (25) sets. The best model (s-COMSA-IVE) led to $q^2 = 0.96$ $s = 0.31$ with five latent components (q^2 test = 0.67) [737]. These results compared well with the CoMFA study ($q^2 = 0.72$, $s = 0.77$ with five components) [406] and CoMSA ($q^2 = 0.77$, $s = 0.71$, with five components) [110]. In fact, charge descriptors yielded a better model than occupancy descriptors. For the whole set, the SOM-4D model led to $q^2 = 0.76$, $s = 0.71$, and the 4D-QSAR-J treatment to 0.80 and 0.65, respectively (the reference structure was the most active compound in the series, contrary to Ref. [110], where it was one of the least active).

With models built on the training set, similar results were obtained. SOM-4D and the sum occupancy descriptor led to $q^2 = 0.53$ and r^2_{ext} (on the test) = 0.61, whereas with the mean charge, SOM-4D yielded $q^2 = 0.74$ and $r^2 = 0.67$. With 4D-QSAR-J, the corresponding values were 0.72 and 0.70. As to the benefit of IVE, results were improved from $q^2 = 0.74$, SDEP = 0.72, and $r^2 = 0.67$, to respectively, 0.93, 0.67, and 0.69 (SOM-4D-QSAR, with charge descriptors).

However, it is clear that the quality of a model depends on the splitting of the molecules between the training and test sets. This prompted the authors to perform numerous additional assays. In SMV, the predictive ability of the approach was verified on a very large number of generated models [738]. For a data set of 50 compounds, it was clearly not possible to examine all the possible splitting into subsets of 33 and 17 samples so, "only" 80,000 of such randomly generated distributions were treated. For the mean charge SOM-4D model, q^2 varied from 0.55 to 0.9. But a region of higher density appeared near $q^2 = 0.7$ and $q^2_{test} = 0.8$ (Figure 9.4). These rather high values suggested that the approach possessed efficient modeling and predictive ability. The authors also observed (on 100 random samplings) that the evolution of q^2 and r^2 (depending on the number of variables eliminated) was a stable and smooth process.

It was also possible to visualize (as in some other 3D-QSARs) those areas with the highest contribution for activity. The point was that these areas were not determined here on a single training, but were defined from the examination of variables simultaneously important for all models. In this series of compounds, the importance of the A, B, but also D rings (previously noted [110,406]) was highlighted. For example, occupancy descriptors near ring D indicated a sterically disfavored area, presumably because of the sterical bulk of the receptor pocket. Near ring A and B, the occupancy descriptors also suggested that substitution would increase activity.

Another example relied on the anti-HIV activity of 107 HEPT derivatives (**9.9**) [363,740], where s-CoMSA-IVE-PLS ($q^2 = 0.86$, $s = 0.58$ on a range of 5.1 log units) outperformed previous studies using varied approaches (CoMFA, BNN,

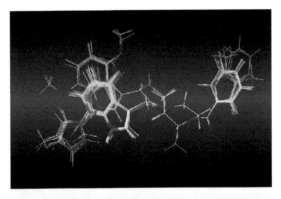

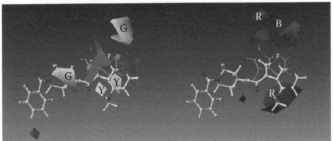

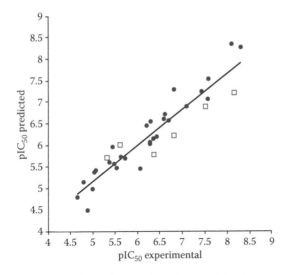

FIGURE 1.5 CoMFA analysis for analogues of BMS-806. (Up) Alignment of 36 ligands' 3D structures; (middle) CoMFA steric (left) and electrostatic (right) contours. Green (G) areas indicate regions where bulky groups increase activity whereas yellow (Y) contours indicate where bulky groups decrease activity. Blue (B) areas represent regions where positive groups enhance activity, whereas red (R) contours indicate where negative groups increase activity. BMS-806 compound is shown with ball and stick representation. Its analogue, which possesses a sulfonamide instead of a ketoamide group, is displayed in wireframe. (Down) Plots of predicted vs. observed pIC_{50} of CoMFA model; square points represent the compounds of the test set. (Reproduced from Teixeira, C. et al., *Eur. J. Med. Chem.*, 44, 3524, 2009. With permission.)

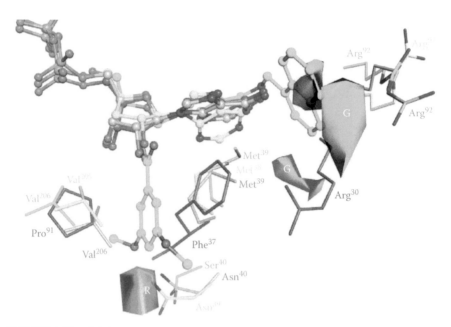

FIGURE 1.10 CoMFA contour plots in selective inhibition of trypanosomatid GADPH. GADPH structures are those of *Leishmania mexicana* (cyan), *Trypanosoma brucei* (yellow), *Trypanosoma cruzi* (gray), and human (magenta). The inhibitor and the cofactor NAD+ are drawn as ball and stick model. In CoMFA contour plots (for *L. mexicana*), the green area (G) corresponds to sterically favored region, and the red one (R) to favorable electronegative area. (Reproduced from Guido, R.V.C. et al., *J. Chem. Inf. Model.*, 48, 918, 2008. With permission.)

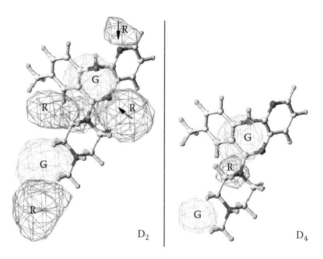

FIGURE 2.3 H-bond acceptor contour plots for pyridobenzodiazepine (**2.2**). Red (R) contours enclose regions in which the presence of a ligand H-bond acceptor decreases (D_2 or D_4) receptor affinities. (Reproduced from Boström, J. et al., *J. Chem. Inf. Comput. Sci.*, 43, 1020, 2003. With permission.)

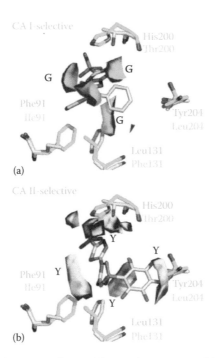

FIGURE 2.5 CoMSIA contour plots and interacting receptor side chains. (a) Green (G) contours correspond to regions where bulky ligand groups increase selectivity toward CA I. Conversely (b) yellow (Y) areas indicate regions where steric bulk increases selectivity toward CA II. (Reproduced from Weber, A. et al., *J. Chem. Inf. Model.*, 46, 2737, 2006. With permission.)

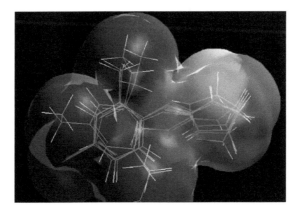

FIGURE 3.10 RSM built (with van der Waals function) from the five more active octopamine agonists (see Section 3.5.3) and colored by hydrophobicity. A (dark) brown area corresponds to a positive contribution of hydrophobicity, (light) gray areas to a negative contribution of hydrophobicity. Note that the phenyl ring and its substituents are hydrophobic whereas the heterorings (imidazolidine and oxazolidine) are hydrophilic. (Reproduced from Hirashima, A. et al., *Internet Electron. J. Mol. Des.*, 2, 274, 2003. With permission.)

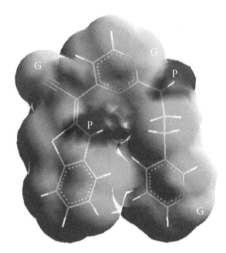

FIGURE 3.11 RSM: mapping of H-bonding propensity. H-bond donor regions are shown in (P) purple while H-bond acceptor regions are shown in (G) cyan color. The compound **3.25** is represented inside. (Reproduced from Shaikh, A.R. et al., *Bioorg. Med. Chem. Lett.*, 16, 5917, 2006. With permission.)

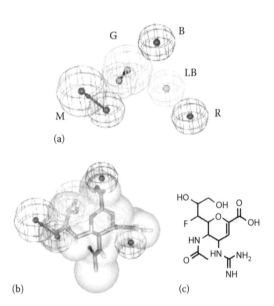

FIGURE 3.19 (a) One of the emerging pharmacophore features. Color code: (LB) Light blue spheres hydrophobic features, (R) red spheres positive ionizable features, (B) blue spheres negative ionizable features, (G) green, and (M) magenta vectored spheres H-bond acceptor and donor features, respectively. (b) inhibitor, (c) $IC_{50} = 0.5\,nM$ fitted against the pharmacophoric features and combined with the corresponding molecular shape. (Adapted from Hammad, A.M.A. and Taha M.O., *J. Chem. Inf. Model.*, 49, 978, 2009. With permission.)

FIGURE 3.23 The best generated pharmacophore model obtained from PHASE. Pharmacophore features are (R) red vectors for H-bond acceptor, (O) orange rings for aromatic groups, and (G) green balls for hydrophobic functions. Two compounds (**3.40**) and (**3.41**) were aligned to the pharmacophore. For molecules, blue indicates nitrogen, red indicates oxygen, yellow refers to sulfur, gray indicates carbon, and white indicates hydrogen. (Reproduced from Zhu, Z. and Buolamwini, J.K., *Bioorg. Med. Chem.*, 16, 3848, 2008. With permission.)

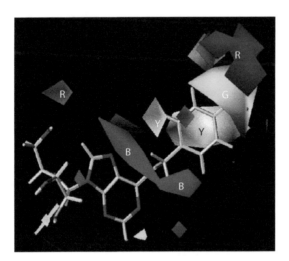

FIGURE 3.24 Contour plots of CoMFA 3D-QSAR model. (R) Red, negative electrostatic potential enhances potency; (B) blue, positive electrostatic potential favors activity; (G) green, bulky substituents enhance potency; (Y) yellow, bulky substituents are unfavorable to activity. (Reproduced from Zhu, Z. and Buolamwini, J.K., *Bioorg. Med. Chem.*, 16, 3848, 2008. With permission.)

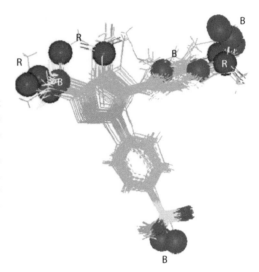

FIGURE 5.7 Map obtained for the set of COX-2 inhibitors (indicator variables for steric and electrostatic contributions, FlexS parameter for hydrophobicity). (R) Red balls indicate regions where an increase of negative charge will enhance affinity. (B) Blue balls indicate regions where more positively charged groups will improve the binding properties. (Reproduced from Kotani, T. and Higashiura, K., *J. Med. Chem.*, 47, 2732, 2004. With permission.)

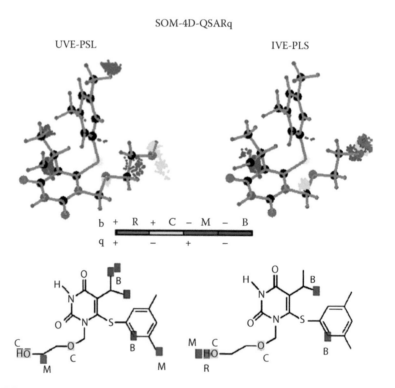

FIGURE 9.5 Molecular coordinates indicated by SOM-4D-QSAR models coupled with variable elimination methods for the compound of the highest activity. Colors code the combination of the mean partial atom charge sign and the sign of the b weight in the model: +/+ (red increases the activity), −/+ (cyan decreases the activity), +/− (magenta decreases the activity), and −/− (blue increases the activity). (Reproduced from Bak, A. and Polanski, J., *Bioorg. Med. Chem.*, 14, 273, 2006. With permission.)

FIGURE 10.1 Steric field vectors originating from the mean envelope surrounding an NK-1 antagonist molecule. Induced-fit is executed along these vectors. (Reproduced from Quasar 5.2 manual. http://www.biograph.ch. With permission.)

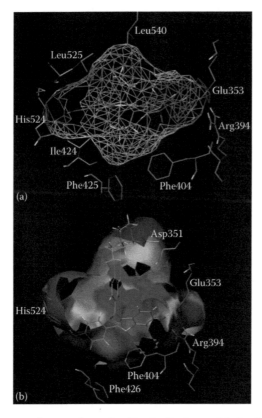

FIGURE 10.5 Common features determined by the Raptor models for the estrogen receptor (hydrophobic fields, (Be) beige; H-bond-donating propensity, (Bl) blue; H-bond-accepting propensity, (R) red, with bound (a) DES and (b) raloxifene). (Reproduced from Lill, M.A. et al., *J. Med. Chem.*, 47, 6174, 2004. With permission.)

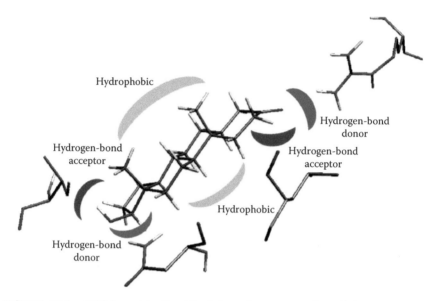

FIGURE 10.6 DHT bound to the AR. Schematic representation of the physicochemical properties of the receptor model and their consistency with the experimental structure. (Reproduced from Lill, M.A. et al., *J. Med. Chem.*, 48, 5666, 2005. With permission.)

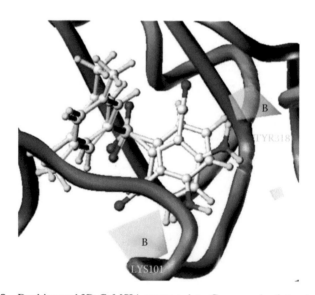

FIGURE 12.8 Docking and 3D-CoMSIA contour plots. Compounds of structure **12.13** with R = di-Me 3,5 on ring B and X = SO and SO₂, and HIV-1RT protein. The cyan area corresponds to favorable H-bond donor fields (and exactly overlaps the residues Lys-101 and Thr-318). Note the flip of the amino substituted phenyl ring. (Reproduced from Hu, R.J. et al., *Bioorg. Med. Chem.*, 17, 2400, 2009. With permission.)

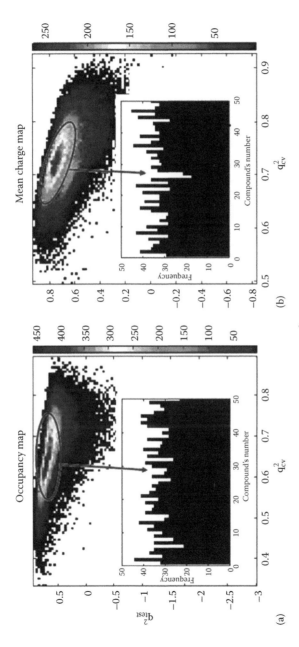

FIGURE 9.4 The density SMV plot illustrating relationships between a q_{cv}^2 value evaluated with the L-O-O CV method for the 33 molecules training set sampled for all 50 molecules against a q_{test}^2 value estimated by the application of the respective L-O-O CV model for the prediction of the activity for the remaining 17 molecules (test set), while using occupancy (a) and charge descriptors (b). Histograms specify a number of the individual compounds appearing in the test set within the most densely populated region that is encircled. (Note that [according to our notations] q_{test}^2 is the determination coefficient calculated on the test set, with SD calculated on the test and not on the mean of the training test.) (Reproduced from Bak, A. and Polanski, J., *J. Chem. Inf. Model.*, 47, 1469, 2007. With permission.)

MLR). For the same series, 4D-QSAR and SOM-4D (plus IVE) gave similar results [740]: the best was SOM-4D (q-IVE) $q^2 = 0.98$, SDEP = 1.68 with nine LV and for 4D-QSAR (J-IVE) $q^2 = 0.92$ SDEP = 1.39 with 10 LV (q and J indicating charge and joint occupancy) (Figure 9.5). For example, CoMFA [240] gave $q^2 = 0.8$, s = 0.53 (80 compounds with one outlier eliminated). Careful validation tests indicated that the s-COMSA method provided reliable and highly predictive models (fulfilling the Golbraikh–Tropsha criterion) [113].

Azo dyes: For 30 heterocyclic monoazo dyes (**9.10** and **9.11**) for which the affinity variations spanned a range of $\Delta(-\Delta\mu^0) = 13.0$ Kj/mol, SOM-4D-QSAR yielded $q^2 = 0.43$ to 0.79, quite consistent with the best CoMFA ($q^2 = 0.44$ to 0.73) [741] and slightly below CoMSA $q^2 = 0.93$ [742].

The treatment indicated that for both SOM-4D and 4D-QSAR, occupancy descriptors led to better models than charge descriptors. This contrasts with CoMFA and CoMSA 3D results that indicated the dominant effect of the electrostatic field [741,742]. So, it was suggested that the incorporation of flexibility (4D models) might lead to different results for nonrigid molecules. Visualization of the important

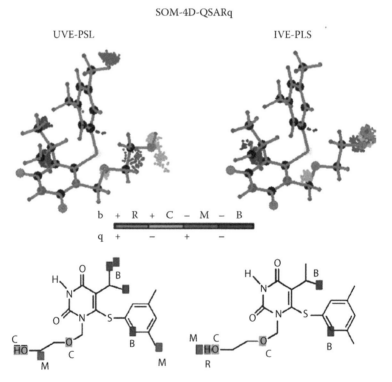

FIGURE 9.5 **(See color insert following page 288.)** Molecular coordinates indicated by SOM-4D-QSAR models coupled with variable elimination methods for the compound of the highest activity. Colors code the combination of the mean partial atom charge sign and the sign of the b weight in the model: +/+ (red increases the activity), −/+ (cyan decreases the activity), +/− (magenta decreases the activity), and −/− (blue increases the activity). (Reproduced from Bak, A. and Polanski, J., *Bioorg. Med. Chem.*, 14, 273, 2006. With permission.)

regions involved more extended areas than in the preceding example of steroids, presumably because of less specific interactions.

For benzoic acids dissociation with 49 molecules in the training set (those studied by Kim and Martin [237]), a CoMFA model (with AM1 charges) led to $q^2 = 0.89$. With SOM-4D, $q^2 = 0.80$, and for the 13 molecules in test, $r_{ext}^2 = 0.65$; SDEP = 0.22 for a single run. This result was obtained with carboxylic group atoms as the only IPEs (the atoms examined in 4D-QSAR) and looked comparable to the two preceding series investigated. But the inclusion of (all) others atoms deteriorated the quality of the model. $q^2 = 0.50$, $r_{ext}^2 = 0.15$, and SDEP = 0.35.

On the other hand, charges on the carboxylic group on a single starting conformer (energy-minimized) gave a good representation of the σ. Here, the SOM-4D model turned out to be much less efficient: The authors argued that the process governed by electronic effects was strictly correlated to a single conformer, and occupancy descriptors were inappropriate. Distortion of the main conformer affected the calculation of charges and decreased the performance. Visualization of the important sites (largest contribution into the activity) focused, as expected, on the carboxylic group. However, the authors stressed the relatively low predictive ability of the model in that series.

9.4   4D-QSAR FROM A SIMPLEX REPRESENTATION

Another approach relying on a system of 3D fragments weighted by conformer probability has been proposed by Kuz'min et al. [743] on the basis of simplexes: tetratomic fragments of fixed nature, chirality, and symmetry. For each molecule, conformers within an energy range of 3 kcal/mol, for example, are described as an assembly of simplexes from 11 base types (Figure 9.6). Atoms are encoded according to their nature, partial charge, lipophilicity, atom refraction, or H-bonding donor or acceptor propensity.

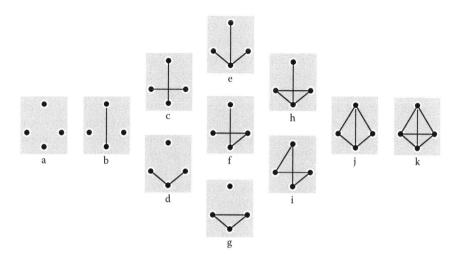

FIGURE 9.6   The 11 base-type of simplexes. (Adapted from Kuz'min, V.E. et al., *SAR QSAR Environ. Res.*, 16, 219, 2005.)

For each of these characteristics, molecular descriptors are calculated as the sum of the structural parameter value for each conformer weighted by the probability of that conformer. The method was presented on the anticancer activity of 31 macrocyclic pyridinophanes and analogues (**9.12**), studied on nine types of tumors. In this application, about 2000 simplex descriptors were calculated. PLS analysis of these descriptors led to good models with r varying between 0.8621 and 0.990. Furthermore, PLS results allowed an estimation of the contribution of individual atoms and also an identification of the fragments that promote or reduce anticancer activity. The model suggested the synthesis of a new compound that proved to be of high activity.

9.12

9.5   DYNAMIC 3D-QSAR: COREPA

The "dynamic 3D-QSAR" techniques proposed by Mekenyan et al. [744] have been mainly used, up to now, for classifying chemicals in various domains of biological activity. But they may also propose the most interesting descriptors for QSAR applications. The approach emphasizes the influence that conformational flexibility may have on the physicochemical descriptors commonly used. For example, conformational changes (in a reasonable energy range of 20 kcal/mol) may induce important variations of HOMO energies (about 1.1 eV for hydroxyflutamide).

The basic idea is that molecules may exist and interact in a variety of conformations (in a context dependent behavior) and that molecules exhibiting similar biological activity must possess a commonality in their stereoelectronic pattern. The problem is therefore to determine a COmmon REactivity PAttern (COREPA) by analysis of the conformational variations of the structural descriptors (such as HOMO, LUMO) potentially associated with the specific biological activity. The method relies somewhat on fuzzy logic, inasmuch as it associates to one chemical, for a given descriptor, a finite set of values. The analysis starts with the generation of a population of conformers by a combinatorial procedure from the 2D formula (3DGEN module) [745] or via a GA, widely exploring the conformational space by means of a fitness function minimizing the similarity between the generated conformers [746].

For each species in the set of selected conformers, each descriptor value (calculated, for example, from the optimized QSAR approach based on a structural index set [OASIS]) [747] is given a probabilistic distribution (Gaussian, Laplacian). This generates a probabilistic distribution of the group of chemicals estimated across a selected descriptor. From subsets of selected active and inactive compounds, the examination of the probability distribution of descriptors identifies the structural requirements for different classes of binding affinity and indicates the descriptors that are more likely to express the relationship between the chemical structure and the observed biological activity. From these profiles, hierarchical rules may be established for the constitution of a decision tree for the prediction of the biological activity in classification tasks. But, the analysis may also propose the descriptors that are important for subsequent QSAR analyses. So, in the prediction of the estrogen receptor binding affinity, the COREPA treatment highlighted the importance of the HOMO energy. It is noteworthy that COREPA avoids the selection of an active conformer, the predetermination of a pharmacophore, and the alignment of conformers (since the conformer distribution is reflected by the descriptor distribution).

The main field of application seems, however, to rely on varied classification problems. Reactivity patterns, based on 3D molecular attributes, have so been established for discrimination of ER agonism and antagonism, an important point with respect to in vivo response [577] or mutagenicity [748]. The approach has potential applications in screening large chemical inventories in a decision-making framework to predict ecologic effects and the environment fate of chemicals [3,749].

10 Induced-Fit in 5D- and 6D-QSAR Models

The 5D and 6D approaches were developed in the framework of a large research project on a "virtual laboratory" that can estimate in silico the harmful effects triggered by chemicals (including drugs) and their metabolites, in line with efforts regarding the limitation of animal experimentation [71,750] (http://www.biograf.ch). Most of the QSAR methods just presented only considered the ligands, without any information about the receptor (unless docking is carried out to refine the orientation of the ligand in the receptor-binding pocket). Apart from this exception, these models have an important limitation. Implicitly, the receptor is considered as a unique rigid average structure. This may potentially bias the ligand alignment (when this phase is needed).

To take an image from Vedani et al. [68], usual QSARs consider that all the samples act simultaneously on a unique receptor (or, more exactly, on a unique surrogate representation of the actual biological receptor), whereas in the actual process each ligand is faced with an individual receptor.

10.1 RECEPTOR ADAPTATION: THE "INDUCED-FIT"

The approaches developed (Quasar and Raptor) aim at a sharper representation of the ligand–receptor interactions and take into account the adaptation of the binding site to the ligand. The physicochemical properties of a ligand are not the only factors intervening in its binding affinity: the interactions with the target protein must also be considered. When a ligand–receptor complex is formed, the solvent is stripped off from the ligand, which induces an energy loss to be compensated by the interactions created in the process. Ligand conformational freedom is also modified, hence a change in entropic terms.

Furthermore, the receptor-binding domain may adapt its shape to accommodate the ligand and solve steric clashes. Various experimental evidences testify for this adaptation process: the "induced-fit." It was observed that the "progesterone receptor tolerates in position 17α larger substituents as compared to the endogenous ligand progesterone" [73]. Similarly, dihydrotestosterone (DHT) with a benzoate group in 17β-position binds androgen receptor (AR), although a rigid structure of the complex (receptor-DHT) does not let enough room and reorientation of some residue side chains seems necessary (as confirmed in molecular dynamics [MD] simulations) [73]. See also Ref. [751] for another example of induced-fit. This adaptation may also change the character hydrophobic and hydrophilic. A flip-flop of the H-bonding capacity may intervene for conformationally flexible residues (such as Ser and Thr). Depending on the ligand, they act as H-bond donor or acceptor. For example, for dimethylstilbestrol binding the AR, the OH group binds the same region as the 3-CO of DHT inducing a flip of the side chain of Gln-711 and engages a H-bond with the amide O atom of this side chain [752].

The important point is that these subtle adaptations depend on the individual ligands. At the basis of the Quasar methodology of Vedani et al. is the construction of a hypothetical quasi atomistic receptor model (an "averaged receptor envelope") that will be subsequently adjusted on each individual ligand by creation of individual receptor surrogates. The shape of this average envelope represents information about steric effects, whereas properties such as hydrophobicity, charge, and H-bonding propensity are mapped onto this surface. Various levels of refinement were successively introduced. First, the approach considered several molecular representations (conformations, orientations, and state of protonation). This constituted the 4D level [714]. Then the models included (in addition to these several possible representations) the possibility for various induced-fit mechanisms with a dynamic interchange: the 5D level [69,70]. At last, at the 6D level, a protocol allows for introducing various solvation processes [73].

A parallel approach Raptor takes into account the changes (induced by the receptor adaptation) in the fields experienced by each individual ligand, thanks to a "dual-shell" representation, stressing the importance of H-bonding and hydrophobicity [751].

10.2 Quasar APPROACH (QUASI ATOMISTIC RECEPTOR MODEL)

10.2.1 Introducing the Induced-Fit in 4D-QSAR

Basically, the sequence starts with after energy minimization, determination of the conformers of low energy (for example, those within 5 kcal/mol from the global energy minimum), and alignment on a template. The following steps are detailed below.

10.2.1.1 Construction of Receptor Envelope and Adaptation to the Individual Ligands

The training set of ligands is surrounded by virtual particles (201 particles of radius 0.8 Å in the treatment of β_2-adrenergic receptor for example [753]) defining, after energy minimization, a van der Waals surface: the "averaged receptor envelope." Then each individual ligand is optimized as a 1:1 complex with this averaged envelope. To simulate the receptor induced-fit, a transient inner envelope featuring the individual ligand is built and the receptor envelope is adjusted to the topology of the very ligand molecule (including a weak positional constraint). A vector is defined between each point of the receptor envelope to the closest point of the inner envelope (Figure 10.1). Then the points of the (mean) receptor envelope are moved along this vector. The magnitude of the fit is linearly (isotropically) or field scaled or determined through energy minimization. The small constraint introduced limits the deformation but allows for relaxing steric clashes between the ligand and the averaged envelope. This process involves rms deviations (between the two envelopes) about 0.4–2.5 Å, corresponding in the subsequent evaluation of the binding energy to variations between 0.2 and 6 kcal/mol (Figure 10.2).

FIGURE 10.1   (**See color insert following page 288.**) Steric field vectors originating from the mean envelope surrounding an NK-1 antagonist molecule. Induced-fit is executed along these vectors. (Reproduced from Quasar 5.2 manual. http://www.biograph.ch. With permission.)

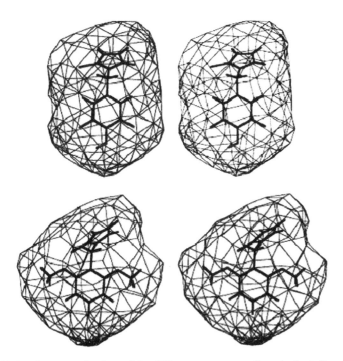

FIGURE 10.2   Stereoscopic view of the different receptor-to-ligand adaptation (simulated local induced-fit) for two dopamine β-hydroxylase inhibitors. Substituents on the phenyl ring are $3,5$-F_2,4-OH (top) and $2,6$-$(OMe)_2$ (bottom). (Reproduced from Vedani, A. et al., *Quant. Struct. Act. Relat.* 19, 149, 2000. With permission.)

10.2.1.2   Generation of an Initial Family of Parent Molecules

The points scattered on the averaged receptor surface are randomly populated with atomistic properties such as hydrophobicity, H-bond propensity, charge, salt-bridge, and possibility of H-bond flip-flop. Regions of the receptor envelope may also be defined as hydrophobic, void, or assigned as representing the solvent. This procedure allows for a multiple representation of ligand topology (conformation, protonation state, and orientation) and a multiple representation of induced-fit depending on the hypotheses. In fact, the induced-fit cannot be evaluated without the knowledge of the biological receptor. So simulations with various hypotheses, as to the mode and magnitude of the fit, must be examined.

An initial population of about 200 models is so defined. In Quasar, six protocols are proposed according to the property selected for modeling adaptation. Energy minimization or a linear scheme can also be used.

10.2.1.3   Evolution of the Models with GA

The selection of the best model(s) is performed with a genetic algorithm (GA) evolving by cross-over and a small probability of random mutation (transcription error). In this process, at each step, the two models with the highest lack of fit are discarded and the process is repeated until a good cross-validated correlation coefficient is obtained. The lack-of-fit function is based on the rms error of $\Delta G^{\circ}_{pred} - \Delta G^{\circ}_{exp}$ (in cross validation) plus three penalty terms that handicap models with many properties mapped, models similar to several others, and models with unspecific selection of conformation or protonation state.

10.2.1.4   Quasar Estimation of Free Energy of Ligand Binding

The evaluation of the free energy of binding includes

- The ligand–receptor interaction energy.
- The entropy loss of the ligand during binding (0.7 kcal/mol for each rotatable bond except for terminal Me group).
- The ligand desolvation energy (<0) to strip solvent off the ligand when binding.
- The change in ligand internal energy upon binding ($E > 0$).
- The energy uptake for adaptation of the receptor: induced-fit (>0).

$$E_{binding} = E_{ligand-receptor} - T\Delta S_{binding} - \Delta G_{ligand\ desolvation} + \Delta E_{ligand\ strain} + \Delta E_{induced\text{-}fit}$$

with

$$E_{ligand-receptor} = E_{electrostatic} + E_{van\ der\ Waals} + E_{H\text{-}bonding} + E_{polarization}$$

When an ensemble of representations is used, the total energy of binding is determined from the contributions of the diverse entities via a normalized Boltzmann distribution. Then, the free energies of binding are obtained by an MLR between experimental $\Delta G°$ and $E_{binding}$ calculated for the training set. The slope and intercept are specific to the selected receptor model and may be applied for the prediction of new ligands.

10.2.2 APPLICATIONS OF QUASAR

In a first application [753], an ensemble of receptor models was generated for the β_2-adrenergic, arylhydrocarbon, cannabinoid, neurokinin-1 (NK-1) and sweet-taste receptors, and for the enzyme carbonic anhydrase (see also Ref. [342]). Obtained q^2 were better than 0.750, but one example (0.62 for the cannabinoid receptor) and prediction on independent test sets were within 0.4–0.8 kcal/mol of the experimental values.

In another publication [714], 47 1-(substituted-benzyl)imidazole-2(3H)-thiones (see compound **3.2**), inhibitors of dopamine β-hydroxylase (that catalyzes the conversion of dopamine to norepinephrine), used in the treatment of disorders related to hypertension, were examined. For these compounds, previously studied by Hahn and Rogers [390], up to 6 conformers per molecule were selected, amounting to a total of 73 conformers for the 34 training ligand molecules and 34 for the 13 test molecules. The treatment led to $r^2 = 0.796$ and $q^2 = 0.759$. In this study, 19 molecules among 34 selected a unique conformation.

A second example [714] concerned 102 polysubstituted aromatics binding the Ah receptor (dibenzodioxins, dibenzofurans, and biphenyls) (see compounds **1.4** through **1.6**). In this example, there was no conformational flexibility but up to four orientations must be considered for unsymmetrical compounds (flip-flop, "up–down," or "right–left") amounting for the whole set to 386 structure representations to consider. The approach yielded $q^2 = 0.839$ ($r^2 = 0.841$). On the 76 ligands of the training set, 42 selected a single orientation but for 11 ligands, 3 or more orientations significantly contributed to the total energy. A similar concern of several simultaneous binding modes was also addressed by Lukakova and Balaz [62].

As with other 4D-QSAR approaches, the bias associated with the choice of the bioactive conformation is reduced. The use of a multiple ligand representation was claimed to be superior to a single-conformer hypothesis. But, in addition, the approach was able to select a small number of active conformations rather than an enlarged set of lesser-contributing species. The study of antagonists of NK-1 receptor system (**10.1**) [68], on a range of 4 log units in binding affinity, offered an example of highly flexible molecules with protonable sites. The NK-1 receptor system is one of a family of neuroreceptors involved in various signal transduction pathways (nausea and vasodilatation).

10.1

The 65 antagonists studied presented a common scaffold of two aromatic groups linked by an amide moiety. Several torsional angles and the protonation state of the two nitrogen atoms (in a piperazyl or two piperidyl fragments) must be examined. Training was carried out on 50 compounds (218 conformers/protomers) and the models were evaluated on a test set of 15 molecules (79 representations). The individual envelopes were generated using a field-based algorithm. From 500 receptor models, GA led to a simulation with $r_{cv}^2 = 0.887$ in an L-10%-O cross validation (the rms for the uncertainty factor on IC_{50} was about 2 for both training and test). The predictive r_{pred}^2 equaled 0.834. The model also worked satisfactorily for newly synthesized and tested compounds. Interestingly, for most compounds, the approach selected a single entity among the conformers/protomers (up to 12) with an identical protonation state.

For other applications of Quasar, see Refs. [363,754–756].

10.2.3 FROM 4D TO 5D

In the preceding approach, a single induced-fit model was chosen and evaluated. The Quasar concept was further extended to a "fifth dimension," allowing for a multiple representation of induced-fit hypotheses [69,70]. Not only the ligands may be represented by an ensemble of conformations, orientations, and protonation states but also different local induced-fit protocols may be evaluated. During the simulation, the ensemble of ligand topology and induced-fit mechanisms are simultaneously available. GA and a Boltzmann criterion select the best combinations. Six adaptation protocols are proposed: scalable linear model, adaptation based on steric, electrostatic, and H-bond fields, respectively, energy minimization along steric field, and adaptation based on lipophilicity potential. During the simulation, each combination of conformers (typically 4–16) and models (2–6) is evaluated. This protocol allows for a dynamic interchange of the induced-fit mechanisms (as frequently observed) and generally during the GA evolution, the simulation converges to a single induced-fit model.

The method was applied to two data sets previously examined at the 4D level [68,714]. For the 75 NK-1 receptor antagonists [68], in 5D with steric field adaptation, $r_{cv}^2 = 0.870$ and $r_{pred}^2 = 0.837$ (values averaged on 500 models and 60 generations), whereas the 4D treatment led to $r_{cv}^2 = 0.887$ and $r_{pred}^2 = 0.834$ (in 80 generations). For polyaromatics binding the Ah receptor [68], $r_{cv}^2 = 0.838$ (0.857 in 4D) for 91 compounds in training (348 orientations) and $r_{pred}^2 = 0.832$ (0.795 in 4D) for 30 compounds in test (113 orientations) using a field energy minimized-based approach.

Although the overall results of the two approaches were not very different, it was worth mentioning that the evolution converged to a single model. As stressed by the authors, the fact that model selection might change during the simulation led to less biased model than 4D-QSAR (where a single mechanism was evaluated at time) and yielded more accurate predictions of new compounds.

Affinity for the AhR was further revisited [750] on 121 chemicals (polyaromatic hydrocarbons and polychlorinated aromatics). With 250 models, Quasar converged to an averaged $q^2 = 0.766$ ($r_{pred}^2 = 0.778$). Reexamination [70] of the 47 ligands of dopamine β-hydroxylase, previously investigated [390,714], also allowed for comparing the successive levels of refinement in the analysis. With 200 models, the 5D treatment for the

34 compounds of the learning set led to $q^2 = 0.892$ versus 0.759 in 4D and 0.641 in 3D. The method was also applied [756] with very good results to 141 inhibitors (**10.2**) of the chemokine receptor-3 (CCR3) treated with two receptor surrogates. Chemokines, low-molecular-weight chemotactic cytokines, are structurally related proteins that play a key role in the action of leukocytes during inflammation.

10.2

10.2.4 To the Sixth Dimension

Whereas the fifth dimension allowed for a simultaneous evaluation of up to six different induced-fit protocols, extension to a sixth dimension includes the possibility to simultaneously consider different solvation scenarii. Solvation terms (ligand desolvation and solvent stripping) are independently scaled in the surrogate family reflecting varied solvent accessibility of the receptor pocket or, alternatively, parts of the surface area may be mapped with solvent properties (position and size optimized by GA). Vedani et al. [73] studied 106 (88/18) agonists binding the estrogen receptor (data from [545,546]). These compounds, from six structural classes, span an activity range of 7 log units for IC_{50}. Compounds were first energy optimized. Then, sampling of energetically feasible arrangements within the binding pocket of the receptor was carried out by a Metropolis Monte Carlo search protocol (based on the X-ray structure of diethylstilbestrol bound to human ERα-binding domain).

For each ligand, up to four arrangements (within 5.0 kcal/mol from the energy minimum) were retained, amounting to 344 representations for the 106 molecules. About 200 models converged in 200 generations to $r_{cv}^2 = 0.903$ (learning set of 88 compounds) and in test $r_{pred}^2 = 0.885$. In this example, where the structure of the actual biological receptor is known, it is noteworthy that the receptor surrogate, built in silico, is quite consistent with it, with H-bond acceptor mimicking Glu-353 and His-524, H-bond donor for Arg-394, and a large central hydrophobic pocket.

The authors also stressed that a 3D treatment would be difficult, since docking identified, for most of the compounds, four binding modes energetically close. Comparison with various simulations using, in 3D, the lowest energy conformer ($r_{cv}^2 = 0.821$), in 4D, a single induced-fit mode ($r_{cv}^2 = 0.810$) or, in 5D, omitting ligand-dependent solvation effects ($r_{cv}^2 = 0.872$), gave inferior results (in 6D, $r_{cv}^2 = 0.903$). Similar results were also found with Raptor (vide infra).

In a subsequent study [71], the approach was applied to the peroxisome proliferator-activated receptor γ (implied in the treatment of diabetes and obesity). For 95 ligands (mainly tyrosine-based compounds), the feasible arrangements (energetically and structurally) in the (known) receptor pocket were determined by flexible docking. Starting from 200 models, 120 generations of GA led to $r_{cv}^2 = 0.832$ (for a training set of 75 molecules) and $r_{pred}^2 = 0.723$ (on a range of 3.5 pK units), whereas previous CoMFA or CoMSIA studies reached only a limited success (q^2 about 0.68).

10.3   Raptor APPROACH

Another approach for taking into account the induced-fit is provided by the Raptor package [750–752], from the same research group.

10.3.1   "DUAL-SHELL" REPRESENTATION

Induced-fit involves steric factors, but it may be also governed by the movements of the residue side chains in the binding pocket, depending on the ligands. These different conformational changes (or H-bonding flip-flop) induced by the ligands generate in turn different fields acting on the ligands. In contrast to Quasar, Raptor methodology expresses these variations of the physicochemical fields, thanks to a "dual-shell" representation [751]. An inner layer maps the fields experienced by a ligand fitting snugly into the binding pocket. For this surface, only the experimentally most affine ligand is taken into account. But additional groups of a ligand lying deeper in the protein may experience different fields generated by the modified binding site. They are evaluated on a second outer level built from all ligands of the training set. Between the two shells, fields are interpolated (Figure 10.3).

Adaptation is calculated as the result of steric adjustment to the ligand topology and a component (attractive or repulsive) between the ligand and the receptor surrogate, which is calculated from H-bond propensity (with directional terms) and

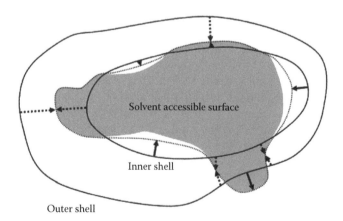

FIGURE 10.3 Dual-shell representation of the receptor surrogate. If the ligand's SAS (solvent accessible surface) lies between the two shells (dashed arrows), the fields generated by the protein are computed by linear interpolation. For ligand points located inside the inner layer, the latter may adapt only in part the receptor surface to the ligand topology (solid arrows; dotted line). (Reproduced with permission from Lill, M.A. et al., *J. Med. Chem.*, 47, 6174, 2004. With permission.)

hydrophobicity. In addition to these terms, solvation effects are explicitly taken into account, but electrostatic contributions are omitted (with the advantage of a model independent of charges). The binding free energy so comprises terms for the topological adaptation ΔG_{IF} (induced-fit), for the change in entropy upon ligand binding ($\Delta G_{T\Delta S}$), and a contribution to binding energy ΔG_{const}, globally due to the loss of the translational and rotational entropy of the ligand and entropy gained by desolvation of the receptor pocket. Additional terms concern hydrophobicity (ΔG_{OH}) and H-bonding (ΔG_{HB}).These terms are weighted by factors (f) characteristic of a receptor model and optimized in the course of the simulation.

$$\Delta G = f_{IF}\Delta G_{IF} + f_f\Delta G_{T\Delta S} + \Delta G_{const} + f_{OH}\Delta G_{OH} + f_{HB}\Delta G_{HB}$$

The ligand molecule is represented by its solvent accessible surface. Fields generated by the ligand are projected on discrete points on this surface. Adaptation between a receptor and an individual ligand depends on the strength of the interaction. It is evaluated with a Fermi function

$$f_{Fermi} = \frac{1}{\left\{1 + \exp\left[a(x-b)\right]\right\}}$$

At surface point a, hydrophobicity of the field receptor model HO_a is:

$$HO_a = \frac{\left(HO_a^\circ\right)}{\{1 + \exp[\delta_{ho,a}(-ho_a - \omega_{ho,a})]\}}$$

where HO_a° is the hydrophobicity of the two layers (inner and outer) linearly interpolated to each ligand surface point a of hydrophobicity ho_a. ho_a is calculated by summation of atom hydrophobic contributors weighted by a Fermi function of the atom-surface point distance. Parameters ω and δ have to be adjusted during the simulation. A rather similar formula holds for H-bonding

$$HB_a = \frac{\left[HB_{Acc,a}^\circ - HB_{Don,a}^\circ\right]}{1 + \exp\left[\delta_{hb,a}\left(hb_a - \omega_{hb,a}\right)\right]} + HB_{Don,a}^\circ$$

$HB_{Acc,a}^\circ$ and $HB_{Don,a}^\circ$, respectively, represent the H-bond donor and acceptor propensities of the two layers linearly interpolated to each ligand surface point a. Remark that this formula explicitly allows for H-bond flip-flop. Energy of interaction is obtained for each surface point as the product of its hydrophobicity ho_a and the hydrophobicity of the field of the receptor model HO_a. A similar formula holds for H-bonding propensity. In a manner similar to that of Quasar, the quality of the model is determined by

$$\sum_{ligands} \left(\Delta G_{cal} - \Delta G_{obs}\right)^4$$

The fourth power weights deviations close to experimental uncertainty. This model implies numerous parameters to optimize (about 5–6). So, rather than changing the values for each surface point, the authors prefer to define "domains" with similar properties (about 50–100). As they stressed, this is not in conflict with the actual nature of biological receptors where properties are not randomly scattered onto the binding site. To avoid overfitting, they also proposed to add variables progressively.

At last Raptor generates a family of models that may be interpreted as featuring different states of the true receptor.

10.3.2   APPLICATIONS OF RAPTOR

Chemokine Receptor CCR-3: Already studied with Quasar [756]. For the 50 compounds (split into 40/10), Raptor [751] on the conformations selected by Quasar (in view of comparison) led, on average on 20 models, to $r^2 = 0.965$ in training and 0.932 in prediction for the test, slightly better than 5D-Quasar [756], with $r^2 = 0.950$ and predictive $r^2 = 0.879$ (Figure 10.4).

Bradykinin B2 Receptor: A G-protein coupled receptor implicated for example in pain, inflammation, and vasodilatation. The study concerned 43 antagonists of the B2 receptor plus nine newly synthesized compounds, split into training (32 + 1) and test (11 + 8) sets. Raptor proposed a family of 10 receptor models, with an averaged $r^2 = 0.949$ (training) and $r^2 = 0.859$ in test.

Estrogen receptor is involved in adverse health effects generated by endocrine disruptors. A total of 116 compounds (93/23) were investigated, using a receptor-mediated alignment (from known structures of ligand ER complexes). With a family of 10 receptor models, $r^2 = 0.908$ and $r^2 = 0.907$. Only 5 ligands (all in the training set)

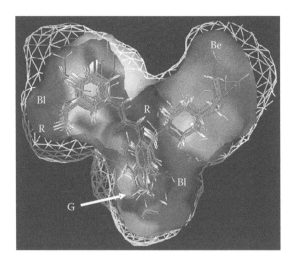

FIGURE 10.4   Dual-shell representation of one of the individual models for the CCR-3 receptor (hydrophobic fields, (Be) beige; H-bond-donating propensity, (Bl); H-bond-accepting propensity, (R); H-bond flip-flop, (G)). The receptor model has a Y-shaped form. The upper part of the right lobe is mostly hydrophobic. The left lobe has mixed hydrophilic/hydrophobic characteristics and the H-bonding areas (donating or accepting) are seen in the central lower part. (Reproduced from Lill, M.A. et al., *J. Med. Chem.*, 47, 6174, 2004. With permission.)

deviate more than 1 log unit for IC_{50}. It is noteworthy that the affinities of charged ligands or neutral molecules are evaluated equally well. Separate models were subsequently proposed for α and β receptors [71].

Another point of interest emphasized by the authors is that Raptor retrieved the essential features of the true biological receptor (bound to DHT): a central hydrophobic part and hydrophilic areas near the phenolic groups were indicated for bound diethylsilbestrol, whereas for bound Raloxifene, an additional interaction (protonated nitrogen-Asp-351) appeared. See also Ref. [750]. These examples pointed out the importance of an anisotropic modeling of individual ligands (Figure 10.5).

Androgen: Raptor was further applied to the binding of 119 ligands to the AR with special emphasis on induced-fit and examination of several possible binding modes [752]. In a first step, the possible binding modes were identified by flexible docking based

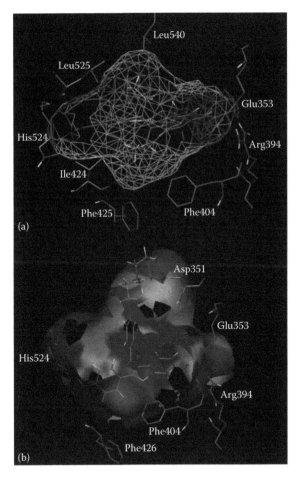

FIGURE 10.5   (**See color insert following page 288.**) Common features determined by the Raptor models for the estrogen receptor (hydrophobic fields, (Be) beige; H-bond-donating propensity, (Bl) blue; H-bond-accepting propensity, (R) red, with bound (a) DES and (b) raloxifene). (Reproduced from Lill, M.A. et al., *J. Med. Chem.*, 47, 6174, 2004. With permission.)

on a Monte Carlo search (allowing for protein adaptation) followed by MD simulation (starting from the most favorable docking scores) to get a better sampling of the configurations for each binding mode. The relative energies of binding were determined by linear interaction energy (LIE) analysis before aligning molecules and carrying out Raptor. Alignment was guided by the known X-ray structure of DHT bound to AR.

This protocol appeared necessary as common (crude) force fields used on static structures are not well suited for a good evaluation of protein–ligand interactions and free energy of binding. In LIE, the free energy, for a given binding mode, is evaluated as the difference (averaged over the simulation time) of electrostatic and van der Waals interaction energies between the ligand bound to the protein and the solvated ligand. See Chapter 12.

The data set encompassed 119 compounds from six structural classes (steroids, phytoestrogens, phenols, diphenylmethanes, di-ethyl-stilbestrol (DES) and related compounds, and organochlorines) on an activity range of 5.5 logRBA units. For each class, a representative compound was submitted to flexible docking and MD simulation to assess the binding mode and guide alignment on the basis of the known structure of the complex AR-DHT. With 10 models, Raptor led to (averaged values) $r^2 = 0.858$ (88 compounds in training) and $r^2 = 0.792$ in test (21 compounds). Similar results were reported in Ref. [750].

In this application also, Raptor retrieved the essential features of the true biological AR, with a central hydrophobic barrel and two H-bonding moieties (H-bond accepting/donating near the 17-OH group mimicking Tyr-877 (acceptor)/Asn-705 (donor) and H-bond donating mimicking Arg-752 near the 3-CO) (Figure 10.6).

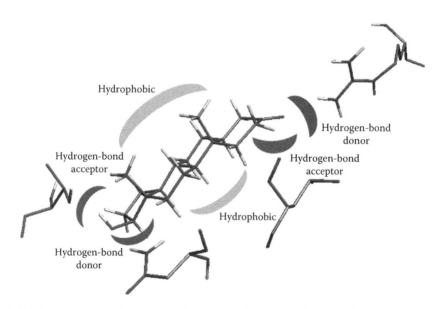

FIGURE 10.6 **(See color insert following page 288.)** DHT bound to the AR. Schematic representation of the physicochemical properties of the receptor model and their consistency with the experimental structure. (Reproduced from Lill, M.A. et al., *J. Med. Chem.*, 48, 5666, 2005. With permission.)

As supplementary check of the efficiency of the treatment, it was observed that the Raptor models were able to satisfactorily predict the activity of compounds in structural classes different from those used in training (polychlorinated biphenyls or diphenyltetrahydrofuran derivatives).

10.4 THE VirtualToxLab: Quasar AND Raptor STUDIES OF ENDOCRINE DISRUPTORS

The VirtualToxLab (in which are integrated, inter alia, Quasar and Raptor) was developed at the foundation *Biographics Laboratory 3R* [71,750] (http://www.biograf.ch). This foundation aims at developing in silico technologies in the field of drug design, with a particular interest for the prediction of adverse effects of environmental chemicals and their metabolites. Particular attention was paid to endocrine disruptors. VirtualToxLab proposes 12 reliable and strictly validated models for ligand binding via an automated technology in real time. The models so encompassed various receptors associated with endocrine disruption: aryl hydrocarbon, estrogen α/β, androgen, thyroid α/β, glucocorticoid, liver X, mineralcorticoid, peroxisome proliferator-activated receptor γ, and enzymes CYP450 3A4 and 2A13.

In early 2009, models established on 662 compounds were validated with excellent results on 191 test compounds (all together). Predictions are within experimental uncertainty. More details can be found in [71,750] and on the site of *Biographics Laboratory 3R*. As to validation, the reliability of the models was established by "y scrambling" or by "poisoning experiments" (for AhR, for example). In this trial, some inactive compounds (with arbitrarily assigned activity) are added to the data set. No good solution could be found. This strongly evidenced that if a good solution was obtained, it actually characterized a common underlying mode of action.

Symposar: In addition to Quasar and Raptor, *Biographics Laboratory 3R* proposed various other modeling tools such as *Symposar* that aims to automatically generate a 3D/4D pharmacophore as input for multidimensional QSAR [755]. The 4D data set built by *Symposar* encompasses the conformations, orientations, or protonation states (up to 64 per ligand) the most similar among the ligands, but some variability is maintained, allowing for some diversity in the bioactive forms. Ligands are superimposed onto one or several template molecules in a two-step process: first a fuzzy-like 2D substructure mapping (with an adapted clique detection module) and then a Monte Carlo search for similarity of physicochemical fields on a 3D grid, using a Gaussian function to score similarity.

The efficiency of the approach was established on the test set previously used to evaluate the FlexS software (from binding data of 272 ligands and 13 proteins) [513] and applied to 186 compounds binding the bradykinin B2 receptor. On the generated data set, Quasar and Raptor gave consistent, satisfactory results.

Part IV

Receptor-Related Models

The preceding approaches only take into account the drug molecules with the implicit assumption that the structural information collected on a set of active compounds is sufficient to evaluate their relative activity in a specific process. Another type of approach explicitly considers both the ligand and its receptor (or more specifically the ligand-binding domain [LBD]) and aims to directly evaluate the interaction energy upon the formation of the ligand–receptor complex. The problem is that this process involves an important participation of the surrounding medium. The formation of the complex implies a partial desolvation of the protein and of the ligand (and also entropic contributions) that is difficult to sharply evaluate.

A first level is limited to determine the best "pose," which is the conformation, position, and orientation of the ligand in the binding domain. This is "Docking." Unfortunately, the associated scoring functions privilege speed over accuracy. As a result, it appears that docking gives good alignment (needed, for example, by CoMFA or CoMSIA treatments) but is not quite successful in evaluating interaction energies [282,757–760].

A precise determination of interaction energies (and of relative affinities) is attainable by more evolved techniques (such as thermodynamic cycle or thermodynamic integration) requiring an extensive sampling of the phase space [61], which is not easily workable on extended populations of chemicals. However, approximation methods such as linear interaction energy (LIE) model [761] and continuum solvation models, with Poisson–Boltzmann/surface area (PBSA) or generalized Born/surface area (GBSA) methods, lead more readily to good-quality results and widely extend the field of such applications [60].

11 Docking and Interaction Energy Calculation

Docking programs aim to determine the best "pose," that is, the preferred conformation, position, and orientation of a ligand binding the receptor pocket (receptor-active site or ligand-binding domain, LBD) by a calculation of ligand–receptor interactions. Three facets were distinguished by Leach et al. [762]:

- Predicting the binding mode of a known active ligand
- Identifying new ligands in virtual screening
- Predicting the binding affinities in a series of known compounds

This approach thus received a continuous interest on account of its prime importance for lead discovery or lead optimization, but it was recently largely boosted up with the development of huge chemical databases and the need for efficient tools in high-throughput virtual screening. Determining, ab initio, the molecules the most able to tightly bind a specific target is of great interest for prioritizing synthesis of the putative most active (potential) new drugs and avoids wasting time and money on compounds that will reveal uninteresting for the practical application looked for.

More marginally, but of prime interest in our concern, the information delivered by docking programs that specify the structure of stable (noncovalent) ligand–protein complexes is highly valuable for establishing 3D-QSAR models, for example, in the choice of the active conformation of a ligand or for the refinement of a CoMFA-type alignment. In a somewhat symmetrical process, in the calculation of ligand–receptor interactions, the introduction of new models for a fast treatment of solvation effects (such as Poisson–Boltzmann/surface area (PBSA) or generalized Born/surface area (GBSA) [763, pp. 502–520]) provides accurate evaluation of binding free energies, at a relatively low cost in computer time. Some treatments have introduced this approach, not only as QSAR models, but also as a higher quality substitute to the scoring functions commonly used in virtual screening. A dual step process may be more efficient where a free energy approach is applied to only one pose per compound (after a first coarse-scoring filter). LIE models were, for example, proposed in lead optimization by Stjernschantz et al. [764].

11.1 DOCKING AND INTERNAL SCORING FUNCTIONS

A lot of docking programs (about 30 according to Leach et al. [762]) have been developed. We will only focus on some of the most used (according to us) in 3D-QSAR. Basically a docking program comprises

- A sampling part that explores the ligand "phase space" (conformation, position, and orientation) with respect to the protein LBD in order to determine the best poses. This may involve a generation step to build up (and examine) flexible ligands.
- A scoring function that evaluates the energetic of binding and that will allow for selecting the best pose(s) or ranking the hits found in screening a database. "Ideally a scoring function should also be able to serve as a docking function because in practice docking and scoring are often inseparable such as in virtual database screening study" [765].

In addition to these (internal) scoring functions implemented in docking programs, other scoring functions were independently developed, and their performance compared to the turnkey tools included in the docking packages. Therefore, important characteristics of docking programs are

- The nature (and completeness) of the force field, particularly regarding the treatment of H-bonds (possibly with directional dependence), solvent effects, and entropic terms
- The degree of flexibility considered (from ligand and receptor fixed to fully flexible entities)
- The method used to sample the system (receptor–ligand)
- The calculation of the scoring function

We will go into more detail about these points, but, generally speaking, it appears that docking programs efficiently separate active compounds (those able to give strong noncovalent complexes with the receptor) from the others and correctly propose (among their first hits) the observed binding mode [513]. This was checked by various simulations where specific active compounds were gathered with decoys (of similar structure but devoid of activity) and randomly selected chemicals. However, if results are often *quite* satisfactory, they may be *quite* disappointing for certain data subsets. Thus, it remains still difficult to guess, for a given target, what method would give acceptable results without some preliminary tests.

On the other hand, docking programs often propose several possible poses (up to some dozens) and the usual scoring functions that give satisfactory (or acceptable) results for virtual screening (with indication of the most potentially active binders) often appear inadequate for the estimation of affinity constants with the accuracy required in QSARs [766]. This is not surprising since, in fact, docking programs were mainly dedicated to an efficient screening of databases. They must be able to treat, as fast as possible, numerous structures, and rapidly give a diagnosis (high-activity predicted and low-activity predicted), not a precise estimate of a binding constant. Hence the choice adopted for the scoring functions.

Furthermore, as stressed by Tirado-Rives and Jorgensen [767], the energy variation interesting (structure/biological activity) correlation only spans a rather limited range (about 4.5 kcal/mol), so that very sharp models are needed. Flexibility of the ligand and limited motions of the protein are now "easily" taken into account in "flexible docking" programs. But difficulties remain to treat protein adaptation (induced fit), and to evaluate entropic terms (immobilization of some ligand rotatable bonds), solvation effects, and removal of mobile water molecules. The rather low degree of sophistication of the force fields included in docking programs is another cause of problems for an accurate evaluation of the various energetic contributions.

Several methodological approaches have been tackled for the generation of the candidate structures and selection of the best pose. For example, DOCK [282] begins with an anchoring fragment and uses a fast shape-matching algorithm. An incremental construction is proposed in FLEXX [768], whereas in LUDI, primarily dedicated to de novo design, the solution is achieved by joining predefined fragments. In the optimization step, very diverse methods were proposed: Tabu Search [769,770], Genetic Algorithm (GA) (in AutoDOCK [343] or Gold [279]), Evolutionary Algorithm [771]. Simulated annealing and Monte Carlo (or biased probability Monte Carlo procedures in ICM [772]) were also developed [273,773,774].

A large diversity also appears in the scoring functions. Some are directly based on force field methods (DOCK [282], GOLD [279]) where interactions are calculated with rather crude force fields. Parameters, easily transferable, have a clear physical meaning and grid techniques allow for a rapid calculation. But some contributions, solvation, and entropy are not always well taken into account. Others (ChemScore [775], Glide [773], SCORE [776], FLEXX [768]) follow a scheme proposed by Böhm in LUDI [777,778]. They decompose the binding free energy as a sum of contributions, such as protein–ligand interactions (van der Waals, electrostatics, H-bonding), internal deformation of the ligand, entropy loss, and desolvation. Empirical fitting parameters are derived from known complexes. On the other hand, knowledge-based functions rely on the analysis of atom–atom distances from the knowledge capitalized from co-crystallized systems, as in potential of mean force (PMF) [779] or small molecule growth (SMoG) [781]. These last approaches are fast and direct, but their parameterization suffers from the reduced number of available experimental data on complexes (particularly regarding energy more than distances). With the increasing number of solved complexes and free energy measurements, the available data pool is continuously extending and will provide these methods with higher accuracy and a wider applicability domain.

Several studies tackled the problem of evaluating the efficiency of docking programs in virtual screening and data base enrichment [757–759,765,766,782–784]. A general opinion is that this is not a trivial task. Many control parameters (and even the definition of the extent of the binding site) may influence the results. Different implementations of a same scoring function may give different results [757]. Performance depends on the selected data sets [758]. Isolated studies established that some algorithms are more efficient than others for certain data sets. But, even if some correspondence may be found between the nature of the binding site and the best docking algorithm, results are difficult to generalize [766].

Thus, Wang et al. [757] suggested dissecting the problem of performance evaluation into distinct modular problems: ability to identify the native binding pose, examination of the robustness of the model to accept small geometrical distortions, and ability to predict binding energy when the correct structure of the complex is provided.

We now rapidly summarize some docking programs and their own specific scoring function used in QSARs studies. External (stand-alone) scoring functions developed for comparison of the docking programs will then be described.

11.2   SOME DOCKING PROGRAMS

11.2.1   DOCK

A pioneering tool in the field is the DOCK approach of Kuntz et al. [282,785]. It addresses small rigid molecules into potentially binding clefts of macromolecular receptors of known structure. It was further extended to include ligand flexibility [786]. The principle is to determine the surface complementarity between the ligand and the receptor by searching a fit (or more precisely some geometrical identity) between the ligand and a "negative" image of the binding site. The shape of the ligand and the negative image of the binding pocket are featured by two sets of imaginary spheres. For the binding pocket, a set of spheres filling the void volume is generated from each surface point. These spheres are drawn as touching the receptor surface and lying outside it. Various criteria are introduced to reduce the number of these spheres to one per atom (the largest from the contact surface point of each receptor atom and the largest from the re-entrant surface of each ligand atom). Overlapping spheres are gathered into clusters associated to cavities, giving a good estimate of the binding sites.

Once both ligand and receptor are represented by sets of spheres, the problem is to fit the set of ligand spheres to the set of receptor spheres. This is carried out by an examination of the internal distances in both ligand and receptor. The algorithm is similar to the Lesk algorithm [787] to identify common 3D substructures. The output consists of short lists of pairs of ligand and receptor spheres having all internal distances matching within a tolerance value. Four pairs of contacts are necessary to ensure docking. Refinement resolves handedness, checks that unmatched spheres do not occupy sterically forbidden positions, and locates the ligand in the receptor pocket thanks to coordinate transformation.

This pioneering approach could be relied on only to shape complementarity without any energy consideration. It was improved by considering the nature of the atoms ("coloring" atoms) and later evaluating interaction energies with a simplified molecular mechanics force field (from AMBER [788]). Computation was made faster by precalculating receptor-dependent terms on a 3D grid mapping the active site [286]. For a recent version, see Ref. [789].

The DOCK package was, for example, used to study inhibitors of α-chimotrypsin [790]. Chimotrypsin catalyzes the hydrolysis of peptide amide bonds on the carboxyl side of an aromatic amino acid residue. The aim of this study was to estimate the performance of the method in database screening (elimination of inactive compounds) rather than identifying all active compounds for a well-known receptor. The binding receptor pocket is a hydrophobic cavity of 40 residues (identified by X-ray crystallography) approximately the size of an anthracene molecule. This binding domain was filled with 21 spheres (diameter 1–4 Å) and the inhibitors docked by sphere center/atom center matching. A rough scoring was based on a soft function of van der Waals contacts (but ignoring electrostatic or H-bond interactions). A total of 103 putative ligands was systematically docked. Although no strict correlation was found between known affinities and docking scores (an observation that will become general in further docking studies), the approach identified 8 potent inhibitors among the top 10 best hits.

Introducing ligand flexibility: the preceding approaches only considered rigid bodies. A first attempt to introduce ligand flexibility was made by DesJarlais et al. [791]. The ligand is divided into smaller rigid fragments (allowing some flexibility at their junction) that will be docked independently. The ligand is then reconstructed, eliminating situations where the fragments cannot be reasonably joined together. Interestingly this method retrieved crystallographic results but also suggested different geometries similar in energy. For example, with methothrexate binding dihydrofolate reductase (DHFR), in addition to the X-ray structure, a solution with the pteridine ring rotated 180° was found (very similar to that supposed for folate).

Leach and Kuntz [792] proposed an incremental construction process of the ligands. An anchor fragment of the ligand is first selected by identification of flexible bonds, and division of the molecule into rigid moieties (more than eight atoms and the most "donor plus acceptor atoms"). This fragment is placed in the receptor by the DOCK algorithm, followed by minimization by a simplex method. Conformations of the remaining parts of the putative ligands are searched by a limited backtrack method (to decrease the number of similar conformations sampled) and minimized to get the most stable conformation.

Shape site points define the receptor docking region. Interactions are calculated on a grid with a force field derived from AMBER, the receptor being assumed to be rigid. In this search, partial energy estimation is performed: only the atoms moved by the current flexible bond motion have to be considered, the interactions with the receptor for the other (fixed) atoms remaining constant. Multistate protonation options are treated simultaneously to determine the best one. The method was tested on 10 ligand–protein complexes and docking to DHFR. The approach was later automated for database searching by Makino and Kuntz [793] and Ewing et al. [785]. A large panel of scoring functions is now offered, including solvent models such as PBSA and GBSA (see Chapter 12) [794].

11.2.2 AutoDOCK

In a different approach, Goodsell and Olsen [273] first proposed to use a Monte Carlo simulated annealing (MCSA) approach to dock a flexible ligand onto a known rigid receptor. The series of AutoDOCK programs (since 1990), dedicated to a fast automatic prediction of geometry and energetics of ligand–protein complexes, now proposes a GA, and more recently a Lamarkian genetic algorithm (LGA) [343], as new options in the search for the best pose. We just consider here the recent version AutoDOCK 4.0 [795]. For new developments see Trott and Olson [796] and Morris et al. [797].

The ligand is considered fully flexible and the receptor is treated as rigid although conformational modeling of selected residues is possible. The empirical force field (calibrated on 30 protein–ligand complexes) involves

- Lennard-Jones (6/12) potential.
- Electrostatics Coulomb law with a distance-dependent dielectric constant.
- H-bonding: only hydrogen atoms potentially able to form H-bonds are defined (hydrogen atoms bonded to O, N, S). A 10/12 potential is used with an angular dependence.
- The docking energy is expressed as the sum of intermolecular interactions in the complex and intramolecular interactions in the ligand.

All these terms are weighted by coefficients (W) determined by the regression analysis of 188 protein–ligand complexes with known 3D structures and binding free energies (the standard error was about 2–3 kcal/mol in cross validation). In addition, a term estimates the change in torsional free energy on ligand binding

$$\Delta G_{tors} = W_{tors} N_{tors}$$

where N_{tors} is the number of rotatable bonds (excluding guanidinium and amide bonds). Desolvation energy (free energy change upon binding) is estimated with

$$\Delta G_{desolv} = W_{desolv} \sum_{i,j} (S_i V_i + S_j V_j) \exp\left(\frac{-r_{ij}^2}{2\sigma^2}\right)$$

where
   S and V denote the solvation term and the atomic fragmental volume for ligand
      atom i and protein atom j at distance r_{ij}
   σ is a Gaussian distance constant (=3.5 Å)

The idea is to evaluate the percentage of the volume around the ligand atoms that is occupied by protein atoms.

For finding the best pose, various optimization algorithms are provided:

- *Monte Carlo simulated annealing (MCSA)*: at each step that randomly changes the position or orientation of the ligand, the variation in energy is calculated and the change is accepted or not according to a Metropolis criterion. The process starts at a high "pseudo" temperature. As the process continues, the "pseudo" temperature is lowered, so that less and less upward changes are accepted. At the end, the best pose is returned.
- *Genetic algorithm (GA)*: a string of digits encodes, into a gene, the degrees of freedom of the ligand. A population of genes is generated at random and submitted to energy evaluation (fitness function). As the genetic operators generate children from pairs of genes, the genes with better scores are selected.
- *Lamarkian genetic algorithm (LGA)*: the difference with GA is that each gene (each state of the ligand) is submitted to energy minimization before scoring and generation of the next population of genes. LGA is faster than the preceding two methods, and allows for docking ligands with more degrees of freedom.
- Local search may be used to optimize a molecule in its local environment.

In order to estimate the free energy of binding, the energy of free (unbound) ligand and protein may be calculated. If it is assumed that the conformations for the two partners are the same as in the bound state, the corresponding contribution to internal energy will be zero. In the conformational search, the ligand is initially located randomly and with a structure nonspecific to the binding domain. The bound state is searched for by exploring translations, rotation, and internal degrees of freedom. Calculations are carried out on a rigid receptor (except specified residues). To speed up the process, the space around the receptor is divided into a grid of nodes (a grid per atom type). At each node, the potential energy of a probe atom is calculated. Receptor is not allowed to move.

For example, Marchand-Geneste et al. [798] carried out a study of endocrine disruptors acting on the rainbow trout estrogen receptor (rt). A model of rtERα was first built by homology with the human estrogen receptor (hERα), of known crystal structure. A high level of sequence conservation was found. Then selected chemicals were docked in the binding site, and the relative free energy of binding was calculated, using AutoDOCK. In this application, $\Delta(\Delta G_{binding})$ values correlates well with the log(RBA); $r^2 = 0.84$ for 14 compounds.

AutoDOCK has been widely used to guide alignments in view of CoMFA treatments as, for example, described in Refs. [172,187,253,275,334,527].

11.2.3 FlexX

The program FlexX developed by Rarey et al. [768] for docking organic ligands into protein active sites takes into account ligand flexibility by a discrete sampling of its conformational space. Basically, as indicated by the authors, FlexX uses the concepts of several preceding approaches developed for structure-based drug design:

- An incremental construction strategy, from Leach and Kuntz [792]
- An interaction model derived from the de novo design tool LUDI [777,778] and a similar scoring function
- A discrete sampling of the conformational state, from MIMUMBA [799], taking into consideration only a predefined set of torsional angles

We now go into further detail. The first step is to define, from the considered ligand, a base fragment (a connected part of the ligand) that will be positioned in the protein active site, the rest of the ligand being, for the moment, ignored. This base fragment will be progressively enlarged by the addition of other fragments. The location of the base fragment in the active site and its extension are controlled by examination of the ligand–receptor interactions (mainly H-bonding or hydrophobic). Three types of interactions are proposed:

H-bond donor	–	H-bond acceptor
Metal	–	Metal acceptor
Aromatic ring center	–	Aromatic ring atom, methyl, amide

To each type of interacting group of the ligand is assigned a geometry (a center and an interaction surface). These interaction surfaces are spheres, part of spheres, capped

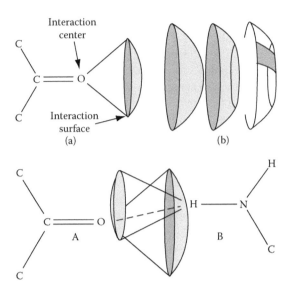

FIGURE 11.1   Interaction geometries. Upper part: interaction surfaces. (a) Interaction center and surface pertaining to a carbonyl group; (b) three of the four different types of interaction surfaces: cones, capped cones, and spherical rectangles. Lower part: geometry requirements. An H-bond between the carbonyl oxygen and the nitrogen. The interaction centers are the oxygen and the hydrogen atom forming the H-bond. They have to fall mutually on the surrounding interaction surfaces. (Reproduced from Rarey, M. et al., *J. Mol. Biol.*, 261, 470, 1996. With permission.)

cones, or spherical rectangles (Figure 11.1). On the receptor, the interaction surface is approximated by a finite set of interaction points. The basic hypothesis is that an interaction occurs if the center of one group approximately lies on the interaction surface of the other and vice versa. Interaction types must of course be compatible.

The problem is to find positions of the base fragment in the active site so that a sufficient number of favorable interactions simultaneously occur. To account for ligand flexibility, only a set of discrete torsional angles around simple acyclic bonds and some (predefined) ring conformations are considered. The location of the base fragment is carried out by mapping three interaction centers onto three receptor interaction points (verifying that angular constraints are satisfied and no overlap occurs with the receptor). Frequently several neighbor solutions are found. They are clustered to define one solution. To avoid long run times, the better selection is a fragment with many potential interaction groups and few conformations. The progressive complex construction starts from the different placements of the base fragment. Since addition of small fragments is more efficient, the ligand is divided at each rotatable acyclic simple bond. The process comes to a tree-search problem to find the leaves that contain placements with favorable binding energies (the nodes of the first level correspond to the different base placements). After each fragment addition to the set of partial ligand currently built, solutions are evaluated by the scoring function and the k-best ones are considered for the next addition. Some discrete flexibility for the receptor has been now introduced.

Scoring function of FLEXX: the scoring function, derived from that proposed by Böhm [778], approximates the free energy of binding, according to

$$\Delta G = \Delta G_0$$

$$+ \Delta G_{rot} * N_{rot}$$

$$+ \Delta G_{nHb} \sum_{nHb} f(\Delta R, \Delta\alpha) + \Delta G_{io} \sum_{io} f(\Delta R, \Delta\alpha) + \Delta G_{aro} \sum_{aro} f(\Delta R, \Delta\alpha)$$

$$+ \Delta G_{lipo} \sum_{lipo} f'(\Delta R)$$

In the second line, N_{rot} is the number of rotatable bonds immobilized in the formation of the complex. The third line takes into account neutral H-bonds (nHb), ionic (io), and aromatic (aro) interactions. The different ΔGs are adjustable parameters (e.g., $\Delta G_{aro} = -0.7\,kJ/mol$). f is a function penalizing deviations from the best geometry. The last term corresponds to lipophilic contact energy. It is calculated as a sum of pairwise atom–atom contacts. The f' function depends on the difference ΔR between the distance of atom centers (R) and an ideal value R_0 (equal to the sum of the van der Waals radii plus $0.6\,\text{Å}$). It is chosen to give a more or less ideal distance (f' = 1 for $-0.2 < \Delta R < 0.2\,\text{Å}$) and forbid close distances.

Applications of FLEXX: the program was tested on 19 protein–ligand complexes of known geometry. The smallest root mean square deviation (RMSD) between prediction and crystal conformation was no more than $0.7\,\text{Å}$, and in 13 cases out of 19 the best prediction was in the three top hits. However, the authors stressed that the approach worked better for medium-sized ligands (up to 17 rotatable bonds). It was suggested that keeping the implicit hypothesis of partial solutions always near optimal might not be possibly satisfied for complex ligands. Although the scoring function considered only a single conformation, solution ranking was good for a wide set of examples, but, as for other programs, the predicted free energy of binding sometimes largely deviated from the experiment. This is a rather common behavior in docking programs owing to chosen force field (vide infra). Another advantage is that the FLEXX approach is fast enough for interactive work and may be used, with a pre-screening step, for database search.

FLEXX was further evaluated [280] on varied tests: first, for 200 protein–ligand complexes, 46.5% of them were reproduced at first rank with an RMSD of less than $2\,\text{Å}$ and the rate reached 70% for the entire generated solution. However, the performance decreased for ligands with more than 15 components. Cross-docking experiments were also carried out: several receptor structures of complexes with identical proteins were used to dock all co-crystallized ligands of these complexes. Results indicated that FLEXX is able to "acceptably dock a ligand in a foreign receptor structure." In a third test, the proteins of 10 complexes were used as targets for a base of 556 molecules (with, among them, 200 ligand structures). The question was to investigate whether FLEXX was able to find the original ligand as one of the most active binders. In eight cases, the original ligand was among the top 7%.

FlexX-Pharm [708], an extended version of FlexX, allows for the introduction of constraints expressing important characteristics of protein–ligand binding modes into a docking calculation. For an example [75], see Section 8.1.2.

11.2.4   FlexS

FlexS proposed by Lemmen et al. [513] may be viewed as an extension and a variation of the docking program FlexX. But while FlexX focused on receptor–ligand complexes, FlexS aims to superimpose pairs of ligands, a (flexible) test ligand and a (rigid) reference ligand, without any knowledge of the receptor structure. This situation is directly related to the alignment problem, either for QSAR models or for a relative ranking of ligands from their similarity, when no receptor structure is available. The speed of the program allows for interactive analyses, or screening databases.

As in FlexX, the general strategy is an iterative incremental construction, adding small and relatively rigid fragments to a user-defined anchor (the base fragment). At each step, the use of a discrete set of torsional angles accounts for conformational flexibility. Directional interactions (H-bond, salt bridge) are modeled in terms of interaction centers and interaction geometries (specifying the position of the expected counter-group) (Figure 11.2). Less directional interactions (e.g., between aromatic groups) are treated as "paired intermolecular interactions" approximated by sets of discrete interaction points. Partial charges, H-bonding potential, and local hydrophobicity are represented by Gaussian functions (generally centered on the atom). Hydrophobic character is determined by rules based on the partial atomic charges.

Scoring function in FlexS: the scoring function is mainly based on the overlap volumes of properties expressed by Gaussian functions (electron density, partial charge, hydrophobicity, H-bond acceptor, H-bond donor character). The scoring function so comprises a sum of terms $\Delta S_x * S_x$, where ΔS_x is a coefficient and S_x the contribution (overlap volume) for the property (x) considered. Three additional terms are respectively associated to

- van der Waals interactions

$$\Delta S_{vdw} * S_{vdw}$$

  (with S_{vdw} corresponding to the overlap of the atomic van der Waals spheres).
- Paired intermolecular interactions, with

$$S_{match} = \sum_{match\ m} S_m(type(m)) * p(\Delta r, \Delta \alpha)$$

  for the three types of matches: uncharged H-bonds, ionic interactions, hydrophobic interactions. The function p penalizes deviations in length (Δr) and angle ($\Delta \alpha$) from an ideal geometry. The value of p is 1 for deviations Δ up to a threshold *th*1 and then it linearly decreases to 0 for Δ between *th*1 and *th*2. The parameters S_m are fixed for the various types of matches.

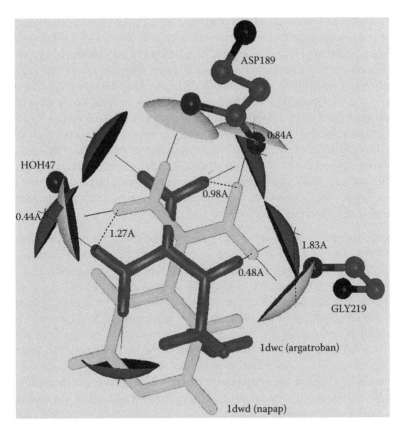

FIGURE 11.2 Alignment where counter groups are nearly coincident but not centers of ligands. Alignment of the benzamidino group of napap (light gray) and the guanidinium group of argatroban (dark gray) as derived by the superposition of corresponding Cα-positions of thrombin. The ligands are shown together with their interaction geometries, main directions, and site points for the main directions (cap, line, and cross in the same color). Two interacting groups of the protein and one structural water molecule are also provided. Neither the interaction centers nor the main directions coincide (distances are indicated by dashed black lines and measured in Å). However, in all cases, the corresponding protein counter group falls close to the intersection of the involved interaction geometries. (Reproduced from Lemmen, C. et al., *J. Med. Chem.*, 41, 4502, 1998. With permission.)

- Matching of subtrees

$$\Delta S_{stm} * S_{stm}$$

where S_{stm} quantifies to what extent subtrees match in the test and reference ligand. The idea is to favor, during the incremental construction, solutions where the size of the remaining fragment of the test corresponds to the respective part of the reference. Coefficients S have been calibrated and change in the diverse phases of the process, as the size of the test ligand increases.

Among the salient points of the program, one can note, for the base placement, a rigid body procedure that optimizes the common volume of the two molecules (according to different physicochemical properties) in a way similar to that of SEAL [151]. This option is best suited for placing large fragments with few directional interactions. It calls for an adaptation of Patterson functions (relying on the phase problem and the molecular replacement technique in X-ray crystallography) in Fourier space, and to Hodgkin similarity measures [474]. This allows for independently optimizing first rotations and then translations for the positioning of the base. On the other hand, in the progressive reconstruction of the ligand, a dynamic selection of the fragments to be added is provided. The point is that in docking, the ligand must lie inside the active site whereas in alignment it may extend beyond the volume of the reference structure. After each partial placement, a list of candidates is built for the next iteration, and FLEXS selects the best one.

Applications of FLEXS: to prove the efficiency of the approach, two tests were carried out. The first one concerned a virtual screening on 984 compounds for fibrinogen receptor antagonists (compounds involved in platelet aggregation properties). The 3D geometry of the active site was not solved, but studies on several rigid cyclic peptides suggested a putative RGD (arginine, glycine, and aspartate) arrangement of the pharmacophoric groups (**11.1**). The screening was performed in two steps: the amidinium and carboxylate groups of a reference (rigid) ligand, considered as essential groups, were separately aligned for all compounds (rigid fit). Only the 50 solutions with the highest combined scores were retained. Then a more elaborate alignment with the entire peptide (base fragment atoms next to the amidinium fragment) was carried out and the best scoring solution retained.

11.1

All known antagonists were found within the 20 top hits and additional interesting candidates were proposed. The approach outperformed a more rapid search based on the 2D-DAYLIGHT fingerprint similarities [800].

In the second test, the authors attempted to reproduce a set of 284, experimentally given, mutual superimpositions (examples were derived from 76 protein–ligand complexes with 14 proteins). The problem was to superimpose, for each ligand, a reference structure corresponding to its crystallographic binding conformation, and, in test, an arbitrarily transformed conformation and orientation. For 60% of the examples, geometries were reproduced with an RMSD below 1.5 Å. In similar validation studies, for docking reproduction, the rate was about 70% with GA [279].

The authors indicated that the approach (such as FLEXX) worked best with small and medium-sized molecules, whereas problems might occur with large systems, or when multiple binding modes might coexist. Similarly, predicted geometries were good but, as observed elsewhere, the docking score did not closely reproduce the ranking of the activities.

11.2.5 FLEXE

Most of the widely used alignment or docking programs use flexible ligands but keep the protein rigid (except some limited movements of terminal H-bonds). The originality of FLEXE [706] is to take into account extended protein flexibility, involving, for example, side chain rotations or adjustment of single loops (commonly referred to as "induced fit"), and even single point mutations. FLEXE includes protein adaptation simultaneously with ligand flexibility during the docking and not as a post-optimization step. Several experimental examples are known where a ligand can only bind an active site with a rotation of a side chain (see Chapter 10) of the protein. Homology modeling may lead to ambiguities in side chain location or uncertainties in loop structure. For example, Alberts et al. [801] reported that peptidic inhibitors bound to the native form of rennin whereas piperidine-based inhibitors bound to a modified receptor structure involving rearrangement of three side chains and opening of the active site flap.

The active site is defined by the union of all ligands. The basic hypothesis, as to protein flexibility, is that the overall structure and the general shape of the active site have to be conserved. The variations in protein structure are represented with a set of discrete conformations that are combined by superposition into a united protein description, the "ensemble," and treated as separate alternatives. FLEXE selects the combination which best suits the given ligand with respect to the scoring function. FLEXE adapts to this "ensemble" approach all the main concepts of FLEXX regarding the interaction scheme, the scoring function, and the incremental construction. FLEXE can treat structures picked up from MD simulations, generated by rotamer libraries, or ambiguous homology modeling. This allows for spanning a much larger conformational space than that covered by the separate structures and provides a more robust approach.

Applications of FLEXE: the method was tested on 10 protein structures corresponding to 105 crystal structures associated with similar backbones but different side chain conformations in complexes with 60 ligands. All ligands were docked in the united protein describing the ensemble. FLEXE, considering the top 10 solutions, retrieved ligand positions within 2 Å RMSD to experiment for 67% of the cases. The percentage was only 63% with FLEXX, in a comparative attempt where all ligands were separately docked into all structures in the ensemble (cross-docking) and a ranking list was built. An interesting example is that of the highly flexible active site of aldose reductase. This enzyme catalyzes the reduction of glucose to sorbitol and is possibly involved in diabetic complications on visual, renal, and nervous systems. FLEXX using just one rigid protein structure would miss in screening some potent inhibitors that could be correctly docked with FLEXE in another member of the ensemble.

11.2.6 GOLD

In the binding of small molecules to macromolecules, another approach to cope with flexibility is afforded by genetic optimization for ligand docking (GOLD) proposed by Jones et al. [279]. The method is based on the development of a GA, already proposed by Jones et al. [802] in the recognition of receptor sites. It explores the full conformational (acyclic) ligand flexibility, with partial protein flexibility. The authors stressed that the method accounted for the displacement of loosely bound water molecules on binding.

In the GA, each chromosome conveys the conformational information about the ligand and the receptor active site into two binary strings: one for the protein, the other for the ligand. Each byte encodes rotation around a rotatable bond. Two other integer strings encode a mapping from possible H-bonding sites of the ligand to those in the protein or internal to the ligands. The fitness function is formed of three terms:

- The H-bond energy determined from precalculated terms (using model fragments and accounting for water displacement) and weighted by a factor depending on both distance and angular constraints.
- The (protein–ligand) complex interaction energy calculated with an atom–atom 4–8 potential (softer than the usual Lennard-Jones potential).
- The ligand internal energy: sum of steric energy (calculated with a Lennard-Jones potential) and torsional energy.

Operators (mutation and crossover) are given a weight and are selected using a roulette wheel based on these weights. A roulette wheel based on fitness score chooses the parents and children (if not already present) to replace the least fit members of the population. Two particularities of the GA may be noted: it uses an "island" model (with several subpopulations and possible migration of individual chromosomes) and "nicheing" to increase population diversity and avoid premature convergence. Two individuals share the same niche if the RMSD between their donor and acceptor atom coordinates is less than $1.0\,\text{Å}$. If there are more than a specified number of individuals in a niche, the new individual replaces the worst member of the niche rather than the worst member of the total population. As an alternative to the fitness function, scoring may be achieved with an adaptation of ChemScore by addition of penalty terms for steric clash and bad conformation in docking [757] (vide infra).

Applications of GOLD: the effectiveness of the method was examined on the complex NADPH-DHFR (intervening in the reduction of folate to dihydrofolate and then onto tetrahydrofolate). The ternary complex of DHFR with NADPH and methotrexate had been solved by X-ray crystallography by Bolin et al. [157]. The test was to dock NADPH into DHFR and compare the results of GOLD to the experiment. It was a difficult problem owing to the high flexibility of NADPH (17 rotatable bonds and 10 free corners for cyclic flexibility). For 20 runs of GA, the best solution led to an RMSD of $1.2\,\text{Å}$ between the predicted and experimental structures. It was observed that the best fit with crystal structure is not always the best GA solution; here the solution closest to the crystal geometry was ranked fourth.

Tests were pursued on a data set of 100 protein–ligand complexes (from PDB) of widely varied structures. Comparison with crystallographic results were ranked good (41%), close (30%), error (9%), or wrong (19%). The authors carried out a very careful analysis of the successes and failures of GOLD. Apart from a poor protein resolution or poorly determined small-ligand geometry, deviations often occur with large and flexible ligands, but the major problems might originate from an underestimation of hydrophobic factors and desolvation.

11.2.7 GLIDE

Grid-based ligand docking with energetics (Glide) [773,774] approximates a complete systematic search of position, orientation, and conformation of the ligand in the receptor binding pocket through a series of hierarchical filters. The receptor is represented on a grid where different fields (associated to its shape and properties) are precomputed, providing progressively more selective scorings of the ligand pose. A set of initial ligand conformations is created (by exhaustive search of energy minima in the ligand torsion angle space). Conformers are clustered, each cluster corresponding to a common core geometry and an exhaustive set of side chain conformations. Each cluster is first docked as a single object. From an initial rough positioning, a scoring phase narrows drastically the search space to a few hundred surviving candidates. Then energy optimization is performed using the van der Waals and electrostatic terms of the all atom optimized potentials for liquid simulations (OPLS-AA) force field [803], precomputed on the receptor grid. At last, the three to six lowest energy poses are refined in a Monte Carlo procedure, adjusting torsion angles and orientation of peripheral groups of the ligand.

These best poses are then rescored using the Glide_Score function, combining the ChemScore function [775] (vide infra) and force field–based terms (ligand–receptor interactions and solvation terms). The best pose is determined by the Emodel that takes into account both Glide_Score and the internal strain of the ligand. Note that van der Waals radii of selected nonpolar atoms (protein or ligand) are scaled to create additional space in the binding pocket.

In the modified ChemScore function

$$\Delta G_{bind} = C_0 + C_{lipo} \sum f(r_{lr}) + C_{hbond} \sum g(\Delta r)\, h(\Delta\alpha) + C_{metal} \sum f(r_{lm}) + C_{rotb}\, H_{rotb}$$

the second term of the right member concerns all pairs (ligand atom–receptor atom) defined as lipophilic

the third term relies on H-bonding interactions, weighted by deviations in distance or angles (functions g and h)

In Glide, the H-bond term distinguishes interaction between neutral and charged partners. For metal–ligand interactions, only the best interaction with anionic acceptor atoms of the ligand is counted. Situations where a polar (non-H-bonding) atom is located in a hydrophobic pocket are also taken into account with a term $C_{pol\text{-}phob} V_{pol\text{-}phob}$. Coulomb and van der Waals interactions between ligand and receptor (with reduced charges) are included, as well as a solvation term considering explicit water molecules docked into the binding domain.

Accuracy of the method was examined by re-docking ligands from 282 co-crystallized PDB complexes starting from optimized ligand geometry (with no memory of the correct pose). Error was less than 1 Å in nearly 50% of the cases and greater than 2 Å for only 30% of the cases. In this example, Glide appeared twice as accurate as GOLD and more than twice as accurate as FlexX for ligands up to 20 rotatable bonds. It also outperformed Surflex of Jain [804]. The ability of Glide to identify active compounds in a database was further assessed in an extensive study of diverse sets of protein receptors [774]. The approach outperformed GOLD, FlexX, and DOCK (see, however, Ref. [760]).

Recently, Extra Precision Glide (XP Glide) incorporated, in addition to water desolvation energy, a model of hydrophobic enclosure for the protein–ligand complex, corresponding to enhanced activity [805].

11.2.8 ICM

ICM [772] works with internal coordinate variables, and uses for energy calculations the Empirical Conformational Energy Program for Peptides-3 (ECEPP-3) and Merck molecular force field (MMFF) partial charges [806]. Five fields characterize the receptor: electrostatic, H-bonding, hydrophobic, and two van der Waals terms. The ligand is fully flexible whereas the protein LBD is kept rigid. Exploration of the torsion and rotational space of the ligand is achieved with a biased probability Monte Carlo (BPMC) algorithm [807]. A conformation is randomly selected and a new position is generated at random, independently of the preceding, but according to a predefined continuous probability distribution. A double energy minimization is carried out (first on analytically differentiable terms, then on the complete energy function). This process allows for incorporating complex energy terms such as those involved in surface-based solvation. Acceptance (or rejection) is decided from a Metropolis criterion on the total energy. Favorable conformations are allocated to a stack (history mechanism) and the process is iterated.

11.2.9 MISCELLANEOUS

Protein flexibility, often ignored, was also taken into account by Alberts et al. [801] who integrated side chain rearrangement into both docking and design modeling. The authors modified the de novo design algorithm Skelgen [808], including an additional transition type to account for side chain mobility in a simulated annealing protocol to dynamically optimize binding energy.

A first example concerned acetylcholinesterase (AChE), a serine hydrolase involved in Alzeimer's disease. Donepezil (yet a potent inhibitor) was not detected as a ligand when a static structure of the nonnative receptor site (1*vot* in PDB) was used whereas including flexibility reproduced the correct binding mode. A second example concerned the design of new ligands to human collagenase MMP-1, a matrix metalloproteinase-1, involved in several pathologies (cancer and arteriosclerosis). A rigid MMP-1 structure allowed for binding only small ligands (one cycle plus few substituents). Introducing side chain mobility exposed the cavity and allowed for binding larger molecules (two rings or fused ring systems plus varied substituents) (Figure 11.3). Such molecules resembled some of the most active known inhibitors.

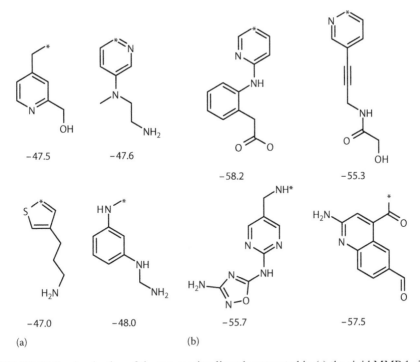

FIGURE 11.3 A selection of the top-scoring ligands generated in (a) the rigid MMP-1 site and (b) the flexible MM-1 site. The attachment point to the anchored scaffold is given by the asterisk. Ligand-binding energies are given beneath each structure. (Extracted from Alberts, I.L., *QSAR Comb. Sci.*, 24, 503, 2005.)

Various other docking programs were proposed for docking or structure-based drug design, such as LigandFit [809] and QXP [810].

11.3 NONSPECIFIC SCORING FUNCTIONS

In many examples, docking scores reproduced the experimentally observed binding modes and have been successfully used for filtering active compounds from those unable to bind the receptor active site. But various examples evidenced that the scoring functions cannot convey the minute activity variations observed in a series of compounds, and so are of little interest for deriving QSAR models, although some success has been reported [811].

Enyedy and Egan [760] examined the relationship between docking scores and pIC_{50} on large data sets. The study concerned the activity of 4300 compounds (measured in the same conditions) toward three kinases. Glide was the program used. For the different series considered, no correlation was found between Glide docking scores and experimental activities ($r^2 < 0.15$). More surprisingly, as to the use of docking scores for enrichment of actives in top scoring compounds, Glide_Scores were not better than the molecular weight (MW) or the partition coefficient ClogP.

The influence on performance of the particular target, the crystal structure, and the precision of the scoring function were also examined.

To tackle this problem a number of scoring functions were proposed. Some of them are examined below.

11.3.1 EMPIRICAL SCORING FUNCTIONS

We first mention, a little different from the common scoring functions, hydropathic interactions (HINT) [192]. This model, developed for computing hydrophobic fields in 3D-QSARs, has been extended to evaluate protein–ligand interaction energy [812]:

$$HINT = \sum_{prot} \sum_{ligand} S_i a_i S_j a_j R_{ij} T_{ij}$$

where

    S_i, S_j are the solvent accessible surface areas
    a_i, a_j are hydrophobic atomic constants (from log P)
    R_{ij} is a distance-dependent function
    T_{ij} ensures a good sign to the expression, so that favored interactions are positive

On the basis of Böhm's model [778] various empirical scoring functions were proposed independently of any docking program. The binding free energy is expressed as a sum of contributions on the general model:

$$\Delta G_{bind} = \Delta G_{motion} + \Delta G_{interaction} + \Delta G_{desolvation} + \Delta G_{configuration}$$

where the different contributions are directly calibrated from a multiple linear regression (MLR) treatment on a set of known protein–ligand complexes. Such approaches present appealing advantages. They are not limited to congeneric series if the training set is diverse enough. Each term in such an empirical function has a clear physical interpretation and the corresponding coefficients in the MLR give some insight about the binding mechanisms.

A first attempt to build an extremely fast, all-purpose estimate is attributed to Eldridge [775], who introduced, in the framework of empirical scoring functions, a four-term expression, ChemScore:

$$\Delta G_{bind} = \Delta G_0 + \Delta G_{H\text{-}bond} \sum_{iI} g_1(\Delta r)\, g_2(\Delta \alpha) + \Delta G_{met} \sum_{aM} f(r_{aM})$$

$$+ \Delta G_{lipo} \sum_{lL} f(r_{il}) + \Delta G_{rot}\, H_{rot}$$

where

    g_1, g_2 are penalty functions for deviations from ideal distance or angles with value
       1 or 0, respectively, for small and too large deviations and a linear variation in
       the border region
    f is a distance-dependent function of similar shape

In addition to the H-bond and metal ion interactions, a lipophilic term is calculated only for pairs of lipophilic atoms on ligand and receptor. The last term corresponds to rotatable bonds frozen upon binding, with a scaling factor (flexibility) taking into account the percentage of nonlipophilic atoms on each side of the rotatable bond to not overestimate the entropy loss. This MLR model adjusted on 82 complexes led to $r^2 = 0.710$, $q^2 = 0.658$ for the full set with an estimated error of 2.2 kcal/mol.

In SCORE, introduced by Wang et al. [765,776], the affinity constant is defined as a sum of 11 terms, scaled by MLR, and characterizing van der Waals interactions, metal–ligand contribution, H-bonding term plus desolvation and internal deformation, whereas overall translational and rotational entropy loss is expressed in the regression constant. The parameters were adjusted on a training set of 120 complexes on a 12 log unit range.

Several original treatments may be noted. Particularly, hydrogen atoms are depleted. The different contributions are evaluated by discontinuous stepwise values and not by continuous explicit functions. So van der Waals effects are expressed by counting pairwise bumps between protein and ligand atoms (other interactions being neglected on account of the competition between ligand and water on going to the protein-bound state). Directional effects (distance and angles) are included for H-bonding: the angular dependence being evaluated from the two angles X–D···A and D···A–Y in a system X–D···(H)···A–Y (Figure 11.4). A distance-dependent term is used for (protein) metal–ligand atom (O, N) terms. Desolvation and hydrophobic effects (loss of lipophilic area upon binding) are treated thanks to a quantitative atomic hydrophobicity scale, and considering only hydrophobic match (a hydrophobic ligand atom in a hydrophobic protein environment from its closest neighbors with $d < 5\,\text{Å}$). At last, entropy effects (immobilization of rotatable bonds) and deformation (enthalpic internal strain of the ligand upon binding) are considered as proportional to the number of rotatable bonds. It may be noticed also that van der Waals radii were adjusted by GA, and that electrostatic interactions are not explicitly evaluated. Another original point is the definition of an atomic binding score that specifies the importance of each atom in the binding process and may be useful for lead optimization in structure-based drug design.

For the 170 complexes of the training set, $r^2 = 0.777$ and $q^2 = 0.743$ (SD = 1.16 log unit). On a test set of 11 endothiapepsin complexes, $r^2 = 0.654$. Robustness and internal consistency of the approach was confirmed in a stepwise evolutionary test where test sets are randomly built with a progressively increasing number of compounds.

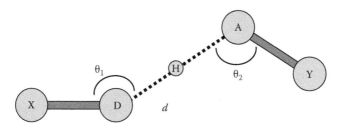

FIGURE 11.4 H-bonding and angular dependence. Evaluation is carried out from the two angles X–D···A and D···A–Y in a system X–D···H···A–Y.

In a further publication, Wang et al. [765] proposed X-CSCORE, in an improved treatment: the binding process was dissected into terms related to van der Waals, H-bonding, deformation, and hydrophobicity (plus a constant term gathering overall translational and rotational entropy loss).

$$\Delta G_{bind} = \Delta G_0 + \Delta G_{vdW} + \Delta G_{H\text{-}bond} + \Delta G_{deformation} + \Delta G_{hydrophobic}$$

The constant term, ΔG_0, implicitly included the translational and rotational entropy loss upon binding. The other terms corresponded to

1. van der Waals ligand–protein interactions, evaluated with a soft 4, 8 potential (only heavy atoms were considered).
2. H-bonding, characterized by the distance between the "root atoms" and angular factors derived from an analysis of ligand–protein H-bonds in the training set. Metal ion–ligand interactions were calculated with the same function.
3. Deformation effects estimating detrimental entropy changes due to the immobilization of some rotors upon binding. In this model, elimination of concerted motions due to synchronized cross rotors avoided the overestimation generally carried out when using the total number of rotatable bonds. The deformation of the protein was not considered.
4. Hydrophobic contribution (nonpolar groups tended to favor each other) and solvation effects, with three options (involving five adjustable parameters weighted by log P) as in SCORE [776].
   - The hydrophobic surface algorithm evaluated the buried surface area (for hydrophobic atoms) of the ligand HS-SCORE.
   - The hydrophobic contact algorithm took into account hydrophobic atom pairs (protein–lignand) with a distance-dependent function HC-SCORE.
   - The hydrophobic matching algorithm as in SCORE [776] considered hydrophobic ligand atoms placed in a hydrophobic proteinic environment HM-SCORE.

The mean of these three models led to a consensus scoring function X-CSCORE. This approach was calibrated with a set of 200 carefully selected protein–ligand complexes (corresponding to 70 different types of proteins). On an independent test set of 30 complexes, r^2 yielded 0.59 with a standard deviation of 2.2 kcal/mol. Potential application of X-CSCORE in molecular docking was also examined on 10 samples corresponding to largely diverse ligands and proteins. Docking was performed with AutoDOCK (GA) and the best 100 conformations reevaluated by X-CSCORE, leading to a largely improved docking accuracy. In the calculation of binding energies, AutoDOCK was clearly outperformed.

Although largely successful for binding affinity prediction, scoring functions suffer some drawbacks: neglect of specific mechanisms such as cation–π interactions, π–π stacking, or the role of water molecules at the protein–ligand interface. An important point is robustness: the capability to tolerate minor uncertainties on atomic positions (avoiding explicit hydrogens, for example, using "soft" potentials are in line with this concern).

11.3.2 POTENTIAL OF MEAN FORCE

In the search for a rapid and general scoring function for evaluating protein–ligand binding affinity, Muegge and Martin [779,780] developed a knowledge-based approach. Information extracted from the Brookhaven database on known protein–ligand complexes was used to define PMF from the distance distribution of specific atom pairs. These empirical potentials were able to lead to the total protein–ligand binding free energy. The interesting point was that all contributions were treated implicitly, a definite advantage, since generally, in free energy calculations, the entropic contributions were difficult to evaluate and the characterization of solvent reorganization time-consuming. Furthermore, no fitting of a training set to experimental results is needed.

The method was mainly devoted to virtual screening. But the good predictions obtained on varied series of ligands (SD from observed binding affinities of 1.8 log units and r^2 about 0.61 for 77 ligand–protein complexes) suggested it might be useful in QSARs. The basic assumptions were that the crystallographic structure of protein–ligand complexes, which represented the optimum placement of the ligands, corresponded to thermodynamic equilibrium in the global free energy minimum and that the distribution of microscopic states obeyed Boltzmann's law and reflected the complex interaction scheme.

For an atom pair (at distance r), the Helmholtz interaction free energy $A(r)$ can be calculated from the pair distribution functions $g(r)$

$$A(r) = -kT \ln [g(r)]$$

with

$$g(r) = \frac{\rho^{ij}(r)}{\rho^{ij}_{bulk}}$$

where
$\rho^{ij}(r)$ is the number density of atom pairs i (protein)–j (ligand) in a spherical shell δr at distance r

ρ^{ij}_{bulk} is the number density when no interaction between i and j occurs (in a sphere of radius R)

In the approach quoted [779], the authors defined a PMF_score (proportional to the absolute binding free energy) as:

$$PMF\_score = \sum_{kl} A_{ij}(r) \quad \text{for } r < r^{ij}_{cutoff}$$

Σ (for all types (k,l) of protein–ligand atom pairs) of the interaction free energy of atom pairs i (protein)–j (ligand). For the calculation of interaction potentials, a slightly modified formula was used, with the introduction of a ligand volume correction $f^j_{v-cor}(r)$

$$A_{ij} = -kT \ln \left[f_{v-cor}^{j}(r) g_{ij}(r) \right]$$

The volume correction factor is introduced to make easier the calculation of the reference bulk density (no interaction). Details may be found in Ref. [780]. The interaction terms were extracted from 697 entries corresponding to 16 proteins and 34 ligand atom types. The efficiency of the approach was examined on six test sets (77 different protein–ligand complexes with large variation ranges on log K_i). Standard deviation from observed binding affinities was of 1.8 log K_i units and $r^2 = 0.61$. On an additional test of 33 inhibitors modeled into the binding site of the L-689, 502-inhibited HIV-1 protease, SD was 0.80 log K_i units, with $r^2 = 0.74$. The soring function was further updated on 7152 proteins-ligand complexes [780].

The authors emphasized the importance of a volume correction term that made the potential deeper at small distances and of a choice of different cutoff values depending on the nature of the atom pairs. They also mentioned that usual scoring functions needed empirical fitting, which made difficult their use for complexes not represented in the training set. Such a limitation did not exist with this PMF approach.

DrugScore of Gohlke et al. [794,813] is also based on knowledge-based potentials (either distance dependent or surface dependent). Diverse other scoring functions have been tested, for example, in the extensive study of Wang [757] (vide infra). For example, it may be noted that commercial packages have incorporated various options of these scoring functions. SYBYL implemented the DOCK function (electrostatics plus van der Waals term with reduced radii to avoid clashes) in Sybyl-Dscore, the GOLD scoring function in Sybyl-G-Score, and the Elridge's ChemScore function, including a term concerning metal ions, in Sybyl-ChemScore. The Sybyl-FScore derived from LUDI and FlexX, whereas Sybyl-PMF score is adapted from the PMF of Muegge et al.

Similarly, Cerius proposed adaptations of LUDI, PMF, a sum of piecewise linear potentials (PLP) of ligand–protein atoms and LigScore involving van der Waals terms, buried surface area with attractive protein–ligand interactions and the square of the buried polar surface in the complex (with both attractive and repulsive interactions).

11.3.3 COMPARED PERFORMANCE OF SCORING FUNCTIONS

Extensive analyses of the merits of these various scoring functions were carried out in several papers from (inter alia) Ewing and Kuntz [814], Sthal and Rarey [815], Kellenberger et al. [816], Bissantz et al. [782], Wang et al. [757,817], Ferrara et al. [818], Cummings [784], Cole et al. [783], Warren et al. [759], and Enyedy et al. [760]. For example, Wang et al. [757,765] studied 11 and then 14 scoring functions. In their subsequent paper, a large set of 800 complexes of small organic molecules with 200 types of proteins were investigated with the concern of prediction of the binding affinity for complexes based upon high-resolution crystal structures. A parallel work was also conducted by Ferrara et al. [818].

A second concern was robustness: can computation tolerate minor inaccuracies on crystal structure? On three most populated proteins in the test set (82 entries for HIV-1 protease, 45 for trypsin, and 40 for carbonic anhydrase II), they examined how well a scoring function can discriminate between low- and high-affinity complexes. Wang X-Score, Drug-score, Sybyl/ChemScore and Cerius2/PLP score gave better correlations with experimentally determined binding constants. These functions were robust enough to accept small variations on crystal structure without optimization of the input crystal structure. On subsets of compounds binding a same protein, more accurate results were obtained with standard deviation as low as 1.4 kcal/mol (1 log unit) but, depending on the subset, performance was case dependent.

All these studies converged to similar conclusions: no clear rule could be laid out as to a "best" scoring function. Results may vary from, more generally, a good recall of experimental poses, about 70% (particularly with consensus approaches) to very bad diagnosis. Furthermore, there is no general relationship between the ability of predicting a good pose and the efficiency in virtual screening. Leach et al. [762] indicated that in some cases, good enrichment was even observed "with poses barely in the binding site." Up to now, reliable rank ordering a list of compounds seemed still beyond the field. However, scoring a homogeneous series binding a same target is more successful.

No doubt these drawbacks would be removed with a more systematic treatment of entropic effects and solvation (some scoring functions already include PBSA and GBSA models). It may be hoped also that the empirical methods or the PMF approach would benefit from an increasing number of resolved complexes. Finally, rigorous physicochemical measurements of fundamental mechanisms (regarding pK, transfer energies, charges) would be of great interest.

11.4 DOCKING: A PREPROCESSING STEP FOR CoMFA-TYPE QSAR MODELS

Alignment is a crucial step in CoMFA, CoMSIA, and related QSAR models. This step, which *must be addressed*, is often time consuming, subjective, and difficult. Docking programs that allow for determining the best geometries (position, orientation, and conformation) for a receptor–ligand complex are therefore privileged tools for a more objective generation of aligned structures. This approach was presented by Sippl [59,274] as very attractive since it combined a receptor-based alignment of high accuracy to guide a computationally efficient ligand-based QSAR model.

In some cases, when examining a data set encompassing several subseries of congeneric compounds, a representative for each subset may be docked in the active site and used as template for the whole corresponding subset. In other applications, all the ligands are docked in the active receptor pocket to determine their best pose. The solutions retained may be directly submitted to field calculations. But it was also proposed to use them as active conformations to be aligned by the usual procedures (e.g., field alignment). An example will be given later [334].

Sippl's study compared the performance of such models to a crude evaluation of interaction energies in a "receptor-dependent" model. The data set investigated (binders to the estrogen receptor) encompassed the 30 compounds previously studied by Sadler et al. [185] on a 4 log unit activity range. Alignment was based on experimental or predicted positions of ligands in the binding pocket. Docking was performed by AutoDOCK [444]. Considering flexible ligands and a fixed protein, interactions were calculated on a grid, similar to that of Goodford [58], with OH, hydrophobic, and methyl probes. A simulated annealing procedure was applied (typically 100 runs of 50 cycles with a maximum of 20,000 steps). Temperature decreased by a factor of 0.95 on each cycle starting from RT = 600 cal/mol. One hundred docked complexes were clustered and energy refined. The first 20 complexes were then energy minimized using a more refined force field, YETI (this program includes directional potential functions for H-bonding, salt linkage, and metal ligand interactions [819,820]).

Comparison with known crystal structures of complexes of estradiol and diethylstilbestrol showed that these structures are well "reproduced" after energy optimization whereas AutoDOCK alone only ranked them in the first five solutions. With the minimum-energy geometries so determined, calculated interaction energies were correlated to binding affinities. The best correlation ($q^2 = 0.570$, $r^2 = 0.617$) was obtained with AM1-ESP charges (charges fitted on the electrostatic potential), protein flexibility (for the atoms nearest to the ligands) being allowed. These results are less satisfactory than those obtained by Sadler in a CoMFA treatment ($q^2 = 0.796$, $r^2 = 0.951$).

A QSAR treatment was then performed by GRID/GOLPE, with a water probe, using as starting point the superimposition of the ligands derived from docking. With variable selection, the best correlation yielded $r^2 = 0.992$ and $q^2 = 0.921$. These results, better than those of Sadler (with a ligand-based alignment), suggested that a model based on the receptor structure gave a better representation of the biological activity. However, the author indicated that the method using a rigid receptor could not be applied if the receptor changed conformation depending on the ligand (this is the problem of "induced fit" addressed by Vedani et al. [see Chapter 10] at least when variations are limited to small changes). The approach was further applied to a test set of 30 new ligands of varied structures, confirming the conclusions of the preceding study [274].

For the training set, three alignment strategies were carried out: one alignment came from the top-scoring YETI orientations, the other two were derived from FLEXS using as fixed reference template either diethylstilbestrol or estradiol and considering flexible ligands. In FLEXS physicochemical properties of molecules to be superimposed are approximated as spatial density distributions in terms of Gaussian functions. Good results were obtained in a 3D-QSAR starting from the refined docked structures, whereas calculation of interaction energies failed to give acceptable results (note, however, that solvation was not included in such a calculation). Similarly, using ligand-based models from two possible alignments proposed by FLEXS led to poor predictivity.

So it was interesting to compare the efficiency of manual and automated alignment. With this concern, Tervo et al. [821] studied the activity of 113 flexible HIV-1 PR cyclic urea inhibitors (**11.2**). HIV-1 PR aspartic protease encoded by HIV plays an essential role in viral infectivity and is one of the primary targets for structure-based drug design against HIV. Manual docking was carried out after energy minimization, on the cyclic urea part, re-minimization, and then adjustment on the cyclic urea ring plus its benzyl and hydroxyl substituents.

11.2

Automatic docking was performed with GOLD [279]. The GOLD score is based on the estimation of H-bonding stabilization energy, van der Waals internal energy, and complex interaction energy. The obtained conformations are in good agreement with existing co-crystallized structures. But, not surprisingly, GOLD rank scores show only a poor correlation with the observed activities ($r^2 = 0.416$), the deviations being more important for low-activity compounds, where secondary factors, not considered in GOLD, may have a larger relative importance.

For both manual and automatic alignments, three models were considered: CoMFA and CoMSIA with steric and electrostatic fields and CoMSIA with H-bond fields only. Generally speaking, consistent and significant results were obtained. The best predictive model (CoMSIA, H-bond fields) yielded $r_{pred}^2 = 0.754$ (on a test set of 20 compounds), $q^2 = 0.649$ (for a learning set of 93 compounds). Automated models were better in test even if slightly inferior in training. However, as previously noted, CoMFA was more sensitive to alignment details with $q^2 = 0.616$ (manual) and 0.523 (automatic) but r_{pred}^2 was only 0.224 (manual) vs. 0.647 (automatic). The authors concluded that automatic alignment could lead to satisfactory predictive QSAR models when a receptor structure is available.

11.5   INTERACTION ENERGY CALCULATION AND RECEPTOR-DEPENDENT MODELS

The situation is somewhat similar to that encountered with scoring functions in virtual screening. But here we are faced with much more limited series of compounds, with generally some structural homogeneity. So, in principle, once the best pose is found, any scoring function already presented might be used to give a rapid estimate of the free energy of binding. However, if van der Waals and electrostatic effects can be relatively easily modeled, the evaluation of entropic terms and the influence of the solvent are more difficult to express. Simplified approaches used relatively crude interaction terms, with no consideration of the solvent and no sampling of the system microstates (as required in accurate calculations). Another difficulty comes from the fact that building a QSAR requires a precise evaluation of small energy differences (remember that 1 log unit variation in pK_i corresponds to an energy change of only 1.4 kcal/mol at room temperature). Owing to the difficulty of accurately and completely computing ligand–receptor binding thermodynamics, the various energy contributions calculated were sometimes treated as structural descriptors submitted to statistical analysis (e.g., MLR and GA) in the general formalism of QSARs.

Among the first attempts, we can cite the model of Goodford [58] with the GRID program, aimed to determine energetically favorable binding sites on biologically important macromolecules. GRID relied on interactions calculated for different (ligand) probes on the nodes of a lattice built around the protein active site and, joined with GOLPE for variable reduction, was the basis of several 3D-QSAR studies in an approach somewhat symmetrical to CoMFA, which uses a lattice built from the ligand.

We have already presented the LUDI program [777,778,822], dedicated to the de novo design of drugs. The free energy of binding is treated as a sum of terms characterizing polar and nonpolar interactions, desolvation, and ligand flexibility. Entropy is estimated by the number of degrees of freedom lost upon binding. These contributions were determined from 82 complexes for which ΔG were known (see also Jain [823] and Eldridge et al. [775]). But the transfer of parameters is difficult.

In a pioneering approach, Ortiz et al. [247] proposed a receptor-dependent model (RD-QSAR) for 26 inhibitors of human synovial fluid phospholipase A2, with their COMparative BINding Energy (COMBINE) method. In the calculation of inter- and intramolecular energies for the ligand, receptor, and their complex, the ligand and receptor are divided into fragments and the partitioned energy terms are considered as structural descriptors for building the QSAR (see Chapter 8).

11.5.1 FREE ENERGY FORCE FIELD

Another approach was the free energy force field (FEFF) 3D-QSAR analysis, proposed by Tokarski and Hopfinger [824]. Basically, enthalpy and entropy contributions for ligand–receptor interaction in a solvent were taken into account and used as independent variables to build a QSAR with a GA. The calculation relied on MD simulations to adjust the force field. Once the force field was parameterized on a set of ligands binding a same receptor, the system allowed for prediction of new compounds. But the so-determined force field was valid only for a specific system (that from which it was calibrated) and no geometrical information could be derived. Due to the scarcity of available absolute ΔG measurements, results were compared to binding or affinity constants (reexpressed in energy terms).

The basic assumption was that energy contributions might be broken into a set of component interaction terms that were further separated into enthalpic and entropic terms (this implying small importance for coupling between degrees of freedom). Furthermore, neglecting the work term $P\Delta V$, enthalpy terms (H) might be represented by internal energy terms (E). Additional hypothesis was low concentration conditions to avoid possible aggregation and/or multiple ligand binding to the receptor. For example, for the complex ligand–protein (LP), P representing the protein receptor:

$$H_{LP} = E_{LP} = E_{LP}(LL) + E_{LP}(LM) + E_{LP}(PP) + E_{LP}(PM) + E_{LP}(LP) + E_{LP}(MM)$$

where the index LP denotes the bound state (the complex). On the right-hand side, the first two terms represent the intramolecular conformational energy of the ligand in the bound state $E_{LP}(LL)$ and its solvation energy $E_{LP}(LM)$; the following two, the corresponding quantities for the receptor, and the last ones the intermolecular ligand–receptor energy $E_{LP}(LP)$ and the solvent organization energy $E_{LP}(MM)$.

A similar sum may be written for the entropic terms, and analogous decompositions hold for the other components involving the unbound receptor (E_{PP}), the free ligand (E_{LL}), the solvation energy terms, and their decomposition into enthalpic and entropic components. For example, for the free ligand:

$$G_L = G_L(LL) + G_L(LM) + G_L(MM)$$

All these quantities may be evaluated when the receptor geometry is known, and they may be used as independent variables (structural descriptors) to construct a QSAR model. The force field used (called an FEFF) is calibrated, with scaling coefficients adjusted while building the QSAR, so as to be transferable to neighboring systems. The authors preferred introducing a decomposition of energy on the basis of changes (upon binding) in individual types of energy contribution. For example, for bond stretching

$$\Delta E_{str} = E_{LP,str} - E_{L,str} - E_{P,str}$$

Similar expressions hold for the various contributions to conformational energy in molecular mechanics, such as bond stretching, angle bending, dihedral torsional energy, plus solvation and entropic terms. The free energy of binding can be expressed as

$$\Delta G = \alpha 1 \Delta E_{str} + \alpha 2 \Delta E_b + \alpha 3 \Delta E_{tor} + \cdots + \alpha T \Delta S$$

Ideally, all α coefficients would be unity. But, since the force field is not absolutely precise, the various energy terms are considered as independent descriptors and must be weighted by the α coefficients, according the usual QSAR methodology.

An FEFF 3D-QSAR was carried out on a series of complexes of Renin with inhibitory peptides. Renin is a member of the aspartic proteases family (also including pepsin, chymosin, and HIV protease). The X-ray structures of human Renin and a complex with a bound inhibitor have been resolved. The thermodynamic binding parameters (K, ΔG, ΔH, and ΔS) have been measured for 13 Renin inhibitory peptides (about eight residues). Ten of them were retained for the study (based on structural diversity and range of activity).

As a preliminary observation, it appeared that binding affinities could not be related to the ΔH variations. A first problem was the choice of an effective size of the receptor model centered at the active site (since all 5177 atoms of Renin could not be considered). All residues farther than 10 Å apart from any bound-ligand atom were discarded. This led to a receptor formed of unconnected fragments, but constraints on masses for the main-chain atoms were introduced to maintain the integrity of the model (and avoid a drift from the X-ray structure in the MD simulations).

The protonation state of the active site (with two aspartates) was also a matter of debate. Various MD simulations and other experimental data led to the choice of a monoprotonated active-site model. Then the FEFF terms were calculated. For the nonscaled energy terms of the bound and unbound states, a series of MD simulations was performed according to an "unbinding process." From a given complex geometry several successive short MD simulations were run at decreasing temperatures (10 ps at 200, 100, 50, 25, 10 K). The lowest energy structure at a given temperature was the starting point of a subsequent simulation at a lower temperature, and so on, down to a user-defined final temperature. The lowest energy structure of the complex was then broken, the ligand being removed unchanged from the complex. Ligand and receptor were then "warmed" into successive independent MD simulations up to the highest temperature (50 ps each, to take into account an increased flexibility). From the lowest energy structure reached at this temperature, MD simulations were performed at each of the intermediate temperatures of the first cooling process. This allowed for the free ligand and receptor to equilibrate at different conformational states. The (unscaled) force field used was based on the AMBER program completed with MM2.

The solvation energy terms were evaluated with the concept of hydration shell (since, as indicated, the use of explicit water molecules would heavily complicate the calculation). A characteristic sphere was centered around each atom or groups of the solute molecule. The intersection with the van der Waals volumes of neighboring solute atoms determined the number of solvent molecules removed. These terms were introduced in the MD simulations, only after energy minimization to avoid energy drifting.

For entropic terms, a group additive property method, previously proposed for polymers [377], was used: the torsion angle unit (TAU) method for the calculation of torsion angle conformational entropy by addition of terms assigned to each torsion angle. Only residues of the receptor in contact with the ligand were considered. So the conformational entropy of the receptor depended on the size and conformation of each (bound) ligand.

Each of the FEFF terms were considered as potential descriptors (independent variables) to be weighted by scaling factors to derive a QSAR (the binding properties being the dependent variable). The problem was that the number of independent variables to scale (about 40) was generally larger than the number of observations, and these variables might be interrelated. To build the model, all terms were considered in a stepwise MLR, using the genetic function approximation (GFA) optimization method [108], to derive the best model(s) according to usual statistical criteria. As frequently observed with GA various models were proposed with nearly equivalent performance.

For ΔG, the best models with one, two, or three descriptors, at the five different temperatures investigated (200–10 K), led to q^2 values from 0.53 to 0.80 (with one exception 0.34). Similar satisfactory correlations were obtained for ΔH and ΔS. Overall, a relatively small number of contributions were the most significant terms since they were frequently encountered in the generation of the best models. They might be considered as dominating the thermodynamics of binding: $E_L(LL)$ intramolecular vacuum energy of the unbound ligand and ΔE_{solv} (change in solvation free energy upon binding). Normalized change in torsion angle conformational entropy of the receptor upon binding and other terms (ligand stretching or bending, intermolecular van der Waals energy) also appeared in some correlations.

As for sampling schemes, energy-minimized and time-averaged energies are less reliable than considering the lowest energy structures sampled from multi-temperature MD. It appeared that assuming a stable alignment by docking, checking for bad steric contacts, and then calculating energy was not correct. It was necessary to allow each participant to assume a relevant stable state in both free and bound states, thanks to a sampling of the phase-spaces of free ligand, unbound receptor, and their complex. To limit calculations, some tricks were suggested: it was better to sample the accessible active-site space with smaller ligands, which were more flexible (and able to explore more phase-space for less computational resources). Conversely, it was more efficient to model the largest active ligand in the active site in order to limit the intermolecular sampling necessary for docking (less geometric exploration). The authors also stressed the fact that the individual FEFFs were not easily transferable to other systems.

An interesting point was the summary of the approximations introduced, the problems they possibly might induce, and the remedies [824]. The following are among the important approximations:

- Using hydration shell model and additive entropy terms
- Pruning the receptor
- Defining the protonation states
- Exploring geometry-energy states by MD simulations

11.5.2 RECEPTOR-DEPENDENT 4D-QSAR MODELS

A new approach, the RD-4D-QSAR model was further proposed by Hopfinger et al. [726] as a modification of the methodology of the receptor-independent 4D-QSAR (RI-4D-QSAR) already developed by the same group [64,700]. Whereas the RI-QSAR models only consider a set of ligands, the RD-4D-QSAR models include information about the structure of the receptor. However, the descriptors are still constituted by cell occupancy (grid cell occupancy descriptors [GCODs]), and the receptor does not explicitly intervene at this stage. But one may infer that the docked conformation of the ligand reflects its interactions with the protein.

The method was presented on a set of glucose analogues binding glycogen phosphorylase b (see compound **4.6**) and comprising 47 compounds in training and 8 in test [726]. This system was the subject of several previous investigations using RI-4D-QSAR models [64,700] and the universal 4D fingerprints [65]. The structure of a complex of glucose bound to the GPb receptor had been solved by X-ray crystallography [825], and it was also known that glucose analogues bind the receptor in a similar mode. The geometry of glucose bound to GPb was used as template and for each inhibitor the minimum-energy conformation was selected for alignment in the binding site.

To determine the GCODs and the corresponding interaction pharmacophore elements (IPE) (as in the RI-4D-QSARs) MD simulations were carried out to define the conformational ensemble profile (CEP) on 2000 frames. But it was necessary to prune the receptor to about 2000 atoms (in order to limit the number of interactions to be calculated) by means of a cutoff at about 12 Å, and introduce constraints to limit flexibility. In RI models, proposed alignments are readily evaluated, whereas in RD models, lining of the active site might lead to bad contacts for inappropriate alignments. In the investigated series, only one alignment was considered. For the diverse CEP, occupancy of the grid cells led to an excessive number of GCODS (more than 60,000). But the previous RI model indicated that the most important GCODs are located very close to the glucose ring. This led the authors to consider various grids of limited size, from 2 to 12 Å. Data analysis was performed by a three-step sequence, GFA—backward MLR elimination—GFA. In the final step, from the conformers within 2 kcal/mol of the absolute minimum, the complex with the highest predicted activity was selected as the active conformation. For each complex, the active geometry is, therefore, composed of a *distinct* active conformation for both the inhibitor and the pruned receptor.

With a lattice size of 8 Å the final result was expressed in a linear relationship between the binding free energy and six GCODs specifying the location of the key sites where appropriate substituents increased activity. This model yielded $r^2 = 0.85$, $q^2 = 0.82$. The performance of this RD model was rather similar to that of the previous RI model for the training set, but higher for the test set, suggesting that the geometrical aspect was not crucial for the training set but important for compounds outside this set. Among the six GCODs retained that defined the pharmacophore, three were attributed to the receptor. Comparing the calculated (ligand–receptor) complex geometry evidenced some induced fit of the receptor, depending on the ligand. So, with NHCOOMe in β position, two H-bonds were formed with the carbonyl group of His-377 and the NH side chain of Asn-284 of the active site. But with more bulky β NHCOPhe, these H-bonds were broken and a new H-bond was formed with the side chain oxygen of Asn-284 and the NH group of the inhibitor (Figure 11.5).

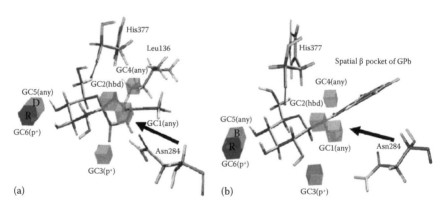

(a)   (b)

FIGURE 11.5 Simplified visual representation of the spatial relationship between two complexes and the GCODs of the best RD-4D-QSAR model. (B) gray cells denote activity enhancing GCODs, (R) dark gray cells denote activity decreasing GCODs. (a) Compound with β = NHCO$_2$Me, α = CONH$_2$; (b) compound with β = NHCOPhe α = H. Remark that, in (a), GC1 (any), indicated by an arrow, is situated on the –CO– of the β substituent of the inhibitor, whereas in (b), it is occupied by the –NH– moiety of this group. (Reproduced from Pan, D et al., *J. Chem. Inf. Comput. Sci.*, 43, 1591, 2003. With permission.)

Interestingly, it appears that the key GCODs selected in the RD and RI models for the same alignment were not strictly equivalent. This was consistent with the fact that they did not encode the same information. The example of receptor adaptation observed in this study also suggested that quantitative models *must take into account* receptor flexibility.

Another study concerned the inhibitory activity of 39 hydroxy-5,6-dihydropyrone analogues (see compound **9.4**) against HIV-1 protease [826]. This data set, spanning 4 log units in inhibitory potency $pIC_{50}(M)$, had been previously investigated with an RI-4D-QSAR analysis [715]. The basis of the study was the crystal structure of a 4-hydroxy-5,6 dihydro-2-pyrone analogue docked into HIV-1 protease. As in the previous example, the analysis first involved pruning the receptor, modeling the ligands, docking them in the active site, selecting the IPEs, and then performing MD simulations of each pruned receptor–ligand complex (after defining constraints). Unlike the glucose GPb-inhibitors analysis, five alignments were tried. The best one was defined by three sites of the lactone part (differing from that selected in the RI model [715]).

After determination of the CEP and calculation of the GCODs, a PLS treatment allowed for reducing these descriptors to the 200 most highly weighted, to form the basis set submitted to GFA-MLR analysis. Six models were extracted, from one to six terms, with q^2 varying from 0.51 (one term) to 0.82. But the correlation matrix of the residuals indicated that they could be reduced to two independent representations (with one or four key GCODs, respectively). It is interesting to see that the one-term model represented about 57% of the variance of the data set. The corresponding site might therefore be attributed a paramount importance for the design of new inhibitors. It is noteworthy that in the RD-4D model, the inhibitors adopted a flat-like active conformation (consistent with crystallographic data) whereas the RI models predicted a fold-like conformation. MD simulations also allowed for detecting H-bonds formed with the ligands (depending upon their nature) and varying over the time course of the bound ligand.

12 Free Energy Calculation: Extensive Sampling and Simplified Models

The preceding approximate methods more or less suffer important limitations such as lack of explicit solvent treatment and lack of sampling of the system configurations. On the other hand, a more acute determination of free energies of binding is attainable via statistical thermodynamics, but by dint of heavier calculations.

12.1 EXTENSIVE SAMPLING

12.1.1 A PINCH OF STATISTICAL THERMODYNAMICS

"Free energy is the energy left once the tax to entropy is paid" [827]

$$\Delta G = \Delta H - T\Delta S$$

In statistical thermodynamics, the Gibbs free energy G of a chemical system, defined at constant number of particles, pressure, and temperature (N,P,T ensemble) can be expressed from the partition function Z

$$G = -k\,T \ln(Z)$$

with

$$Z = \sum_i \exp(-\beta E_i)$$

Σ over the configurations i of the system. $\beta = (k\,T)^{-1}$ with k = Boltzmann constant. "Configuration" is here synonymous with "state" and has nothing to do with the stereochemistry of an asymmetrical carbon. Note that the Helmotz free energy would correspond to (N, V, T) conditions. In addition to free energy, the partition function Z plays a key role in statistical thermodynamics since it leads to

Internal energy	$U = \langle E \rangle = \partial[\ln(Z)]/\partial\beta$
Pressure	$P = kT\,\{\partial[\ln(Z)]/\partial T\}_{N,T}$
Expectation value of a quantity O	$<O> = (1/Z)\sum_i O_i \exp(-\beta E_i)$

Remark that this approach relies on the ergodicity hypothesis (a link between micro-scopic and macroscopic scales): the average value of a quantity or a property for an ensemble of individuals may be obtained taking the average value over time of this property for one individual. Hence, evaluation of Z implies

- Sampling of microstates of the system (e.g., Monte Carlo (MC) or molecular dynamics (MD) simulations)
- Calculation of the energy of theses microstates (e.g., molecular mechanics)

Unfortunately, free energy, as other "thermal" properties (entropy, chemical poten-tial) cannot be accurately directly determined with a standard MC or MD simulation because such processes do not adequately sample high-energy regions of the phase space. 3N coordinates for the positions and 3N components for the velocities are necessary to define the state of a system of N particles, as a point in a phase space of dimension 6N). On the other hand, evaluating the free energy difference between two states is easier and can be approached by standard MC or MD simulations via various types of sampling.

12.1.2 ZWANZIG FORMULA

Although the thermodynamics aspects were known for a long time, practical appli-cations were not possible until powerful computers became available. A fundamen-tal formula was due to Zwanzig [828] as early as 1954. According to Zwanzig, the free energy difference between two states A and B is

$$\Delta G = G_B - G_A = -kT \ln \left(\frac{Z_A}{Z_B} \right)$$

or

$$\Delta G = -kT \ln \left\langle \exp \left(-\frac{\Delta V}{kT} \right) \right\rangle_A$$

where $\langle\ \rangle_A$ corresponds to an ensemble average of $\Delta V = V_B - V_A$, sampled using the V_A potential (generation carried out by MC or MD, at constant T and P).

An important point is that free energy is a state function. Its variations between two states only depend on these two states and not on the (reversible) way followed between them. It even may be, as we will see, a nonphysical path, an "alchemical transformation." Let us take a simple example: suppose we wish to calculate the free energy difference between CH_3F and CH_3Br in a common solvent S. We can sample the different states of CH_3F surrounded by S (e.g., in an MD simulation of CH_3F in a box of molecules of S) and then evaluate the free energy of CH_3Br, with, for each configuration of the system, the parameters of Br in place of those of F (with the same organization of the solvent, that evaluated with CH_3F). Of course, the configurations sampled with the V_A potential should have a reasonable probability to also occur with V_B: in other words, the thermally accessible regions of the two potentials should have a reasonable degree of overlap (in the phase space). To fulfill this condition, variations between the two systems must not be too large and a multistep approach is normally used, comparing in each step "reasonably" neighbor systems.

Two methods, for exact sampling, have been proposed: free energy perturbations and thermodynamic integration.

12.1.3 FREE ENERGY PERTURBATIONS

A set of intermediate potential energy functions is introduced between the initial (A) and final (B) state. In intermediate state m

$$V_m = (1 - \lambda_m)V_A + \lambda_m V_B$$

the coupling parameter λ_m varying from 0 to 1. If energy is (as usual) calculated by MM, the various parameters of the force field (force constants of stretching, angle bending, torsion, charges, ε and σ of the van der Waals potential), are given by linear combination. For example, for bond i in state m, the stretching force constant is

$$k_{i,\lambda_m} = (1 - \lambda_m)k_{i,A} + \lambda_m k_{i,B}$$

For each step of the simulation, the energy is accumulated as

$$-\beta^{-1} \ln \langle \exp[-\beta(V_{m+1} - V_m)] \rangle_m$$

The total change in free energy is the sum of these free energy changes for λ_m varying from 0 to 1. This corresponds to forward sampling. The calculation may be also carried out by backward sampling, from B to A (λ_m decreasing from 1 to 0).

This process is somewhat approximate, since with the conformation of the solute, the organization of the solvent in the λ_m state is used for the $\lambda_{(m+1)}$ state. So, it may be wise to check that forward and backward sampling lead to the same results. The divergences observed ("hysteresis") give an indication of the accuracy of the simulation. A variant, the double wide spacing, considers for each λ_m the steps to $\lambda_{(m+1)}$ and $\lambda_{(m-1)}$, hence a better efficiency.

12.1.4 THERMODYNAMIC INTEGRATION

For small variations of λ_m (independently of the preceding hypothesis of a linear interpolation),

$$V_{m+1} - V_m = \left(\frac{\partial V_m}{\partial \lambda_m}\right) \Delta \lambda_m$$

so that

$$\Delta G = G_B - G_A = -\beta^{-1} \ln\langle \exp(-\beta\Delta V)\rangle_A$$

can be written as

$$\Delta G = -\beta^{-1} \sum_m \ln\left\langle \exp\left[-\beta \frac{\partial V_m}{\partial \lambda_m} \Delta\lambda_m\right]\right\rangle_m$$

For small steps, a Taylor expansion gives

$$\Delta G = \sum_m \left\langle \frac{\partial V_m}{\partial \lambda_m}\right\rangle \Delta\lambda_m$$

And, if $\Delta\lambda \to 0$

$$\Delta G = \int_{0,1} \left\langle \frac{\partial V(\lambda)}{\partial \lambda}\right\rangle_\lambda d\lambda \quad \text{for } \lambda \text{ running from 0 to 1}$$

ΔG is evaluated as the area under the graph of $\langle(\partial V(\lambda)/\partial\lambda)\lambda\rangle$ vs. λ. If a linear interpolation is admitted

$$\Delta G = \sum \langle\Delta V\rangle_{\lambda m} d\lambda$$

This method was sometimes called "slow growth," with λ varying at each step of the MD simulation [61]. It was criticized because of suffering of "Hamiltonian lag" [763, pp. 495].

The relative qualities of these two approaches were discussed by Brandsdal et al. [61]. As to the simulation methods, MC is not well suited for systems with high conformational flexibility but works better with more rigid small molecules. Alternate steps of MC and MD have been proposed to improve the sampling of physical configurations: MC for a faster treatment of the solvent and MD for modeling the ligand and receptor. The chemical Monte Carlo/molecular dynamics method (CMC/MD) of Pitera and Kollman [829] corresponds to a quite different approach (vide infra).

12.1.5 THERMODYNAMIC CYCLE

A frequent problem in QSAR models is the determination of the variation in the binding energies of various ligands to a common receptor active site, for example, for ligands L and L′ binding the active site of Protein P (the "receptor"). A direct solution would be to start from ligand and receptor far away and to gradually form the complex. This would require conformational changes in the ligand, the receptor, and reorganization of the solvent, which makes a correct sampling very difficult and time

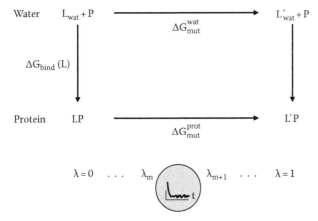

FIGURE 12.1 Thermodynamic cycle. L and L′ are the ligands, P the receptor. Superscripts wat and prot indicate an aqueous medium and the bound state to solvated protein P. The index mut denotes mutation of L to L′. The various steps of MD simulation (with coupling parameter λ are indicated in the lower line.

consuming [763, pp. 489]. An easier solution proposed by Tembe and McCammon [830] was to consider the following cycle illustrated in Figure 12.1.

Since the free energy is a state function, its variation round such a cycle is null. So,

$$\Delta\Delta G_{bind} = \Delta G_{bind}(L') - \Delta G_{bind}(L) = \Delta G_{mut}^{prot} - \Delta G_{mut}^{wat}$$

The last term corresponds to variation of free energy in the mutation of L to L′ either free or bound to P. Note that these mutations need not to be a transformation really possible but only "alchemical mutations." Depending on the problem, various organizations may be adopted for such thermodynamic cycles. See, for example, Ref. [831] considering the free energy of binding of R- and S-propanolol to wild type and F483A-mutant cytochrome P450 2D6 from MD simulations.

Such mutations often imply creation or annihilation of atoms. In the single topology method, this is carried out via dummy atoms. These dummy atoms do not interact with the other atoms in the system, but are progressively transformed when the simulation goes on. An alternative choice is the dual topology method where the two species coexist during the entire simulation but do not "see each other" (no interaction between them), the total energy being, for example, a linear combination calculated from the coupling parameter λ. Problems sometimes associated with the end points, when atoms appear or disappear ("end-point catastrophe") have been discussed by Pitera and van Gunsteren [832].

Thermodynamic cycle was used, for example, by Oostenbrink et al. [833] to calculate the relative free energy of binding 17β-estradiol, diethylstilbestrol (see Figure 12.2), and genistein (compound **12.1** in Figure 12.4) to the active site of estrogen receptor α (ERα) (Figure 12.2). The simulations were carried out in a dual topology process. The two species (DES + E2 or DES + GEN) always share a common phenolic ring, the other atoms being changed into dummy or created from nihil. For nonbonded interactions, a soft-core potential was used to avoid instabilities (vide

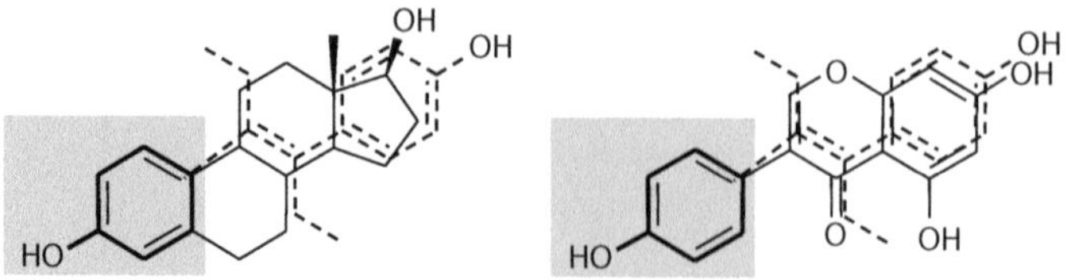

FIGURE 12.2 Dual-topology ligand used for TI calculations between DES and E2 and between DES and GEN. The phenolic ring is common to both end states and does not change. In the mutation DES to E2 or GEN, DES-specific atoms (linked by dotted lines) are turned into dummy while the E2- or GEN-atoms (linked by straight lines) come into existence. (Adapted from Oostenbrink, B.C. et al., *J. Med. Chem.*, 43, 4594, 2000.)

infra). Simulations were carried out in a box of water molecules: about 1200 for the free ligand, 15,000 (and 6 Na⁺ ions to ensure a net charge of 0) for the complex. The initial structure was that solved by X-ray crystallography. Solvent organization was determined with protein and ligand positions restrained by a harmonic interaction. 11 λ values were used, with for each, a 20 ps MD simulation (forward and backward).

12.1.6 SINGLE STEP PERTURBATION

Such approaches are heavily time consuming, due to the need for multiple steps to limit structural changes. However, the choice of an adequate reference state ("nonphysical" reference state) allows for carrying out a single step perturbation methodology [224,833,834].

In the thermodynamic cycle indicated, the relative free energy of binding $\Delta\Delta G$ between two ligands L and L' was evaluated by two mutations (L → L' bound to the receptor or free in water). The problem was to get a correct sampling for L' using a sampling performed on L (remember the Zwanzig's formula). The trick here is to choose the reference compound R so that the ensemble of states of R shows considerable overlap with the ensemble that would be generated for L and L'. If this condition is fulfilled, the relative free energy change may be estimated with only two simulations on R (in water and bound to the receptor in water) (Figure 12.3). One advantage of this process is that, with these two simulations only, several ligands (say L', L'', L''' and so on) may be studied with limited computational effort.

This approach performs better if a "soft potential" is used for van der Waals interactions. At distance r, this potential is expressed by

$$\left[\frac{C_{12}}{(\alpha+r^6)}-C_6\right]\left(\frac{1}{\alpha+r^6}\right)$$

where C_{12} and C_6 are the usual Lennard-Jones parameters for the atom pair considered, the offset parameter α being set, for example, at $\alpha = 0.3775$ (C_{12}/C_6) [224]. This soft potential deadens van der Waals interactions so that atoms may occasionally overlap with "soft atoms" of the reference without prompting computational singularities. This allows for sampling configurations for ligands of varied shapes with only a single simulation.

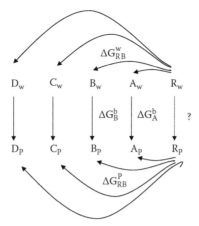

FIGURE 12.3 Thermodynamic scheme used to calculate relative free energies ΔG of binding for compounds A and B from two simulations of the reference ligand indicated by R. (Reproduced from Oostenbrink, C. and van Gunsteren, W.F., *Proteins: Struct. Funct. Bioinf.*, 54, 237, 2004.)

The process was applied [833] to GEN and eight aromatic diols (chrysene, benz[a] anthracene, benzo[a]pyrene derivatives **12.2** through **12.4** in Figure 12.4). In this approach, only ΔΔG were calculated (by respect to the nonphysical reference state). These values cannot be directly compared to the observed $\Delta G_{binding}$. But relative values between different ligands can. A fairly good agreement is observed for the five compounds with known $\Delta G_{binding}$ with a maximum deviation to experiment of 0.8 kcal/mol, suggesting possible applications to larger-scale screening.

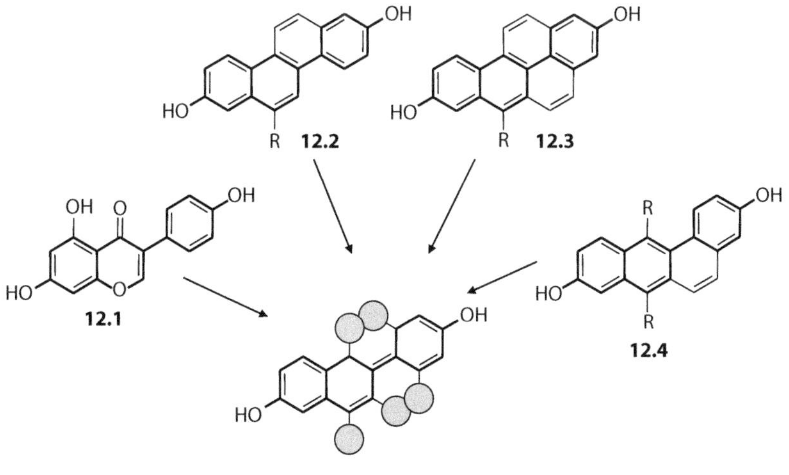

FIGURE 12.4 Soft-core reference state used for single step perturbation calculations. Soft van der Waals interaction sites are drawn as spheres. (Reproduced from Oostenbrink, B.C. et al., *J. Med. Chem.*, 43, 4594, 2000. With permission.)

Another study [224] concerned 17 hydroxylated polychlorinated biphenyls binding ERα receptor. The mean absolute difference to experimental binding free energies was 0.8 kcal/mol, but only 0.5 kcal/mol (less than kT) if excluding the three largest ligands. The accuracy of the calculation competed well with the other types of approach. In addition, by comparison to classical FEP treatments, needed computertime was largely reduced. Furthermore, since the heavier simulations were carried out once (for the reference compound only) additional ligands can be treated with limited work (calculation of interactions of the ligand atoms with their surroundings only).

An interesting point is that, in this study, the evolution of ΔG (for the bound ligand) during the simulation clearly indicated that a ligand bound to the receptor could not be represented by only a single conformation.

12.2 APPROXIMATION AND LINEAR INTERACTION ENERGY MODEL

Free energy perturbations or thermodynamic integration are heavily time consuming. They better apply to the study of solvation processes of ions or organic compounds differing only by few substituents or to the calculation of the binding free energies of small ligands to host molecules. But they are not well suited for a systematic exploration in series of compounds where the perturbations or transformations required may be too drastic [60]. Another approach more accurate than usual scoring functions, less time consuming than free energy perturbation methods and more suited to the study of rather similar ligands [61,835,836], relies on the expansion of free energy differences between two states (A and B) in a Taylor series around one of these states [837].

$$\Delta G_{AB} = \langle \Delta V \rangle_A - \frac{\beta}{2} \left\langle \left(\Delta V - < \Delta V >_A \right)^2 \right\rangle_A + \frac{\beta^2}{6} \left\langle \left(\Delta V - \langle \Delta V \rangle_A \right)^3 \right\rangle_A \cdots$$

where $\Delta V = V_B - V_A$ and $\langle \Delta V \rangle_A$ is a (MC- or MD-generated) ensemble average of ΔV, sampled using the V_A potential. A first approximation, with the truncation after the second term, corresponded to the assumption of a linear response of the state A to a perturbation. A truncation after the first term was proposed by Gerber et al. [838] for binding free energies of different ligands to DHFR. But it was indicated [61] that there is no guarantee that the method is general. In the hypothesis of a linear response, free energy can be deduced from two simulations for states A and B, respectively:

$$\Delta G_{AB} = 0.5 \left[\langle \Delta V \rangle_A + \langle \Delta V \rangle_B \right]$$

If the states A and B correspond to end points of a process where a charge is created or annihilated, one state (A for example) is "nothing" and the formula reduces to

$$\Delta G_{AB} = 0.5 \left[\langle \Delta V \rangle_B \right]$$

A single simulation is sufficient. For examples, on charge rearrangement, see Refs. [834,837]. One interest of this approach is that the method uses a single reference state and so different end points (different ligands) can be easily studied (sampling corresponding to the unique reference state)

12.2.1 LINEAR INTERACTION ENERGY

In the linear interaction energy (LIE), the basic idea is that the binding energy of a ligand corresponds to the change in free energy when this ligand is transferred from solution to the solvated receptor binding site. The hypothesis is that this absolute binding free energy is composed of a polar and nonpolar contribution, and will be calculated considering only the corners of the thermodynamic cycle, and not spending to much time in the mutation process (in which many configurations correspond to unphysical mixtures) [61,835,836]. This approach was more accurate than usual scoring functions used with docking programs but less time consuming than FEP or TI methods.

Electrostatic contribution is evaluated by the linear response approximation. Nonpolar contributions are estimated with empirically derived parameters scaling van der Waals contributions from MD simulations (in line with the observation that average solute solvent van der Waals energies correlate with solute accessible surface area).

$$\Delta G_{bind} = \beta \Delta \langle V^{el} \rangle + \alpha \Delta \langle V^{vdW} \rangle$$

with

$$\Delta \langle V^{el} \rangle = \langle V^{el} \rangle_R - \langle V^{el} \rangle_W$$

Δ denotes the changes in the ensemble average energies for the bound and free states; $\langle V \rangle$ represents MD or MC averages of intermolecular electrostatic or van der Waals energies for the ligand with its surroundings (water or solvated receptor). In this formalism α and β are parameters empirically adjusted on a training set (caution β is no more $[kT]^{-1}$). We maintain here the notion of Åquist et al., α corresponding to the van der Waals contribution [835]. To determine absolute free energy of binding, two simulations are carried out: one for free ligand in solution, the other when bound to the solvated receptor [836]. At the beginnings, β was determined as 0.5 [835]. But in further studies it was considered as an adjustable parameter [839,840] and a constant term was added, leading to

$$\Delta G_{bind} = \beta \Delta \langle V^{el} \rangle + \alpha \Delta \langle V^{vdW} \rangle + \gamma$$

The term in α implies a linear dependence of the van der Waals energies (ligand-surroundings) and the size of the ligand, characterized by its solvent accessible surface area (SASA). This term may be completed by a contribution expressing the difference in shape of the ligand between its free and bound states something like strain energy. But this supplementary term, related to the differences in SASA of the ligand, seemed unnecessary since SASA is strongly correlated with the intermolecular van der Waals energy [841].

Current values are $\alpha = 0.18$ (based on simulation of 18 protein–ligand complexes) and β between 0.33 and 0.5 (see Ref. [842]). Wang et al. [843] established that in various simulations $\alpha = 0.16$ and $\beta = 0.5$ gave good results, although α might depend on the nature of binding site for different proteins and they proposed a relationship between α and the weighted desolvation nonpolar ratio representing the hydrophobicity of the binding sites.

12.2.2 APPLICATIONS

Various examples of applications are reported in Refs. [61,836]. This model was tested by Stjernschantz et al. [764] on four receptors of differing characteristics and extended sets of ligands. The aim of the study was to examine the efficiency of LIE method as intermediate between fast docking and scoring functions for series of compounds on one hand and on the other hand elaborate and time-consuming MD simulations on a few ligands. Particular interest was devoted to the importance of a good starting structure for MD simulations, and to the improvement to be expected by respect to scoring functions (that usually give rms error about 2.5 kcal/mol, whereas about 1 kcal/mol may be expected from LIE). The applicability of LIE method to a systematic lead optimization was also examined. The results were compared to diverse docking–scoring evaluations using Glide SP3.5 for docking [773,774,844] and various scoring functions [279,282,757,765,768,776,780,817,845]. In their study [764], Stjernschantz et al. get the best results with a simplified formula:

$$\Delta G_{bind} = \beta \, \Delta \langle V^{el} \rangle + \alpha \, \Delta \langle V^{vdw} \rangle + \delta \, \Delta \left(\langle V^{el}_{1-1} \rangle + \langle V^{vdw}_{1-1} \rangle \right)$$

where the term with the subscript 1−1 may correct for a difference in free and bound conformation of the ligand (intramolecular strain).

This study encompassed seven ligands to retinoic acid receptor γ (RARγ), 54 for matrix metalloprotease 3 (MMP-3), 36 for estrogen receptor α (ERα), and 41 for DHFR. Ligands were docked to a crystal structure to define a starting structure for MD simulations. The authors indicated that the MMP3 and ERα receptor binding sites were strikingly different: one was large, open toward the solvent, and containing a Zn ion (MMP-3) whereas the other (ERα) was small, closed, typically binding rigid hydrophobic molecules. For DHFR, two different sets of ligands were considered.

An originality of the approach was an automated setup of the MD simulations of the docked compounds after only a manual preparation for the protein (determination of protonation state, hydrogen addition to PDB files, restrained minimization). MD simulations were carried out in a solvated simulation sphere, at 300°K with coupling to a thermal bath. The net charge of the protein was set to zero with a correction term for neutralized charges. Restraints on specific residues might be applied.

For RARγ, two simulations were performed starting either from known crystal structures for seven ligands or from ligand conformations docked to one of the crystal structures. The two models gave similar results (RMSE 0.7 and 0.9 kcal/mol), with a slight advantage for the crystal structure-based model.

For MMP 3, errors on training and test sets were about 0.9 and 1.0 kcal/mol. respectively. Similar values were obtained with the best scoring function but ranking (Spearman rank correlation) was worse than that of the LIE model.

For ERα ligands (previously studied by Sippl [59,274]), the LIE model was qualified of "satisfactory" (RMSE = 0.6 and 1.3 kcal/mol), whereas the best docking score among 10 scoring functions (HS-score) was 1.0 and 1.54 kcal/mol in training and test, respectively. On this example, it appeared that different docking poses between compounds, or for a same compound, corresponded to different interaction energies. Dividing the data set into two groups (to take this into account) improved the results but limited the high throughput character of the approach.

For DHFR ligands, docking was carried out on the co-crystal structure of *Escherichia coli* DHFR, NADPH, methotrexate. For both approaches (docking or LIE), scores were rather mediocre and even worse for another set of 11 compounds extracted from a high-throughput screening. These poor results (on a low-activity range of about 3 kcal/mol) have been attributed to the uncertainties of docking poses, perhaps due to changes in protein conformation not seen during the MD simulation.

The authors concluded that, in this study, LIE method performed generally better than the best scoring functions (selected among 10). LIE gave information about solvation effects, flexibility, and ligand strain. They also stressed that LIE is not a turnkey method but parameter adjustment is necessary. So, LIE is not, up to now, applicable to virtual screening of a large amount of compounds, but is still usable in lead optimization on a more limited and homogeneous set of chemicals. MC simulation was also proposed (in place of MD) assuming a rigid protein backbone (moves of the side chains in the ligand binding domain were, however, allowed) [846]. Speed was largely increased, but suppressing some protein motions might be in some cases questionable. In this study, for six sulphonamide inhibitors of human thrombin on a range of 6 kcal/mol the average error was 0.8 kcal/mol.

MC simulations were also used by Rizzo et al. [841] for activity prediction of non-nucleoside inhibitors of HIV-1 RT. Each inhibitor was modeled as free molecule in water and complexed with the solvated receptor. The originality was to extract structural descriptors (to be subsequently treated in multilinear regression) during the MC simulation. The protein backbone was fixed and the ligand fully flexible. MC simulation (with several millions configurations) involved moves of the water molecules, less frequently of protein side chains and (still less frequently) of the inhibitor. The treatment concerned 200 NNRTIs representing eight chemotypes. Different MLR models were developed for the chemical families, but results were gathered in a "master regression" with $r^2 = 0.60$ and an average error of 0.69 kcal/mol. The model was successfully extended to 27 new compounds with a similar accuracy.

Interestingly, using only ligand descriptors significantly diminished the predictive ability. The most commonly intervening descriptors are the van der Waals interaction energy ligand–protein (reflecting shape complementarity) and the change in the number of hydrogen bonds on binding. Other useful descriptors depended on the chemical families: They involved changes (upon binding) in hydrophobic SASA, in weakly polar SASA or in aromatic SASA; changes in the dipole moment; or the number of H bond donated to a protein π system, or water bridges.

The LIE approach has been recently applied to a set of 43 benzylpyridinone derivatives (**12.5**), inhibitors of HIV-1 RT [761].

12.5

The crystal structure of one of these inhibitors in complex with HIV-1 RT was used as reference starting structure in docking and MD simulations. For each inhibitor, automatic docking was carried out by GOLD but the scoring function reflected very poorly the relative affinities ($r^2 = 0.02$!!!). This was in line with the observation that docking algorithms frequently identified the correct binding mode, but were not able to rank the inhibitors by affinity. LIE analysis with standard parameterization reproduced experimental results with an average absolute error 0.8 kcal/mol ($r^2 = 0.70$ for 39 compounds), four compounds significantly larger or smaller than the reference being discarded. A detailed analysis also enabled to characterize the few residues most important for binding, an important observation as to reactivity of mutant strains.

Recent work [764,847] indicated that "LIE (with automated generation of force field parameters and MD simulations), could be used to estimate binding affinities for thousands of potential inhibitors." The success of affinity prediction largely relied on the choice of a correct binding mode. This was usually guided by the known crystal structure of a ligand–receptor complex. However, if a new compound is to be studied with a receptor structure determined with a ligand from another (different) chemical scaffold, difficulties may arise. In other words, what is the impact of using different receptor models in affinity prediction?

The study of Nervall et al. [848] concerned 34 HIV-RT inhibitors (**12.6** and **12.7**) that were docked into two crystal structures of HIV-RT (one of them being a non-native crystal structure). Two distinct clusters were found, one compatible with an existing crystal structure, the other corresponding to a 180° flip of the heterocyclic part (Figure 12.5). Several scoring functions (ChemScore [776], X-SCORE [765], or PMF [780]) gave similar results with, for the two clusters, no significant correlation with experimental data. Hence, they did not differentiate the correct binding mode. Even performing MD simulations for sampling different conformations did not improve the results. On the contrary, an LIE treatment led to a good correlation ($r = 0.79$; mean error = 0.57 kcal/mol) and, more important, the correct binding mode was predicted for each ligand. From the analysis of the interactions it was suggested that, when two conformations fitted equally well the binding site, electrostatics determined the binding mode.

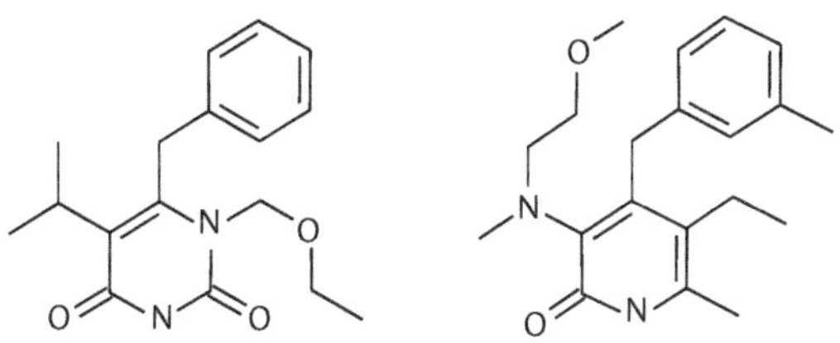

12.6   **12.7**

FIGURE 12.5 Two ligands in their respective crystal structure. The main difference is the ether moieties protruding from the pyridinone core in opposite directions. (Adapted from Nervall, M. et al., *J. Med. Chem.*, 51, 2657, 2008.)

The authors indicated that the scoring function lacks sufficient negative contributions for polar mismatches and unfavorable desolvation. These results were in agreement with Warren et al. [759] concluding that docking methods were generally able to generate poses close to crystallographic conformation but had difficulties in ranking the correct conformation in top position. In this application, MD sampling gave no improvement, at the opposite of the observations of Guitiérrez de Terán [849] in ranking malaria protease inhibitors.

12.3 CONTINUUM-SOLVENT MODELS: POISSON–BOLTZMANN AND GENERALIZED BORN SURFACE AREA METHODS

One important aspect of free energy calculations is the treatment of solvation. Rigorous treatments as FEP or TI explicitly take into account the solvent molecules, (in a box or a sphere surrounding the entities under study), but by dint of heavy time-consuming calculations. On the other hand, a distant-dependent dielectric constant is only a crude way to express the influence of the medium. Approximate methods were thus developed, still relying on an analysis of MD trajectories, but using an implicit solvent representation which integrates all solvent contributions [850]. These approaches are less accurate than traditional free energy methods but offer the advantage to be broadly applicable to systems which differ substantially in structure. This calculation also avoids considering intermediate states.

Such treatments have been used for determination of binding free energies, in place of the rigorous TI or FEP methods. But, on the other hand, their speed and low computer-resource demand make them particularly attractive to rank-docking results, in place of the usual scoring functions. Indeed, these scoring functions were mainly designed for speed, but they hardy labor to rank the hits.

12.3.1 EVALUATING THE DIVERSE CONTRIBUTIONS

Basically, in these approximate treatments, the average free energy of a molecular specie is evaluated (from a MD simulation) as

$$G = E_{MM} + G_{solv} - TS_{MM}$$

where E_{MM} is the average MM energy (intramolecular bond, angle, torsion, electrostatic, and van der Waals contributions). The second term includes all solvent influences [60,851] and the third one is the solute entropy (with contribution from vibrations, rotations, and translations). And for ligand-binding to protein

$$\Delta G = G_{complex} - G_{protein} - G_{ligand}$$

A major contribution is the reorganization of water molecules around hydrophobic species. Water molecules form a cage around them and are more ordered in that region than in the bulk. When a complex is formed with the receptor site, water molecules are liberated and the solvent reorganizes itself. For the evaluation of ligand–receptor binding energies, two options exist:

- For the three entities (ligand L, protein P, complex ligand–protein LP), G is evaluated on separate trajectories
- One trajectory only is run (for the complex) and on snapshots, the ligand or the receptor are in turn removed to characterize the other entity

In fact, the first option is not easily tractable for protein ligand studies (too large computing time for both receptor and complex). Unless the structures L and P change on binding, no intramolecular term intervenes (an assumption not present in the LIE method, thanks to weight coefficients).

Evaluation of entropic terms $T\langle\Delta S\rangle$ may be carried out by normal mode analysis or quasi-harmonic analysis of atomic fluctuation matrix from the MD snapshots of the trajectory [852–854]. Wang et al. stressed some difficulties of the methods: normal mode analysis does not account for anharmonic contributions, and low frequency modes leading to important displacements. Anharmonicity is considered in quasi-harmonic analysis, but requires a careful sampling [855]. Alternatively, a quasi-Gaussian approach may be used [856]. This evaluation may be difficult, and is was also indicated that carrying out a normal mode analysis of a ligand L from a simulation of the complexes LP, might be considered as a drastic simplification for estimating ligand entropies, although good results have been obtained [61]. Brandsdal et al. also reported that ligand binding a receptor in an extended conformation might in solution arrange itself so as to limit the water-exposed hydrophobic surface, making the situation in water more favorable than calculated from removal of the extended ligand from the complex. Fortunately, if the problem is only to compare binding affinities of structurally similar ligands of comparable size to a same receptor, a common assumption is to consider the entropic contributions as constant (although the various ligands may present some conformational freedom) [850].

The main problem is the evaluation of the solvation term ΔG_{solv}. The basic assumption is that this term ΔG_{solv} is composed of an electrostatic term and a nonpolar one:

$$\Delta G_{solv} = \Delta G_{solv}^{elec} + \Delta G_{solv}^{npol}$$

The nonpolar part involves van der Waals terms (favorable interaction solute–solvent) plus detrimental contributions, related to the cost of breaking the structure of the solvent around the solute. It encompasses a "cavity" term (free energy required to form the solute cavity within the solvent), an entropic term (corresponding to solvent reorganization), and the work against solvent pressure. These nonpolar contributions are evaluated as proportional to the SASA:

$$\Delta G_{solv}^{npol} = \sigma \; SASA + \beta$$

where σ and β are constants.

The linear form of the equation results from the fact that van der Waals forces are short range. So, like also the cavity term and the entropy penalty, they mainly depend on the number of solvent molecules in the first solvation shell, which is approximately proportional to the SASA of the solute. Constant β is commonly set to zero, and σ is about $6 \, cal/(mol*Å^2)$. An extra contribution may be added for localized H-bonds between solute and solvent. The variation of solvation free energy upon forming the receptor ligand complex is

$$\Delta G_{desolv} = \Delta G_{comp,solv}^{elec} - \left(\Delta G_{lig,solv}^{elec} + \Delta G_{prot,solv}^{elec} \right) + \sigma \left(SASA_{comp} - (SASA_{lig} + SASA_{prot}) \right)$$

For the evaluation of the electrostatic component ΔG_{solv}^{elec}, two ways were used:

- MM-Poisson–Boltzmann surface area approach (MM-PBSA)
- MM-generalized Born surface area approach (MM-GBSA)

The PBSA approach is the more time consuming, since the Poisson–Boltzmann equation may be solved for each conformational change, whereas the GBSA model is computationally more efficient. These two methods are available, for example, in the AMBER package [788].

12.3.2 MM-PBSA Models

In this approach, the solvation electrostatic contribution to free energy is calculated using a numerical solution of Poisson–Boltzmann equation and a surface area-based estimate of nonpolar free energy [851].

12.3.2.1 Poisson–Boltzmann Equation

The Poisson–Boltzmann model was used for evaluating electrostatic contribution to solvation or formation of intermolecular complexes [763]. The solvent is treated as a continuum of high dielectric constant ($\varepsilon = 80$ for water) and the solute as a body of low dielectric constant ($\varepsilon = 1$). The electrostatic contribution to solvation-free energy corresponds to the change in electrostatic energy for transfer from vacuum ($\varepsilon = 1$) to the aqueous solvent ($\varepsilon = 80$). For a charge q_i in a potential V this transfer corresponds to an energy Vq_i.

For a set of charges in a medium of fixed dielectric constant, Coulomb law allows for evaluating the potential. If the dielectric varies, Coulomb law must be replaced by the Poisson equation for calculating the potential V created by a charge density distribution ρ in a medium (ε). In reduced units, omitting the $4\pi\varepsilon_0$ factor:

$$\nabla^2 V(r) = \frac{-4\pi\,\rho(r)}{\varepsilon}$$

ρ is the charge density

or

$$\nabla(\varepsilon(r)\nabla V(r)) = -4\pi\rho(r)$$

When mobile charge are present, their motion under the influence of V modifies V itself, leading to a Boltzmann distribution (repulsion between charges avoids accumulation in region of extreme electrostatic potential). In the Poisson–Boltzmann model, the solute is represented as a set of point charges z_i (with a number density c_i far from the solute) and the potential V(r) is obtained solving

$$\nabla(\varepsilon(r)\nabla V(r)) = -4\pi\rho(r) - 4\pi \sum_i z_i c_i \, \exp\left[-z_i \frac{V(r)}{kT}\right]$$

$$\nabla(\varepsilon(r)\nabla V(r)) - \kappa' \, \sinh[V(r)] = -4\pi\,\rho(r)$$

where κ' is related to the Debye–Hückel inverse length

$$\kappa' = \varepsilon^{0.5} \left\{ \frac{8\pi N_A e^2 I}{1000\,\varepsilon\,kT} \right\}^{0.5}$$

with N_A Avogadro's number and I ionic strength of the solution.

Expanding the sinh term in a Taylor series, and retaining only the first two terms leads to the linearized Poisson–Boltzmann equation:

$$\nabla(\varepsilon\nabla V(r)) - \kappa' \, V(r) = -4\pi\,\rho(r)$$

This equation may be solved numerically, for example, with a finite difference method. Elements of the calculation (potential values, charge density, dielectric constant, ionic strength) are determined on a lattice including the solute and the surrounding solvent. The potential on a node depends on the potential values on neighboring nodes and iterations are carried out up to convergence. For the dielectric constant, a high value is chosen for points outside the molecular surface, a low-one for points internal. Details can be found in Leach [763, pp. 502–520] and in Ref. [857]. See also Ref. [855] for a re-parameterization of the model, and [858] for a statistical mechanics study relying the change in free energy upon binding to MD trajectories.

12.3.2.2 Applications of MM-PBSA

Various fields of application are presented in Ref. [60]: stability of DNA, RNA, duplex DNA or RNA, hairpin RNA, protein folding, protein–RNA interactions, etc.

Biotin derivatives: As to ligand–receptor interactions, Kuhn and Kollman [859] applied MM-PBSA approach to the study of analogues of biotin-binding avidin (a tetrameric glycol protein, probably antibiotic and inhibiting bacterial growth). The free energies spanned a 15 kcal/mol range. The correlation of calculated ΔG with experimental values led to $r^2 = 0.92$ and an average absolute error $\Delta\Delta G = 1.7$ kcal/mol, whereas the LIE method only yielded $r^2 = 0.55$ and $\Delta\Delta G = 2.3$ kcal/mol. The authors emphasized that the approach is less accurate than traditional free energy methods but is broadly applicable to systems which differ substantially in structure. Some approaches used only a continuum model for the calculation of free energies, but it was shown that including in the model van der Waals and torsional energies largely improved results. A caveat was also put on the use of continuum models when water molecules formed critical hydrogen bonds or when ions were involved.

Inhibitors of HIV-1 RT: We already mentioned the interest of non-nucleoside inhibitors of HIV-1 RT (NNRTIs) because their binding site is unique to the HIV-1 RT and therefore, less adverse side effects may occur. In the paper of Wang et al. [855], 12 TIBO-like inhibitors (see compound **1.10**) (with ΔG_{bind} varying over 4 kcal/mol) were treated by MM-PBSA. Starting from the crystal structure of a protein–ligand complex, MD trajectory (on 200 ps) was analyzed in 100 snapshots. A water cap (20 Å) was added to the binding domain. Water, ligand, and receptor side chains within 20 Å of the center of the ligand were considered flexible.

For each snapshot, ΔE^{npol} was set proportional to the SASA and ΔE^{elec} plus solvation energy ΔE^{PB} were calculated. For the 12 inhibitors, the treatment led to an average error of 0.97 kcal/mol, and the predicted rank order was satisfactory. Analysis of the different contributions indicated that electrostatic ($\Delta E^{elec} + \Delta G^{PB}$) and van der Waals hydrophobic ($\Delta E^{vdW} + \Delta G^{SA}$) interactions were most important.

The same approach (MM-PBSA combined with normal mode calculation) was applied to the prediction of the binding mode of evafirenz (**12.8**), a promising inhibitor for which no ligand protein crystal structure was published. Five binding modes suggested by DOCK were submitted to MD simulations and MM-PBSA treatment. A binding mode 7 kcal/mol more favorable than the others, was proposed as the correct one and led to a calculated ΔG_{bind} in fair agreement with experiment. Furthermore, the last snapshots of MD simulation indicated a structure very close to the crystal structure of efavirenz-HIV-1 complex. These consistent results testified of the efficiency of the method for ranking binding modes and determining free energies of binding.

12.8

12.3.2.3 Chemical Monte Carlo/Molecular Dynamics Approach

The chemical MC/MD (CMC/MD) method developed by Pitera and Kollman [829] combines MD to sample coordinate space with MC to sample among the various chemical states of the system. The approach shows some analogy with "umbrella sampling" [860] and "λ-dynamics" [861].

Rather than jumping between different Cartesian configurations (as do the previous hybrid MC/MD models) MC steps are here used to jump between ligands and adjust the potential function representing the interaction of different ligands with the receptor. Furthermore, if an estimate of ΔG_{solv} is attainable (by MD), the relative free energy of binding in a series is:

$$\Delta\Delta G_{bind} = -k\mathrm{T} \ln \left\langle \exp\left(-\frac{(\Delta E_{recep} - \Delta G_{solv})}{k\mathrm{T}}\right)\right\rangle$$

The solvation energy of each ligand is included as a biasing potential in the MC steps to focus sampling toward the best binding ligands. More precisely, all ligands (up to 10 in the binding site) are simultaneously included in the system but, at a time, only one is "real" and "seen" by the protein and the others are "ghosts." For the MC step, a ligand is chosen at random. The potential is changed and the move is accepted or not depending on a Metropolis criterion. It was shown that this potential switching (between two states A and B) corresponds to some form of configurational sampling in proportion that reflects the free energy difference between states A and B. In practice, a set of additional coordinates must be created, one per chemical state, with weights 0 or 1 according to whether the state is real or not.

12.3.2.4 Pictorial Representation of Free Energy Changes

The PROFEC approach from Radmer and Kollman [862] aims to estimate the average cost in adding a particle around the inhibitor in the protein complex and in the solution. It was used to suggest what changes to introduce in the ligand to get a better affinity. These approaches were used by Eriksson et al. [863] to rank 13 TIBO derivatives using PBSA or CMC/MD. For the eight compounds with experimentally determined activity, good agreement with experiment was obtained (average error 1.0 kcal/mol for MC/MD and 1.3 by PBSA). In this study, one proposal was predicted to bind 1–2 kcal/mol better than the most potent known inhibitor. These conclusions were supported by the rigorous methods TI or FEP. The authors concluded that this new strategy allowed for new derivatives to be tested in a reasonable time, focusing chemistry on the most promising ones.

Prediction of binding affinities for TIBO inhibitors of HIV-1 reverse transcriptase was also successfully tackled by Smith et al. [864] using MC simulations in a linear response method (with also introduction of several empirical parameters). However, Wang et al. [855] argued that LIE supposed that binding free energy was a combination of weighted electrostatic and van der Waals interactions. This later term largely depended on the systems, which would perhaps make difficult the treatment of very varied compounds. A comparison was also carried out with the Protein Dipole/Langevin Dipole (PDLD) model [858].

MM-PBSA was also used to calculate the free energies of seven inhibitors of Cathepsin D (an aspartyl protease implicated in Alzheimer disease and some cancers [440]). A good correlation (r = 0.98, average error 1 kcal/mol) was obtained between calculated ΔG and observed activities. Whereas the scoring function of DOCK (based on intermolecular van der Waals and electrostatic interactions, with a distance dependent dielectric constant) poorly reproduced the trends (see also Ref. [849]).

12.3.2.5 Comparison of PBSA Approach with Scoring Functions

As already stated, scoring functions cannot be applied to accurate large scale comparison due to severe approximations such as lack of protein flexibility, inadequate solvation treatment, simplistic force field. On the other hand, thermodynamic integration (TI) or free energy perturbation (FEP) is too heavily time and resource demanding. It was thus tempting to examine the potentialities of more rapid, approximate but valuable methods such as MM-PBSA (or MM-GBSA—vide infra).

To evaluate the capabilities of the PBSA approach, an extended study was carried out by Kuhn et al. [865] on eight proteins and a large number of ligands. The results of the scoring function FlexX/ScreenScore was compared to the usual MM(D)-PBSA approach (applied to averaged snapshots of MD simulations) and a simplified single structure MM-PBSA model (where the Poisson–Boltzmann formalism is applied to a single relaxed complex structure). A very crude model of solvation thanks to a distant-dependent dielectric constant was also considered. The study concerned 8 biotin derivatives binding avidin, 12 inhibitors of p38 MAP kinase, and a second set of 16 congeneric p38 MAP kinase binders.

In these different examples, MM-PBSA applied to a single relaxed complex structure was judged adequate and sometimes more accurate than standard free energy averaging over MD snapshots. The approach outperformed FlexX/ScreenScore and the distance-dependent dielectric model (that however gave good results for biotin derivatives). The single structure PBSA approach was very efficient in categorizing strong and weak binders, but was unable to express small free energy differences (TI or FEP remaining the best, although costly, methods).

It was also established that MM-PBSA might act as an efficient post-docking filter on top-ranking compounds from docking runs in virtual screening of focused databases (100–1000 compounds) and might also be used to select interesting substituents in advance of chemical synthesis in the search for new leads. The efficiency (compared to usual scoring functions) came from a better treatment of solvation effects. An example is given in Figure 12.6. FRED/ChemScore ranks favorably the

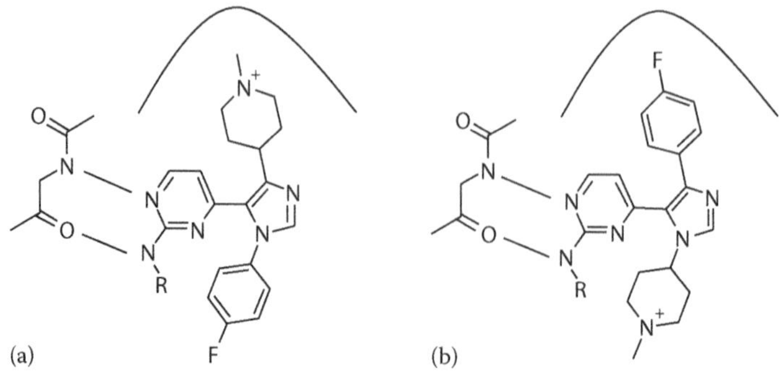

FIGURE 12.6 Illustration of handling of solvation effects. (a) Binding mode suggested by FRED/ChemScore (Rank 7); corresponding MM-PBSA result (Rank 122, ΔG_{bind} = +7.8 kcal/mol). (b) X-ray binding mode obtained by rotation around the pyrimidine-imidazole bond (MM-PBSA: Rank 57, ΔG_{bind} = −2.6 kcal/mol). R = CH_2-phenyl. (Reproduced from Kuhn, B. et al., *J. Med. Chem.*, 48, 4040, 2005. With permission.)

inhibitor, despite of geometry where the charged piperidine cycle lies in a lipophilic pocket whereas MM-PBSA gave a less favorable score. Rotation around the imidazole-pyrimidine bond would lead to a better solution.

As to the, rather unexpected, efficiency of single structure MM-PBSA, the authors suggested that "MD simulations introduce additional structural uncertainties which are often not favorable and averaging over a number of snapshots does not add sufficient accuracy to compensate for these effects."

A comparative evaluation of MM-PBSA and XSCORE was carried out by Obiol-Pardo and Rubio-Martinez [866] on seven XIAP–peptide complexes. XIAP is an apoptosis protein whose inhibition can lead to new drugs against cancer. XSCORE takes into account van der Waals, H-bonding, hydrophobic effects, and deformation penalty (frozen rotatable bonds). It was able to predict binding free energies on 30 protein–ligand complexes with a deviation of 2.2 kcal/mol [765,817]. In this study, XSCORE predicted binding energies with a maximum error of 3 kcal/mol, but a very low correlation coefficient with experimental values. It was also unable to identify good- and bad-peptide inhibitors in a large range of activity. This might be due to the high mobility of the peptide side chains not captured in XSCORE that only considered conformations in a close energy range. On the other hand, MM-PBSA, on a representative set of structures generated by MD simulations, appeared as a more robust approach able to identify slight binding differences (max error 1.7 kcal/mol and correlation coefficient r = 0.86).

12.3.3 GENERALIZED BORN MODEL: THE MM-GBSA APPROACH

In the framework of continuum solvent models; the generalized Born solvation approach (MM-GBSA approach) appeared as a very attractive alternative to TI or FEP since it gave accurate results and could handle more structurally dissimilar ligands at reduced cost. It was also well adapted to MD simulations as requiring less calculation than MM-PBSA that involves iterative calculations to solve the Poisson–Boltzmann equation. The starting point is the Born model (1920) [867] of electrostatic contribution to the free energy of solvation of an ion. We just consider here the application of the generalized Born model in MM and dynamics. For extension to quantum mechanics calculations, see for example, Cramer and Truhlar [682] and Constanciel and Conteras [868].

12.3.3.1 Generalized Born Model

The generalized Born model [763, pp. 508] treats the electrostatic interaction between two charges as a function of their distance and their effective solvation radii that somewhat depend on their exposure to solvent. As in PBSA model, hydrophobic effect is accounted for in terms of changes in the SASA. For an ion, represented by a sphere of radius a, and charge q_i, the electrostatic component of its solvation-free energy is equal to the work to transfer it from vacuum to the medium (ε). This can be evaluated by removing the charge in vacuum and adding it back in a continuum solvent environment. In vacuum, this work is $q^2/2a$, and in medium ε, $(q^2/2a) * (1/\varepsilon)$. So

$$\Delta G_{elec} = -\frac{q^2}{2a}\left(1-\frac{1}{\varepsilon}\right)$$

A system of particles is considered as a set of atoms i (radii a_i and charge q_i). The interior of the atom is considered as a material of dielectric constant 1, surrounded by a solvent of high dielectric constant (for water 80). The total electrostatic free energy in a medium of dielectric constant ε is

$$G_{ele} = \sum_{i=1,N}\sum_{j=1,N}\frac{q_iq_j}{\varepsilon r_{ij}} - 0.5\left(1-\frac{1}{\varepsilon}\right)\sum_{i=1,N}\frac{q_i^2}{a_i}$$

In the right part of this equation, the first term denotes the Coulomb energy and the second one represents the Born free energy of solvation. The variation of G between vacuum and the medium (ε) is

$$\Delta G_{elec} = -\left(1-\frac{1}{\varepsilon}\right)\sum_{i=1,N}\sum_{j=i+1,N}\frac{q_iq_j}{r_{ij}} - 0.5\left(1-\frac{1}{\varepsilon}\right)\sum_{i=1,N}\frac{q_i^2}{a_i}$$

The second member of the equation can be expressed with pairwise terms by

$$-0.5\left(1-\frac{1}{\varepsilon}\right)\left(\sum_{i=1,N}\sum_{j=i+1,N}\frac{q_iq_j}{f(r_{ij}a_{ij})}\right)$$

where function f depends on the interatomic distances r_{ij} and Born radii a_i. Various expressions can be selected for the f function. Still et al. [869] proposed

$$f(r_{ij},a_{ij}) = \left[r_{ij}^2 + a_{ij}^2 \exp(-D)\right]^{0.5}$$

with
$$a_{ij} = (a_i a_j)^{0.5} \quad \text{and} \quad D = \frac{r_{ij}^2}{(2a_{ij})^2}$$

Justifications for the choice of this function are highlighted in Leach [763, pp. 508)]

for $i = j$ the expression comes to the Born formulation;
if r_{ij} is small, to the Onsager formula;
if r_{ij} is large, to the sum of Coulomb and Born expressions.

Born radius ("effective Born radius") corresponds to the radius that would return the electrostatic energy of the system in Born equation if all other atoms are uncharged (and intervene only to displace the solvent). In Still force field, the OPLS atomic radii are empirically corrected by -0.09 Å to define the dielectric boundary.

a_i is somewhat a measure of the solvent exposure of the atom, approximating the average distance from the atomic charge center to the boundary of the dielectric medium [870]. a_i depends on the positions of all other atoms of the solute but is independent of the charge distribution. For an atom totally exposed to the solvent, a_i equals its atomic radius. For an atom at the center of a molecule, it would be equal to the radius of the molecule.

Electrostatic energy of an atom is calculated numerically in a series of concentric spherical shells until the entire van der Waals surface is included. This model implies calculation of the surface areas of the shell that are exposed to the solvent. Wodak and Janin [871] proposed a fast approximate evaluation:

$$a_i = S_i \prod_j \left[1.0 - \frac{b_{ij}}{S_i} \right]$$

S_i total accessible surface area of atom i of radius r_i defined with a solvent probe of radius r_s. b_{ij} amount of surface area removed in the overlap with atom j at distance d_{ij}

$$S_i = 4\pi (r_i + r_s)^2$$

$$b_{ij} = \pi(r_i + r_s)(r_i + r_j - 2r_s - d_{ij}) \left[1.0 + \frac{(r_i - r_j)}{d_{ij}} \right]$$

It was also proposed to calculate the effective Born radii a_i from the volumes (v) of elementary cells occupied by the molecule embedded in a grid space:

$$\frac{1}{a_i} = \frac{1}{a'_i} - \frac{V}{4\pi} \sum \frac{1}{r_{ik}^4}$$

where a'_i is the van der Waals radius of atom i and r_{ik} the distance between atom i and the (occupied) k^{th} cell. Other approximations are indicated in Ref. [873].

12.3.3.2   Modification of GBSA Formalism

Zou et al. [872] modified the initial formalism to take into account unoccupied regions between protein and ligand, regions where the effective dielectric constant is not as high as for the bulk solvent but presumably as low as for ligand and protein. They also express the nonpolar contribution as the sum of a van der Waals attractive term (receptor–ligand), a term proportional to the total surface area plus, for cavity, a term proportional to the nonpolar SA. This model implemented in DOCK provided a "free energy score" that superseded the current DOCK scoring function on two test systems: dihydrofolate reductase-methotrexate and benzamidine-trypsin, see also Ref. [286].

Rather than the van der Waals surface, Yu et al. [874] introduced a Gaussian surface that better mimics the boundary surface of biomolecules and presents the advantage of being analytically differentiable.

12.3.3.3   Applications of GBSA Model

MM-GBSA approach was compared to Glide scoring function [773] for sets of inhibitors of four kinases (p38, Aurora A, Cdk-2, Jnk-3, **12.9** through **12.12**) [875]. The method, which clearly outperformed Glide scoring, was judged as "very successful in getting relative ranking and also high correlations between observed and predicted activities."

12.9

12.10

12.11

12.12

Application of GBSA for rescoring docking poses was recently investigated. It may be expected that a more refined model led to improved enrichment in virtual screening and superior correlation between calculated binding affinities and experiment. Guimarães and Cardozo [870] examined several targets, Cdk2, Factor Xα, Thrombin, and HIV-RT. In all cases, good correlations were obtained between pK_i or pIC_{50} and predicted GBSA score (r^2 about 0.75), whereas Glide scoring functions are quite unable to correctly rank congeneric series of compounds. In this study, a single conformer representation was sufficient for the bound state, but better results are obtained if an ensemble of conformers is generated for the unbound ligand (otherwise, intramolecular and desolvation penalties were underestimated). It was also observed that the scoring penalty (0.65 kcal/mol per rotatable bond) applied by Glide overestimated conformational entropy penalty of the ligand upon binding compared to a Boltzmann estimate.

The MM-GBSA approach was used by Huang et al. [876] to improve inhibitor enrichment in virtual screening of extended databases. It is expressed by the enrichment factor, fraction of binders found in a subset with respect to random selection:

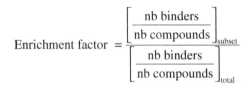

$$\text{Enrichment factor} = \frac{\left[\dfrac{\text{nb binders}}{\text{nb compounds}}\right]_{\text{subset}}}{\left[\dfrac{\text{nb binders}}{\text{nb compounds}}\right]_{\text{total}}}$$

(see Chapter 8). The process relied on rescoring: a first step identified the best poses (the simplified scoring functions being sufficient). In a second step (for single poses on the top 25% of the base) the more refined MM-GBSA treatment was used with geometry optimization. The study was carried out on 9 enzymes and a database of about 100,000 drug-like decoys. A significant improvement was observed within the first 1% of the base (the first thousand compounds). This early gain was mainly attributed to a more realistic treatment of desolvation and possibly to energy minimization.

12.4   A CASE STUDY: 2-AMINO-6-ARYLSULFONYLBENZONITRILE ANALOGUES, INHIBITORS OF HIV-1RT

The study of Hu et al. [102,334,877] on inhibitors of HIV-1RT constitutes an interesting example where the same data set was investigated by various nonlinear 2.5D and 3D techniques including ligand-based and receptor-dependent approaches. The dataset was constituted of 68 2-amino-6-arylsulfonylbenzonitriles and thio- or sulfinyl-congeners (**12.13**). Activity data encompassed (for 64 compounds), pIC_{50} anti-HIV-1 activity, and (for 51 compounds) pIC_{50} HIV-1 reverse transcriptase binding affinity. These compounds were the subject of several QSAR studies using "classical" descriptors such as molecular connectivity, E-state parameters, hydrophobicity, Hammett's σ [878,879]. Other models were derived from multivariate image analysis [880] or several nonlinear models, such as a mixed radial basis function network-based transform for SVM [613] with descriptors taken from Cerius^2 3.5 [56].

12.13

12.4.1 2.5D ANALYSIS

In a first step, this dataset was revisited with both introduction of 2.5D-descriptors generated by CODESSA (including quantum-chemical and geometrical descriptors) and use of various nonlinear models MARS, PPR, RBFNN, GRNN, and SVM (for details about the methods, see Appendix A).

A selection of CODESSA descriptors was first performed using forward stepwise regression on the initial descriptor pool (about 600) and the break-point technique: In a plot of r^2 or q^2 versus the number of descriptors (in MLR), the curve grows up monotonically before a sudden reduction of the slope. This point can be chosen as indicating the best number of descriptors, as a trade-off between precision and complexity of the model. In this application, for anti-HIV-1 activity, six descriptors where selected: three topological (two Kier et Hall connectivity indices and Kier shape descriptors), one geometrical (shadow index, also related to the molecular shape), and two quantum chemical descriptors (relying on nucleus repulsion and electron resonance energy).

The selection of training and test sets (48 and 16 compound, respectively) was achieved by the principal component analysis from the two first components that represent about 70% of the overall variance. The best models were searched for in leave-one-out (L-O-O) cross-validation on the training set and the performance checked on the test set. For anti-HIV-1 activity, SVM ($r^2 = 0.831$) and PPR ($r^2 = 0.890$) significantly outperformed MLR (0.793) and the other nonlinear methods used. A similar treatment for HIV-1 binding affinity, led to neighbor results, PPR remaining the best approach ($r^2 = 0.843$, $r^2_{pred} = 0.843$ vs. 0.738 and 0.750 for MLR) (Figure 12.7). Values of r^2_m or r^2_0 [881] suggested good external predictability.

12.4.2 CoMFA AND CoMSIA ANALYSES

In the second step, 3D-QSARs were carried out for this series of compounds. The whole data set was manually split into training (51 chemicals) and test (13). Four compounds for which experimental activity were not available constituted an external prediction test. The "probable" active conformations of inhibitors were selected by docking them (AutoDOCK4) into the receptor pocket isolated from a known crystal structure of an inhibitor–receptor complex. The reliability of the procedure was checked in a satisfactory comparison (for that inhibitor) of its experimental geometry and that derived from docking. Docking was then performed on all compounds to select the active geometry. For each compounds, 100 runs were performed, the solutions clustered and the most populated (or failing that, the lowest energy) representative in each group is selected.

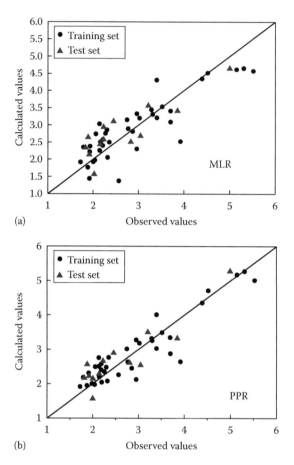

FIGURE 12.7 Calculated values vs. observed values of affinity using MLR (a) and PPR (b) modeling for HIV-1 RT binding affinity data set. The diagonal is the y = x line. (Reproduced from Hu, R. et al., *Eur. J. Med. Chem.*, 44, 2158, 2009. With permission.)

In a first treatment, the docked geometries were chosen for alignment (a "receptor-guided" model). In an alternate approach, a two-step procedure was used ("receptor and ligand based," the authors said). First, for each chemical family (thio-, sulfonyl-, or sulfinyl-derivatives) a representative was chosen as template for alignment, and then the three families were joined in the same data set.

Interestingly, the binding modes are found different in the three chemical families; as exemplified in a comparison between compounds with X = S (thio), X = SO (sulfinyl), and X = SO$_2$ (sulfonyl) all bearing Me groups in 3-, 5-positions of the B ring, that devoid of the nitrile group. Compounds X = S and X = SO$_2$ have a rather similar binding mode. For X = S, two H-bonds are formed from the amino hydrogens to the carbonyl of Lys-101 and to the nitrogen atom of Lys-103. For compound X = SO$_2$, one amino hydrogen is engaged in competitive H-bonds with either oxygen or nitrogen of Lys-101. By contrast, in docked compound X = SO, there is a flip of the phenyl ring

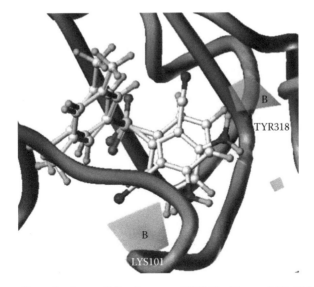

FIGURE 12.8 **(See color insert following page 288.)** Docking and 3D-CoMSIA contour plots. Compounds of structure **12.13** with R = di-Me 3,5 on ring B and X = SO and SO$_2$, and HIV-1RT protein. The cyan area corresponds to favorable H-bond donor fields (and exactly overlaps the residues Lys-101 and Thr-318). Note the flip of the amino substituted phenyl ring. (Reproduced from Hu, R.J. et al., *Bioorg. Med. Chem.*, 17, 2400, 2009. With permission.)

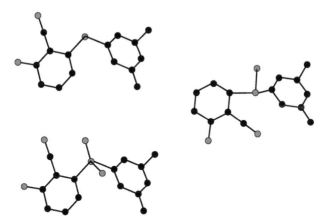

FIGURE 12.9   Different binding modes of thio-, sulfinyl-, and sulfonyl-derivatives.

bearing the amino group, leading to an H-bond with the hydroxyl oxygen of Tyr-318 (Figures 12.8 and 12.9).

The receptor- and ligand-guided model gave better results than the receptor-guided one. CoMSIA with steric, hydrophobic and H-bond donor and acceptor fields only, led to $q^2 = 0.760$, $r^2 = 0.959$; $r_{pred}^2 = 0.520$ in a PLS treatment (eight components). CoMFA with 4 LV only yielded slightly lower results ($q^2 = 0.723$, $r^2 = 0.868$; $r_{pred}^2 = 0.483$).

From the CoMFa and CoMSIA contour plots, it was concluded that bulky and hydrophobic groups in the 3-, 5-positions of the B phenyl ring increased activity.

12.4.3 MOLECULAR DYNAMICS AND BINDING FREE ENERGY

To get a more refined insight about binding modes and protein–ligand interaction energies, a MD simulation was performed on the preceding three molecules, chosen as representatives of the involved chemical families. From the known crystal structure, a simplified model of the binding site was determined, limited to residues with at least one atom at a distance lower than 12 Å from the ligand (acetyl and amino caps were introduced for isolated terminal residues). A water box of about 4000 TIP3P water molecules was created with counter ions to ensure neutrality. Long-range interactions were treated with the Particle Mesh Ewald (PME) method. MM optimization was carried out first on water and counter ions only, and then the whole system was relaxed. It was checked that the receptor model did not deviate from the original protein. MD simulation (N, V, T) was performed at 300°K during 5 ns with a time step of 2 fs, with a smooth restraint applied to the farthest protein-model atoms. Averaged ligand structures were determined during the last 2 ns. From ligand conformations generated every 1 ps, the conformation closest to the average structure was extracted. Interestingly, for compound X = SO a significant conformational change occurred at 2 ns. H-bond patterns from MD simulations were in agreement with docking conclusions.

Binding energies directly estimated from AutoDOCK4 did not satisfactorily rank the three molecules. Hence, more refined treatments of binding free energies were carried out with Poisson–Boltzmann (MM-PBSA) or generalized Born (MM-GBSA) approximations. Results are in good agreement with the ranking of experimental activities (for both anti-HIV-1 activity and RT inhibitory activity). The higher activity of compound X = SO_2 reflected differences in gas-phase electrostatic term and polar contribution to solvation.

	Molecule X		
Model	S	SO	SO_2
Experimental rank[a]	3	2	1
MM-PBSA[b]	−48.4	−45.5	−52.3
MM-GBSA[b]	−52.8	−53.11	−58.6

[a] Most active first. Anti HIV-1 activity.
[b] Energies in kcal/mol.

In conclusion, the authors stressed that the higher activity of molecule X = SO_2 (and higher stability of its complex) originated mainly with interactions (including H-bond) to Lys-101, and not as a general property of sulfone derivatives. For a more detailed analysis of the contributions of individual residues and solvent environment, see [877].

Concluding Remarks

About half a century after the modern QSARs were introduced by Hansch and Fujita [23] and 30 years after the inception of CoMFA, the archetype of 3D models, a question may be posed: Have QSARs kept their promises [882]?

Indeed, QSAR models sometimes were given a cool reception. A rather general feeling of disillusionment appeared, if not going with acerbic comments or criticisms. This was not quite unjustified owing to a lot of published spurious models, bringing QSARs studies into disrepute. As stressed by Doweyko [883], QSARs were expected to "predict activity of unknowns, illustrate regions where structural changes may significantly modify activity, and give some insight on possible interactions." But, too often, "predictions were illusive," graphical displays subject to "vagaries of alignment," and interaction mechanisms "dependent on the eye of the beholder."

This feeling of frustration and disappointment must not let us forget the countless successes of these approaches (only a few of them are quoted in this book). However, the question deserves comments. An extensive, and well-documented, paper of Cronin and Schultz [884] reviewed some of the possible reasons of pitfalls and highlighted the fact that QSARs belong to a multidisciplinary field, hence a large variety of possible problems.

Clearly, some critical points (common in fact to all QSAR/QSPR applications) have been carefully examined, and some caveats (not always strictly observed in the past, but now largely accepted by the community) defined what "good practice" must be. Clearly

- Data quality is (obviously) crucial ("garbage in, garbage out"). This is true not only for experimental results but also for some descriptors calculated from models (log P is a well-known example).
- Robustness and generalization capability of the models are crucial. The applicability domain and the limits of a "reasonable" extrapolation must be strictly examined.
- Mathematical or statistical treatment must be correctly applied, avoiding overfitting and, for certain models, overtraining.
- Cross-validated L-O-O q^2 is not an automatic criterion for a predictive correlation, and some other tests have been proposed. Generally speaking, the predictive ability is better established on an external test set (not "known" beforehand by the model). Note, however, that Horwell et al. [483] remarked that discarding some data in order to define a test set came to loose a part of information.
- Transparent, easily interpretable model in terms of interaction mechanisms is highly desirable.

As to, more specifically, 3D-QSARs, the stumbling block is clearly the choice of the active conformation. The delicate step of alignment needed by some methods is now often made easier with the increasing number of protein structures solved by X-ray crystallography, NMR, or homology modeling.

These methodological points can be—relatively easily—taken into account. But, on a more fundamental aspect, Maggiora [885] suggested that some QSARs were not as good as expected because the response activity surface is not as smooth as supposed. In addition to the presence of such activity cliffs, Johnson [882] wondered how we arrive to wrong models (given the wide panel of available tools), and why we accept these wrong models. For this latter point, the suggested reason would be that a blind confidence in statistical tests has replaced a true scientific critical sense (rational thought, controlled experiments, and personal observation). A similar concern was also shared by Benigni et al. [537].

With the huge number of descriptors and the diversity of statistical methods, one is now able to propose a variety of models of comparable quality. Some are neighbor, but some may correspond to different hypotheses. Selecting a "best model" from a statistical test on only some observations, moved apart beforehand from the data set, is not a guarantee. The model ought to be validated on selected molecules specially designed to check the consistency of the hypotheses. In other words, if getting numerical results is pleasant, it is better to obtain meaningful results.

Practitioners have now at their disposal a great choice of approaches that constitute, despite their limitations and if well applied, invaluable tools. If users are conscious of these limits, and keep some critical judgment, no doubt QSARs will play a still more important role in the future.

As to the foreseeable future of QSARs (and 3D-QSARs), two elements may already look as positive indications:

1. From a political point of view, environmental problems take a crucial importance on account of the outburst of new chemicals widely scattered all over the earth. Faced with such a worrying question, international policies (the REACH project for example in Europe) recommend the use of QSARs models (duly designed) in hazard identification and risk assessment as a reliable and efficient tool.

2. As to methodology, we witness a close symbiosis between applications traditionally relevant of QSARs and methods associated with the domain of molecular modeling and database management. These domains also take advantage of the continuously growing computer resources.

   • With the increasing number of ligand–receptor complexes solved by X-ray crystallography and the advances in homology modeling, docking methods offer an efficient alternative solution to the delicate alignment phase.

   • Calculation of thermodynamic parameters (and not only comparisons between *variations* of activity and *variations* in structural descriptors, more or less well selected) becomes easier. Whereas in the past, determination of binding free energies required intensive calculations, and was only tractable for few selected examples, approximate methods combining speed and good accuracy are now available for series of molecules. This particularly concerns the estimation of solvent influence (partial desolvation upon binding). Adaptation of the receptor to individual ligands (induced fit) can be also treated. This would make easier the comparison of non-congeneric series.

- Incorporation of the dynamic nature of the molecules is also a major challenge. Molecules are no longer considered as an assembly of rigid hard spheres, described in their minimal energy state, but as adaptable entities. Generalizing methods "beyond 4D models" will give a more realistic picture, including a sharper evaluation of solvation effects.
- Various examples exist where QSAR models are now efficiently substituted (after pre-filtering) to scoring functions in data mining.

Today, faced with the multiplicity of mechanisms possibly monitoring the biological activity, it seems difficult, as stressed by Manchester and Czerminski [886], to assess the merits of current QSARs from the popular benchmarks. We have to search for convergence and complementarity of several methods until the multiplication of solved examples makes the choice of the good approach unequivocal.

Appendix A: The Tool Kit

In this appendix, we would only give few schematic indications about some methods frequently used in QSARs and quoted in the main text. As previously stated, QSARs have no specific methods but frequently call for techniques or tools of more general use and also intervening in diverse other fields. Our aim is not to present extensive developments about the latest options exposed but rather sketch out the fundamental principles of these techniques. For example, we will generally present only basic algorithms even if they are not the most efficient in a particular application. This survey will surely appear fragmentary and over-simplified, if not simplistic. But we hope it may provide beginners or people entering in the discipline, some starting points and open some leads before going about more in-depth books or reviews.

In this appendix, we will consider

- Firstly, optimization, a general concern intervening in energy minimization, docking, variable selection. Genetic algorithms will be also included on account of their wide use in the later point.
- Secondly, modeling problems with empirical force field in molecular mechanics (MM) and molecular dynamics (MD).
- At last, some mathematical (or statistical) models used to correlate experimental and calculated (or predicted) activities, with special emphasis on nonlinear approaches.

A.1 OPTIMIZATION

Many scientific problems come down to optimization processes, i.e., finding the extremum (maximum or minimum) of an objective function (e.g., potential energy, fitness function). If we restrict ourselves to common questions intervening in the field of QSARs:

- A very usual problem is the choice of the "active conformation" of a drug. In this conformational analysis problem, one has thus to determine (depending on the hypotheses) the minimum energy structure or, more generally, the low-energy structures in a range of about 5–10 kcal/mol above the absolute minimum, in order to derive a conformational energy profile.
- In docking, the best pose (position, orientation, conformation) of the ligand in the receptor binding site corresponds to a maximum interaction energy, taking also into account the adaptation of the ligand and possibly of the receptor.
- In an alignment process (superimposition of two or several molecules), a possible solution may be obtained determining the minimum of the sum of distances between corresponding atoms. (Other approaches look at a similarity criterion in the fields generated.)

- To build QSARs, one has to select the relevant descriptors (those conveying the information), and choose a mathematical model. The best descriptor set may be determined searching for the maximum of the (cross-validated) correlation coefficient between observed and calculated values or the minimum of the root mean square error (RMSE).

Optimization techniques can be schematically divided into three main types: calculus-based, enumerative, and guided random search methods. We will just indicate the basic principles of these methods [887], using as privileged example the search for the minimum energy conformation of a molecule. For a more in-depth presentation, see [461, pp. 140–144], [763, pp. 211–233], and [887, pp. 274–334].

A.1.1 CALCULUS-BASED TECHNIQUES

These techniques look for local extrema of an objective function. Among the most widely used methods, we retain the simplex, gradient, and Newton methods.

A.1.1.1 Simplex Method

A very safe method (but sometimes slow) is the *simplex*. It always gives the solution without need to calculate derivatives. In the 3D space, a simplex is a set of $3 + 1 = 4$ configurations of a system, represented in the solution space by four points. In the current step, the least favorable point as to the objective function to optimize (cost function) is replaced by its symmetrical by respect to the plane formed by the three other points (possibly with expansion or compression), and the process is iterated (Figure A.1).

Another method that does not require derivative calculation is the *Conjugate Directions* method.

A.1.1.2 Gradient Methods

Other methods of multidimensional minimization carry out (successive) one-dimensional-minimizations and rely on the derivatives (first or second) of the

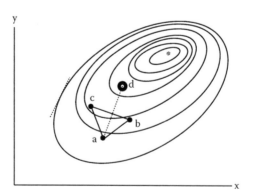

FIGURE A.1 A simplex in 2D z = f(x,y). Given three points, the worst a is replaced by d, its symmetrical with respect to bc. The new simplex becomes b, c, and d. The star indicates the minimum.

objective function. Remember that for a function of two variables (say x and y) it is not sufficient to minimize first along x and then along y and stop. The same process must be continued until the actual minimum is found.

Given a function f(**x**), of vector variable **x**, for a displacement **δ** around a starting position $\mathbf{x}_0$, in a Taylor expansion

$$f(\mathbf{x}_0 + \boldsymbol{\delta}) = f(\mathbf{x}_0) + \sum_i \left(\frac{\partial f}{\partial x_i}\right)_0 \delta_i + 0.5 \sum_{i,j} \left(\frac{\partial^2 f}{\partial x_i \partial x_j}\right)_0 \delta_i \delta_j + \cdots$$

the index 0 indicating that derivatives are calculated at point $\mathbf{x}_0$
δ_i is the displacement along coordinate i

or

$$f(\mathbf{x}_0 + \boldsymbol{\delta}) = f(\mathbf{x}_0) + \mathbf{g}(\mathbf{x}_0)\boldsymbol{\delta} + 0.5\boldsymbol{\delta}' \, \mathbf{G}\boldsymbol{\delta} + \cdots$$

where
g is the gradient, Λf of components $(\partial f/\partial x_i)$
G is the Hessian matrix, elements $(\partial^2 f/\partial x_i \partial x_j)$

In *indirect gradient method*, the optimum may be found by solving the equations resulting from setting to zero the gradient, first derivative of the function. The *direct gradient method* works on the gradient itself (calculated either analytically or numerically). At the minimum **x**, the gradient is null.

An energy minimum for example is sought for moving in the direction opposite to the gradient $\mathbf{g} = \partial f/\partial \mathbf{x}$, **x** representing the vector variable of the atomic coordinates defining the objective function f to minimize, the greater the gain in stability, the greater the displacement. It is the *steepest descent method* (Figure A.2). This method

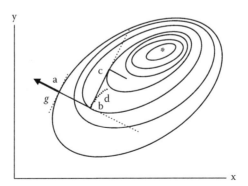

FIGURE A.2 Minimization in 2D z = f(x,y). In Steepest descent method, starting from point a (where the gradient is g) down to b (the minimum in gradient direction); then along the new gradient direction to the minimum c and so on. The star indicates the absolute minimum. Conjugate gradient moves along bd (dotted line).

is reliable. But it may be slow: successive motions are along perpendicular directions and may zigzag a lot in a shallow energy basin [888].

The search corresponds to a (mono-dimensional) line minimization that may be achieved via the parabola method ("bracketing" the minimum) or using a predefined step-size iteratively expanded or contracted depending on the variations observed.

After the minimum is found along direction $\mathbf{u}$, choosing a new direction of minimization ($\mathbf{v}$), so as not to spoil in the current step the gain of the preceding one, leads to elect a direction so that the new gradient direction ($\mathbf{g}_j$) stays perpendicular to $\mathbf{u}$. Thus

$$\mathbf{u}\mathbf{g}_j = 0$$

$$\mathbf{u}\delta(\Lambda f) = \mathbf{u}\mathbf{G}\mathbf{v} = 0$$

$$\mathbf{u}\mathbf{v}_j = 0$$

The directions $\mathbf{u}$ and $\mathbf{v}$ are said conjugates. For a quadratic form, minimization along two conjugate directions will lead to the minimum.

Conjugate gradient method speeds up the process, keeping trace of the gradient in the preceding step to determine the current move, a technique previously proposed by Engler et al. [888] in their "pattern search."

At iteration k, minimization is carried out along direction

$$\mathbf{s}_k = -\mathbf{g}_k + \beta\mathbf{s}_{k-1} \quad \text{with} \quad \beta = \frac{|\mathbf{g}_k|^2}{|\mathbf{g}_{k-1}|^2}$$

in the Fletcher-Reeves method. A neighbor variant exists, which is strictly equivalent for a quadratic form, the Polak–Ribiere method with

$$\beta = \frac{(\mathbf{g}_k - \mathbf{g}_{k-1})\mathbf{g}_k}{|\mathbf{g}_{k-1}|^2}$$

A.1.1.3 Newton's Methods

The *Newton–Raphson* method works on the second derivatives, the $\mathbf{G}$ Hessian matrix with elements $\partial^2 f/\partial x_i \partial x_j$.

$$\mathbf{g}(\mathbf{x} + \delta) = \mathbf{g}(\mathbf{x}) + \mathbf{G}\delta$$

At the minimum

$$\mathbf{g}(\mathbf{x} + \delta) = 0 \quad \text{so that} \quad \mathbf{G}\delta = -\mathbf{g}(\mathbf{x})$$

The minimum is directly determined for a quadratic function, reached in few iterations for more complex functions. In such a case, the step is

$$\delta = -\mathbf{G}^{-1}\mathbf{g}(\mathbf{x}) \qquad \mathbf{x}_{k+1} = \mathbf{x}_k + \delta$$

The Newton–Raphson method is largely used, but it may be less efficient if the starting point is far from the minimum. Problems also occur when the Hessian matrix is not positive definite. In the *restricted step* method, geometrical changes are shortened to avoid an increase in energy, according to

$$(G + \lambda I)\delta = -g(x)$$

where
   I is a unit matrix (m*m, with m number of parameters to optimize)
   λ is a scalar chosen, so that the new matrix is positive definite

Quasi-Newton methods: The drawback is the calculation of the Hessian matrix and its storage at each iteration. Variants have thus been proposed to speed up the process. *Block-diagonal Newton–Raphson* method, for example, considers that only one atom moves at each iteration. This reduces the calculation to the inversion of a 3*3 matrix. But concerted motions in cyclic systems may pose problems.

   On the other hand, Davidon–Fletcher–Powell (DFP) or Broyden–Fletcher–Goldfarb–Shanno (BFGS), use approximate expressions of the Hessian, thanks to information gained in the preceding iteration. The trick is to define iteratively a matrix H that is a good approximation of G^{-1} and tend to G^{-1} as iterations go along.
   At step k + 1, the new position $x_{k+1} = x_k + \delta$ is calculated by

$$\delta = -H_k g_k$$

and H_k is updated. See also Refs. [889–891] for the introduction of constraints.
   An important point is that these methods stop at the first extremum (all derivatives null) encountered, which may only be a local minimum. To find the absolute extremum, the exploration must be carried out from several starting points of the configurational space. The absolute extremum is often the most interesting point, but local extrema may be sometimes considered; for example, to define the different molecular conformations possibly coexisting in flexible molecules.

A.1.2   ENUMERATIVE METHODS

These methods examine the objective function at every point (one at a time) within the given space, as for example the *systematic search* in conformational analysis: energy is calculated at the nodes of a grid. This approach is simple but slow, and so not very efficient. In simple cases, when the conformational energy only depends on two dihedral angles, conformational maps may be drawn from such systematic search, as the well-known (φ, ψ) Ramachandran maps for amino acids.

A.1.3   GUIDED RANDOM SEARCH METHODS

They rely on enumerative methods but with the help of supplementary information to guide the search. This may be viewed as a restless random walk from point to point on the potential energy surface; but such a random walk cannot give better results than an exhaustive enumerative search. Furthermore, strictly speaking, the user is never sure to have found the best solution.

Two main avenues may be distinguished:

- Monte Carlo methods and simulated annealing
- Evolutionary algorithms

A.1.3.1 Monte Carlo Search

It is often associated with a *simulated annealing* process. In conformational analysis for example, starting from an initial structure of the system, a random change is performed on one of the parameters defining the geometry, generally a dihedral angle. Indeed torsional motions, involving weak energy changes, are more likely to be involved in conformational equilibria. In basic Monte Carlo method, the energy of this new configuration is calculated. The change is accepted if the energy decreases, that is if you go nearer to an energy minimum, otherwise it is rejected. And the process goes on (random walk). For the next kick, you can restart from the initial geometry, from the new one generated in the preceding change, or from the energy minimum the less frequently found. After each kick, it is also possible to refine the search and locate the (nearest) energy well. The process ends when no new minimum is found after a fixed number (500, for example) of changes. See, for example, the random incremental pulse search (RIPS) of Ferguson and Raber [154]. This procedure is efficient to find the attraction basins of the various energy minima, but the process may be accelerated, if you only want the absolute minimum, thanks to simulated annealing.

The term "simulated annealing" was adopted by analogy with the so-called physical process. At a high temperature, in a sample, molecules are free and disordered moves occur. If the temperature slowly decreases, molecules tend to fall in order and finally organize themselves in a crystal. If temperature decreases rapidly, a glass or a metastable state is formed. In optimization, the principle of simulated annealing is to introduce a new parameter: a "pseudo" temperature that will be slowly decreased. We prefer the term "pseudo" temperature since in many optimization processes it has nothing to do with the real temperature of the experiment. It is only an easy way to introduce Boltzmann probability.

At "pseudo" temperature T, a change is always accepted if the energy decreases ($\Delta E < 0$). If energy increases ($\Delta E > 0$), the change is not always rejected. It may be accepted with a probability proportional to $\exp(-\Delta E/RT)$: Metropolis criterion [515]. At a high "temperature," many changes are accepted. It is possible to largely explore the conformational landscape and escape from local minima. As temperature decreases, less and less upward changes are accepted. The process converges to the absolute minimum. To verify the acceptance condition, a number (n) between 0 and 1 is randomly selected. The change is accepted if

$$n < \exp\left(\frac{-\Delta E}{RT}\right)$$

A.1.3.2 Miscellaneous Methods

At last, we should mention MD. MD is not per se an energy minimization method. It only describes how the energy of a molecular system evolves with time (e.g., conformational energy for a single molecule or interaction energy in a protein–ligand complex). Although some limitations (difficulty to cross high-energy barriers), MD is an efficient way to draw conformational energy profiles and characterize the low-energy states of a molecular systems and its conformational transitions.

On the other hand, evolutionary methods (genetic algorithm and evolutionary programming) apply the concepts of natural selection and genetic evolution. In view of many other applications (such as variable selection, building up a receptor model, determining pharmacophoric features, docking), we will treat separately these methods.

A.2 EVOLUTIONARY METHODS: GENETIC ALGORITHM AND EVOLUTIONARY PROGRAMMING

Evolutionary approaches such as genetic algorithms (GA) (which involve both mutations and recombination), described by Holland [892,893] and evolutionary programming (EP) (which prefer mutations) [85], mimic the Darwinian natural evolution and selection process. They are now recognized as efficient approaches to solve search and optimization problems. In addition to the development of QSAR (QSPR) models, our main concern here, they have also been used to perform variable selection, conformational analyses, flexible docking simulations, or pharmacophore search [84–86,155,343,410].

In real life, the transmission of the heredity is carried out by the genes contained in the chromosomes that store the information determining the individuality of the organism. Chromosomes are replicated, possibly modified by genetic operations, and passed onto the next generation. The organisms best suited to their environment survive and mate, whereas others waste away and disappear. In a population of chromosomes, better the fitness (how "good" the chromosome is), greater is its probability to be transmitted to the offspring. Evolution over many generations leads to optimal (or near optimal) population, that is the best adapted individuals. It is to be noted that the process operates on a population of individuals and works on a series of intermediate solutions to reach the optimal one(s).

A.2.1 Main Steps

GA and EP reproduce this evolution and selection process, but differ in the way an offspring is created from the current generation: basically, sexual reproduction (in GA: children have two parents) vs. asexual reproduction (in EP, a child has only one parent). Schematically, GA and EP address problems where [393,394]

- A possible solution can be encoded via a string of characters
- A given solution may be evaluated quantitatively

The approach involves as major components: generation of the chromosomes, fitness assessment, and deletion. We will here develop it in the context of QSARs. It must also be kept in mind that many different options of GA or EP are offered and that the implementation of the method largely depends on the problem to solve (regarding encoding, choice of the population, evaluation of the fitness, generation of the offspring). We summarize here only the fundamentals of these approaches. For a more detailed presentation, the reader can refer to specialized papers [48,49,84,487,488].

The successive steps are

- *Encoding*: The molecules are described by a set of structural descriptors (genes) called "chromosomes." All kinds of descriptors can be used: similarity indices, indication of the presence/absence of a given fragment, physicochemical properties. But, traditionally they are handled via a binary string (1: the corresponding descriptor is included in the model; 0: it is discarded). Continuous variables are treated as integers defined in a specific range and encoded as a fixed number of binary bits. Rather than a binary encoding Gray code presents the advantage that successive integers differ only by a single bit. For example, integers 11 and 12 become 1011 and 1100 (binary), but 1110 and 1010 (Gray). For the conversion binary ↔ Gray code, see Ref. [894]. So, GAs work with an encoding of the descriptors not with the descriptors themselves.

  Other encoding schemes are possible such as
  1. Node encoding: the string represents a route, as the order of the journey in the traveling salesman problem. The values of the string are not independent.
  2. Delta encoding: the change with respect to the current template is indicated.

  Of course, the choice of an encoding scheme has an effect on processing the analysis, but we only consider here the case of binary encoding.

- *Creation of a population of chromosomes*: GA works on a population of individuals (chromosomes). For example, in variable selection, starting from the huge number of descriptors available from several packages (such as CODESSA, DRAGON) subsets are selected at random, forming the chromosomes. The common number of variables (genes) forming the chromosomes, the number of chromosomes, that is the extent of the population, must be fixed depending on the problem. In this pool of chromosomes, two individuals or two descriptors in an individual must not be identical.

  Schematically, a large number of chromosomes reflects a large structural diversity but may imply longer time to converge to the optimal solution. Too small this number increases the risk of rapidly converging to a suboptimal (local) solution. Similarly, the size of the chromosomes (the subset of descriptors) reflects the complexity of the model and must be examined in terms of parsimony or, conversely, overfitting.

- *Definition by the user of a fitness function*: Objective function to optimize. For variable selection, a correlation model (MLR, PLS, ANN) is generally used to indicate how well a given chromosome (subset of descriptors) is able to calculate (and predict) the property under scrutiny. As already mentioned, the fitness function may be the RMS deviation between

calculated and experimental values, or the L-O-O cross-validated correlation coefficient q².

   This fitness function directs the selection process: high fitness individuals are more likely to be selected as parents for the next generation, and at
last the fitness function indicates the quality of the final solution(s).

• *Genetic operations*: There are a lot of possible strategies for updating the
   current generation or generating a new population from the current one.
   The principle is to increase the number of chromosomes with better fitness.
   We just summarize here the basic processes. Other options may be found
   in Refs. [84–86,487].

   In most cases, a first selection of the initial population is carried out with
the creation of a mating pool, where the bit strings encoding the chromosomes
are copied (*replacement*). As in natural selection, where the more fit individuals have more chances to survive, chromosomes are reproduced with a probability depending on their fitness function, according to the roulette wheel
selection for example. However, a danger exists: super-individuals are selected
more often and may become predominant, inducing premature convergence.

   If, at a time, all individuals share a common value for a given position
in the genetic string, it will be impossible to introduce another value at
this position. This corresponds to *focusing* [86]. A remedy may be using
islands (creating several subpopulations with possible migrations of individual chromosomes) and *nicheing*: if too many individuals are very similar ("sharing the same niche"), the new individuals generated replace the
least fit in the niche rather than in the whole population. Such a solution
was, for example, adopted in GOLD [279,802]. On the other hand, chromosome of low fitness would never be able to transmit part of their gene pool.
So, it is possible to include in the mating pool all the chromosomes of the
initial population plus extra copies of the best fit individuals.

   During the generation phase, new chromosomes are created thanks
to genetic operators, the most important being crossover and mutation
(Figure A.3).

1. In *crossover* (sexual generation), two parents exchange parts of their chromosomes, cut at a randomly selected point, and generate two children. Generally,
   the first parent is chosen at random. The second one may be selected at

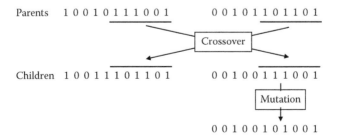

FIGURE A.3 Examples of genetic operations: crossover (two parents) and mutation (one parent).

random or the choice may be based on its fitness (the most fit chromosomes being chosen more often). Two points crossover may be also carried out.

2. *Mutations* only imply the random change on one (or several) genes. It may protect the system against irrecoverable loss of some useful genetic material. For example, a "1" in a given position of the chromosome (inclusion of the corresponding descriptor in the model) may be important, but unfortunately it may be lost in crossover).

   In GA, mutations are given a low probability ($P_{mutation}$ about 0.01) far below $P_{crossover}$ (current values are about 0.5); whereas in EP, $P_{crossover} = 0$ and $P_{mutation} = 1$. So, in standard EP the current population is placed in a mating population, with very fit members placed more often than others. Each parent is sequentially chosen and allowed to produce a single offspring through mutation.

Crossover and mutations are important operations since replacement (copy) does not change the structure of the chromosomes, but only their proportions in the population. Creating new strings extends the search procedure to other regions of the search space.

Other genetic operators are

1. *Replacement* (as already seen), just the copy of (some) individuals of the mating pool.
2. *Elitism*, an artificial operator without biological reference. It guarantees that the best individuals are transferred without mutation, and so are not lost.

   Many options may be defined as to the fate of offspring. It is not mandatory that the number of children equals that of the parents (many offsprings can be created from two parents) or that all children are kept.

   Children may replace a member of the preceding generation, for example, its weakest parent or (more often) the weakest in the population, or the more similar. They can also be placed in another (new) generation either alone or combined with the preceding generation, favoring the best solutions. From this general framework, many variants have been reviewed by Luke [86]. Particularly,

   - In the *standard GA*, at least one parent is chosen from its fitness and the offspring replaces the weakest member of the population.
   - In *Evolutionary Programming*, one child is created per parent and the next generation combines parent plus offspring with a probabilistic replacement.

These operations: genetic manipulation of chromosomes of the current population, evaluation of the fitness of the new individuals, generation of a new population (by partial or total replacement) are repeated until optimal or near optimal solution(s), or according to some other termination criteria. For mathematical details, see Ref. [894].

Typically, in variable selection 100 or 200 chromosomes are formed with, each, about 20 variables (genes). GA works on a population and so may be considered as a parallelized approach. Generally, it does not lead to a unique solution, but to several solutions of similar quality. This allows for getting different views of the same problem. Users can then choose among these models according to, for example, interpretability, ease of extension to other systems.

A.2.2   GENETIC FUNCTION APPROXIMATION

The genetic function approximation (GFA) of Rogers [109] is an extension of the simple genetic algorithm (SGA). In simple GA, an internal model (such as MLR, PLS) is chosen and genetic operators change the descriptors forming the chromosome. In GFA, the chromosomes are constituted by a set of mathematical functions acting on predefined descriptors and the genetic operations are performed on these functions.

Extending the usual multilinear relationships, QSAR models are often expressed as a sum of terms:

$$Activity = \sum a_k \Phi_k(\mathbf{x}) + b$$

Where $\Phi_k(\mathbf{x})$ are called the basis functions (such as square, log), the vector function $\mathbf{x}(x_1, x_2, \ldots)$ representing the structural descriptors. A very simple example is Hansch-type equation where intervene terms in π and π^2.

From a set of descriptors considered as potentially useful, and user-defined basis functions genetic models are first created as random sequences of these basis functions. Then, repeated crossover operations are performed:

- Models are chosen as parents, proportional to their fitness.
- Children are generated. Cuts occur between basis functions, not inside an individual basis function.
- Children may suffer mutations.
- This new model replaces the worst one of the preceding generation.

The process ends when the average fitness of the population stops increasing. Another original character in the treatment of Rogers is the use, for evaluating the fitness, of the lack-of-fit function (LOF function) introduced by Friedman and Silverman [895]. In LOF, the complexity of the model is penalized

$$LOF(model) = \frac{1}{N} \frac{LSE(model)}{[1 - (c + 1 + (d*p))/N]^2}$$

where

LSE is the least squares error $\sum (y - y_{pred})^2$

c is the number of nonconstant basis functions
p is the total number of parameters of the model
N is the total number of samples
d is a smoothing parameter generally set to 2

Basis functions used in Roger's model are truncated power splines also used in multivariate adaptive regression splines (MARS)—vide infra. These functions, when used in a regression, allow for introducing nonlinearity. Particularly, the knot indicates where discontinuity occurs, making interpretation easier. Other possible basis functions are polynomials. But a caveat must be put on more intricate functions where the physical meaning (if any) is difficult to extract.

An example of nonlinear modeling using GFA concerns the activity of dopamine β-hydroxylase inhibitors, previously studied by Burke and Hopfinger [368], see Section 3.1.3.

Suffice here to indicate that the best model, built with two splines, is nearly of the same quality ($r^2 = 0.808$, N = 45) as the polynomial models of Burke and Hopfinger involving six or three parameters ($r^2 = 0.828$ [six terms, N = 47] and 0.810 [three terms, but with two compounds discarded]).

$$\mathrm{Log}\left(\frac{1}{\mathrm{IC}_{50}}\right) = 3.762 + 0.296\langle -10.203 - \mathrm{E_{int}}\rangle + 0.089\langle 26.855 - \mathrm{E_{inside}}\rangle$$

where $\mathrm{E_{int}}$ and $\mathrm{E_{inside}}$ are interaction energies calculated by a receptor surface model. As to nonlinearity, note that this formalism indicates a linear variation with $\mathrm{E_{int}}$, if $\mathrm{E_{int}} < -10.203$, but a null contribution for $\mathrm{E_{int}}$ superior to this knot value.

A.2.3 DERIVED MODELS AND APPLICATIONS

We just mention some applications related to variable selection. Basically, GA is used to propose some (relevant) descriptors to an objective function, which is here a model for associating observed and calculated activity. The simplest function is MLR, alternately PLS or neural networks were used.

A.2.3.1 MUSEUM

Neighbor to evolutionary programming, the MUtation and SElection Uncover Model (MUSEUM) proposed by Kubinyi is dedicated to variable selection in regression and PLS analyses. Stepwise regression (upward or backward) has been largely used, and appears appropriate for a reduced number of variables, but may lead to local solutions if too many variables are involved [896].

Basically, in MUSEUM, a random model (any combination of the data set) is first created. One or few variables are added to or eliminated from this model. If fitness increases, the new model replaces the current one. If no improvement appears after a certain number of mutations, mutation rate is enhanced (leading to more exchanges). Again, if no improvement, all variables inside and outside are systematically added or eliminated, one at a time. If the best model is stable, all variables inside are checked for significance and possibly eliminated. But, the system keeps a record of the intermediate models.

In variable reduction, it was found that the results depend on the criterion: using Fisher test, few variables are selected, whereas with the standard error, too many variables are maintained. So, Kubinyi introduced, as a good compromise, the FIT criterion

$$\text{FIT} = \left[\frac{r^2}{(1-r^2)} \right] \left[\frac{(n-a-1)}{(n+a^2)} \right]$$

where
    a number of variables
    n number of objects

The treatment was carried out on 11 data sets (data from real life or simulated). Most often, results are better than with PLS analysis including all variables. According to the author, "the variables retained in the best MUSEUM models are suitable subsets for PLS and some of these are better than the best regression results."

Another advantage is that MUSEUM, as a generally evolutionary approach, leads to several solutions. The user can choose among them the most interpretable in terms of chemical or biological behaviors. However, MUSEUM is not suited much for treating the numerous descriptors generated by molecular field analyses (such as CoMFA).

A.2.3.2 Hybrid GA

Leardi [105] stressed that GA is able to largely explore the variable space and find several local extrema but is not very efficient for a sharp analysis in their neighborhood. On the other hand, the usual method of stepwise regression operating by single variable addition from a simple model is a deterministic, conservative approach. But one drawback is that, each choice affects the following ones. If a variable is selected, all models that do not contain it are eliminated. Furthermore, only one combination is obtained with no choice for the user.

According to Rogers, if the fitness landscape shows local minima, classical techniques such as stepwise regression would be unable to reach the absolute minimum (unless it starts in its attraction basin) and remain stuck on the nearest local minimum. On the contrary, GA can recombine separately discovered partial solutions and give a population of solutions. The Hybrid GA of Leardi alternates generations of GA with cycles of stepwise selection (forward or preferably backward).

On the example of the anesthetic activity of 14 halogenated ethyl methyl ethers [164], a previous PLS analysis with 41 descriptors (log P, molecular volume, charges) led to a two-component model explaining 86% of the variance for the whole data set (75% in cross-validation). However, a cluster analysis showed that the descriptors could be gathered in nine groups, indicating strong similarities between them. A hybrid GA analysis in a L-20%-O cross-validation, only selected six variables (belonging to five different clusters) and yielded, with five components, a global predictive ability at least comparable to that obtained with all the variables. Another example concerned the activity of antifilarial antimycin analogues of the Selwood data set [897].

A.2.3.3   Genetic Partial Least Squares

In the genetic partial least squares (GPLS) approach [108,120], the efficiency of genetic function approximation (GFA) to generate models is combined with PLS analysis that is able to treat many more variables. PLS is used in place of MLR to determine the best coefficients in linear combinations of basis functions. Varied applications have been published. See, for example, Rogers and Hopfinger [108], LIV [364], RSM [366], or FEFF [824] models.

A.2.3.4   Genetic Neural Network

In a somewhat similar approach, GFA was combined with artificial neural networks, owing to the ability of BNNs to model nonlinear phenomena. Karplus et al. carried out various comparisons of performance for models implying GA or EP, using classical (substituent constants) [488], 2.5D descriptors [487,610] or similarity indices [48,49]. This latter study was developed in Chapter 5.

The first work concerned the well-known Selwood data set [897]. Thirty-one antifilarial antimycin analogues characterized by 53 physicochemical 2.5D descriptors (such as partial charges, superdecocalizabilities, volume, surface area, log P). Several QSARs were already published: forward-stepping MLR with three parameters (Selwood [897]), BNN (Wikel and Dow [898]). GFA led to several linear models of improved performance (Roger and Hopfinger [108]); and Luke [85] found, by evolutionary programming, some good models missed by GFA.

These studies focused on three-descriptor MLR. In a first step using the same descriptors, substituting BNN to MLR gave only a marginal improvement, but damaged the predictive power. According to the authors, this might be due to the fact that a linear decision function is reluctant to select descriptors involved in a nonlinear correlation with activity [487].

In a second step, EP or GA is used for selecting the best three descriptors to be input in a 3–3–1 BNN. This number maintained a good ratio (number of connections)/number of data points. GFA-NN and EP-NN discovered the same best model at the same evolution time. However, the average fitness was lower for GA-NN; this was explained by the elitist algorithm that retained the best model and eliminated during the evolution other relevant models. The best result obtained with both GA-NN and EP-NN yielded cross-validated correlation coefficient q = 0.866, slightly better than that obtained in the best MLR 0.849.

GA offers no guarantee that the best solution is found. But a comparison with an exhaustive search, carried out on more than 20,000 combinations of three descriptors among 53, established the efficiency of the EP-NN approach: the top-10 models being discovered from the 14th generation. A fitness function taking into account the error in prediction (on a very limited test set or by L-O-O) confirmed the efficiency of the approach.

As a last remark, GA often gives multiple models of comparable quality. Merging the predictions of high scoring models (consensus model) as proposed in Ref. [108] may in some cases (but not here) improve the results.

The GNN approach was extended [488] to a set of 57 1,4 benzodiazepin-2-ones binding GABA-R. To build the QSAR model, a back propagation neural network was used. However, to avoid a slow convergence and the risk of stopping in local minima, the usual steepest descent algorithm was replaced by the scaled conjugate gradient (SCG) method, which gave more homogeneous results in training and improved speed and stability [899]. Selection was carried out by Evolutionary Programming (EP) on a population of 200 individuals, evolving during 50 generations (a full L-O-O validation being performed only in the last round). Using 6 or 10 descriptors gave nearly similar results (q > 0.93), outperforming the previous results of Maddalena and Johnston [900].

To guide the search for new active compounds, it was interesting to specify the influence of a given substituent. This is not quite easy with BNN since units of different layers are strongly interconnected. Using the procedure due to Andrea and Kalayeh [94] (all but one input blocked at their value in a template, the remaining one scanning its variation range) suggested to the authors new active compounds by adequate replacement of substituents [488].

More extended comparative studies were further carried out on a set of 56 steroids binding in vitro the progesterone receptor (on a 2.8 log unit scale in RBA) characterized by 52 2.5D structural descriptors [609,610]. In a preliminary study, CoMFA outperformed a 10–10–1 BNN (that was clearly overfitted) [608]. In subsequent publications, diverse methods for descriptor selection were compared, on the same data set. The best results were obtained combining GFA regression for descriptor selection and BNN for model building.

GFA was started with 250 randomly generated equations (of length = 7). A total of 20,000 generations were performed. Basis functions (the genes) included not only linear, square functions, but also splines and squared splines functions. The back propagation neural network consisted of 10 input units, a single hidden layer with 1–10 units and one output layer of 1 unit. To avoid overtraining, a L-O-O cross-validation procedure was implemented. Variable selection was first tempted via stepwise regression, or PLS, but with bad results, particularly regarding prediction ($r^2_{pred} < 0.1$). However, performance is better if two compounds with unique structural features are discarded: $r^2_{pred} = 0.42$ (MLR) and 0.59 (PLS).

With GFA, 250 nonlinear models were obtained (for the best one $r^2 = 0.64$, but $r^2_{pred} = 0.49$ [for 11 compounds], which is only slightly better than the statistical techniques MLR or PLS). The authors concluded that one advantage of GFA was its ability to select relevant properties, even if they are not linearly correlated with the RBA values.

Comparison was pursued [610] on the same data set, but with only eight descriptors to maintain a 6/1 ratio between the number of patterns and descriptors. Several methods were used for descriptor selection. In addition to subjective selection from known models and forward-stepping regression, genetic algorithm methods were examined. In GFA regression, a genetic algorithm selected the descriptors input in a nonlinear regression (with, inter alia, splines). In two other models, data were mapped from a neural network, and descriptors selected either by simulating annealing (generalized simulated annealing, GSA), or by a genetic algorithm (genetic neural network, GNN). It was then established [610] that a 8–2–1 BNN combined to

GSA or genetic algorithm gave the best results, $q^2 = 0.717$ and 0.635 and $r^2 = 0.880$ and 0.860, respectively. The more limited performance of GFA may result from its difficulty to treat nonlinearity. However, GFA ($q^2 = 0.626$, $r^2 = 0.722$) or stepwise regression ($q^2 = 0.535$, $r^2 = 0.721$) may be useful in preliminary studies since easier to implement.

On the other hand, genetic algorithms often give several models of comparable quality, making possibly the interpretation less easy. It may be noticed that, here, five descriptors were common in the selection by GSA and GFA and that half of the descriptors retained concerned the "key" positions 11, 13, 17 of the steroid scaffold. Niculescu and Kaiser [611] compared these results to the performance of a probabilistic neural network (PNN) with the same descriptors. On learning, PNN ($r^2 = 0.863$) was comparable to GSA ($r^2 = 0.860$) and slightly lower than GNN ($r^2 = 0.880$) but a little better in test (0.769 vs. 0.488 and 0.610 on 10 compounds).

A.2.3.5 Genetically Evolved Receptor Models

This approach is a bit different since here GA selects interaction points and not structural descriptors. The method [393–395] aims to produce, at atomic level, models of receptor sites based on a small set of known structure–activity relationships. But it may be also used to derive QSAR from a set of ligands of known activity. The approach relies on evaluation of interaction energies between the ligands and a model of the receptor constituted by atoms of variable nature in the neighborhood of aligned ligands, see Section 3.5.5.

A.3 EMPIRICAL FORCE FIELDS

The techniques of empirical force fields applied in MM and MD constitute an appealing approach for determining the geometrical and energetic features of a molecule or a system of interacting molecules, in all problems where there is no explicit need for specifying electronic features. Indeed, the methods of quantum chemistry are of course quite suited for calculating or predicting the geometrical, electronic, and energetic features of a molecular system. However, many applications in molecular modeling are too large to be treated by quantum mechanical methods even at the approximate level of semiempirical methods. With carefully adjusted parameters, the empirical force fields give now high-quality results as to low-energy geometries and conformational preferences, with fast, nonintensive, calculations, whereas MD carries out dynamic conformational analysis, characterizing the evolution along time of energetic or geometrical features.

A.3.1 MOLECULAR MECHANICS

A.3.1.1 Foundations of Molecular Mechanics

Schematically, in empirical force fields, molecules are considered as composed of balls and springs (according to an idea traced back to Andrews from 1930 [901]). MM methods may be viewed as the other facet of the Born Oppenheimer separation of nuclear and electronic motions. Quantum mechanics looks at the electron repartition in the field generated by fixed nuclei. In MM, one examines the location of the nuclei in the field generated by the electrons. These are not explicitly considered but represented by an "effective potential" treated according to classical mechanics. The fundamental assumption of MM is that the potential energy of a molecule (near an equilibrium position) may be expressed as a sum of terms expressing (with respect to the equilibrium position) the deformations of bond lengths, valence angles, diehedral torsions, respectively, plus van der Waals interactions and, possibly, specific terms for H-bonds or electrostatic interactions. This relies on an ensemble of notions not too far from the usual chemist's intuition.

An important notion is that of "atom type." Different parameter values are given to atoms depending on their hybridization, their chemical functions or direct environment. For example, a carbonyl carbon is considered as different from a sp^3 carbon. This allows for reproducing the structural characteristics of a molecule without explicitly introducing electronic features. But more subtle distinctions are also possible. For example, junction-cycle sp^3 carbons differ from those of acyclic systems.

Schematically, the internal energy of a system will be evaluated (in internal atomic coordinates) by a Taylor's expansion around an equilibrium position. The *steric energy* calculated from the sum of these terms represents the additional energy associated with deviations from an ideal situation where all geometrical elements would be in a reference state and nonbonded interactions null. Numerical values for the necessary parameters are extracted from physicochemical studies (IR spectroscopy particularly) of small molecules, come from database-mining (interatomic distances, angular factors in H-bonded systems), or are derived from quantum calculations on small systems [242,902].

Aiming to reproduce or predict the different structural features of a molecular system with some a priori (and somewhat arbitrary) decomposition of the interaction mechanism and the use of simplified mathematical expressions obviously requires approximations. But various solutions may be elected depending on the importance attached to given physicochemical properties and the nature of the chemical species investigated. Note, however, that, over years, since Lifson and Wharshel [903] many efforts were dedicated to the development of a "consistent force field" that is a "universal" force field able to model the entire periodic table [904].

Several research groups or laboratories develop their own force fields and MM softwares with very variable diffusion. But, in addition to several simplified force fields incorporated in virtual screening or docking programs (see Chapters 8 and 11) various softwares are now available for extensive MM calculations since the pioneering work of Allinger [905,906] with the MM suite of programs MM2 to MM4 [907–910] and the extension MM+ [911,912].

Among other seemingly most widely used packages, we can cite: the TRIPOS force field [913], AMBER [788] with its various extensions (MD, QM-MM methods), CHARMM [914], YETI [819], CFF [902], OPLS [803], MMFF94 [242,915–918], or JUMNA [919]. For a critical review of some of these force fields and their applications to protein and nucleic acids, see Varma [920]. AMBER looks well adapted for condensed phase simulations and for the study of proteins and nucleic acids. It also (as CHARMM) easily incorporates MD applications. JUMNA is more oriented toward DNA derivatives study. Note that for the same package several force fields may be found for differing chemical systems In AMBER, GAFF [921] is applicable for small ligands and FF03 for proteins or macromolecular systems [922]. On years, different parameterizations were also proposed: see Ponder and Case, Cheatham and Young or Hornak et al. [923–925].

It must therefore be kept in mind that the force field (the ensemble of the necessary parameters) is calibrated to give a good result for the energy and the geometry of the system as a whole. It does not mean that each term (such as bond stretching, angle bending) is individually well evaluated. So caution is necessary when discussing of minute variations in a particular component of the force field!

Among promising developments (although not yet actually commonly used in QSARs) the mixed method QM/MM offers a powerful tool for studying chemical and biochemical processes, combining accuracy of QM and speed of MM. A localized region (something like the "reaction center") is treated with precise QM techniques but taking into account the influence of the surrounding (polarization effects, geometric restraints) via a MM treatment. This partition is made possible thanks to appropriate boundary conditions. The interest is that the heart of the system is treated at a high level (but taking into account more distant influences) by dint of a limited computational work [926,927].

A.3.1.2 Basic Components of the Force Field

As a very simple example, let us consider a diatomic molecule treated as a classical-mechanics system. For a small variation of the internuclear distance r around the equilibrium position r_0, the potential energy is expressed by

$$V(r) = V_0 + \frac{dV}{dr}(r - r_0) + \frac{1}{2}\frac{d^2V}{dr^2}(r - r_0)^2 + \cdots$$

If we choose a null value for the minimum V_0, and since at equilibrium the first derivative is zero, it comes (neglecting higher order terms)

$$V(r) = \frac{1}{2}\frac{d^2V}{dr^2}(r - r_0)^2$$

This corresponds to the largely used harmonic expression (Hooke's law in spectroscopy)

$$V(r) = \frac{1}{2}k(r - r_0)^2$$

where k is the stretching bond constant.

As previously stated, all MM force fields encompass inevitable terms associated to bond length, valence angle, torsion dihedral deformations plus nonbonded atom interactions, electrostatic contribution, and possibly specific terms (H-bonding, solvation). However, the corresponding mathematical expressions, the associated numerical parameters, and even the atom types may vary from one package to another one. We prefer here to limit ourselves to a general presentation not bound to a particular package, but indicate, some time, specific features.

$$V = V(r) + V(\theta) + V(\Phi) + V(nb) + V(elec) + (\text{specific terms})$$

Bond stretching is represented by the Hooke's law

$$V(r) = \frac{1}{2} \sum_i k_i (r - r_0)^2$$

where
   summation over all bonds i of reference-bond-length r_0
   k_i is the stretching force constant

Note that r_0 is not strictly the equilibrium bond length (as determined by RX or spectroscopy), but the (very neighbor) length value that will give the good geometry when all terms in the force field are evaluated.
   For *valence angle bending*, a similar harmonic expression is generally retained

$$V(\theta) = \frac{1}{2} \sum_\theta k_j (\theta - \theta_0)^2$$

sum over the j valence angles. For highly constrained cycles, this second order Taylor expansion would be insufficient. The harmonic formula is generally maintained but with adequate parameters k_j and θ_0.
   Bond torsion effects can be reproduced using a Fourier series

$$V(\Phi) = \sum_n \frac{V_n}{2} [1 + \cos(n\Phi - \delta)]$$

Generally, three terms are sufficient when nonbonded repulsions are taken into account.

$$V(\Phi) = \sum \left[\frac{V_1}{2}(1 + \cos\Phi) + \frac{V_2}{2}(1 - \cos 2\Phi) + \frac{V_3}{2}(1 + \cos 3\Phi) \right]$$

where Φ represents the dihedral angle, i.e., for a four-atom fragment ABCD, the angle between planes ABC and BCD (Figure A.4). In a Newman projection, along the central bond BC, Φ appears as the angle between the projection of the bonds AB and CD. According to the convention proposed by Klyne and Prelog, a positive value corresponds to a clockwise rotation of the first-named bond to bring it along the second one (it takes the same value when looking from one side or the other one).

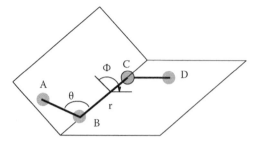

FIGURE A.4 Diheral angle (Φ) for four-atom chain.

For a symmetrical simple bond, the third term is sufficient and inclusion of the first and second ones only slightly improves the results. For a double bond (as in ethylene), the second term is preponderant. Note that this term alone privileges an eclipsed situation, without explicit treatment of the π electrons (Figure A.5). This three-term expansion allows for reproducing complex situations, as, for example, the torsional barrier in a O−C−C−O fragment (Figure A.6).

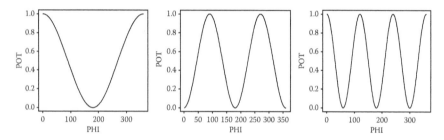

FIGURE A.5 Angular variation of torsional barriers of order 1, 2, 3.

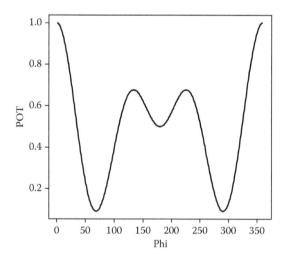

FIGURE A.6 Torsional barrier of a O−C−C−O fragment: Angular variation.

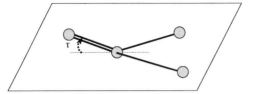

FIGURE A.7 Out-of-plane deformation.

The coplanarity around a double bond is also reproduced with quadratic term of *out-of-plane bending* (Figure A.7). Its inclusion is important in small strained cyclic systems (as cyclobutanone) to retain exocyclic atom in the plane of the cycle. Similar terms have been also introduced for *improper torsion* in nonsequential arrangement of four atoms, sometimes useful for maintaining the stereochemistry around a chiral center.

van der Waals interactions (between nonbonded atoms) are generally represented with a Lennard-Jones potential (Figure A.8)

$$V(nb) = \sum 4\varepsilon_{kl} \left[\left(\frac{\sigma_{kl}}{r_{kl}} \right)^{12} - \left(\frac{\sigma_{kl}}{r_{kl}} \right)^{6} \right]$$

where σ is the collision diameter (distance for which the energy is null). An alternative expression is

$$V(nb) = \sum \varepsilon_{kl} \left[\left(\frac{r_m}{r_{kl}} \right)^{12} - 2 \left(\frac{r_m}{r_{kl}} \right)^{6} \right]$$

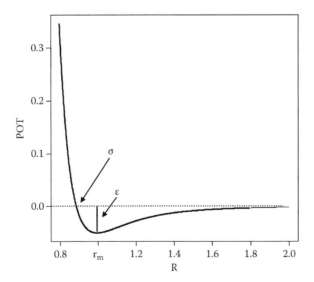

FIGURE A.8 Lennard-Jones-type potential energy curve for 1–4 interaction.

where

r_m corresponds to the energy minimum

ε_{kl} to the depth of the potential well, at the minimum $r_m = 2^{(1/6)} \sigma_{kl}$

Or alternatively

$$V(nb) = \sum \left\{ \frac{A}{r_{kl}^{12}} - \frac{C}{r_{kl}^{6}} \right\}$$

sum over all pairs of atoms (k,l) except those bonded to a same atom or a same bond (since already treated with angle bending and torsion).

For any pair of atoms, k, l, σ_{kl} is the half sum of the collision diameters or of the van der Waals radii of atoms k and l,

$$\sigma_{kl} = \frac{1}{2} (\sigma_{kk} + \sigma_{ll})$$

alternately

$$r_{m(kl)} = r_{m(kk)} + r_{m(ll)}$$

and

$$\varepsilon_{kl} = (\varepsilon_k \varepsilon_l)^{0.5}$$

where ε_k and ε_l represent the "hardness" of atoms k and l.

For large molecules, there would be many pairwise terms to calculate, often insignificant if atoms are far apart. Introducing a *cutoff* (eliminating the evaluation of nonbonded interactions at a long distance) makes the calculation faster. For neutral species, a cutoff of about 12 Å is sufficient, but with ions interactions are still sizeable at longer distances and specific methods may be used (Ewald series), see Section A.2.2.4.

On the other hand, in the *united atoms* option (as in AMBER), hydrogens (except those involved in H-bonds) are omitted. The heavy atom with its attached hydrogens is treated as a single entity with an adapted van der Waals radius.

Electrostatic interactions are expressed with a Coulombic potential

$$V(elec) = \sum \frac{q_k q_l}{4\pi\varepsilon_0 r_{kl}}$$

Various empirical schemes have been proposed for the derivation of partial atomic charges, for example, from a partial orbital atomic electonegativity equalization of

Gasteiger and Marsili [158–161]. Orbital electronegativity (a concept generalizing the atom electronegativity) is expressed by a polynomial function

$$\chi_{\mu A} = a_\mu + b_{\mu A} q_A + c_{\mu A} q_A^2$$

where coefficients a, b, and c have been determined for common elements from their usual valence state. The method of partial equalization of orbital electronegativity relies on a partial electron transfer from the less electronegative atom (A) to the more electronegative (B). The charge transferred is (at iteration k)

$$q^{(k)} = \alpha^k \left[\frac{\left(\chi_B^{(k)} - \chi_A^{(k)} \right)}{\chi_A^+} \right]$$

Where α is a damping factor and χ_A^+ is the electronegativity of the cation A⁺. With the damping factor, the charge transferred is reduced at each iteration, and convergence is rapidly obtained.

In MMFF94 [242], empirical charges are calculated as the sum of formal atomic charges, plus bond charge increments describing the polarity of bonds adjacent to the atom under scrutiny. Charges may be also derived from quantum mechanics calculations at a semiempirical level (AM1, PM3, for example, [162,163]). For a better estimation of electrostatic interactions, atom-centered point charges may be fitted on the electrostatic potential evaluated from the wave-function (the use of an extended basis set such as 6-31G* has been recommended). Least-squares adjustment is carried out on nodes on a grid embedding the molecule or (in AMBER) on spherical shells surrounding it at 1.4, 1.6, 1.8, and 2.0 times the van der Waals radii. The restrained electrostatic potential (RESP) variant [928,929] limits the overestimated influence of buried atoms. These RESP charges are conformation-dependent (although less than ESP charges). So, it is wise to use multiple conformations in the fitting process [930].

For a better representation, the AM1-BCC method combines charges calculated by AM1 and a bond charge correction (BCC) aiming at fitting the charges obtained with a training set of molecules studied at the higher HF/6–31G* ab initio level. This process associates the speed of semiempirical calculations and the accuracy of ab initio methods [931,932]. For extended systems, charges may be obtained from more limited subsystems. For polypeptides, it was thus proposed from the first versions of AMBER to calculate charges on residues (or preferably on dipeptide moieties better reflecting the immediate environment) and patch them together assuming additivity.

A.3.1.3 Specific Terms: H-Bonding and Polarizability Effects

Hydrogen bonding plays a preeminent role in processes involving biological molecules. Its effects are more or less expressed with the usual van der Waals and electrostatic components of the force field. But, for a better modeling, supplementary terms were proposed with a modified Lennard-Jones-type potential

$$V(hb) = \frac{A}{r^{12}} - \frac{B}{r^{10}}$$

to express interactions between donor and acceptor atom. The directional character of H-bonds was introduced in Goodford's force field GRID [58] as soon as 1985,

$$V(hb) = \left(\frac{C}{d^6} - \frac{D}{d^4} \right) \cos^m \theta$$

θ being the angle donor-H-acceptor. Directional effects have been also taken into account in the YETI software of Vedani [819] and in some docking and scoring programs (see chapter 11). *Metal ions* interactions (that intervene for example with Zn digits) are also considered in YETI with directional terms.

At least a contribution may be introduced for *polarizability* effects (distortion of the electronic distribution due to the field created by other molecules or other parts of the molecular system). Atom-centered dipole polarizabilities may be included in AMBER. See also refinements of fields in CoMFA models (Chapter 1). Considering, in MM treatments, polarization effects (that reduce electrostatic interactions) is also the subject of various developments regarding macromolecules and proteins.

Solvent effects may be treated with explicit solvent molecules (see "solvent box" in MD) or with a continuum model, using, for example, a distance-dependent dielectric constant $\varepsilon = r$ or with more refined models such as MM-PBSA or MM-GBSA. See Chapter 12.

A.3.1.4   Refinements of the Force Field

Applications to QSAR models or docking processes have been generally limited to such simplified force fields, which looks rather sound when examining *variations* within a homologous series of congeners. However, more complex functions have been proposed for a sharper description of the molecular structure. We shall not develop these aspects, a little apart from our concern.

For bond stretching, the Taylor expansion included *cubic* and *quartic* terms (although with a cubic, potential may tend to minus infinite if the starting point is too far from equilibrium). A *Morse potential* gives a more realistic representation of bond stretching, as evidenced in spectroscopic applications.

For valence angle deformation, terms up to $(\theta - \theta_0)^6$ have been proposed. Lennard-Jones potential can be replaced by *Buckingham* or *Hill* potentials (with an exponential repulsive part) [461, pp. 131], or by a "buffered 7–14" potential in MMFF94, which avoids V(nb) to grow up infinitely at very short distances (one recognizes a classical 7–14 formula if the terms r_{kl}^* are omitted in denominator).

$$V(nb) = \sum \varepsilon_{kl} \left[\frac{1.07\, r_{kl}^*}{(r_{kl} + 0.07\, r_{kl}^*)} \right]^7 \left[\frac{1.12\, r_{kl}^{*7}}{(r_{kl}^7 + 0.12\, r_{kl}^{*7})} - 2 \right]$$

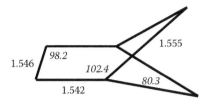

FIGURE A.9 Bonds lengthen when the valence angle decreases. Bond lengths in Å, valence angles (italics) in degrees.

Cross terms were introduced to correct the hypothesis of independent and additive contributions to the expression of potential function. For example, a term

$$\frac{1}{2}k_{i,j,\theta}(k_i\Delta r_i + k_j\Delta r_j)(\theta_{ij} - \theta_{ij,0})$$

where Δr_i denotes the length variation of bond i, accounts for the lengthening of bonds (r_i, r_j) adjacent to a decreased valency angle θ_{ij} (Figure A.9).

For *extended π systems*, the MMP2, MM3, and MM4 programs proposed to adjust force field parameters (stretching, angle deformation, torsion barriers) on bond orders determined by a crude quantum mechanics calculation on a planar model. However, with the increased performance of modern computers, a direct quanto-chemical calculation at a semiempirical level would frequently seem to be the best solution, if the size of the problem, is not too high.

Steric energy and heats of formation: Summation of the various energy terms gives the steric energy that characterizes the influence of geometrical deformations. This is sufficient for comparing conformations of a same molecule or *cis trans* isomers. But steric energy cannot be directly used for molecules with a different topology (and so with different atom types).

For a comparison with experimental data, *heats of formation* can be evaluated in some packages (as the MM programs) adding to the steric energy supplementary terms: Bond enthalpy increments for groups CH_3, CH_2, etc. (plus possibly specific terms for branched motifs), and a contribution of 2.4 kcal/mol for the translation/rotation term (contribution from partition function). Influence of low frequency torsional motions or of high energy conformers may be accounted for with an increment of 0.36 kcal/mol per rotational barrier lower than 7 kcal/mol.

A.3.1.5 Energy Minimization, Conformational Analysis, and Interaction Energy

The main applications, in connection with QSAR models, are conformational analysis (for isolated ligands) or determination of interaction energies ligand–receptor binding domain for example in docking. As to conformational preferences, one can note that torsional motions are significantly less energy-demanding than deformation of valence angles and, by far, modification of bond lengths. So, for QSAR applications, one can only examine variations of dihedral angles (possibly among predefined values followed by an ultimate refinement of the solutions). In docking, the problem is to determine the best "pose" that is the best position, orientation, and conformation of the ligand in the receptor site. Energy contributions involve the internal steric energy of both entities in the complex (taking into account adaptation of the ligand and possibly of the receptor) and their interaction energy. Desolvation effects during complex formation can be also considered with models such as MM-PBSA or MM-GBSA (see Chapter 12).

These two aspects come to an optimization problem: minimum steric energy in one case, maximum stabilization on the other one. Note that usual minimization programs stop at the first minimum encountered, which may be only a local minimum. So caution is necessary to find the true extremum. For most complex cases, grid searches (often slow working) may be replaced by stochastic methods (with or without minimization after each draw) or, if a good starting point is known, by MD simulation which gives a conformational profile and may indicate conformational transitions along time.

More generally speaking, MM was also used in very varied applications relevant of basic physicochemical problems. Stability of Anti-Bredt olefins, determination of conformational filiation or stabilomers, study of carbenium ions rearrangements or solvolysis reactivity constitute some of the applications. For more details see [461, pp. 151–167].

A.3.2 MOLECULAR DYNAMICS

MM evaluates, on a static point of view, the energy of a system due to internal deformations by respect to a reference state. MD focuses on the evolution with time and so generates successive conformations of a molecule (or a molecular system), which constitute a "trajectory." MD is thus well suited to study conformational transitions or molecular flexibility. See van Gunsteren and Berendsen [933] for a general overview of MD, and the review of Rognan [934] for applications in the protein field.

At the basis of MD, Newton's equation relies the acceleration γ_i of a particle i (mass m_i, position $\mathbf{x}_i$, velocity $\mathbf{v}_i$) to the applied force $\mathbf{f}_i$. These forces are intramolecular (isolated molecule) and intermolecular (molecule in a solvent or ligand interacting with a receptor).

$$\frac{d^2\mathbf{x}_i}{dt^2} = \gamma_i = \frac{\mathbf{f}_i}{m_i} \quad \text{with} \quad \mathbf{f}_i = -\frac{\partial V}{\partial \mathbf{x}_i}$$

where the potential V can be obtained by a MM force field. We do not consider here, in the framework of QSAR models, "quantum dynamics."

The evolution of the system is described by the finite difference method. Remember that velocity is the derivate of the position and acceleration (thus force) the derivate of the velocity or the second derivate of the position. Integration of preceding equation is carried out by successive short steps (time step δt). The state of the system at time $t + \delta t$ is calculated from its state at t. The force and other quantities, such as acceleration and velocity, are assumed to be constant during the time step. Schematically,

$$\mathbf{v}(t+\delta t) = \mathbf{v}(t) + \gamma \, \delta t \quad \text{and} \quad \mathbf{x}(t+\delta t) = \mathbf{x}(t) + \mathbf{v}(t) \, \delta t$$

At each step of the simulation (t to t + δt), the potential V(t) is calculated and its partial derivatives give the forces acting on the atoms (and so the accelerations). Then the new velocities (and the kinetic energy) and the new positions at the end of the time step are obtained. The potential at V(t + δt) is again evaluated. And this cycle is repeated again and so on.

So, a trajectory is generated, specifying how the characteristics of the system evolve with time. During the period of production (vide infra) one can store (and examine) the values of selected quantities (such as distances, angles). A sampling of the conformational space is so obtained allowing for characterizing *conformational flexibility* and creating a conformational energy profile (CEP) (see Chapter 9). Alternately, *conformational transitions* are picked up: a sudden variation of an interatomic distance or angle may indicate the formation of a hydrogen bond. At the end of the simulation, the *time-average of a property* P (taking into account for example the fluctuations in conformation or in solvent organization) is calculated by numerical integration

$$\langle P \rangle = \frac{1}{M} \sum_{i=1,M} P_i$$

where

M is the number of time steps

P_i is the property value at time step i

A.3.2.1 Integration and Time Step

Various integration methods are available, based on Taylor series expansions. We just develop here the *Verlet algorithm* [935] (probably, the most widely used in such simulations).

At time t + δt

$$x(t+\delta t) = x(t) + v(t)\delta t + \frac{1}{2}\gamma(t)\delta t^2 + \cdots$$

Similarly, at t − δt

$$x(t-\delta t) = x(t) - v(t)\delta t + \frac{1}{2}\gamma(t)\delta t^2 + \cdots$$

So that:

$$x(t+\delta t) = 2x(t) - x(t-\delta t) + \gamma(t)\delta t^2$$

Note that the velocity does not intervene. But it can be calculated by

$$v(t) = \frac{x(t+\delta t) - x(t-\delta t)}{2\delta t}$$

To start the algorithm (t = 0), it would be necessary to know $\mathbf{x}(t - \delta t)$. This may be guessed, using the same Taylor expansion, with

$$\mathbf{x}(0 - \delta t) = \mathbf{x}(0) - \mathbf{v}(0)\delta t + \gamma(0)\delta t^2$$

However, in Verlet algorithm, the position at $(t + \delta t)$ is obtained adding a small term $\gamma(t)\delta t^2$ to the difference of two large terms, and velocities are not directly attainable. An improvement is the "leap frog" algorithm [936].

$$\mathbf{x}(t + \delta t) = \mathbf{x}(t) + \mathbf{v}\left(t + \frac{\delta t}{2}\right)\delta t$$

$$\mathbf{v}\left(t + \frac{\delta t}{2}\right) = \mathbf{v}\left(t - \frac{\delta t}{2}\right) + \gamma(t)\delta t$$

The velocities at $(t + \delta t/2)$ are first calculated from velocities at $(t - \delta t/2)$ and acceleration at time t. Then, positions at $(t + \delta t)$ are obtained from positions at t and velocities just obtained. For the calculation of kinetic energy, $v(t)$ is given by

$$v(t) = 0.5\left[\mathbf{v}\left(t + \frac{\delta t}{2}\right) + \mathbf{v}\left(t - \frac{\delta t}{2}\right)\right]$$

This algorithm explicitly delivers the velocity and does not imply differences between large numbers. Unfortunately, velocities and positions are not synchronized (hence, the name of "leap frog").

The Velocity Verlet [937] directly uses

$$\mathbf{x}(t + \delta t) = \mathbf{x}(t) + \mathbf{v}(t)\delta t + \frac{1}{2}\gamma(t)\delta t^2$$

and

$$\mathbf{v}(t + \delta t) = \mathbf{v}(t) + \frac{1}{2}\delta t[\gamma(t) + \gamma(t + \delta t)]$$

Several other algorithms are available, see, for example, [763, pp. 315].

Choice of the time step is of crucial importance. Intuitively, it is clear that a short time step would allow a more realistic trajectory since the configuration of the system (potential energy, positions, and velocities of the atoms) will be frequently updated. A long time step makes a truncated Taylor expansion inexact and leads to less accurate trajectories. Too long a time step may also lead to instabilities and even "collisions." Remember that during the time step velocities are not updated. For example, for two particles i, j (atom i of the molecule under scrutiny and atom j of a solvent molecule), their positions at t + δt are determined by the initial positions x_i, x_j and the calculated displacement v_i δt and v_j δt. A long time step may bring (wrongly) the two atoms close together, hence a very high potential energy: the system "blows up."

On the other hand, to be able to describe the evolution of the system over a period long enough (!) to have a good picture of its flexibility or its interactions with a solvent, very short time steps would cause very long computation times.

For the applications considered here (small drug molecules alone, or possibly interacting with a receptor pocket), a common rule of thumb is to choose a time step about one order of magnitude smaller than the shortest motions. For organic molecules, it corresponds to C–H vibrations, with a wave number about $3000\,cm^{-1}$ that is a period about 10^{-14} s. So, a convenient time step δt would be 10^{-15} s = 1 fs. A trick to limit the computation time is to "freeze out" such vibrations (and let the other motions occur). This can be achieved, for example, with the SHAKE option for the Verlet algorithm, allowing time steps of, e.g., 2 fs [938].

A.3.2.2 Running an MD Simulation

MD simulations are usually performed in the *microcanonical* ensemble or constant NVE ensemble, i.e., constant number of particles, volume, and energy. An MD simulation generally encompasses three phases: preparation, evolution, and production.

The first part of the simulation corresponds to the *preparation* of the system: an initial configuration of the system is defined. For example, to avoid too strong repulsions, one may start from a minimum energy conformation (for an isolated molecule, possibly in a solvent box) or a "pose," ligand-schematized receptor, obtained by docking.

At temperature T, the total kinetic energy, for a system with N atoms, is

$$K = \sum_{i=1,N}\left[\frac{\left(|\mathbf{p}_i|^2\right)}{2\,m_i}\right] = \frac{kT}{2}(3N - N_c)$$

where

N_c is the number of constraints limiting the total number of degrees of freedom
$\mathbf{p}_i$ is the momentum of atom i

Initial velocities may be chosen from a uniform distribution or at random using a Maxwell-Boltzmann distribution. For atom i (mass m_i), the probability to have velocity v_{ix} in direction $\mathbf{x}$ is

$$p(\mathbf{v}_{ix}) = \left(\frac{m_i}{2\pi k_B T}\right)^{0.5} \exp\left[-0.5\frac{m_i v_{ix}^2}{kT}\right]$$

Initial velocities may be chosen so that for an isolated system (simulation in vacuo) the total translational momentum and the total angular momentum are null. This corresponds to subtraction of 6 degrees of freedom. In a simulation with boundary conditions (in a water box, for example [vide infra]) only the center of mass is constant (3 degrees of freedom): the translational momentum remains null, but not the angular momentum. To introduce kinetic energy, in other words, to give velocities to atoms; it is often recommended to "heat" the system progressively.

Then comes the *evolution* (or *equilibration*) phase. Velocities assigned to atoms do not generally correspond to a realistic situation. It is necessary to let the system come to an equilibrium state. This will be determined observing the evolution of characteristic properties. The total energy (kinetic plus potential) must remain fixed in an isolated system (but the kinetic and potential energy contributions will individually fluctuate), a drift indicating a bad equilibration. Velocities should have a Maxwell-Boltzman distribution along the three axes. To maintain a given temperature, velocities may be scaled. This equilibration phase is also mandatory when a molecule is surrounded by solvent molecules.

When the system is stable, the *production* phase begins, according to the same updating cycle. At regular intervals, for example, every 500 time steps, various quantities such as positions, energies are stored for further analyses. Depending on the problem, properties (or characteristics) of the system are then calculated, regarding, for example, conformation, existence of H-bonds, interacting residues in a ligand–protein complex. The stability of the system can be checked looking at the root mean squares mean deviation (rmsd) of the molecule (or the complex) with respect to the reference starting structure.

Typically, with limited computer resources, for a small drug in a receptor binding site, a simulation during 5 ns with a time step of 2 fs implies 2.5×10^6 calculation steps. Picking up information every ps during the last half of the trajectory gives 2500 snapshots on which property values will be extracted. It is remarked that such a duration would be quite insufficient to characterize slow motions of proteins.

A.3.2.3 Controlling Temperature

It is sometimes useful to monitor the temperature, for example, in a *simulated annealing* process (to obtain an absolute energy minimum without getting trapped in a local minimum), or to observe how a property may change with temperature. In an NVE simulation, it is often recommended to adjust the temperature at the desired value (but no adjustment will be carried out in the production phase and the temperature becomes a calculated property).

A possible way is to scale the velocities. The time-average kinetic energy (for N_f degrees of freedom) is

$$\langle K \rangle = \frac{1}{2} N_f k_B T$$

At each step, velocities are scaled by

$$\left[\frac{T_{new}}{T(t)} \right]^{0.5}$$

where $T(t)$ is the current temperature calculated from $\langle K \rangle$ at time t.

Another way is to "couple" the system with a *thermal bath* allowing for heat exchanges. The change in temperature at step t is given by, according to a formal, first-order kinetics:

$$\frac{d[T(t)]}{dt} = \tau^{-1}[T_{bath} - T(t)]$$

so that the change in temperature during step (t → t + δt) is

$$\Delta T = \frac{\delta t}{\tau}[T_{bath} - T(t)]$$

specifying the velocity-scaling factor. A small rate (τ) is equivalent to velocity scaling. A high τ (weak coupling) would lead to large T fluctuations. With a time step of 1 fs, τ = 0.4 ps has been suggested [763, pp. 343].

A.3.2.4   Incorporating Solvent

In the field of QSARs, the most common situation corresponds to a ligand or a complex ligand–protein immersed in water. The simplest way to take into account solvent influence (implicit solvent) is to use the dielectric "constant." A distant-dependent dielectric (ε = r), although without any physical meaning, allows for reducing long-range effects. A sigmoidal variation from a weak value ε = 1 or 4 in the vicinity of the molecule to a high value (about 80 for bulk water at long distance) have been proposed, for example, in JUMNA [919]. A more evolved, but more intensive treatment corresponds to "explicit solvent" with *periodic boundary conditions* and the *minimum image convention*.

- *Water models*: The first step is the definition of a model for water molecules. Various solutions have been proposed. The simplest is the TIP3P model of Jorgensen et al. [939]. The SPC/E model of Berendsen et al. [940] is somewhat similar. In TIP3P, the water molecule is supposedly rigid and schematized by three sites (Figure A.10). Only oxygen intervenes for the evaluation of van der Waals' terms, whereas the three sites are involved in electrostatic interactions. More sophisticated models have been also proposed, involving a negative charge slightly shifted from the oxygen location (TIP4P) on the bisector of the HOH angle, or replaced by two fictive charges in the direction of the lone pairs (model ST2).

- *The solvent box*: Schematically, the molecule is embedded in a (say, cubic) box filled with water molecules at an adequate density, with possibly some counter-ions to maintain the neutrality of the system. Simulation takes into account forces internal to the solute molecule (or the complex) and interactions solute–water molecules. For the initial configuration of the system, molecules are scattered at random

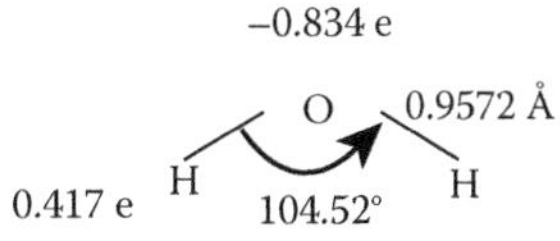

−0.834 e

O   0.9572 Å

H   H
0.417 e   104.52°

$10^{-3}A = 582.0$ Kcal Å^{12}/mol
$C = 595.0$ Kcal Å^6/mol

FIGURE   A.10   The   TIP3P water model. A and C denote the Lennard-Jones coefficients.

or reproduce the organization of common crystal lattices (such as face-centered cube). In the first phase, this solvent box is equilibrated. Then the solute is added (water molecules too close to the solute are removed). And the simulation itself begins with equilibration and then production phases.

- *Periodic boundary conditions*: This is a trick to simulate the influence of a bulk solvent with only a limited number of water molecules. Translating the solvent box in each dimension (x, y, z) gives image boxes reproducing the bulk. Note that if a molecule leaves the box during the simulation another one enters by the opposite face. The number of particles in the (central) simulation box is thus constant. Other box shapes than cubic may be used (chopped octahedron, for example). The condition is that they fill all space by translation (Figure A.11).

Looking at ligand–protein interactions would lead to a huge number of atoms to consider. So, it is wise to examine only the receptor binding site. Only residues closer to 10 or 12 Å from the solute will be treated (this corresponds to a limit for sizeable nonionic interactions). Atoms further apart may be discarded (cutoff) or, in order to maintain the cohesion of the receptor, constrained to their initial position (adding to the force field a supplementary potential or artificially increasing the mass of the concerned atoms).

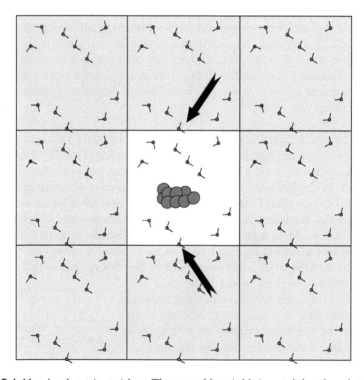

FIGURE A.11 A solvent (water) box. The central box (white) containing the solute, is surrounded by its images. Arrows indicate a water molecule entering the box (upper side), while another one goes out on the lower side. Note that in usual conditions, a box of 50 Å edge contains about 4000 water molecules.

- *Cutoff and minimum image convention*: Nonbonded (van der Waals) terms are, by far, the more time-consuming steps on account of the number of pairwise interactions, but many of these terms are negligible, a consequence of the r^{-6} dependence of the attractive part of the van der Waals function. An easy way to avoid calculating many negligible terms is to fix a nonbonded cutoff value, with the minimum image convention. Each atom sees at the most every other atom or its image in the box (but never the two at a time). Interactions for distances larger than the cutoff values are set to 0.

  These conditions imply that the cutoff value must not be larger than half the box edge. An edge of 30–50 Å is often considered as sufficient, for neutral systems. But ion–ion interactions (force varying in r^{-2}) fade more slowly with distances and a larger box may be necessary, which would multiply the computation time.
- *Particle Mesh Ewald (PME) method*: PME summation aims to provide an easy way to rapidly evaluate interaction energy between particles in a system of periodic symmetry [941,942]. We will take the example of the "solvent box" just presented. The basic idea is to separate the contributions into
  1. A short-range term, easily evaluated
  2. A long-range term, corresponding to interactions of charges in the central unit cell with the charges in the surrounding cells

The trick is that this term can be evaluated from the Fourier transforms of the potential and of the charge density field and that these transforms converge rapidly (limiting the number of terms to calculate).

Note that PME method [941] requires a neutral system. Counter-ions must therefore be added. Other methods have been proposed, such as reaction field, image charge methods or cell multipole method. For more details, see Leach [763, pp. 294–302].

A.4 THE MATHEMATICAL-STATISTICAL TOOL KIT

The most widely used mathematical model in QSARs, to associate a biological activity to a structural description of a set of molecules, is undoubtedly the MLR. However, problems appear when collinearity between variables occurs or when the number of variables exceeds the number of samples. In such situations, solutions exist: Principal component regression (PCR), and the neighboring method partial least squares (PLS) regression. In addition, nonlinear methods were further introduced as, for example, the parabolic relationship proposed by Hansch and Clayton [27]. Beyond some successes these polynomial-type relationships were not widely used and a large number of more flexible representations were introduced. Neural networks and their different architectures are model-free tools. Support vector machines (SVM) or projection pursuit regression (PPR) operate by projection in spaces of enlarged or restrained dimensionality. Multivariate adaptive regression splines (MARS) or locally weighted scatterplot smoothing (LOWESS) correspond to local fitting devices.

Generally speaking, for a population of compounds I, activity y_i (log $K_{binding}$, pIC_{50}), the biological response (the quantity to "explain" or predict) is expressed in a vector $\mathbf{y}$ (component y_i), whereas molecular descriptors are ordered in a matrix $\mathbf{X}$ (elements x_{ir} descriptor r for compound i) where columns are associated with a type of descriptors and rows to an object (a molecule). $\mathbf{y}$ is the vector of the dependent variable and $\mathbf{X}$, the matrix of the independent variables (or predictors). We suppose that the $\mathbf{X}$ matrix (as often done) has been submitted to column auto-scaling, so that, for each column r (descriptor r), the mean is 0 and the variance is unit, which gives the same importance to all descriptors. We suppose also that the response y (the biological activity) is uni-variate (a unique response per compound), a common situation in QSARs.

In the following pages, matrices are represented with bold upper-case letters with their elements described as (row, column), vectors by bold lower-case letters and scalars by usual characters.

A.4.1 MULTILINEAR REGRESSION

A.4.1.1 Proportionality Relationships

MLR is the simplest of the statistical models currently used in QSARs. Suppose, a population of n structures i (i = 1 … n), each represented by m variables $(x_{i,1}; x_{i,2}; x_{i,3} \ x_{i,m})$, for which we search for a linear relationship

$$y_i = b_1 x_{i,1} + b_2 x_{i,2} + b_3 x_{i,3} + \cdots + b_m x_{i,m} + e$$

that can be written as

$$\mathbf{y} = \mathbf{Xb} + \mathbf{e}$$

where
   $\mathbf{y}$ is the column vector of activities
   $\mathbf{X}$ is the matrix (n,m) of descriptors
   $\mathbf{b}$ is the column vector of the coefficients (weight or slope)
   $\mathbf{e}$ is the residual (error vector)

If the matrix is not normalized, the same formula applies, provided a column of 1 is added to the descriptor matrix to account for an intercept.

If n = m, one exact solution exists. If n > m, there is no exact solution. An approximate solution is searched for by minimizing the square of the residual vector $\mathbf{e} = \mathbf{y} - \mathbf{Xb}$, and the solution is

$$\mathbf{b} = (\mathbf{X'X})^{-1}\mathbf{X'y}$$

The symbol $'$ indicating the transpose matrix. The estimated values for $\mathbf{y}$ are given by

$$\mathbf{y}_{est} = \mathbf{Xb}$$

(in the following, we will some time use y_{fit} for values calculated for the training set during the model is built up and y_{pred} for values predicted for unknowns, not considered in building the model).

If **X** is singular (or nearly singular), that is if it contains variables (closely) intercorrelated, the method fails in the inversion of ($X'X$).

A.4.1.2   Quality of a Model

The aim of QSARs is ultimately to propose a calculation or a prediction (for an unknown) of the biological activity from a set of structural descriptors This comes to a comparison of estimated values (y_{est}) vs. the experimentally observed one (y_{obs}). The crucial question is therefore, what trust may we have in the calculated values? And what about validation of the model?

The concept of quality of a model encompasses varied facets.

1. On the first point, the *accuracy* of the representation. This may be easily quantified with usual statistical criteria (correlation coefficients, Fisher test and so on) and, for the end-user, the precision of the prediction.
2. But other important questions are
   a.   Sure, the model correctly represents the data it treated, but what about *predictions* on unknown compounds, not used in building up the model?
   b.   The model is built with a given data set. Is it *applicable* only for other congeneric compounds or may it be *extended* to structurally different chemicals?

We just consider here the statistical criteria and discuss further the last point. Among the most commonly used criteria to testify of the quality of the correlation are

- The *correlation coefficient* r (more precisely its square, the determination coefficient r^2)

$$r = \frac{\sum (y_{i,obs} - y_{obs,mean})(y_{i,fit} - y_{fit,mean})}{\left\{ \left[\sum (y_{i,obs} - y_{mean})^2 \right] \left[\sum (y_{i,fit} - y_{fit,mean})^2 \right] \right\}^{0.5}}$$

summation running on the n compounds, where $y_{i,fit}$ and $y_{i,obs}$ are the calculated and the observed property values respectively in the training set. An alternative expression is [487]

$$r = \frac{\sum (y_{i,obs} - y_{obs,mean})(y_{i,fit} - y_{fit,mean})}{\left[\left(\sum y_{i,obs}^2 - ny_{mean}^2 \right) \left(\sum y_{i,fit}^2 - ny_{fit,mean}^2 \right) \right]^{0.5}}$$

- The leave-one-out (L-O-O) *cross-validated correlation coefficient* (one compound is in turn left out. The model is built on the remaining compounds and is applied to predict the activity of the compound currently discarded). In this way, each molecule is predicted once as though the system had never seen it before

$$q^2 = 1 - \frac{\text{PRESS}}{\text{SSD}}$$

with PRESS (predicted residual sum of squares)

$$\text{PRESS} = \sum (y_{i,\text{pred}} - y_{i,\text{obs}})^2$$

and

$$\text{SSD} = \sum (y_{i,\text{obs}} - y_{\text{mean}})^2$$

$y_{i,\text{pred}}$ is the value predicted for the compound left out and ignored when building the model.

Remark that q^2 may be negative (!!!). This means that a "no model," taking the mean of the values, gives better predictions than the proposed model.

When a fraction of the data set (e.g., 10%, 20%) is left out and used as an external set to test the model defined without them (Leave-Some-Out processes [L-S-O]), the "external" determination coefficient r^2_{ext} is calculated, as r^2, with SSD determined on the external set considered. In another formulation, r^2_{pred} is evaluated, as q^2, with SSD measured with respect to the mean y value for the *training set*.

- The root mean squared *standard error of estimation* (or calibration) in training

$$\text{RMSEC, possibly called SPRESS (or } \sigma) = \left[\frac{\text{PRESS}}{(n - a - 1)} \right]^{0.5}$$

a is the number of adjustable parameters, latent variables etc. (degrees of freedom frozen by the model). It is a root mean squared error *corrected* by the complexity of the model

- But, in prediction, the (often-called) "Standard Error in Prediction" RMSP (or RMSEP, or SDEP) is, in fact, the root mean squared error of prediction

$$\text{RMSP} = \left[\frac{\text{PRESS}}{\text{nt}} \right]^{0.5}$$

where nt is the number of compounds of the test set.

However, Polanski introduced the *cross-validated standard error*

$$RMSECV = \left[\frac{PRESS}{(nt-a-1)} \right]^{0.5}$$

where the complexity of the model is taken into account.

- The *Fisher test*

$$F = r^2 \frac{[n-a-1]}{a(1-r^2)}$$

is easily calculated from r^2, n, and a values.

- Other metrics have also been proposed to assess the quality of a model. So, Peterson [141] proposed a composite score function:

$$Score = 5r^2_{pred} + 4q^2 + 2r^2$$

A similar function was used by Schultz et al. [169], see also Kolossov and Stanforth [943]. Another quality criterion was formulated by Akaike: The Aakaike information criterion (AIC) weights differently the goodness of the fit and the number of parameters introduced [659,944,945].

$$AIC = RSS \frac{(n+a)}{(n-a)^2}$$

In variable reduction, it was found that the results depend on the criterion: using Fisher test, few variables are selected, whereas with the SD, too many variables are maintained. So, Kubinyi [106,107] introduced, as a good compromise, the FIT criterion

$$FIT = \frac{r^2}{(1-r^2)} \frac{(n-a-1)}{(n+a^2)}$$

We would, however, stress that, in addition to these well-known statistical criteria, and on a very pedestrian point of view, it may be crucial to compare the "error" of the model to the *activity range* spanned by the data set and the *experimental uncertainty*.

A.4.1.3 q² or Not q²? What Criterion for Prediction?

It is well accepted that r^2 measures how well a model fits the data it was built from. But L-O-O cross-validation has been frequently questioned as a criterion of predictive ability. It is obvious that a low value of q^2 means poor predictions. A q^2 value significantly lower than r^2 also clearly indicates an overfitted model [483]. But a high q^2 is only a necessary but not sufficient condition for a good prediction. In their famous paper "beware of q²," Golbraikh and Tropsha [113] concluded that "often, a high value of this statistical characteristic ($q^2 > 0.5$) is considered as a proof of the high predictive ability of the model …, this assumption is generally incorrect." See also Aptula et al. [114]. It is now generally admitted that q^2 corresponds to some *interpolation* inside the data set, and characterizes the *robustness* of the model rather than its ability to prediction. Note also that s_{pred} values are not a reliable measure of predictive ability.

To define better criteria, Golbraikh and Tropsha considered regression lines passing through the origin for the pairs (y_{cal}, y_{obs}), and they defined as criteria of high predictive power:

- A high value of r^2 (q^2)
- A r^2 value close to 1 and at least one (but better both) correlation coefficient for regressions through the origin (y_{pred} vs. y_{obs} or y_{obs} vs. y_{pred}) close to r^2
- At least one slope of the regression lines through the origin close to 1

Recently, Roy and Roy [881] proposed to examine the difference between r^2 and r_0^2, the correlation coefficient for the regression through origin (y_{pred} vs. y_{obs}).

$$r_m^2 = r^2[1 - (r^2 - r_0^2)^{0.5}]$$

Good values correspond to $r_m^2 > 0.5$.

However, although these criticisms, q^2 values are very often reported, since these values, which perhaps are not the best criteria, are strictly reproducible [458], and allow for easy performance comparisons (which is not actually true with different L-S-O assays, vide infra).

Some authors proposed thus to carry out cross-validation by L-S-O (rather than L-O-O); and randomly split the data set into a training and a test set. However, it is not always without problems. If too many objects are left out, the training data no longer span the entire data space and some information is lost to build the model. If too few objects are left out, overfitting may occur since the construction data and the validation data are too similar. It was observed that overfitting can occur in two different ways: on the one hand, uninformative variables are included in the model and on the other hand, too many latent variables are selected. This situation may induce "chance correlation," an artifact highlighted as early as 1972 [77]. In the case of overfitting, results can be excellent in training but poor in prediction.

Note that Hawkins et al. stressed that, when the analyzed data set is small, holding a portion of it back in order to create an "external" test set is wasteful. *Cross-validation is better, if done properly.* The hold-out compound must not be somehow reflected by one of the remaining compounds used in the model [946].

On the other hand, as stated by Polanski, such a random splitting (training/test) on limited data sets may lead to largely differing and confusing conclusions. The authors revisited the steroid benchmark and examined the reliability and consistency of the process (training/test in the ratio 21:10 as commonly performed). This would amount to 44,352,165 possible different experiments. Considering only each one of five possible assays reduces this number to 8,870,433 combinations for which q^2 vary from −0.16 (!) to 0.95 [363]. On the same data set, SOM-COMSA shows less instability ($q^2 = 0.42$–0.98).

To avoid chance correlation (which may occur in L-O-O) Baumann [116,433,434] proposed a Leave-50%-Out with a random splitting into training and test sets, but with 3n assays if n is the number of compounds. In a plot (observed values vs. calculated values), for each compound, the spread of the 3n/2 predicted values in the various assays, and the position of this segment with respect to the correlation gives a good visual indication on the quality of the model [116].

An interesting approach as to external validation was that of Waller et al. [173]. They studied 55 compounds binding the estrogen receptor (on a 6.5 log unit range in RBA) and belonging to eight structurally diverse subsets (phenols, DDTs, DESs, PCBs, phthalates, phytoestrogens, steroids, pesticides). External predictive ability was examined excluding in turn a structural family and training the model on the other ones. The good results obtained suggest that the training sets already encompassed a sufficient structural diversity to be able to treat new families.

In randomization tests [48,472] (shuffling, scrambling), the elements of the response vector are shuffled by about 100 random exchanges, and models are sought for with the re-ordered responses. If satisfactory correlations are obtained, the significance of the QSAR must be suspected: the method would be potentially able to model any kind of data. Of course, the point is not to test the model obtained from the original data, but to entirely rebuild models from the permuted data.

It must also be clear that a model is built and validated on compounds spanning a defined part of the chemical structural space. But nobody can assess that this model would be acceptable for new compounds lying out of this subspace. At the most, it can be hoped that, beyond structural interpolation, a limited extrapolation will be reasonable.

As to splitting a unique data set into training and test, how to avoid luck or bad luck? A structural analysis of the data set may give some clues. It is wise that different chemical families are represented in both training and test sets. This can be ascertained by clustering compounds from their structural descriptors. Various classification methods can be used such as principal component analysis (PCA), k-nearest neighbors (k-NN), or self-organizing maps (SOM).

A.4.2 PROJECTION METHODS: PCA, PCR, AND PLS

MLR cannot work if there are more descriptors (independent variables) than compounds (y_{exp} dependent variable). Furthermore, if there is some approximate collinearity between the descriptors, the coefficients in the MLR are meaningless as to interpretability. To cope with data defined in a high-dimensional space [y, ($x_1, x_2, ..., x_m$)] a possible solution is to project the initial space onto a lower dimensional space with well-chosen axes. Principal component analysis (PCA) and the derived PCR or PLS methods belong to such approaches.

A.4.2.1 Principal Component Analysis

PCA is typically an unsupervised classification method (qualitative SAR), which, inter alia, allows for selecting the more relevant variables (neglecting details and noise) and so constitutes a possible solution for descriptor selection. It is also at the basis of PCR. From a qualitative point of view, PCA may be considered as a mean to translate a complex data matrix in a simpler representation retaining the main information [947].

Basically, the elements $x_{i,r}$ of **X** can be viewed as the co-ordinates of points i in a space defined by m axes $(x_1, x_2, x_3, \ldots, x_m)$. PCA aims to replace this original reference axes by a reduced set of new axes: the *principal components* and characterizes the data points by their projections onto these principal components. These new axes are built as a linear combination of the old ones, and are chosen mutually orthogonal, eliminating problems of collinearity between descriptors. The first principal component PC_1 is defined in the direction of the maximum variance of the whole data set, PC_2 as the direction describing at best the remaining variance in the subspace orthogonal to PC_1, etc. [166] (Figure A.12). Each new axis is generated into such an order that it explains at best the residual variance not yet explained by the preceding PCs. Clearly, if the number of PCs considered is equal to the number of descriptors, the process only amounts to changing the reference coordinate axes. But the interest is that, in many cases, the relevant information is mainly concentrated in the first PCs, the following ones referring only to details, and noise. The number of PCs to retain for accounting for experimental data depends on the application. This will be discussed later.

The search for these principal components comes to a problem of eigenvalues, eigenvectors. Let $X(i,r)$ the auto-scaled data matrix (n,m) where each of the n compounds is defined by m descriptors. The data matrix **X** multiplied by its transpose **X′** gives the (square) correlation matrix **C**

$$\mathbf{C} = \mathbf{X'X}$$

The search for the eigenvalues and eigenvectors of **C** leads to

$$\mathbf{CP} = \lambda\mathbf{P}$$

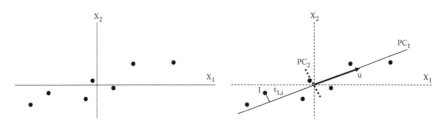

FIGURE A.12 Principal component analysis. 2D scheme. The proximity between points can be exactly determined in the X_1, X_2 system (left). A good approximation may be obtained (right) looking only at the projection of the points onto the direction of maximum variance (PC_1). If the second PC is considered (dotted bold line) it would come to a change in axes. $t_{i,1}$ is the first score for point I. The loadings for PC_1 are the angle cosines of the direction vector u with respect to X_1 and X_2.

where
  λ is the vector of the eigenvalues
  $\mathbf{P}$ is the eigenvector matrix

$$\mathbf{P}^{-1}(\mathbf{X}'\mathbf{X})\mathbf{P} = \lambda$$

Since the eigenvectors are orthogonal, this is equivalent to

$$\mathbf{P}'\mathbf{X}'\mathbf{X}\mathbf{P} = \lambda$$

or

$$\mathbf{T}'\mathbf{T} = \lambda$$

with

$$\mathbf{T} = \mathbf{X}\mathbf{P}$$

So, the original data matrix can be written as

$$\mathbf{X} = \mathbf{T}\mathbf{P}^{-1} \quad \text{or} \quad \mathbf{X} = \mathbf{T}\mathbf{P}'$$

The original data matrix $\mathbf{X}$ has been factored in two matrices:

- $\mathbf{T}(n,m)$ the "score matrix" characterizing the objects (compounds) in the new coordinate system
- $\mathbf{P}^{-1}(m,m) = \mathbf{P}'$, the "loading matrix" whose columns describe the new axes by respect to the old axes

This decomposition is strictly exact (it is just a change in the reference axes). But the trick is that the relation $\mathbf{X} = \mathbf{T}\mathbf{P}'$ may often be written in a simplified form by deleting rows from $\mathbf{P}'$ (and columns from $\mathbf{T}$); and retaining only a few components (those associated with the highest eigenvalues). Indeed, it can be shown that the eigenvalues λ indicate the relative importance of the rows in $\mathbf{P}'$ and so suggest how many rows can be eliminated without leading to large errors. This simplification allows for neglecting unimportant variables, and noise but, possibly, by dint of a loss of information. A simple rule of thumb is that the number of rows to retain corresponds to the number of independent factors needed to reproduce the initial data within a given error limit (e.g., the experimental uncertainty).

$$\mathbf{X} = \mathbf{T}_{(i,s)}\,\mathbf{P}'_{(s,m)} \quad \text{with } s < m$$

In QSARs, the first principal components may be considered as a good representation of $\mathbf{X}$. The loading vectors indicate what factors are important (large loadings).

As to score vectors, they contain information about the relative proximity of the compounds (similarity or dissimilarity).

A.4.2.2 Principal Component Regression

The basic idea of PCR is to perform a linear regression of $\mathbf{y}$, but using the score matrix $\mathbf{T}$ to represent the original data matrix of descriptors $\mathbf{X}$, and to retain only a limited number of components of this score matrix. The property $\mathbf{y}$, a column vector with one component, activity, per compound is so predicted using a limited number of principal components of the descriptor matrix

$$\mathbf{y} = \mathbf{Tb} + \mathbf{e} = b_1' \mathbf{t}_1 + b_2' \mathbf{t}_2 + \cdots + \mathbf{e}$$

with no intercept if data are auto-scaled. $\mathbf{e}$ is the error vector. As in MLR, the solution is

$$\mathbf{b} = (\mathbf{T'T})^{-1} \mathbf{T'y}$$

PCR replaces the original descriptors by new ones with better properties (orthogonality) and that span the multidimensionality of $\mathbf{X}$. Inversion de $\mathbf{T'T}$ now poses no problem (since scores are orthogonal). It may be interesting to express the regression in terms of the original variables ($\mathbf{X}$)

$$\mathbf{y} = \mathbf{Xb}_{PCR} = \mathbf{Tb}$$

With $\mathbf{T} = \mathbf{XP}$, one solution for $\mathbf{b}_{PCR}$ is therefore $\mathbf{Pb}$. But $\mathbf{X'X}$ can be close to singular, the solution for $\mathbf{b}$ is hardly unique, which may create problems [166]. See also Ref. [948] for a neighboring situation.

A.4.2.3 Nonlinear Iterative Partial Least Squares

We addressed PCA using diagonalization of $\mathbf{C}$, the correlation matrix. But other methods exist as single-value decomposition, or nonlinear iterative partial least squares (NIPALS) method. We just indicate here the principle of NIPALS, from Ref. [165], since it is not so far from the very widely used PLS approach. NIPALS calculates the scores and loadings sequentially.

The data matrix X will be written as the sum of matrices of rank 1

$$\mathbf{X} = \mathbf{M1} + \mathbf{M2} + \mathbf{M3} + \mathbf{Mr}$$

Each of these matrices is the outer product of two vectors: scores $\mathbf{t}_h$ (column vector of size n*1) and loading $\mathbf{p}_h'$ (row vector, size 1*m)

$$\mathbf{X} = \mathbf{TP'}$$

$\mathbf{P}'$ is made of the $\mathbf{p}'$ as rows and $\mathbf{T}$ from the $\mathbf{t}$ as columns. Starting from a vector $\mathbf{x}_j$ (size 1*m) the algorithm calculates the first terms $\mathbf{P}'_2$ and $\mathbf{t}_1$ from $\mathbf{X}$. Then the product $\mathbf{t}_1\mathbf{p}'_1$ is subtracted from $\mathbf{X}$ and the residue ($\mathbf{E}_1 = \mathbf{X} - \mathbf{t}_1\mathbf{p}'_1$) used to calculate the second terms $\mathbf{t}_2\mathbf{p}'_2$ and so on. Schematically, in the current step, NIPALS uses the following sequence:

One column vector of $\mathbf{X}$ is chosen $\mathbf{x}_j$

Let

$$\mathbf{t}_h = \mathbf{x}_j \qquad \mathbf{t}_h(n*1)$$

$\mathbf{P}'_h$ is calculated as

$$\mathbf{p}'_h = \frac{\mathbf{t}'_h\mathbf{X}}{\mathbf{t}'_h\mathbf{t}_h}$$

and normalized

$$\mathbf{p}'_{h,new} = \frac{\mathbf{p}'_{h,old}}{\left\|\mathbf{p}'_{h,old}\right\|}$$

$\mathbf{t}_h$ is calculated

$$\mathbf{t}_h = \frac{\mathbf{X}\mathbf{p}_h}{\mathbf{p}'_h\mathbf{p}_h}$$

If the value is the same as that chosen, the process stops (convergence). If not, one goes to the second step, calculation of $\mathbf{P}'_h$, and so on. After the first components ($\mathbf{t}_1$, $\mathbf{p}_1$) are obtained, $\mathbf{X}$ is updated and replaced by its residual

$$\mathbf{E}_1 = \mathbf{X} - \mathbf{t}_h\mathbf{p}'_h$$

As stressed by Geladi [165], things are easier to understand if we remark that $\mathbf{t}'_h\mathbf{t}_h$ or $\mathbf{P}'_h\mathbf{P}_h$ and $\left\|\mathbf{p}'_h\right\|$ are scalars (that can be combined in one general constant C).

From

$$\mathbf{t}_h = \frac{\mathbf{X}\mathbf{p}_h}{\mathbf{p}'_h\mathbf{p}_h} \quad \text{and} \quad \mathbf{p}'_h = \frac{\mathbf{t}'_h\mathbf{X}}{\mathbf{t}'_h\mathbf{t}_h}$$

it comes

$$(\mathbf{CIm} - \mathbf{X}'\mathbf{X})\mathbf{p}_h = 0$$

And similarly

$$(\mathbf{C'In} - \mathbf{XX'})\mathbf{t}_h = 0$$

where **Im** and **In** are the identity square matrices of dimension m*m and n*n, respectively. These two equations are the eigenvalues/eigenvectors equations. At the end of the process:

$$\mathbf{X} = \mathbf{TP'} \rightarrow \mathbf{T} = \mathbf{XP}$$

where
    T is the matrix of scores
    P is the matrix of loadings

Another presentation can be proposed using the *singular value decomposition* technique [166], **X** is decomposed in three matrices:

$$\mathbf{X} = \mathbf{USV'}$$

where **U** and **V** are orthogonal eigenvector square matrices, and **S** a diagonal matrix containing the singular values (square root of the eigenvalues). **US** is the score matrix **T**, and **V** corresponds to the loading matrix **P**

$$\mathbf{y} = \mathbf{Xb} \qquad \mathbf{b} = \mathbf{VS^{-1}U'y}$$

A.4.2.4   Partial Least Squares

PLS is a neighboring method very largely used in QSARs, since applicable when there are a large number of predictors, and particularly when there are more predictors than experiments (as, for example, in CoMFA). We consider here only PLS1: the response **y** is a vector (n,1).

In PCR, the principal components of **X** are used to predict **y**. This is carried out using the components that best explain **X**. But it is not sure that these PCs would be the best ones to explain the dependent variable **y**, which is the actual problem on hand. In PLS, the dependent variable **y** is taken into account for the decomposition of **X** and a new matrix of weights **W** for the independent variables is introduced with the constraint that the components retained explain as much as possible the covariance between **X** and **y**. This condition is often designed as "inner relation" linking **y** and **X**, whereas the decomposition of **X**, and possibly of **Y** (when the response is multiple) are the "outer relations." Then the decomposition of **X** is used to predict **y** (Figure A.13).

Various algorithms have been proposed for the "classical" approach [78,165,166,949]. The following version corresponds to Wold's algorithm presented by Smilde [949]. Furthermore, several developments (increasing speed) or extension to arrays of data (N-way PLS) have been published [143,144,312,950]. Similar to the NIPALS algorithm, loading and scores used to predict **y** from **X** are determined iteratively. The independent-variable matrix **X** (n,m), of elements $x_{i,j}$, is expressed as

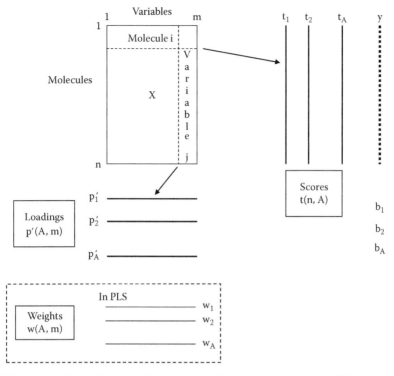

FIGURE A.13 PCA, PCR, and PLS: decomposition of the data matrix **X** in a space of dimensionality A (<m). Each row of **X** (a molecule) is projected onto an element of each of the score vectors **t**. Each column of **X** (a variable) is projected onto an element of each of the loading vectors **p′**. Weight vectors **w** intervene only in PLS. (Adapted from Geladi P. and Kowalski B.R., *Anal. Chim. Acta*, 185, 1986.)

the outer product of the score vector **t** (n*1) and the weight vector **w** (m*1) of length one. **w** is chosen *so that **t** has a maximum covariance with **y***. The model (with one component) takes the form

$$\mathbf{t} = \mathbf{Xw}$$

$$\mathbf{y} = \mathbf{tb} + \mathbf{f}$$

where
   f is the error vector
   w is the **X**-weight vector
   b is the **y**-loading vector

The sequence can be summarized as follows [949]:
   Beginning with the first component, the problem is to find a weight vector **w** satisfying

$$\text{Max}_w[\,\text{cov}(\mathbf{t},\mathbf{y})\big|\mathbf{Xw} = \mathbf{t} \quad \text{and} \quad \|\mathbf{w}\| = 1]$$

which can be written

$$\text{Max}_w \left[\sum_i t_i y_i \middle| t_i = \sum_j x_{ij} w_j \middle| \quad \|\mathbf{w}\| = 1 \right]$$

equivalent to

$$\text{Max}_w \left[\sum_j \sum_j y_i x_{ij} w_j \qquad \|\mathbf{w}\| = 1 \right]$$

Or, defining **z** by

$$z = \mathbf{X'y}$$

$$\text{Max}_w \left[\sum_j z_j w_j \qquad \|\mathbf{w}\| = 1 \right]$$

this condition comes to

$$\mathbf{w} = \mathbf{z} \left(\|\mathbf{z}\| \right)^{-1}$$

that is

$$\mathbf{w} = \mathbf{X'y} \left(\|\mathbf{X'y}\| \right)^{-1}$$

The algorithm is therefore

- Determination of the first weight loading vector **w** and loop start

$$\mathbf{w} = \frac{\mathbf{X'y}}{\|\mathbf{X'y}\|}$$

$$\mathbf{t} = \mathbf{Xw}$$

$$\mathbf{p} = \frac{\mathbf{X't}}{\mathbf{t't}}$$

$$b = \frac{t'y}{t't}$$

- Update: deplete $\mathbf{X}$ and $\mathbf{y}$

$$\mathbf{X} \leftarrow \mathbf{X} - \mathbf{tp}'$$

$$\mathbf{y} = \mathbf{y} - \mathbf{t}b$$

- Repeat the loop to determine the next weight loading vector

The process continues until a sufficient number of components or "latent variables" (say A) is calculated. The number of components is chosen according to usual statistical criteria for the calculation of the response $\mathbf{y}$.

New objects ($\mathbf{X}_{new}$) are predicted as

$$y_{pred} = \mathbf{b}_{pls}\mathbf{X}_{new}$$

The coefficient $\mathbf{b}_{pls}$ is

$$\mathbf{b}_{pls} = \mathbf{W}(\mathbf{P}'\mathbf{W})^{-1}\mathbf{b}$$

where

$$\mathbf{W} = [\mathbf{w}_1, \mathbf{w}_2, \ldots, \mathbf{w}_A], \quad \mathbf{P} = [\mathbf{p}_1, \mathbf{p}_2, \ldots, \mathbf{p}_A], \quad \text{and} \quad \mathbf{b} = (b_1, b_2, \ldots, b_A)$$

This algorithm is said to be orthogonalized since scores and weights are orthogonal:

$$\mathbf{t}'_r\mathbf{t}_s = 0 \quad \text{and} \quad \mathbf{w}_r\mathbf{w}_s = 0$$

If all columns of $\mathbf{X}$ are used, the result is equivalent to that of MLR.

Kaneko et al. [948] indicated that $\mathbf{y}_{calc}$ can be expressed as

$$\mathbf{y}_{calc} = \frac{(\mathbf{XX}'\mathbf{yy}'\mathbf{XX}'\mathbf{y})}{\|\mathbf{y}'\mathbf{XX}'\mathbf{XX}'\mathbf{y}\|}$$

where $\mathbf{y}'\mathbf{XX}'\mathbf{y}$ and $\mathbf{y}'\mathbf{XX}'\mathbf{XX}'\mathbf{y}$ are scalar.

$$\mathbf{y}_{calc} \text{ is then}: \mathbf{y}_{calc} = \mathbf{X}c\mathbf{X}'\mathbf{y}$$

with

$$c = \frac{(\mathbf{y}'\mathbf{X}\mathbf{X}'\mathbf{y})}{(\mathbf{y}'\mathbf{X}\mathbf{X}'\mathbf{X}\mathbf{X}'\mathbf{y})}$$

and

$$\mathbf{b}_{PLS} = c\mathbf{X}'\mathbf{y}$$

An alternative formulation was proposed by Indahl [951] who mentioned that vector **w** maximizing the covariance between **y** and **Xw** can be expressed as

$$\mathbf{w} = k[\text{cov}(\mathbf{y},\mathbf{x}_1), \ldots, \text{cov}(\mathbf{y},\mathbf{x}_m)]$$

or

$$\mathbf{w} = k \, \text{std}(\mathbf{y})[\text{corr}(\mathbf{y},\mathbf{x}_1)\text{std}(\mathbf{x}_1), \ldots, \text{corr}(\mathbf{y},\mathbf{x}_m)\text{std}(\mathbf{x}_p)]$$

Another algorithm is by Martens and Naes [952] and gives the same results but the scores are no longer orthogonal. This algorithm does not imply the calculation of the extra regression vectors **p** and the coefficients b must be calculated simultaneously. Note that **t**, **b**, and **X** (after updating) are different from Wold's algorithm.

$$\mathbf{b} = (\mathbf{T}'\mathbf{T})^{-1}\mathbf{T}\mathbf{y}$$

Updating **X** is carried out by

$$\mathbf{X} = \mathbf{X} - \mathbf{t}\mathbf{w}'$$

(compared to the corresponding Wold's relationship $\mathbf{X} = \mathbf{X} - \mathbf{t}\mathbf{p}'$).

A.4.2.5 Variants of PLS

- *Kernel PLS*: since the inception of the classical PLS method, several variants were proposed with the main aim of increasing computational speed. For a review, see Lindgren and Rånnar [950] and Bastien [953]. The starting point is the remark that PLS can be reformulated as an eigenvalue/eigenvector problem and that the various scores and weights can be determined as the eigenvectors of a set of variance/covariance matrices. For example, **t** and **b** can be determined as the eigenvectors of the square matrices **XX'yy'** and **yy'XX'**. An important observation is that it is also possible to deflate only **X** or **y**, which is interesting if the response is of low dimensionality.

The basic idea is to work on matrices of lower dimensionalities and the proposed solutions have been optimized depending on the relative number of variables and observations. Hence, the term "kernel PLS" where "kernel" stresses the notion of lower dimensionality rather than projection in a different space (as in SVM for example). In QSARs (and particularly 3D-QSARs), there are usually many more variables than molecules ($m \gg n$).

- Universal partial least squares (UNIPALS) proposed by Glen in 1989 [954,955] was based on the m*m matrix $\mathbf{y'XX'y}$. For applications, see Dunn et al. [310,375,376].
- SAMple-distance partial least squares (SAMPLS) of Bush and Naschbar [144], rather than examining individual values of descriptors, reduces explanatory data to pairwise distances among molecules or equivalently to a n*n covariance matrix $\mathbf{C} = \mathbf{XX'}$. Another originality is that only $\mathbf{y}$ is deflated. This algorithm, which operates on one $\mathbf{y}$ response, is well suited for CoMFA or CoMSIA analyses and is particularly efficient for L-O-O cross-validation.
- In statistically inspired modification of the PLS (SIMPLS) method [143], the scores are calculated from a direct combination of the $\mathbf{X}$ matrix by a constrained optimization instead of deflating $\mathbf{X}$. The matrix $\mathbf{T} = \mathbf{Xw'}$ is obtained maximizing $cov^2(\mathbf{y}, \mathbf{Xw'})$ with constraints of $\mathbf{w'}$ normed, $\mathbf{t} = \mathbf{Xw'}$ orthogonal. In case of a unique response $\mathbf{y}$, results are identical to classical PLS (in other cases only the first component is identical, but differences remain small).

- *Nonlinear PLS*: PLS has been also extended to deal with nonlinear phenomena. The simplest models introduce square terms in the descriptor matrix as in the implicit nonlinear latent variable regression (INLR) method [956] or separate the $\mathbf{X}$ matrix in blocs in the *serial extension of multiblock PLS* [957]. It is also possible to introduce a nonlinear function in the inner relationship linking the scores $\mathbf{t}$ to $\mathbf{y}$. A quadratic function was used by Funatsu et al. [145]. For example, with PLS1

$$\mathbf{X} = \mathbf{tp}$$

$$\mathbf{y} = \mathbf{ub} \quad \text{with} \quad \mathbf{b} = d_0 + d_1\mathbf{t} + d_2\mathbf{t}^2$$

Splines or neural networks have also been used.

The GIFI-PLS aims to model nonlinearities (even severe) and discontinuities (occurring, for example, when examining simultaneously clusters of compounds significantly separated) [958]; GIFI-PLS is based on binning quantitative x variables into categorical variables.

- *GA-PLS*: PLS could also benefit the capacity of GA to select the most relevant variables. The aim is to obtain a PLS model with high internal predictive ability using a small number of variables that convey the essential structural information. This approach was explored by Hasegawa et al.

[959,960] using a modified GA where the most informative chromosomes are "protected." The best chromosomes using a given number of variables are selected unless a chromosome gives better internal predictive ability with a lower number of variables and is considered as more informative.

A.4.2.6   Self-Organizing Regression and Multicomponent SOR

Self-organizing regression (SOR) was proposed as a built-in correlation tool in SOMFA and its performance has been questioned by Tuppurainen et al. [347,348]. Recall that in SOMFA (a grid-based, alignment-dependent method) the descriptors for each object of the training set are combined to form a master grid. The data, descriptors ($\mathbf{X}$), and property ($\mathbf{y}$) are mean centered, which (for $\mathbf{y}$) tend to give less importance to compounds with an activity close to the mean value. From the master grid, a master vector is created, with the same dimensionality as that of descriptors

$$\mathbf{mv} = \sum_{i=1,n} \mathbf{x}_i \mathbf{y}_i$$

where i refers to the objects (rows of the grid). For calculating the property, a new predictor vector is created.

For object i

$$\mathbf{p}_i = \mathbf{x}_i \mathbf{mv}'$$

and a MLR is sought for between $\mathbf{y}_i$ and $\mathbf{p}_i$.

Compared to NIPALS or SIMPLS, it was demonstrated [348] that, with adequate normalization conditions, the result of SOR is strictly equivalent to the first component of NIPALS and SIMPLS. For example, for the first PLS component, the weight vector $\mathbf{w}$ is identical to $\mathbf{mv}'$ in SOR and $\mathbf{b}_{SOR}$ to $\mathbf{b}_{PLS}$. The fact that SOR considers only this first component was advocated as the reason of poor performance when modeling complex phenomena, for example, in the analysis of xenoestrogen activity [347].

Multi component SOR (MCSOR) is a direct extension of SOR, according to a process somewhat similar to that of PLS. The first component is derived as in SOR. Then for the determination of the other components, successive master vectors, regression coefficients, and $\mathbf{y}_{pred}$ are stored. The calculation is iterated replacing $\mathbf{y}$ by the residual $\mathbf{y} - \mathbf{y}_{pred}$ of the preceding step.

An extended comparison of MCSOR with PLS and derived methods SIMPLS, SVDPLS, powered PLS (PPLS) was carried out by Tuppurainen et al. [961] on various extended data sets. PPLS [951] operates by stepwise optimization of a set of candidate loadings and weights by taking powers of the $\mathbf{y}$–$\mathbf{X}$ correlations and $\mathbf{X}$ SDs. The question was mainly to examine the performance of MCSOR in cases where the number of descriptors is larger than that of molecules, and variables present high collinearity. Physicochemical properties (boiling, melting points, log P), biological affinity (Xα inhibition), or multivariate calibration tasks were examined.

Although the descriptors involved (derived from SMILES codes) do not convey any 3D-information (except some MOE values), it is interesting to observe that, schematically, MCSOR favorably competes with PLS (whereas SVDPLS or PPLS is often slightly inferior). The authors stressed that MCSOR is relatively insensitive to the number of components retained (an important point to avoid overfitting). However, the components of the model are more difficult to interpret (compared to PCR or PLS) and, up to now, the analysis is only applicable to univariate response (a unique **y** vector). Transferability of the model to other data sets was also questioned.

A.4.2.7   Independent-Component Analysis

Although up to now very scarcely introduced in Chemistry, mainly for analysis of NIR spectra [962,963], it seems worth mentioning the independent-component analysis (ICA) method. This approach, largely used in signal-processing studies, relies on the "blind source separation" problem, sometimes referred to as the "cocktail party problem." Several persons speak simultaneously and the sound is transmitted by several microphones situated at different locations. The problem is to recover individual sources from the mixed signals received, which are a combination of these individual sources.

This approach may be considered as an extension of PCA, but operates on components which are determined as *independent* rather than *orthogonal*. Schematically, from the descriptor matrix **X** an independent-component matrix **S** is extracted. Then the process is similar to the derivation of PCR or PLS from the classical **X** matrix. In QSPR, the method was successfully used by Kaneko et al. [948] on solubility data.

A.4.3   BACK PROPAGATION NEURAL NETWORKS

Beside MLR and PLS that were for years the basis of QSAR models, other approaches tended to account for the nonlinear behaviors often encountered with biological data. In the first attempts, one can note the incorporation of *quadratic* (or higher power) terms in the framework of usual MLR. On the other hand, nonadditive behaviors in polysubstituted compounds have been treated with the inclusion of *pairwise interaction terms*. But these treatments (although some successes) were difficult to generalize, since they implied an a priori choice of the polynomial formula to apply and/or a selection of the interaction terms to retain. Quadratic terms have also been introduced in PLS treatments [145].

A more efficient approach was developed with *Artificial Neural Networks* (ANN), which possesses the double advantage of learning from information and to be model free. ANN are computer models derived from a very simplified concept of the brain and supposed to mimic its functioning. After a period of doubts, due to the inability of the first systems (perceptrons) to solve relatively simple problems, NN were definitively accepted in the 1980s as very powerful tools, now largely widespread in a lot of domains [87,88]. Several books and articles reviewed the numerous applications of ANN in drug design and QSARs [48,49,83,87–92,401,487,488].

Their success came from their ability to learn from examples and generalize to new cases, particularly when they are faced with incomplete or noisy data. One of their main interests is also to find complex relationships (e.g., in QSAR, between the structure of a molecule and its physiological action) without the need to use a predefined model, or when the underlying phenomena are still not well known.

Various architectures of NN have been proposed, defining in the field of QSARs a wide panel of applications. Back propagation neural networks (BNN) are seemingly the most widely used layered networks in QSARs. As just mentioned, they have been largely applied with the numerous 2D and 2.5D descriptors easily available from various packages (CODESSA, DRAGON). But they were also introduced, coupled with a GA, for the treatment of similarity matrices [48,49] and constitute an integral part of CoMSA and COMPASS 3D approaches [95,367,397].

The network comprises a set of identical elementary units (the formal neuron), organized in layers. All units of a layer are connected to all units of the adjacent upstream and downstream layers, but there is no connection between units within the same layer. Connections are unidirectional, from input to output. The most commonly used BNN in QSARs are formed of three layers:

- One input layer (with a number of units corresponding to the number of structural descriptors) receiving the information.
- One (or rarely more) hidden layer with a number of neurons to be determined.
- One output layer delivering the calculated response. The number of output units depends on the property. One output unit is sufficient for a unimodal (scalar) response, such as a biological activity, but several output units may be introduced if the response is a vector $y_i(y_{i1}, y_{i2}, ..., y_{im})$ (Figure A.14).

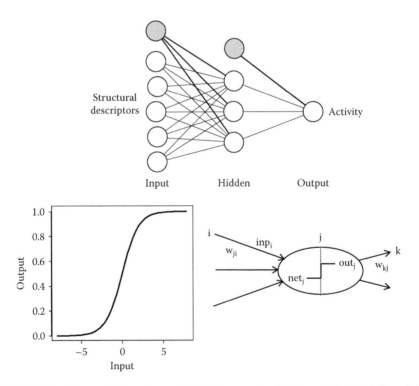

FIGURE A.14 Scheme of a three-layered ANN. Upper part: architecture of the ANN. Gray disks correspond to biases. Lower part: the elementary unit (neuron) and the sigmoid transfer function.

For a given pattern (a compound), the input units (**i**) are fed with the structural descriptors (Inp_i). The information, weighted by connection weights w_{ji}, between units i of the input layer and j of the hidden layer, is transmitted to the neurons of the hidden layer (j). For each hidden unit, the incoming information (from all upstream units) is summed up, corresponding to the Net input of unit j (Net j), and transmitted to the output unit (k) thanks to a transfer function (often, a sigmoid).

$$sf(u) = \frac{1}{(1 + \exp(-u))}$$

$$Net_j = \sum_i inp_i * w_{ji}$$

$$Out_j = \frac{1}{[1 + \exp(-\alpha Net_j + \theta_i)]}$$

where θ_i is the bias. This can be written in a more synthetic way, considering the bias as a supplementary unit with an input fixed to 1 and a "weight" θ_i and incorporating this term with Net_i

$$Out_j = \frac{1}{1 + \exp\left(-\sum_i inp_i * w_{ji}\right)}$$

(coefficient α may be skipped with adequate scaling).

After summation of all the signals transmitted from the hidden layer, the output unit delivers the response of the network

$$Net_k = \sum_j out_j * w_{kj}$$

$$Out_k = \frac{1}{[1 + \exp(-Net_k)]}$$

In other words, for the object (pattern) n, the input x_n (x_{n1}, x_{n2}, x_{n3}, ...), plus the biases, leads to the response $\ddot{y}_n$

$$\ddot{y}_n = sf\left\{\sum_j\left[sf\left(\sum_i x_{n,i} w_{ji}\right)\right] w_{kj}\right\}$$

where sf is the sigmoid function. Two parameters, determined by the learning algorithm, are important:

- Bias (θ) that allows for adjusting the input in a region where the transfer function is more efficient for the performance of the method
- Weights (w_{ji}, w_{kj}) that scale the information transmitted from one layer to the following one

Note that, due to the shape of the sf, it is recommended to scale y in the range (say) 0.1–0.9.

A.4.3.1 Learning and Testing

The BNN functions in two steps: learning (or training) and test. In the *learning* step, a set of pairs (input descriptors, associated response) is given to the networks. The BNN adjusts its parameters (weights and biases) so as to reproduce as output, the correct response. Then weights and biases are frozen and, in the *test* phase, when new compounds are submitted, the network is able, using the so-determined weights, to give the appropriate response from the descriptors submitted. The key step in network adjustment is therefore the correction of weights and biases. This is currently carried out by the algorithm of *back-propagation of errors* (from output to input layer).

A.4.3.2 Weight Adjustment: Back Propagation of the Error

At the beginning of the training step, all weights and biases are initialized to small random numbers (say [−0.3, 0.3]) to avoid saturation of the weights and possible problems. Then each object (pattern) of the training set is in turn presented to the network, and an output is generated. For each submitted pattern, weights are iteratively adjusted (however, vide infra the possibility to correct the weights after each "*epoch*," that is once all patterns have been considered).

The basis of the calculation is schematized in Figure A.15. The problem is to determine the relationship between the change in a weight and the resulting change at the output of a layer. The adjustment of the weights is carried out layer by layer working upstream from network output to input. The terms upstream, downstream refer here to the usual data flow, and not to the propagation of the error in the opposite direction.

The process that corresponds to an approximate gradient optimization, is in line with the *Delta rule*, which considers that the correction on a unit is proportional to both its input and the error (δ) it delivers at output. Note that the correction at a given layer depends on the status of the unit under scrutiny, the status of the adjacent upstream units and of the downstream one (or the property value).

The process is therefore the following: in the current step, the status of the units in the different layers is determined. From the error at output, Δw_{kj} (connections to output unit) is calculated, and then the connections of the upstream layers. Then, another pattern is submitted. Working pattern after pattern may lead to instabilities (depending on the order of the objects) and it seems better to store the information for all patterns before starting the iterative adjustment.

The *learning rate* is an important parameter: too low a value makes the process long and increases the risk of stopping in a local minimum. Too high a value may

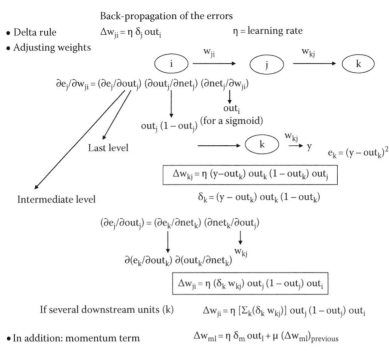

- Delta rule $\quad \Delta w_{ji} = \eta\, \delta_j\, out_i$ $\qquad$ η = learning rate
- Adjusting weights

$\partial e_j/\partial w_{ji} = (\partial e_j/\partial out_j)\,(\partial out_j/\partial net_j)\,(\partial net_j/\partial w_{ji})$

$out_j\,(1-out_j)$ (for a sigmoid)

Last level

$e_k = (y - out_k)^2$

$\Delta w_{kj} = \eta\,(y-out_k)\,out_k\,(1-out_k)\,out_j$

Intermediate level $\qquad\qquad\qquad \delta_k = (y - out_k)\,out_k\,(1-out_k)$

$(\partial e_j/\partial out_j) = (\partial e_k/\partial net_k)\,(\partial net_k/\partial out_j)$

$\partial(e_k/\partial out_k)\,\partial(out_k/\partial net_k)$

$\Delta w_{ji} = \eta\,(\delta_k\, w_{kj})\,out_j\,(1-out_j)\,out_i$

If several downstream units (k) $\qquad \Delta w_{ji} = \eta\,[\Sigma_k(\delta_k\, w_{kj})]\,out_j\,(1-out_j)\,out_i$

- In addition: momentum term $\qquad \Delta w_{ml} = \eta\,\delta_m\,out_l + \mu\,(\Delta w_{ml})_{previous}$

FIGURE A.15 Back propagation of the error. The calculation is exemplified for a three-layer network with only one output unit (k), and a sigmoid transfer function. η is the learning rate. Note that the status of a unit only depends on the connections adjacent to it (in and out) and that the output of a unit is the input of the downstream unit (scaled by the connection weight).

induce oscillations, which can be avoided by including a momentum μ (adjustment at step n + 1 keeps some memory of step n). Typical values may be $\eta = 0.05$, $\mu = 0.9$.

Various modifications of the BP algorithm have been proposed. See Refs. [488,899] for the Scaled Conjugate Gradient and [887,964] for the Levenberg-Marquardt algorithm.

Some particular aspects of BNN, particularly regarding comparison with, say, MLR, deserve some comments.

- The first remark is that the process is not deterministic. Starting from weights randomly initialized, various solutions of comparable performance may be commonly obtained. It may be interesting to perform a *consensus* study using an ensemble of networks (about 100 networks, for example) and average the results. *Associative NN* (vide infra) constitute a possible alternative solution [964].
- There are no theoretical rules for the choice of the best number of hidden neurons. It seems logical to keep it as reduced as possible. This number must be sufficient to ensure generalization capabilities, and to limit the risk of getting stuck in local minima, but if too large, it will increase the number of weights to adjust, leading to an *overfitted* model.

This poses the problem of the ratio: number of samples (patterns) in the training set/number of adjustable connection weights. A ratio about 2 is generally considered as correct. In some cases, it may be interesting to reduce the number of descriptors by pre-processing: Principal Component Analysis (PCA), for example, allows for replacing the original descriptors by a reduced number of composite variables obtained by linear combination. GA has also been widely used [48,49,487,488,610].

Duprat et al. [965] remarked that neural networks apply a parsimony process and need for equivalent performance less parameters than MLR or PLS.

- Another potential difficulty is *overtraining*. Too long a training reduces the error (possibly below realistic values): the network adjusts its weights very sharply, even on unimportant details or noise. Beyond time wasting, the network, faced with new compounds, shows poor generalization performance: this behavior has been called "learning by heart."

  One possibility to avoid overtraining is to divide the data set equally into a learning set and a validation (or "control") set. When the training goes on, the error on the learning set continuously decreases whereas on the validation set it starts decreasing but after a time, the error begins to increase: this is a clue for overtraining. Stopping the training at this point, "early stopping," is a good recipe to avoid overtraining and not damage the prediction ability [966].

- An important point is also that BNN are often considered as "black-boxes" since the interpretation of the weights is not at all straightforward. However, to guide the search for new active compounds, it would be interesting to specify the influence of a given substituent, or a given descriptor. This is not quite easy with BNN since units of different layers are strongly interconnected. A possible way to get some insight on this point was proposed by Andrea and Kalayeh [94]. All but one input descriptors are kept at the value they have in a template compound (generally among the most active) and one examines how the results are modified when the selected descriptor scans its variation range [94,488]. The resulting plot gives the functional dependence of the activity on this variable. The process was used by So and Karplus [488] in their study of 1,4 benzodiazepin-2-ones, binding GABA receptor for suggesting new possibly active compounds.

A.4.3.3 Applications of BNN

The papers of Ayoma et al. [93,967] on anticarcinogenic activity of carboquinones are among the first (if not the first) to propose BNN models in QSARs, using usual substituent constants. They were followed by a huge number of papers proposing structure/activity (or property) correlations: for a review see for instance Refs. [83,89–92]. Input structural descriptors were mainly at that time, 2D or 2.5D descriptors. Some years later, BNN were incorporated as fitness function in a GA model for variable selection in QSAR approaches based on various types of descriptors [48,49,487,488,608–610].

A.4.3.4 Associative Neural Networks

In addition to a consensus approach, a possible solution to get more stable and improved results is offered by the ASsociative Neural Networks (ASNN) proposed by Tetko [964]. ASNN associates a layered NN (though in the original paper, the back-propagation algorithm was not used) and a k-NN approach. The initial remark was that in a layered network, the information obtained from the input data is stored in the weights (after training) and no more considered. So, it cannot be used to correct or refine the output results. On the contrary, by using k-NN method one can take advantage of the output results of the training objects to improve the prediction for an unknown. The correction is calculated from the errors observed on nearest training objects. The originality of the method is that the correction is determined in the space of the output values and not in the initial descriptor space. For applications in QSPRs, see, for example, Ref. [968].

A.4.3.5 Volume Learning Algorithm

The basic idea of VLA [969] is to decrease the dimensionality of input data space (often thousands of inputs) by clustering the input-to-hidden neuron weights. The method was presented on the example of aminoalkyl indoles agonists of the cannabinoid receptor. CoMFA or CoMSIA treatments led to a huge number of field values that are generally analyzed by PLS. Using a BNN to do this task would be extremely inefficient on account of the number of weights to adjust with, as consequences, a high computer-time and, more importantly, overfitting. The VLA approach combines an unsupervised network (Kohonen map) with a supervised network (BNN) in a recurring iterative process. A Kohonen self-organizing map clusters the field values of CoMFA. The mean values in the cells are then analyzed with the BNN and the dimensions of the Kohonen map are progressively decreased. An interesting point is that a very efficient pruning can be carried out on the weights (input-hidden layers) of the BNN during the process.

A.4.4 OTHER NEURAL NETWORKS

More recently, new layered networks have been proposed as more efficient than BNN. Particularly, for correlation, radial basis function neural networks (RBFNN) [96] and general regression neural networks (GRNN) [98] both present the advantage of easy settings, rapid training, and guarantee to find the global minimum of the error surface. New architectures have been also proposed. Hopfield networks were used for pattern recognition. Self-organizing Kohonen maps (SOMs) allow for variable reduction or for selection of test and training sets. At last, counter-propagation neural networks (CPNNs), adding a supplementary layer to a Kohonen network, perform classification and correlation tasks.

A.4.4.1 Radial Basis Function Neural Networks

They are also three-layer feed-forward structures (input, hidden and output layers). The principle relies on the problem of exact interpolation of a set of data points in a multidimensional space that is to the projection of every input vector $\mathbf{x}_i$ onto the corresponding target y_i [96]:

$$y_i = f(\mathbf{x}_i) \quad \text{for each i object}$$

This can be achieved using a nonlinear radial function Φ, function of the distance between object $\mathbf{x}_i$ and the center $\mathbf{x}_j$ of the function.

$$f(\mathbf{x}_i) = \sum_{j=1,n} w_j \Phi\left(\left\|\mathbf{x}_i - \mathbf{x}_j\right\|\right)$$

In matrix notations

$$\mathbf{y} = \Phi\mathbf{w}$$

so that

$$\mathbf{w} = \Phi^{-1}\mathbf{y}$$

Going to RBFNN, some modifications are introduced. The number of basis function is inferior to the number of inputs. Their centers are no longer constrained to the input data vectors but are determined during training, as well as the width of each Gaussian. A bias is also introduced (Figure A.16).

Architecture of a RBFNN: The input layer does not process the information. It only distributes the input vectors (structural descriptors of the patterns) to the hidden layer. This hidden layer consists of a number of RBF units (n_k) and bias (b_k). Each of these units employs, as a nonlinear transfer function, a RBF operating on the input data. Gaussian functions are generally used

$$\Phi\left(\left\|\mathbf{x}_i - \mathbf{x}_j\right\|\right) = \exp\left(\frac{-\left\|\mathbf{x}_i - \mathbf{x}_j\right\|^2}{2\sigma^2}\right)$$

The jth RBF is characterized by its center ($\mathbf{c}_j$) and width (r_j). The RBF operates by measuring the Euclidian distance between input vector ($\mathbf{x}_i$) of pattern i, and the RBF center ($\mathbf{c}_j$) and performs a nonlinear transformation

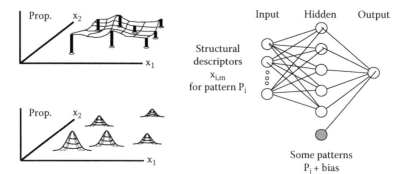

FIGURE A.16 Radial basis function neural network. Left: for objects defined by descriptors $X\{x_1, x_2\}$, the property is represented by vertical bars. The problem is to describe the response surface by adding Gaussian functions centered on selected points. Right: architecture of a RBFNN. Gaussian functions are centered on selected points (i).

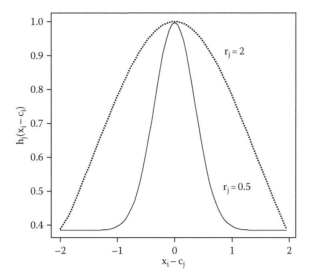

FIGURE A.17 Examples of radial basis Gaussian functions for $r_j = 2$ (outer curve) and 0.5 (inner curve).

$$h_j(\mathbf{x}_i) = \exp\left(\frac{-\|\mathbf{x}_i - \mathbf{c}_j\|^2}{r_j^2}\right)$$

where h_j represents the output of the jth RBF unit of the hidden layer (Figure A.17).
The operation to the output layer is linear, according to

$$y(\mathbf{x}_i) = \sum_{j=1,nh} w_{oj}h_j(\mathbf{x}_i) + b_o$$

where
   scalar y is the output for the input vector $\mathbf{x}_i$
   w_{oj} is the connection weight between the output unit (o) and the jth hidden layer unit
   b_o is the bias of unit o

Designing a RBFNN involves selecting centers (number and width) of hidden layer units and weights. The overall performance is evaluated in terms of the RMSE between predicted and observed values. This performance had to be optimized by adjusting the values of

n_h number of hidden units
centers c_j and width r_j of each RBF h_j

The widths r_j of the RBF can be chosen the same for all units (as in the examples quoted here) or different for each unit. For selecting the centers $\mathbf{c}_j$, various options exist such as K-means clustering, orthogonal least squares learning algorithm, and RBF-PLS. The forward subset selection method proposed by Orr [970,971]

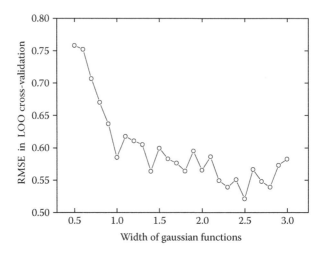

FIGURE A.18 RMSE versus width of RBFNN on L-O-O cross-validation.

simultaneously selects the number and the centers of the hidden units. The weight adjustment between hidden and output layers is performed by a least squares method after this selection. The optimal width is determined from a number of trials looking at the minimum error on a learning set in L-O-O cross-validation (Figure A.18). The property can then be calculated in a second step.

For applications of RBFNN, and comparison with other nonlinear models, see, for example [97,99,102,972].

A.4.4.2 General Regression Neural Networks

Introduced by Specht in 1991 [973], GRNN are fed forward networks with four layers: input, hidden, summation, and output. They are nonparametric estimators that calculate a weighted average of the target values [98]. Each unit of the input layer receives a descriptor of the input patterns. The hidden layer comprises as many units as training patterns. Their activations for a given pattern are calculated from the Euclidian distance of these units to the input pattern. The property value predicted by the network is a weighted average of the property values of the training patterns close to the input pattern under scrutiny, determined from a probability density function using Parzen's nonparametric estimator [974] (Figure A.19).

This function is calculated as a weighted sum of kernel functions represented by normalized Gaussian functions with a common width r_j. Each training pattern is weighted exponentially according to its distance to the unknown pattern and a smoothing factor r_j. The output y associated to the input vector variable $\mathbf{x}$ is given by

$$y(\mathbf{x}) = \frac{\displaystyle\sum_{i=1,n} y_i \exp\left[-(\mathbf{x}-\mathbf{x}_i)'(\mathbf{x}-\mathbf{x}_i)/2r_i^2\right]}{\displaystyle\sum_{i=1,n} \exp\left[-(\mathbf{x}-\mathbf{x}_i)'(\mathbf{x}-\mathbf{x}_i)/2r_i^2\right]}$$

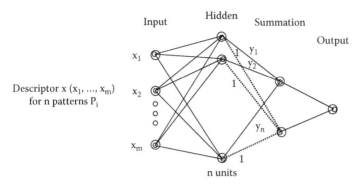

FIGURE A.19   Architecture of a GRNN.

where x_j, y_j correspond to the structural descriptors and the property value of the hidden node corresponding to training pattern j. The numerator and denominator of this expression are evaluated on the two units of the summation layer (connection weights being y_j and 1, respectively). The output unit calculates their quotient, the predicted property value y_j^{pred}. The common width r_i of the Gaussian function are determined in L-O-O cross-validation (Figure A.20).

So, RBFNN and GRNN present some analogies with the k-NN approach. In all cases, the property is calculated from the averaged values from the neighbors. In GRNN, all "neighbors" are considered, whereas in RBFNN some are selected by the algorithm and are used for all predictions. In k-NN, few neighbors are selected but vary with the pattern to predict.

A.4.4.3 Kohonen Neural Network and Self-Organizing Maps

KNN is one of the neural network techniques to perform unsupervised learning [975–977]. Its main interest is its ability to generate a projection of objects from a high-dimensional space onto a two-dimensional space, while conserving the

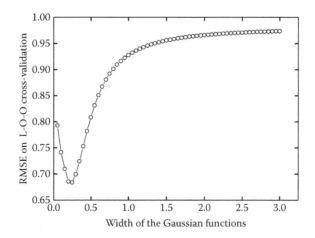

FIGURE A.20   GRNN: RMSE versus width of the Gaussian functions on L-O-O cross-validation.

topology of information as much as possible [83,88,400,401,978]. In these aspects, KNN may be viewed as bearing some analogy with human brain processes that tend to compress and organize information. In other words, samples, which are close to each other in the multidimensional descriptor space remain neighbors on KNN map at the end of the learning process. Thanks to this ability to conserve topology, KNN are very useful tools for visualization of multivariate data: color encoding the neurons of the network output layer (the Kohonen map or SOM) allows for global perception, at a glance, of properties such as charges, electrostatic potential, lipophilic character, much more easily than looking at various 3D slides [400,401]. In QSARs, they provide efficient tools for classification (intervening for instance in variable selection). They also allow for easier quantitative analyses on 2D maps of 3D data scattered on the molecular surface.

Architecture and training: Typically, KNN comprises a single active output layer of neurons generally arranged in a two-dimensional rectangular or hexagonal grid of nodes (neurons) giving the SOM. Typically, 20×20 to 50×50 unit maps are used. But for the representation of molecular surfaces (continuous surfaces) the topology is usually defined as "torus." "Torus" means that the right (and top) edge of map is continued to its left (and low) edge, respectively, and vice versa. Then the torus is cutoff along two perpendicular directions to get a planar map. An important notion is that of topology. On the map formed by the units, one distinguishes for a particular neuron (the central neuron) its first neighbors, second neighbors, etc.

Each sample (s) in the data set is considered as a vector $\mathbf{x}$ of m components (the descriptors $\mathbf{x}_{si}$). Each neuron (j) of the active layer (the Kohonen map) is characterized by a weight vector $\mathbf{w}$ of m components w_{ji}. The values of weight vector are randomly initialized (but other options exist [977]). All the neurons of the active layer receive the same input ($\mathbf{x}$) and competitive learning is carried out. Then Euclidean distance d_{js} is calculated between each component of the input vector x_{si} (i^{th} descriptor of sample (s)) and each component of the weight vector w_{ji} (i^{th} component of neuron j of the output layer) according to

$$d_{js} = \sum_{i=1,m} (x_{si} - w_{ji})^2$$

The neuron j having the shortest distance to the input vector $\mathbf{x}_s$ (and only this neuron) is declared the winner (c). "Winner takes it all."

$$Out_c = min\left[\sum_{i=1,m} (x_{si} - w_{ji})^2 \right]$$

Alternately, one can choose the neuron giving the largest output $\sum_{i=1,m} w_{ji} x_{si}$ (Figure A.21).

After the winning neuron, denoted by c, is found, the adjustment of weight vector starts an iterative self-organization phase. The weight vectors of the winner node

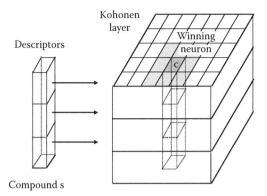

FIGURE A.21 Kohonen neural network. c (dark gray) is the winning neuron for compound s. The first neighbors of c are shown in light gray.

(and the neighboring nodes) are iteratively modified in order to make the central node even closer to the current object. The weights of the neighboring units are decreased by an amount vanishing with increasing topological distance with unit c. This enhances the difference between the winner and its neighbors.

$$w_{ji}^{new} - w_{ji}^{old} = \alpha * h(d_{cs} - d_{js}, t) * (x_{si} - w_{ji}^{old})$$

where

α is the learning rate

t is the iteration number

$h(d_{cs} - d_{js}, t)$ is the neighborhood function which determines the modification degree for neighboring neurons

A triangular function or a "Mexican hat" function may be used to encode the influence of the distance to the winner unit. The change also vanishes with the iteration number t.

Applications: After training on an input data set, the weights form a basis for the projection of new data. SOM working by unsupervised learning received many applications in classification. Particularly, since mirroring the similarities between compounds (from their descriptors) they may be used (as principal component analysis) to select, for building QSAR models, representative *training* and *test sets* spanning the available structural space. Similarly, SOM could *select descriptors* using a transposed data matrix where indices indicating objects or descriptors are exchanged. From the map where units are occupied with descriptors, one or two descriptors from each neuron may be selected.

A major interest of SOMs is the possibility to *map 3D objects* onto a planar representation. This appeared particularly interesting for displaying values of atomic densities or representing molecular properties calculated on the 3D molecular surface, such as electrostatic or lipophilic potentials. See Refs. [401,402]. In addition, the information collected on a 3D surface, but organized in the SOM, can be introduced in quantitative 3D or 4D schemes as in comparative molecular surface analysis (CoMSA) and related methods [95,110,111,363,399,404,735–737,740,742]. For such applications, comparative maps are drawn, taking advantage of the fact that the weights of a trained network contain information about the molecular shape. A network trained on a reference compound is used to process a property recorded on the molecular surface in a series of molecules ("template" and "counter-template" CoMSA approach) [404,736].

At last, a direct extension of KNN is the counter-propagation NN (CPNN), see Chapter 5.

A.4.4.4 Hopfield Neural Networks

Hopfield neural networks [290,291] are generally known as "Content_Addressable Memory" able to retrieve incomplete or noisy information [979]. Various stimuli (input information), defined by the activation S_i of the neurons, are presented to the network; and the connections between the units are defined by summation. After training, if incomplete information is presented, the network is nevertheless able to deliver the good answer. However, Hopfield networks can also be used for solving combinatorial optimization problems, and inter alia, pattern matching [292,296].

In a Hopfield network, neurons (or "units") are arranged in a single layer and are fully interconnected in a symmetrical manner (see Figure 1.11): if w_{ij} is the strength of the connection (weight) between neurons i and j, $w_{ij} = w_{ji}$ and $w_{ii} = 0$. By contrast to layered BNN or KNN, connection strengths between units are here *fixed, depending on the problem to solve*. Activation of neurons only takes binary values 1 or 0 (alternatively 0, −1 may be used).

First, the network is initialized: neurons take at random values 0 or 1. Then the network evolves, and each neuron is updated using Hebb's rule

If

$$\sum w_{ij} S_j - \theta_i > 0 \rightarrow S_i = 1$$

Otherwise

$$S_i = 0$$

where
  the right part of the expression represents the input to neuron i coming from the other neurons j
  θ_i is the threshold value
  S_i is the activation of neuron i
  w_{ij} is the weight of the connection between units i and j

Various schemes may be adopted for this process. For example, periodically each neuron is randomly updated (with a mean common frequency). Alternatively, all neurons may be updated simultaneously, or one after another, in a defined order.

To the network, is associated an energy function

$$E = -\left(\frac{1}{2}\right)\sum_i \sum_{j\neq i} W_{ij}S_iS_j - \sum_i \theta_iS_i$$

The expression of W_{ij} and θ_i depends on the problem. When the network evolves according to Hebb's rule, it was shown that the energy always decreases or remains stationary. This property can be used to minimize symmetrical quadratic forms of binary variables: The problem is mapped onto a Hopfield network, weights corresponding to the coefficients of the quadratic form.

Alignment using a Hopfield network: The pattern matching problem comes to establish a correspondence between subsets of atoms of the target molecule (T), that under scrutiny, and of the reference template or substructure (R). This can be represented by a correspondence matrix where lines and columns represent the atoms of the template and of the target molecule, respectively, coefficients being 1 if the two atoms correspond or 0 otherwise. A candidate atom–atom correspondence is accepted or not depending on selected properties on potentially corresponding atoms [289,294]. As such, there is a close analogy to the *traveling salesman problem*: given a set of cities, how to find the shortest path passing one time and only one by each city? The solution is represented by a correspondence matrix between the cities and their rank in the travel, the objective function to minimize being the length (Figure A.22).

For the pattern matching (finding a substructure of m atoms in a molecule of n atoms) the problem is mapped on a Hopfield network of m*n binary units (1 for match, 0 otherwise). The objective function to minimize consists of four terms

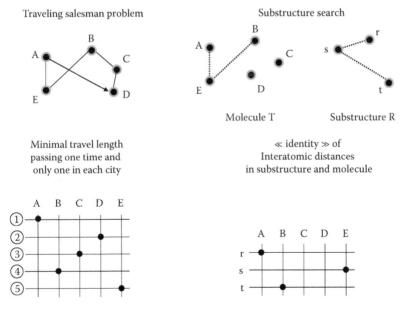

FIGURE A.22 Analogy between the traveling salesman problem and the substructure search: a correspondence matrix solution.

$$E = T_1 + T_2 + T_3 + T_4$$

The first two terms are

$$T_1 + T_2 = \sum_{i,m}\sum_{I,n}\sum_{j,m}\sum_{J,n} S_{iI}S_{jJ}[A\delta_{IJ}(1-\delta_{ij})+B\delta_{ij}(1-\delta_{IJ})]$$

where
   i,j refer to atoms of the substructure
   I,J refer to the molecule
   m,n are the number of properties assigned to molecule T and substructure R
   δ is the Kronecker delta function (1 if i = j, 0 otherwise)

These terms impose that there is, at the most, one «1» in each line and column.
After minimization of the energy function, each property of the atoms of the target
may correspond to the property of one, or zero, atom of the reference, and vice versa.
   The third term

$$T_3 = C\left[\min(m,n) - \sum_{i,m}\sum_{I,n} S_{iI}\right]^2$$

maximizes the number of correspondences.
   The fourth term (T_4) evaluates the distance between corresponding properties
(here interatomic distances in the molecule and the substructure)

$$T_4 = \frac{1}{2}D\left[\sum_{i,m}\sum_{I,n}\sum_{j,m}\sum_{J,n} S_{iI}S_{jJ}(1-\delta_{ij})(1-\delta_{IJ})\left|dT(ij)-dR(IJ)\right|\right]$$

In our pattern recognition approach, the constants A, B, C, and D were given values
of A = B = C = 2 and D = 1 (if distances are expressed in Å) [294]. The weights and
biases are determined from the objective function:

$$W_{iI,jJ} = -2A\delta_{IJ}(1-\delta_{ij})-2B\delta_{ij}(1-\delta_{IJ})-2C-D(1-\delta_{ij})(1-\delta_{IJ})|dT(ij)-dR(IJ)|$$

$$\theta iI = -2Cm$$

Once the connections are fixed, activations of the units are set at random and the network evolves until a stable configuration (the solution) is found. After the correspondence is found; the real molecular alignment can be performed by usual methods.
   An important improvement is the possibility to add a fifth term expressing supplementary constraints, for example, the difference in atom character (e.g., H-bond
donor or acceptor) or in an atomic property value (Pr). Evaluating the consistency of
each kind of property for candidate atoms is expressed by

$$T_5 = F\sum_{i,m}\sum_{I,n} S_{iI}\left|Pr(i)Pr(I)\right|$$

According to the relative values of constants D and F, more or less importance may be put on geometrical concordance or property similarities. Examples may be found in Ref. [289].

> *Remark*: In the work of Arakawa et al. work, the properties considered were [287,288]

- Hydrophobic group (HY): aromatic ring, CF_3, aliphatic chain of three atoms or more
- Hydrogen bonding donor (HD): NH in amine, amide, amidine, guanidine
- Hydrogen bonding acceptor (HA): oxygen of a carbonyl or sulfone group
- Hydrogen bonding donor/acceptor (HAD): O of OH, S in thiols

And the score matrix was defined according to Table A.1.

> *Boltzmann machine and mean field annealing*: One drawback of Hopfield neural networks is that they often remain stuck in a local mimimum (see, for example, Ref. [289]). But a new algorithm relieves this problem [296]. Alternatively, we proposed to use a *Boltzmann Machine* combining Hopfield network and Simulated Annealing with Metropolis Monte Carlo criterion.

In the study on hand, neurons are updated stochastically. The corresponding change in energy ΔE when updating a neuron is always accepted if the energy decreases. If ΔE is positive, the change is not automatically rejected; it is accepted if: the Boltzmann factor B

$$B = \exp\left(\frac{-\Delta E}{T}\right)$$

is higher than a random number in the interval [0,1], T being a pseudo-temperature.

TABLE A.1

Scores between Chemical Properties

	HY	HD	HA	HAD
HY	−3	Inf	Inf	Inf
HD	Inf	−2	Inf	−1
HA	Inf	Inf	−2	−1
HAD	Inf	−1	−1	−1

Source: Extracted from Iwase K. and Hirono S., J. Comput. Aided Mol. Des. 13; 499, 1999.

Note: Inf means infinite, and strongly penalizes the objective function (too differing properties cannot be put in correspondence); whereas negative signs are beneficial, since lowering the objective function.

The idea is to accept some moves increasing energy so that the system is able to escape from a local minimum. The "pseudo" temperature T is gradually lowered. At high "temperature," many changes in the neurons activations are accepted. But when T is low, less and less detrimental moves $\Delta E > 0$ are allowed and the system goes to the global energy minimum.

An alternative to the Boltzmann machine may be the so-called *mean field annealing* proposed by Bilbro et al. [295] and related to Ising spin Hamiltonian. The basic idea is to consider that the activation of a neuron i depends on the "mean field" expressing the action of the other neurons it feels, and a self-consistent configuration of the network is iteratively searched in the framework of a simulated annealing process. For more details see Ref. [289,294].

A.4.5 OTHER NONLINEAR METHODS

In parallel to the use of neural networks as correlation tools, various recent statistical methods relying on rather different concepts were introduced in QSAR models, in order to estimate complex regression surfaces. To our knowledge, their use is not yet widely generalized. But recent examples evidenced that these approaches may be useful in the context of 2.5D-QSARs and may constitute interesting alternatives to neural network approaches. See Ren [103], Panaye et al. [99], Yao et al. [97], Doucet et al. [101], and Hu et al. [102,877]. We only give some examples: projection pursuit regression (PPR) working in spaces of smaller dimensionality, support vector machine (SVM) operating in a higher dimension space, and multivariate adaptive regression splines (MARS) using smoothers. For other models also relying on smoothers, generalized additive model (GAM), least absolute shrinkage and selection operator (LASSO), localized weighted regression scatter plot smoothing (LOWESS), see Ren [103].

A.4.5.1 Projection Pursuit Regression

PPR is a nonparametric method developed by Friedman et al. [980–983]. The principle is to project high dimensional data onto lower dimensional (usually 1 or 2) spaces and then apply smoothing functions. These projections may provide more revealing views than the full-dimensional data. Particularly nonlinear effects are easily recognized in a low-dimensional visual representation of the data density. There are an infinite number of projections from a higher dimension to a lower dimension, and more than one projection may be interesting. It is thus important to have a technique to *pursue* a finite sequence of projections that can reveal the most interesting structures of the data; hence, the idea to combine both projection and pursuit in the so-called projection pursuit.

Given a trial direction vector $\alpha'_k[\alpha_{k1}, \alpha_{k2}, \alpha_{k3}, \ldots, \alpha_{kp}]$, the descriptor matrix $\mathbf{X}$(m descriptors * n compounds) is projected to matrix $\mathbf{Z}$ (p*n) with p < m

$$\mathbf{Z} = \alpha' \mathbf{X}$$

and a smooth curve is searched for to estimate the residual from $\mathbf{Z}$. The important elements are the direction of the projection ($\boldsymbol{\alpha}$) and the functions f. They are iteratively estimated. PPR models the response variable by a linear combination of K predictor functions of the projection $f_k(\alpha'_k \mathbf{X})$. The estimated value $\hat{y}$

$$\hat{y} = y + \sum_{k=1,K} \beta_k f_k(\alpha'_k \mathbf{X})$$

with $y = E(y) =$ mean of y_i; $E(f_k) = 0$, $E(f_k^2) = 1$, and $\sum_{j=1,p} \alpha_{kj}^2 = 1$

where E represents the expectation value.

The best projection has to be determined, according to its "interestingness"; for example, with f fixed, the optimal direction of projection will be that minimizing the residuals. The process is repeated forward working on the residuals (a backward fitting possibly adjusting the previous fitted pair). This procedure allows for removing the structure present in the solution and preserves the multivariate structure not yet captured. The process stops when the residual sum of squares is less than a predetermined value. For curve fitting, Friedman used the supersmoother function, but Hermitian polynomials or splines have also been used. Different smoothers for f_k may be found in the R Documentation [984].

As an example, it was indicated [983] that the (typically nonlinear) function

$$y = X_1 * X_2$$

could be modeled by

$$0.25(X_1 + X_2)^2 - 0.25(X_1 - X_2)^2$$

corresponding to $\alpha_1 = (0.7, 0.7)$ and $\alpha_2 = (0.7, -0.7)$. Of course, the analytical form of the data was not a priori known, and the model was extracted from the raw $\{y, X_1, X_2\}$ values.

PPR model can be used to approximate a large class of function by suitable choices of α_i and f_i. Using univariate regression functions leads to simple and efficient estimations, avoiding the curse of dimensionality. PPR and BPNN show some formal similarity in their basic expressions invoking transformed hidden variables

$$y(x_i) = \sum_{k=1,K} \beta_k f_k(\alpha'_k \mathbf{X}) y(x_i) \quad \text{or} \quad sf \left\{ \sum_h sf \left(\sum_i x_i w_{ih} \right) w_{ho} \right\}$$

sf representing the sigmoid transfer function. However, fitting is carried out successively for each term in PPR, simultaneously in BNN [985].

Up to now, PPR was not widely used in QSARs, but some applications have been published [102,103,877,986].

A.4.5.2 Multivariate Adaptive Regression Splines

MARS was explored by Friedman [987] and introduced in Chemistry by De Veaux et al. [988]. This is an adaptive regression procedure well suited to problems with a large number of predictor variables. MARS is a generalization of stepwise linear regression, but the regression is fitted using a series of basis functions, and in this aspect bears some analogy to the process of recursive partitioning in classification. The basis functions consist of one single spline function or two (or more) functions (Figure A.23).

For example

$$b_q^-(x-t) \quad \text{and} \quad b_q^+(x - t) \text{ with}$$

$$b_q^-(x-t) = [-(x-t)]_q^+ = \begin{cases} (t-x)^q, & \text{if } x < t \\ 0, & \text{otherwise} \end{cases}$$

$$b_q^+(x-t) = [+(x-t)]_q^+ = \begin{cases} (x-t)^q, & \text{if } x > t \\ 0, & \text{otherwise} \end{cases}$$

t is called the "knot" and the subscript "+" indicates that the positive part of the argument is considered (and a value of "0" for negative values of the argument).

The different basis functions are combined in one multidimensional model, which describes the response as a function of the explanatory variables. The result is a complex nonlinear model of the form

$$\hat{y} = a_0 + \sum_{m=1}^{M} a_m B_m(x)$$

where
    $\hat{y}$ is the predicted value for the response variable
    a_0 is the coefficient of the constant base function
    M is the number of base functions
    B_m and a_m are the mth base function and its coefficient

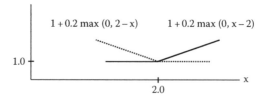

FIGURE A.23 A pair of splines with a knot at x = 2.0.

In MARS analysis, first a stepwise forward procedure selects, at each iteration, the best pair of basis functions in order to improve the model. All possible predictors and knot locations (for each predictor) are evaluated. This building process continues until a user-defined maximum number of basis functions is reached. This global model usually overfits the training data. A backward pass prunes the model and eliminates the least effective terms (possibly only one side of the paired splines). Interactions may also be introduced at the end of each iteration in the forward pass for a better representation.

A pioneering application was proposed by Massart et al. [989] who investigated the activity of 208 NNRTIs (TIBO- and HEPT-like molecules). The descriptors were constituted by the Coulomb and van der Waals interaction energies with the side chains and backbone parts of the 93 residues forming the binding pocket, resulting in an initial data matrix of 208*(93*2*2) values, reduced to 238*54 for the treatment. With 23 functions, q^2 reached 0.716.

A.4.5.3 Locally Weighted Scatterplot Smoothing LOWESS

LOWESS and its variant locally weighted scatterplot smoothing LOWESS (LOESS) may be viewed as an approach to regression analysis by local fitting [990,991]. Around each data point, a low-degree polynomial carries out a local fitting on a subset of weighted data points whose descriptors are neighbor to those of the point examined, the weighting scheme giving them an importance decreasing with their distance to this center.

LOWESS does not need any a priori choice of a function to fit the data (only the degree of the polynomial and the smoothing parameter are needed), and is thus very flexible. However, it is rather computationally intensive, needs large data sets and (more importantly) it does not deliver the answer as a mathematical formula.

A.4.5.4 Generalized Additive Model

GAM combines additive models and generalized linear models. It extends the concept of MLR but replacing the linear function of predictors by summation of nonparametric smoother functions of the predictors $(x_1, x_2, \ldots, x_m)$ [992,993]. Furthermore, the dependent variable (y) is connected to this linear form by a nonlinear transformation function, the "link function" g.

$$g(\hat{y}) = b + f_1(x_1) + f_2(x_2) + \cdots + f_m(x_m) + \varepsilon$$

where $\hat{y}$ is the fitted y value. Different functions may be introduced to ensure local smoothing: running mean, cubic splines, LOWESS tri-cube weight function or kernel functions with weights inversely proportional to the distance between the target and the points considered, for example. Note that parametric functions may also be added to the nonparametric terms.

Such an expression gives to GAM an extended flexibility. However, the model may lead to overfitting and the results are often difficult to interpret.

A.4.5.5 Least Absolute Shrinkage and Selecting Operator

LASSO proposed by Tibshirani [994] builds a linear model ($\mathbf{y} = \mathbf{xb'}$) where the vector of regression coefficients ($\mathbf{b}$) is estimated by minimizing the residual sum of squares subject to a constraint on the L^1-norm of the coefficient vector.

Minimize

$$\text{Min}_{\beta,1\beta m} \frac{1}{2} \sum_{i=1,n} \left(y_i - \sum_{j=1,m} x_{ij}\beta_j \right)^2$$

subject to

$$\sum_{j=1,m} \left| \beta_j \right| \leq s$$

where (s) is the tuning parameter. If large, it has nearly no effect; the result is similar to that of MLR. But, for smaller (s) values, shrunken versions are obtained, with some coefficients put to zero, which comes to a sort of variable selection. Choosing (s) is more or less as choosing the number of components in the model.

A good way to perform the calculation is the "least angle regression algorithm" (neighbor to the "forward stepwise regression algorithm"). Schematically, one starts with the variable x_j the most correlated with $\mathbf{y}$. Then the slope b_j is increased until another variable x_k better correlates with the residue ($y - y_{calc}$). The step is repeated again with the two variables x_j, x_k until another variable x_l better correlates with the new residue; and so on.

A.4.5.6 Support Vector Machine

The most widely used ANN layered networks working with back-propagation of errors (BNN) lead to a lot of successful applications. However, they suffer from some limitations: (a) architecture and parameter setting are difficult to fix, without any systematic method; (b) convergence of the algorithm may be slow, with the risk of getting stuck in a local minimum; (c) poor generalization capability may result from overfitting and overtraining, with a lack of robustness (nearly similar networks may give widely different results due to random initialization of the weights).

To overcome these difficulties, other artificial intelligence techniques have been introduced. Among them, SVM (initially proposed for classification problems and later extended to regression applications) has attracted attention, due to its remarkable generalization performance. It has now gained extensive applications in pattern recognition and regression problems where it seems very promising for nonlinear problems. As stated by Zernov [995], the main advantages of SVM are: (a) results are stable, reproducible, and largely independent of the optimization algorithm; (b) solution is guaranteed to be optimum, the quadratic programming approach, avoiding to get stuck at local minima; (c) a simple geometric interpretation is attainable; (d) any complex classifier can be built with the introduction of kernels for nonlinear decisions; (e) few parameters have to be adjusted like the regularization parameter (C), the nature and the parameters of the kernel function; (f) the result is built on a sparse subset of training samples. SVM avoids the "curse of dimensionality" in high-dimension problems.

SVM, developed by Vapnik et al. [100] appears to be a very promising classification and regression method. The basic principle is to privilege the capacity of generalization of the model over a very sharp recall of a training set: in other words, SVM prefers to minimize the structural risk rather than the empirical risk. SVM was the subject of several books and tutorials, and its applications in varied fields of chemo- and bioinformatics have been recently reviewed [100,101,996–999]. Information can also be found on several Internet sites:

http://www.csie.ntu.edu.tw/~cjlin/libsvm
http://www.kernel-machines.org; http://www.cs.columbia.edu/~bgrundy/svm
http://www.isis.ecs.soton.ac.uk/isystems/kernel, http://www.ncrg.aston.ac.uk/
  netlab

The basic idea is to map the input vectors $\mathbf{x}$ (structural descriptors) into a higher dimensional space, the *feature space* where a linear model can be applied.

$$\mathbf{x_i} \rightarrow \Phi(\mathbf{x_i})$$

This mapping is achieved thanks to a kernel function. The trick is that the kernel function makes it possible the projection of the descriptor vectors onto a space of high arbitrary dimension and allows for the calculations to be made from the initial space avoiding a painful explicit calculation in that high dimensional space. In the feature space, a *linear model* is built, generated by a *nonlinear transformation* of the initial data.

The desired values y_i will be approximated by z_i

$$z_i = \sum_{i=1,m} w_i \Phi(\mathbf{x_i}) + b$$

where $\Phi(\mathbf{x})$ represents a nonlinear transformation mapping $\mathbf{x}$ onto the feature space. The originality of the treatment is that the function $\Phi(\mathbf{x})$ does not need to be explicited. It is enough to know the kernel function $K(\mathbf{x_i},\mathbf{x_j})$ such as

$$K(\mathbf{x_i},\mathbf{x_j}) = \Phi(\mathbf{x_i}) * \Phi(\mathbf{x_j})$$

The coefficients ω and b are the hypotheses and have to be optimized during training. The quality of the model is measured by a loss function L(y, z). SVM regression uses a new type of loss function, proposed by Vapnik, a "ε-insensitive loss function"

$$L_\varepsilon(y,z) = |y - z| - \varepsilon$$

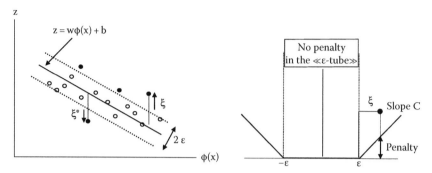

FIGURE A.24 Support vector regression. Left: with ε-insensitive loss function, only points deviating from more than ε are penalized with slack variable ξ or ξ*. Right: penalty associated with a slack variable ξ.

$$\text{If } |y - z| - \varepsilon \geq 0$$

and 0 otherwise,

so that, deviations between the model prediction (z) and the observed y values are ignored if the error remains inferior to ε. In other words, in a tube of diameter 2ε around the output function, the cost of errors is zero (Figure A.24). This allows for protecting about overfitting and for using sparse data points to represent the decision function.

SVM aims to perform a linear regression in the high dimensional feature space, using ε-insensitive loss function and at the same time it tries to reduce the complexity of the model. This comes to minimize

$$\text{Min} \left[\frac{1}{2} \|w\|^2 + C \sum_{i=1,m} L_\varepsilon (y_i, z_i) \right]$$

The first term, a measure of function flatness, corresponds to reducing model complexity (minimization of the norm of the normal vector ω); the second one represents the empirical risk, m being the number of samples and C a regularization constant.

To take into account deviations of training samples out of the "ε-tube" (the ε-insensitive zone) (nonnegative) slack variables ξ_i, ξ_i^* (with i = 1 ... m) are introduced.

So, SVM regression is built as minimization of

$$\text{Min} \left[\frac{1}{2} \|w\|^2 + C \sum_{i=1,m} (\xi_i + \xi_i^*) \right]$$

Subject to constraints

$$y_i \leq z_i + \varepsilon + \xi_i$$

$$y_i \geq z_i - \varepsilon - \xi_i^*$$

$$\xi_i, \xi_i^* \geq 0 \qquad i = 1...m$$

(the symbol * distinguishes deviations above and below the regression line).

This optimization problem can be transformed into a quadratic programming problem, introducing Lagrange multipliers $\boldsymbol{\alpha}$ and $\boldsymbol{\alpha}^*$ (dimension m) [1000]. These multipliers are determined by minimizing the Lagrangian with respect to $\mathbf{w}$, b, ξ (and maximizing it with respect to $\boldsymbol{\alpha}$, $\boldsymbol{\alpha}^*$ and the Lagrange multipliers $\boldsymbol{\upsilon}$ of the slack variables)

$$\frac{1}{2}\|\mathbf{w}\|^2 + C\sum_{i=1,m}(\xi_i + \xi_i^*) - \sum_{i=1,m}(\upsilon_i\xi_i + \upsilon_i^*\xi_i^*) - \sum_{i=1,m}\alpha_i(\varepsilon + \xi_i + z_i - y_i)$$

$$- \sum_{i=1,m}\alpha_i^*(\varepsilon + \xi_i^* - z_i + y_i)$$

with $\alpha_i, \alpha_i^*, \upsilon_i, \upsilon_i^* \geq 0$. Differentiating over $\mathbf{w}$, b, ξ and ξ^* and substituting, we have to maximize F by respect to $\boldsymbol{\alpha}$, $\boldsymbol{\alpha}^*$. This leads to

$$F(\alpha_i, \alpha_i^*) = \sum_{i=1,m}y_i(\alpha_i - \alpha_i^*) - \varepsilon\sum_{i=1,m}(\alpha_i - \alpha_i^*) - \frac{1}{2}\sum_{i,j=1,m}(\alpha_i - \alpha_i^*)(\alpha_j - \alpha_j^*)\,K(x_i, x_j)\cdots$$

Subject to

$$0 \leq \alpha_i, \alpha_j \leq C \quad \text{and} \quad \sum_{i=1,m}(\alpha_i - \alpha_j^*) = 0$$

The summation would run over the m data points. However, only a limited number of coefficients $\alpha_i - \alpha_i^*$ have nonzero values. The associated data points are called the "support vectors" (points with prediction error larger than ε). Once these Lagrange multipliers have been determined, the regression weight coefficients can be written

$$w = \sum_{sv}(\alpha_i - \alpha_i^*)\,\Phi(\mathbf{x}_s)$$

The summation running only on the support vectors (SV) $\mathbf{x}_s$. And the solution becomes

$$\mathbf{z}(\mathbf{x}) = \sum_{sv}(\alpha_i - \alpha_i^*)K(\mathbf{x}_s, \mathbf{x}) + b$$

where
   $\mathbf{x}_s$ are, here, the support vectors (components of the training set)
   $\mathbf{x}$ is the current point

The regression weights are a function of the training objects and of the Lagrange multipliers that also indicate the relative importance of each training object. Note that only the support vectors contribute to define the final function and the solution is an expansion on these vectors only.

The generalization performance of SVM depends on a good setting of the following parameters: C regularization constant, ε of the ε-insensitive loss function, the kernel type with its specific parameters.

Various functions may be currently elected as kernels (linear, polynomial, Gaussian or sigmoid). The necessary and sufficient condition for a symmetric function to be a kernel is to be positive, semidefinite, that is with no negative eigenvalues (Mercer's condition [997,1001]). However, for SVM applications in chemoinformatics, Gaussian RBF kernels were largely preferred (ability to treat nonlinear cases, smaller number of hyperparameters influencing the complexity of the model selection, less numerical difficulties). It was also mentioned that linear kernels are a particular case of RBF kernels and that sigmoid kernel behaves like RBF for certain parameter settings [1002].

Let us also remark that Czerminski [1003] pointed out that the SVM formulation (with a Gaussian kernel) is quite similar to a regularized RBFNN, where the decision function is a sum of Gaussian functions. The support vectors and hidden units are comparable. Whereas in RBFNN different strategies may be used to select the centers of the RBF expansions, these centers, in SVM, are restricted to the support vectors (a subset of the training points) and are selected by the algorithm itself.

More recently, a universal Pearson VII function (PUK) or the Tanimoto index were also proposed [1004,1005] and graph kernels directly related to the organization of the molecular skeleton were also introduced [1006,1007].

Selection of the kernel function is particularly important, because it implicitly determines the distribution of the training set samples in the feature space and so in the linear model to be applied. For such a kernel,

$$k(u, v) = \exp(-\gamma |u - v|^2)$$

the most important parameters are the spread γ of the Gaussian and the limit ε for tolerating deviations to the model. As to the regularization parameter C, it controls the trade-off between maximizing the margin and minimizing the training error. If C is too small, insufficient stress is placed on fitting the data. If C is too large, deviations from $|\varepsilon|$ count heavier, the algorithm would overfit the training data. For ε, the optimal value depends on the level of noise present in the data which is generally unknown. Even if some knowledge of the noise is available, a practical consideration is primordial: the value of ε affects the number of support vectors used to construct the regression model. The bigger ε is, the fewer support vectors are selected, with the risk of badly representing the data.

Selection of the proper values for C, g, and ε is usually carried out in a systematic grid search; the model with the best performance in L-O-O cross-validation being

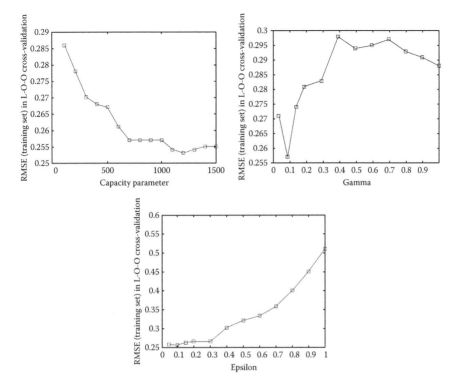

FIGURE A.25   An example of the impact of the various parameters (C, ε, and γ) on SVM results.

retained for further analysis. Influence of the choice of the capacity factor C, epsilon or width of the Gaussian kernel have been illustrated in Ref. [97] (Figure A.25).

In further developments (υ SVM), ε becomes a variable to optimize [1008]. In least squares SVM [1009,1010], the inequality constraint is replaced by an equality: LS-SVM fits a linear model in a kernel-induced space

$$y_i = \mathbf{w}'\Phi(\mathbf{x}_i) + b + \xi_i$$

w and b being determined by

$$\text{Min}\left[\frac{1}{2}\|\mathbf{w}\|^2 + \frac{1}{2}\gamma\sum_{i=1,m}\xi_i^2\right]$$

subject to

$$y_i\left[\mathbf{w}'\,\phi(\mathbf{x}_i) + b\right] = 1 - \xi_i$$

The solution is expressed by

$$z = \sum_{sv} \alpha_i\, K(\mathbf{x}_i, \mathbf{x}) + b$$

LS-SVM is claimed to be less sensitive to overfitting. See Ref. [1011] for an application.

Applications: In addition to numerous developments in classification (in Chemistry as well as in Bioinformatics) [101], support vector regression (SVR) was introduced as a correlation tool (in place of MLR, PLS, or CPANN) in various applications using 2D classical and/or 2.5D descriptors [101,524]. In was also introduced in the SAMFA approach [360], in data set mining [76] or in the search for supplementary descriptors with CoMFA [233].

Some comparisons: To our knowledge, only limited comparisons (on restricted data sets) of these various nonlinear regression methods are available, and have been quoted in the main text. In addition to Hu et al., results on 64 2-amino-6-arylsulfonylbenzonitriles and congeners as to anti-HIV-1 activity and HIV-1 RT binding affinity (for 51 derivatives) [102] seen in Chapter 12, we can give one more example.

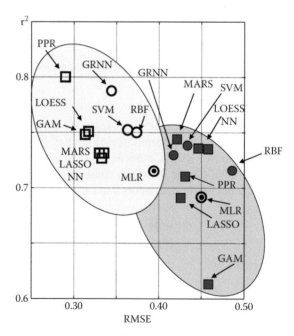

FIGURE A.26 Performances of the different models on toxicity of 202 nitro- and cyano-aromatics to *Tetrahymena pyriformis* in the response surface methodology: activity = f(K$_{ow}$, maximum acceptor superdelocalizability). Light gray: training set of 162 compounds; dark gray: test set of 40 compounds. Squares: Results from Ren S., *J. Chem. Inf. Comput. Sci.*, 43, 1679, 2003.; Circles: Results from Panaye A. et al., *SAR, QSAR Environ. Res.*, 17, 75, 2006.

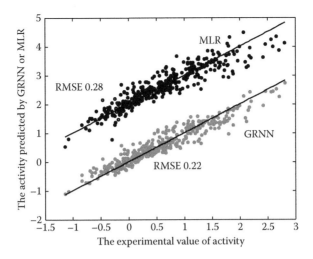

FIGURE A.27 Comparison of GRNN and MLR models in response surface methodology: toxicity of 384 aromatics to *Tetrahymena pyriformis*. For the sake of clarity, the MLR correlation has been shifted 2 units up.

The toxicity of 202 cyano- and nitro-aromatics to *Tetrahymena pyriformis* (a common test in eco-toxicology) was studied by Ren [103] and Panaye et al. [99]. Experimental data span a range of about 3.5 log 1/IGC$_{50}$ (50% growth inhibition concentration).The study was carried out by the response surface methodology, a general avenue in the field, referring to two basic mechanisms of action, uptake of the chemical into the biophase, transport and distribution, roughly speaking *penetration*, and *interaction* with the site of action. These two factors are approached using hydrophobicity descriptors K$_{ow}$ and, at 2.5D level, electophilic character (represented by the LUMO energy or, here, the maximum acceptor superdelocalizability).

The study of Ren [103] concerned six nonlinear models: GAM, LASSO, LOESS, MARS, PPR, and BNN. This study was completed by the use of RBFNN, GRNN, and SVM [99] and the results carefully validated using an external test set or 10-fold cross-validation (Figure A.26). Suffice here to indicate that all the methods investigated appear *better than MLR*, a non-unexpected result and that RBFNN, GRNN, or SVM competes favorably with the nonlinear methods previously used. It is interesting to remark that all these approaches, yielding here comparable performance, look rather different in their concepts: some (LOESS, MARS, PPR) use smoothers, BNN, RBFNN, and GRNN employ nonlinear transformations and SVM operates via a linear model in a feature space of enlarged dimensionality. As stated by Ren [103], this would suggest, on this example, that "the choice of the modeling method may simply be the personal preference of the model developer." Similar results were obtained on a more extended population of 384 aromatics (experimental data from the work of Schultz et al. [1012]) (Figure A.27).

Appendix B: The Steroid Benchmark

Data set of 31 steroids with affinity to CBG- and TBG-receptors has been extensively studied in the design of 3D models. For years, they constituted a reference benchmark. In most applications, compounds **s-1** through **s-21** form the training set and compounds **s-22** through **s-31**, the test set. However, compound **s-31** often deviated from the proposed models, and a reduced nine compound test set was also used. Note that this choice has been criticized. The activity range (less than 3 log units) is rather limited, the molecular skeleton does not suffer high flexibility, and, more importantly, the training set does not correctly span the structural space. Another splitting has thus been proposed by Kubinyi (see main text). Anyway, we gathered in Table B.1 some results obtained by various models.

The structural formulas and their CBG binding affinities (log K) are the following [48,126,396]:

s-1: 6.279

s-2: 5.000

s-3: 5.000

s-4: 5.763

s-5: 5.613

s-6: 7.881

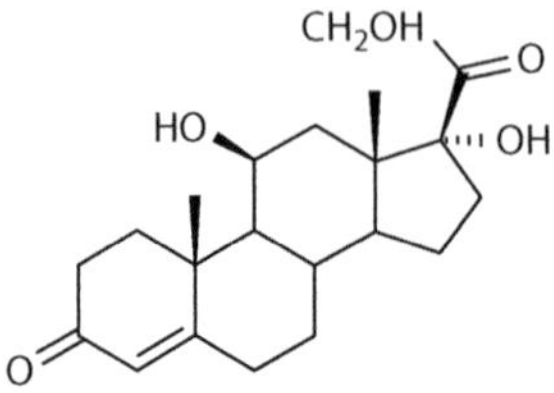

s-7: 7.881

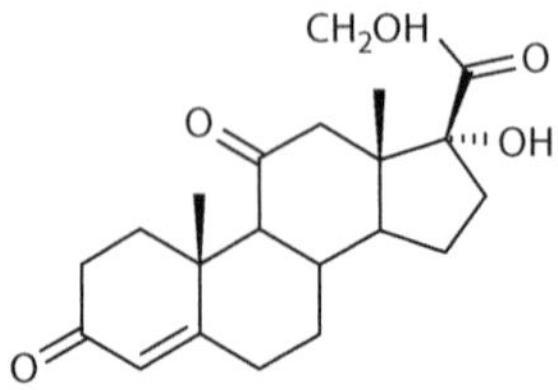

s-8: 6.892

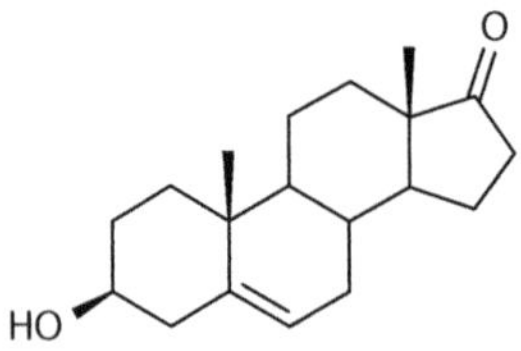

s-9: 5.000

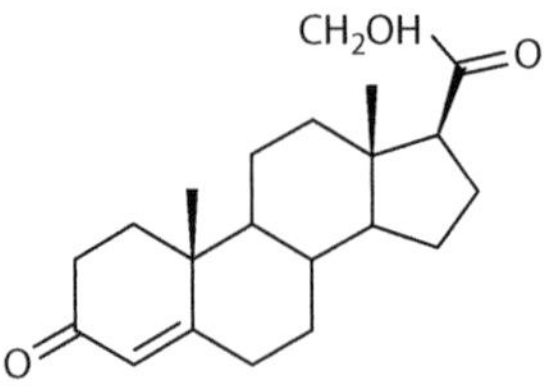

s-10: 7.653

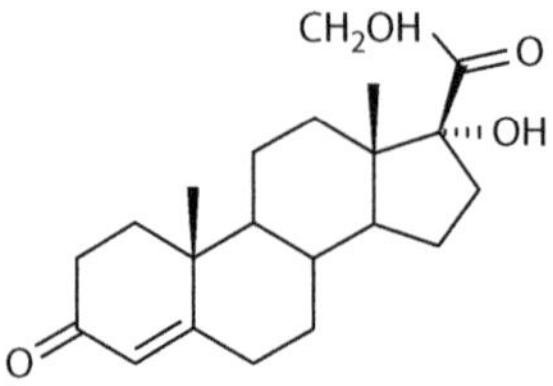

s-11: 7.881

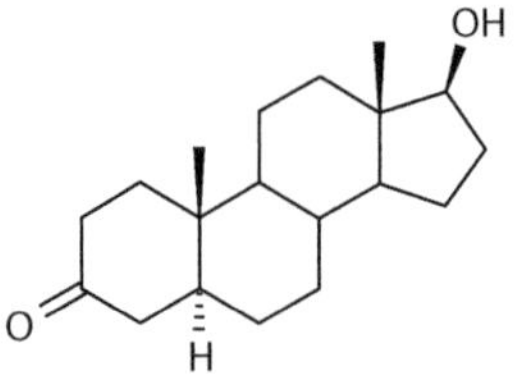

s-12: 5.919

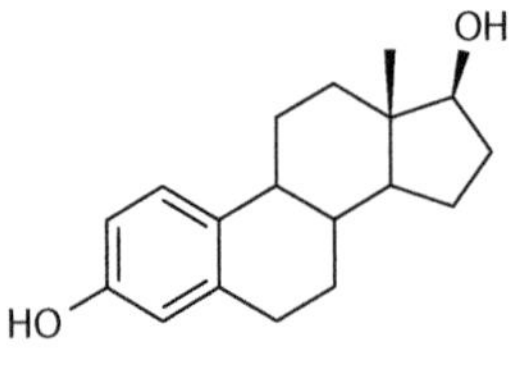

s-13: 5.000

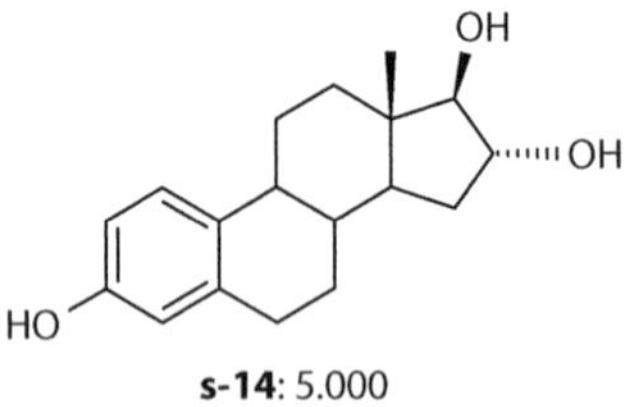

s-14: 5.000

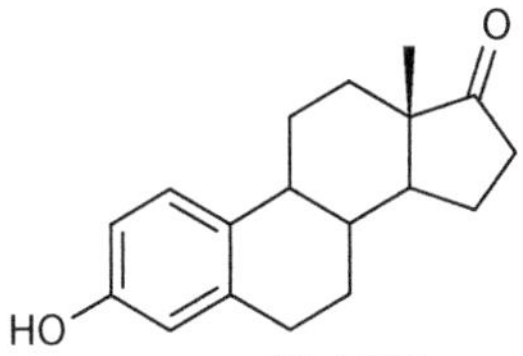

s-15: 5.000

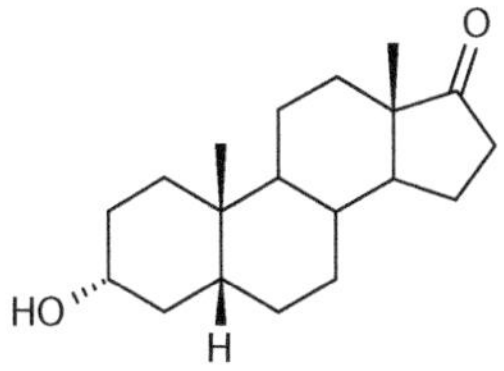

s-16: 5.225*

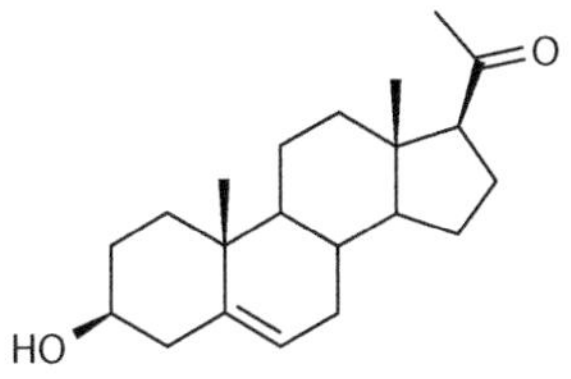

s-17: 5.225*

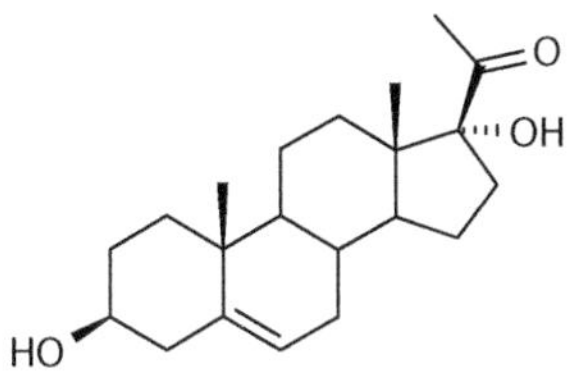

s-18: 5.000

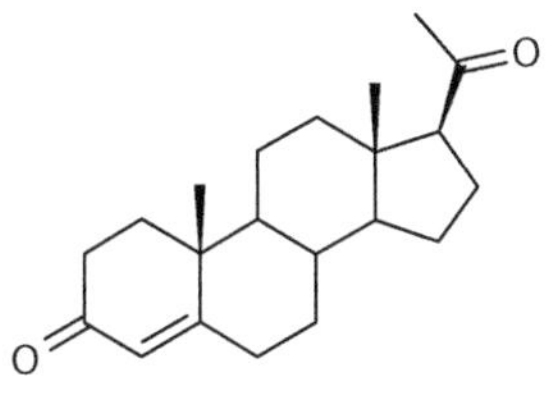

s-19: 7.380

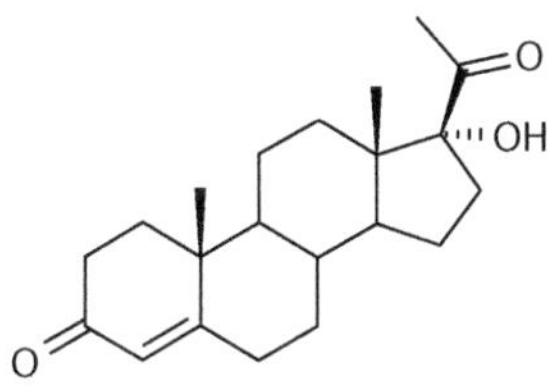

s-20: 7.740

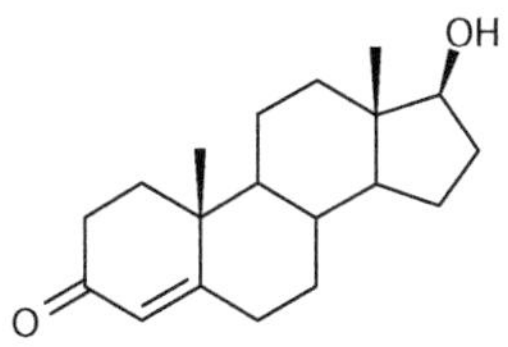

s-21: 6.724

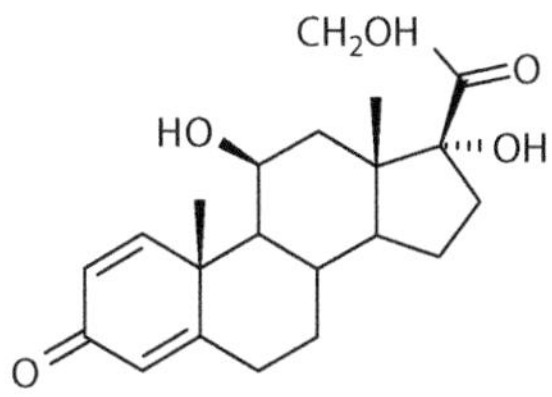

s-22: 7.512

s-23: 7.553

s-24: 6.779

s-25: 7.200

s-26: 6.114

s-27: 6.247

s-28: 7.120

s-29: 6.817

s-30: 7.668

s-31: 5.797

---

* For compound s-16 and s-17, Coats [126] reported log K = 5.255. The difference is of no significant consequence. See Ref. [146].

TABLE B.1
Samples of Steroid Activity (CBG-Receptor)

Methods	Full Set r^2	31 Compounds. LV	31 Compounds. q^2	Training 21 Compounds r^2-21	LV	q^2-21	Test-10 $r^2_{pred}-10$	Test-9 $r^2_{pred}-9$	Test-10 $r^2_{ext}-10$	Test-9 $r^2_{ext}-9$	References	
CoMFA-Cramer						0.665	0.31	0.45	0.05	0.12	[38,126, 390,664]	Re-evaluated in [126]
				0.873 s 0.45		0.75 0.75	sdep 0.8	sdep 0.75				Mean 4 CV runs
CoMFA, default				0.94	3	0.7 sdep 0.7	0.31 sdep 0.8	0.8 sdep 0.45	0.21	0.68	[182]	
CoMFA, best settings				0.95	2	0.84 sdep 0.49	0.64 sdep 0.58	0.91 sdep 0.3	0.34	0.73	[182]	
CoMFA, Robinson				0.896 rmse 0.39		0.662 rmse 0.71	0.24 sdep 0.84	0.77 sdep 0.49	0.14	0.58	[344]	
CoMFA, Ferguson				0.93 s 0.32	2	0.87 rmse 0.45	0.45 sdep 0.71	0.84 sdep 0.4	0.25	0.75	[530]	
CoMFA, Norinder					2	0.791	0.56				[171,191]	
CoMFA, Norinder					2	0.903	0.478				[191]	Variable select.
CoMFA, Bravi				0.719	3	0.84	0.24	0.77	0.14	0.58	[664]	

(continued)

TABLE B.1 (continued)
Samples of Steroid Activity (CBG-Receptor)

Methods	Full Set r²	31 Compounds. LV	31 Compounds. q²	Training 21 Compounds r² – 21	LV	q² – 21	Test-10 r²pred – 10	Test-9 r²pred – 9	Test-10 r²ext – 10	Test-9 r²ext – 9	References	
CoMFA-FFD, Bravi				s 0.84	3	spress 0.51 / 0.729	sdep 0.84 / 0.44	sdep 0.49 / 0.87	0.16	0.65	[664,530]	Train: 16 comp.
CoMFA-shape, Good					2	spress 0.61 / 0.76	sdep 0.72	sdep 0.36			[396]	Steric field only
CoMFA-q2 G.R.S.					2	s 0.60 / 0.79	sdep 0.76	sdep 0.42			[184]	
CoMFA, Coats					3	0.734 / s 0.66	0.47 / sdep 0.7	0.85 / sdep 0.39	0.13	0.71	[126]	
CoMFA, Kubinyi				0.897 / s 0.4		0.662 / spress 0.72	0.31				[115]	
CoMFA-SEAL, Kubinyi				0.947 / s 0.3		0.598 / spress 0.83	0.36				[115]	
CoMFA-HINT				0.973 / s 0.22	5	0.739 / s 0.63					[192]	3 fields

(continued)

COMFA, Kotani FLUFF-BALL (FIX)	0.719	2	0.662	0.894	3	0.758	-0.1	0.53	0.14	0.56	[43] [356]
CoMSIA				*s 0.4* 0.937 *s 0.33*	4	*spress 0.6* 0.665 *spress 0.76*	*sdep 1.0* 0.18	*sdep 0.69* 0.64			[40]
CoMSIA, Kubinyi				0.941		0.662					[115]
CoMSIA-SEAL				*s 0.32* 0.937 *s 0.33*	4	*spress 0.76* 0.665 *spress 0.76*	0.4				[115]	3 fields
CoMSIA, Kotani SOMFA	0.763	4	0.662	0.689							[43]
SOMFA							0.63 *sdep 0.58* 0.58	0.82 *sdep 0.36* 0.69	0.2	0.62	[344]	Test: 20 comp.
SOMFA COMMA (ab initio)		6	0.689		3	0.828	0.58	0.69			[344] [489]
MEDV-13	0.812 *rmse 0.46*		0.762 *rmse 0.52*	0.882 *rmse 0.52*		0.79 *rmse 0.52*	0.54 *sdep 0.65*	0.66 *sdep 0.59*	0.45	0.57	[686]
MEDEV GA	0.843 *rmse 0.42*		0.778 *rmse 0.50*								[687]
EEVA				0.97		0.84	0.64	0.85	0.36	0.58	[554]

TABLE B.1 (continued)
Samples of Steroid Activity (CBG-Receptor)

Methods	Full Set r^2	31 Compounds LV	31 Compounds q^2	Training 21 Compounds r^2 – 21	Training 21 Compounds LV	Training 21 Compounds q^2 – 21	Test-10 r^2_{pred} – 10	Test-9 r^2_{pred} – 9	Test-10 r^2_{ext} – 10	Test-9 r^2_{ext} – 9	References	
EVA				0.96 *spress* 0.24	2	0.8 *spress* 0.59	0.69 *sdep* 0.58	0.74 *sdep* 0.4	0.36	0.34	[44,51, 530,539]	
SAMFA-PLS			0.69	*s* 0.24		0.55 *s* 0.55	*sdep* 0.53	*sdep* 0.51			[360]	
TSAR				0.545		0.505	*sdep* 1.05				[422]	
3D-TBD				0.75	3	0.558	*sdep* 0.58				[422]	
Autocorrelation	0.98 *s* 0.18		0.63 *s* 0.65			0.689*	0.64				[146,*422]	
Autocorrelation	0.82		0.63								[1013]	
TDQ, grid				0.89	2	0.8	0.24	0.55			[398]	Nonbonded + electrostatic
TDQ, surface				0.9	2	0.68	0.01	0.82			[398]	shape + H-bond
COMPASS						0.89	0.46 *sdep* 0.71	0.89 *sdep* 0.34	0.16	0.69	[368]	
SM-SEAL, 3 fields		2	0.722		2	0.768	0.66				[115]	
SM-shape, Good	0.72	3	0.68		3	0.633 *s* 0.82	*sdep* 0.71	*sdep* 0.4			[396,551]	Steric only

Method											Ref	Comment
SM-ESP	0.764	0.734									[47]	
SM, Richards							0.56 sdep 0.64	0.86 sdep 0.38	0.12	0.51	[344]	
SM-GNN, Karplus	0.951	0.903	6								[48]	Electrostatic mean 50 runs
SM-GNN, Karplus		0.941									[48]	Electrostatic + shape
SM-Elec., Parretti	0.764	0.734	1	0.819	0.78	1					[1013]	
PARM				0.83	0.806 s 0.47		0.45 sdep 0.71	0.45 sdep 0.74	0.33	0.3	[1014]	
MFTA							0.9 sdep 0.3	0.9 sdep 0.31	0.87	0.82	[555]	
TQSI	0.837 s 0.43	0.775	3				0.58 sdep 0.62	0.61 sdep 0.63	0.31	0.27	[616]	
MQ-SM	0.833	0.759	5	0.939	0.832	6	0.6 sdep 0.6	0.84 sdep 0.41	0.17	0.43	[506]	Coulomb
MQ-SM	0.781 s 0.53	0.705	3								[616]	
TQSAR	0.828	0.779	4			6	0.37 sdep 0.76	0.69 sdep 0.56	0.16	0.36	[506]	Coulomb + overlap
TQSAR	0.909*	0.872*	5	0.939	0.832	6					[499,506]	*s-31 excluded
	0.903	0.842	6									Coulomb
QS-SM, fragments	0.917	0.866	4				0.68 sdep 0.54	0.76	0.36	0.22	[499]	
RSM				0.702	0.646		0.84 sdep 0.38	0.696 sdep 0.49	0.65	0.55	[390]	s-23 excluded
RSM open				0.664	0.628		0.84 sdep 0.38	0.85 sdep 0.39			[390]	

(continued)

TABLE B.1 (continued)
Samples of Steroid Activity (CBG-Receptor)

Methods	Full Set r^2	31 Compounds. LV	q^2	Training 21 Compounds r^2-21	LV	q^2-21	Test-10 $r^2_{pred}-10$	Test-9 $r^2_{pred}-9$	Test-10 $r^2_{ext}-10$	Test-9 $r^2_{ext}-9$	References	
COSASA	0.8	3	0.73								[551]	30 comp.
COSA	0.8	3	0.78								[351]	30 comp.
CoSCOSA-COSY	0.84	3	0.74								[351]	30 comp.
HE, E-state	0.98	3	0.8	0.98	3	0.8					[221]	
E-state	0.96	3	0.79								[221]	
QSAR/E-state	0.82	3	0.78								[551]	
SESP				0.94 *s 0.32*	4	0.87 *rmse 0.41*	*sdep >1*	0.81 *sdep 0.46*			[429]	
MaP	0.91 *s 0.36*		0.88 *s 0.38*	0.92 *s 0.38*		0.89 *s 0.39*		0.81 *sdep 0.44*			[458]	20/9 comp.
Grind	0.83	2	0.76	0.82	2	0.64 *s 0.67*		0.93 / 0.26		0.88	[45]	20/9 comp.
k-NN-MFA, GA						0.93	0.85 *sdep 0.37*	0.87 *sdep 0.36*	0.79	0.8	[171]	
s-COMSA + IVE			0.75 *s 0.54*			0.9	*sdep 0.77*				[363]	

COMSA	1	0.76 s 0.54		2	0.88 s 0.42	0.47 sdep 0.7	0.81 sdep 0.44	0.09	0.41	[95]	
SOM-4D	4	0.68 s 0.65		3	0.86 s 0.47	0.75				[735]	Occupancy
Grid-4D	2	0.69 s 0.63		2	0.84 s 0.5	sdep 0.83				[735]	Occupancy
Grid-4D-IVE					0.94 s 0.38	sdep 0.77				[363]	Occupancy
QUASAR		0.982			0.9					[363]	
COMASA	4	0.798	0.929	2	0.779 / 0.31	sdep 0.58				[43]	Different models
MS_WHIM	2			2	0.631 spress 0.75	0.41 sdep 0.74	0.74 sdep 0.54	0.24	0.48	[664]	
MS_WHIM-FFD	2			2	0.605 spress 0.8	0.52 sdep 0.66	0.87 sdep 0.41	0.28	0.63	[664]	Train: 13 comp.
WHIM	3			3	0.667 spress 0.74	-2.3 sdep 1.75	-1.6 sdep 1.6	0.16	0.3	[664]	s-23 outlier
WHIM-FFD						-1.9 sdep 1.65	-1.5 sdep 1.6	0.17	0.34	[664]	Train: 7 comp. / s-23 outlier
Excluded volume					0.766					[365]	Mean 10 assays

Some of the QSAR results on CBG-binding steroids.

Each model corresponds to two rows of the table. The upper line concerns the determination coefficients, the lower one the errors. The first three columns concern the whole data set (31 molecules) and specify the determination coefficient r^2, the number of components of the model (latent variables [LV]), and the L-O-O cross-validated determination coefficient q^2. The next three columns refer to the traditional training set (steroids **s-1** through **s-21**) and similarly give r^2, LV, and q^2 for this set. The last two columns relate to the test sets: molecules **s-22** through **s-31** (10 compounds) or **s-22** through **s-30** (nine compounds, excluding **s-31** that often appears as an outlier). In the lower lines, unless otherwise specified, the quality of the model is indicated (in italics) by the standard error (also referred to as *SPRESS* in C-V), which takes into account the complexity of the model (number of independent variables). In some cases, the *RMSE* (root mean squared error) is mentioned. For the test sets, *SDEP* (which only considers the number of compounds and not the complexity of the model) is reported.

Cramer's original result [38] is only indicated as a rough guide (several structures are inexact, there was no L-O-O cross validation). Corrected structures have been published by Wagener et al. [146] in 1995 and works after this date generally (but not always) used the good set. For a detailed discussion, see Coats [126]. In the Table B.1, the reported values either correspond to the results from original papers or if not available, have been recalculated from the published data.

Recently, Cherkasov et al. [1015], carried out docking and QSAR studies on ligands with affinity for sex hormone-binding globulin (SHBG), either steroidal or nonsteroidal compounds. This led the authors to propose an alignment differing from the traditional one, to be used in CoMFA-CoMSIA treatments of the classical benchmark of CBG binding steroids. However the 2 models yielded nearly identical results despite a quite different field distribution. So that a caveat was put on possible "misleading hypotheses concerning novel compound design".

References

1. H. Kubinyi, *From narcosis to hyperspace: The history of QSAR*, Quant. Struct. Act. Relat. 21 (2002), pp. 348–356.
2. T.W. Schultz, M.T.D. Cronin, J.D. Walker, and A.O. Aptula, *Quantitative structure-activity relationships (QSARs) in toxicology: A historical perspective*, J. Mol. Struct. (Theochem) 622 (2003), pp. 1–22.
3. P. Schmieder, O. Mekenyan, S. Bradbury, and G. Veith, *QSAR prioritization of chemical inventories for endocrine disruptor testing*, Pure Appl. Chem. 75 (2003), pp. 2389–2396.
4. P.K. Schmieder, G. Ankley, O. Mekenyan, J.D. Walker, and S. Bradbury, *Quantitative structure–activity relationship models for prediction of estrogen receptor binding affinity of structurally diverse chemicals*, Environ. Toxicol. Chem. 22 (2003), pp. 1844–1854.
5. R.F. Rekker, *The history of drug research: From Overton to Hansch*, Quant. Struct. Act. Relat. 11 (1992), pp. 195–199.
6. J. Parascandola, *To bond or not to bond: Chemical versus physical theories of drug action*, Bull. Hist. Chem. 28 (2003), pp. 1–8.
7. O.F. Güner, QSAR and QSPR at the 222nd ACS Meeting, Chicago IL, October 2001, Newsletter, 12, http://qsar.org
8. A.C. Brown and T.R. Fraser, *On the connection between chemical constitution and physiological action, Part I—On the physiological action of the salts of the ammonium bases, derived from Strychnia, Brucia, Thebaia, Codeia, Morphia, and Nicotia*, Trans. R. Soc. Edinb. 25 (1869), pp. 151–203.
9. A.C. Brown and T.R. Fraser, *On the connection between chemical constitution and physiological action, Part II—On the physiological action of the ammonium bases, derived from Atropia and Conia*, Trans. R. Soc. Edinb. 25 (1869), pp. 693–739.
10. A. Gallegos Saliner and X. Gironés, *Topological quantum similarity measures: Applications in QSAR*, J. Mol. Struct. (Theochem) 727 (2005), pp. 97–106.
11. J.N. Langley, *On the physiology of the salivary secretion. Part II. On the mutual antagonism of Atropin and Pilocarpin, having especial reference to their relations in the submaxillary gland of the cat*, J. Physiol. 1 (1878), pp. 339–369.
12. E. Fischer, *Einfluss der configuration auf die wirkung der enzyme*, Ber. Dtsch. Chem. Ges. 27 (1894), pp. 2985–2993; Part II, pp. 3479–3483.
13. F.W. Lichtenthaler, *100 years "schlüssel-schloss-prinzip": What made Emil Fischer use this analogy*, Angew. Chem. 106 (1994), pp. 2456–2467; Angew. Chem. Int. Ed. Engl. 33 (1994), pp. 2364–2374.
14. G. Folkers, Ed., *Lock and key—A hundred years after*, Emil Fischer Commemorate Symposium, Pharm. Acta Helv. 69 (1995), pp. 175–269 (special issue).
15. D.E. Koshland Jr., G. Némethy, and D. Filmer, *Comparison of experimental binding data and theoretical models in proteins containing subunits*, Biochemistry 5 (1966), pp. 365–385.
16. M.C. Richet, *Note sur le rapport entre la toxicite et les propriétés physiques des corps*, Compt. Rend. Soc. Biol. (Paris) 45 (1893), pp. 775–776.
17. H. Meyer, *Zur theorie der alkoholnarkose*, Arch. Exp. Pathol. Pharmakol. 42 (1899), pp. 109–118.
18. E. Overton, *Studien über dir narkose, zugleich ein Beitrag zur allgemeinen, Pharmakologie*, Gustav Fischer, Jena, Germany, 1901, pp. 1–195.
19. L.P. Hammett, *The effect of structure upon the reactions of organic compounds. Benzene derivatives*, J. Am. Chem. Soc. 59 (1937), pp. 96–103.

20. R.W. Taft Jr., *Linear free energy relationships from rates of esterification and hydrolysis of aliphatic and ortho-substituted benzoate esters*, J. Am. Chem. Soc. 74 (1952), pp. 2729–2732.

21. R.W. Taft Jr., *Polar and steric substituent constants for aliphatic and o-benzoate groups from rates of esterification and hydrolysis of esters*, J. Am. Chem. Soc. 74 (1952), pp. 3120–3128.

22. J. Ferguson, *The use of chemical potentials as indices of toxicity*, Proc. R. Soc. London Ser. B 127 (1939), pp. 387–404.

23. C. Hansch and T. Fujita, *ρ-σ-π analysis. A method for the correlation of biological activity and chemical structure*, J. Am. Chem. Soc. 86 (1964), pp. 1616–1626.

24. J.A. McPhee, A. Panaye, and J.E. Dubois, *Steric effects. 4. Multiparameter correlation models. Geometrical and proximity site effects for carboxylic acid esterification and related reactions*, J. Org. Chem. 45 (1980), pp. 1164–1166.

25. J.E. Leffler and E. Grunwald, *Rates and Equilibria of Organic Reactions*, John Wiley & Sons, New York, 1963.

26. A. Verloop, W. Hoogenstraaten, and J.Tipker, *Develoment and application of new steric substituent parameters in drug design*, in *Drug Design*, Vol. VII, E.J. Ariens, Ed., Academic Press, New York, 1976, pp. 165–207.

27. C. Hansch and J.M. Clayton, *Lipophilic character and biological activity of drugs. II: The parabolic case*, J. Pharm. Sci. 62 (1973), pp. 1–21.

28. C. Hansch, *A quantitative approach to biochemical structure–activity relationships*, Acc. Chem. Res. 2 (1969), pp. 232–239.

29. C. Hansch and A. Leo, *Exploring QSAR. Fundamentals and Applications in Chemistry and Biology*, ACS Professional Reference Book, American Chemical Society, Washington, DC, 1995.

30. S.M. Free Jr. and J.W. Wilson, *A mathematical contribution to structure–activity studies*, J. Med. Chem. 7 (1964), pp. 395–399.

31. T. Fujita and T. Ban, *Structure–activity relation. 3. Structure–activity study of phenethylamines as substrates of biosynthetic enzymes of sympathetic transmitters*, J. Med. Chem. 14 (1971), pp. 148–152.

32. T.C. Bruice, N. Kharasch, and R.J. Winzler, *A correlation of thyroxine-like activity and chemical structure*, Arch. Biochem. Biophys. 62 (1956), pp. 305–317.

33. J.E. Dubois, *Principles of the DARC topological system*, Entropie 27 (1969), pp. 1.

34. J.E. Dubois, J.P. Doucet, A. Panaye, and B.T. Fan, *DARC site topological correlations; Ordered structural descriptors and property evaluation*, in *Topological Indices and Related Descriptors in QSAR and QSPR*, J. Devillers and A.T. Balaban, Eds., Gordon & Breach Science Publishers, Amsterdam, the Netherlands, 1999, pp. 613–673.

35. H. Wiener, *Structural determination of paraffin boiling points*, J. Am. Chem. Soc. 69 (1947), pp. 17–20.

36. J. Devillers and A.T. Balaban, Eds., *Topological Indices and Related Descriptors in QSAR and QSPR*, Gordon & Breach Science Publishers, Amsterdam, the Netherlands, 1999.

37. R. Todeschini and V. Consonni, *Handbook of Molecular Descriptors*, Wiley-VCH, Weinheim, Germany, 2000.

38. R.D. Cramer III, D.E. Patterson, and J.D. Bunce, *Comparative molecular fields analysis (CoMFA). 1. Effect of shape on binding of steroids to carrier proteins*, J. Am. Chem. Soc. 110 (1988), pp. 5959–5967.

39. R.D. Cramer III and J.D. Bunce, *The DYLOMMS method: Initial results from a comparative study of approaches to 3D QSAR*, Pharmacochem. Libr. 10 (1987), pp. 3–12.

40. G. Klebe, U. Abraham, and T. Mietzner, *Molecular similarity indices in a comparative analysis (CoMSIA) of drug molecules to correlate and predict their biological activity*, J. Med. Chem. 37 (1994), pp. 4130–4146.

41. A.J. Hopfinger, *A QSAR investigation of dihydrofolate reductase inhibition by Baker triazines based upon molecular shape analysis*, J. Am. Chem. Soc. 102 (1980), pp. 7196–7206.

42. A.M. Doweyko, *The hypothetical active site lattice. An approach to modelling active sites from data on inhibitor molecules*, J. Med. Chem. 31 (1988), pp. 1396–1406.

43. T. Kotani and K. Higashiura, *Comparative molecular active site analysis (CoMASA). 1. An approach to rapid evaluation of 3D QSAR*, J. Med. Chem. 47 (2004), pp. 2732–2742.

44. D.B. Turner and P. Willett, *The EVA spectral descriptor*, Eur. J. Med. Chem. 35 (2000), pp. 367–375.

45. M. Pastor, G. Cruciani, I. McLay, S. Pickett, and S. Clementi, *GRid-INdependent Descriptors (GRIND): A novel class of alignment-independent three-dimensional molecular descriptors*, J. Med. Chem. 43 (2000), pp. 3233–3243.

46. A.G. Maldonado, J.P. Doucet, M. Petitjean, and B.T. Fan, *Molecular similarity and diversity in chemoinformatics: From theory to applications*, Mol. Divers. 10 (2006), pp. 39–79.

47. A.C. Good, S.J. Peterson, and W.G. Richards, *QSAR's from similarity matrices. Technique validation and application in the comparison of different similarity evaluation methods*, J. Med. Chem. 36 (1993), pp. 2929–2937.

48. S.S. So and M. Karplus, *Three-dimensional quantitative structure–activity relationships from molecular similarity matrices and genetic neural networks. 1. Method and validations*, J. Med. Chem. 40 (1997), pp. 4347–4359.

49. S.S. So and M. Karplus, *Three-dimensional quantitative structure–activity relationships from molecular similarity matrices and genetic neural networks. 2. Applications*, J. Med. Chem. 40 (1997), pp. 4360–4371.

50. R. Carbó, L. Leyda, and M. Arnau, *How similar is a molecule to another? An electron density measure of similarity between two molecular structures*, Int. J. Quantum Chem. 17 (1980), pp. 1185–1189.

51. D.B. Turner, P. Willett, A.M. Ferguson, and T. Heritage, *Evaluation of a novel infrared range vibration-based descriptor (EVA) for QSAR studies. 1. General application*, J. Comput. Aided Mol. Des. 11 (1997), pp. 409–422.

52. R. Bursi, T. Dao, T. van Wijk, M. de Gooyer, E. Kellenbach, and P. Verwer, *Comparative spectra analysis (CoSA): Spectra as three-dimensional molecular descriptors for the prediction of biological activities*, J. Chem. Inf. Comput. Sci. 39 (1999), pp. 861–867.

53. A.R. Katritzky and V.S. Lobanov, CODESSA, version 5.3, University of Florida, Gainesville, FL, 1994.

54. R. Todeschini, V. Consonni, and M. Pavan, Dragon Software. http//www.talete.mi.it

55. ALMOND, v.3.3.0, Molecular discovery via stoppani, Perugia, Italy, 2000, http://www.moldiscovery.com

56. Cerius², version 4.10, QSAR, Accelerys, January 2005.

57. A.J. Stuper and P.C. Jurs, *ADAPT: A computer system for automated data analysis using pattern recognition techniques*, J. Chem. Inf. 16 (1976), pp. 99–105. http://research.chem.psu.edu./pcjgroup/ADAPT.html

58. P.J. Goodford, *A computational procedure for determining energetically favorable binding sites on biologically important macromolecules*, J. Med. Chem. 28 (1985), pp. 849–857.

59. W. Sippl, *Receptor-based 3D QSAR analysis of estrogen receptor ligands—Merging the accuracy of receptor-based alignments with the computational efficiency of ligand-based methods*, J. Comput. Aided Mol. Des. 14 (2000), pp. 559–572.

60. P.A. Kollman, I. Massova, C. Reyes, B. Kuhn, S. Huo, L. Chong, M. Lee et al., *Calculating structures and free energies of complex molecules: Combining molecular mechanics and continuum models*, Acc. Chem. Res. 33 (2000), pp. 889–897.

61. B.O. Brandsdal, F. Österberg, M. Almlöf, I. Feierberg, V.B. Luzhkov, and J. Åqvist, *Free-energy calculation and ligand binding*, Adv. Protein Chem. 66 (2003), pp. 123–158.

62. V. Lukacova and S. Balaz, *Multimode ligand binding in receptor site modeling: Implementation in CoMFA*, J. Chem. Inf. Comput. Sci. 43 (2003), pp. 2093–2105.

63. A.J. Hopfinger, S. Wang, J.S. Tokarski, B. Jin, M. Albuquerque, P.J. Madhav, and C. Duraiswami, *Construction of 3D-QSAR models using the 4D-QSAR analysis formalism*, J. Am. Chem. Soc. 119 (1997), pp. 10509–10524.

64. P. Venkatarangan and A.J. Hopfinger, *Prediction of ligand–receptor binding free energy by 4D-QSAR analysis: Application to a set of glucose analogue inhibitors of glycogen phosphorylase*, J. Chem. Inf. Comput. Sci. 39 (1999), pp. 1141–1150.

65. C.L. Senese, J. Duca, D. Pan, A.J. Hopfinger, and Y.J. Tseng, *4D-Fingerprints universal QSAR and QSPR descriptors*, J. Chem. Inf. Comput. Sci. 44 (2004), pp. 1526–1539.

66. K. Hasegawa, M. Arakawa, and K. Funatsu, *3D-QSAR study of insecticidal neonicotinoid compounds based on 3-way partial least squares model*, Chemom. Intell. Lab. Syst. 47 (1999), pp. 33–40.

67. K. Hasegawa, S. Matsuoka, M. Arakawa, and K. Funatsu, *Multi-way PLS modeling of structure-activity data by incorporating electrostatic and lipophilic potentials on molecular surface*, Comput. Biol. Chem. 27 (2003), pp. 381–386.

68. A. Vedani, H. Briem, M. Dober, H. Dollinger, and D.R. McMasters, *Multiple-conformation and protonation-state representation in 4D-QSAR: The neurokinin-1 receptor system*, J. Med. Chem. 43 (2000), pp. 4416–4427.

69. A. Vedani and M. Dobler, *5D-QSAR: The key for simulating induced fit?* J. Med. Chem. 45 (2002), pp. 2139–2149.

70. A. Vedani and M. Dobler, *Multidimensional QSAR: Moving from three- to five-dimensional concepts*, Quant. Struct. Act. Relat. 21 (2002), pp. 382–390.

71. A. Vedani, A.V. Descloux, M. Spreafico, and B. Ernst, *Predicting the toxic potential of drugs and chemicals in silico: A model for peroxisome proliferator-activated receptor γ (PPARγ)*, Toxicol. Lett. 173 (2007), pp. 17–23.

72. J.J. Sutherland, L.A. O'Brien, and D.F. Weaver, *A comparison of methods for modeling quantitative structure-activity relationships*, J. Med. Chem. 47 (2004), pp. 5541–5554.

73. A. Vedani, M. Dobler, and M.A. Lill, *Combining protein modeling and 6D-QSAR. Simulating the binding of structurally diverse ligands to the estrogen receptor*, J. Med. Chem. 48 (2005), pp. 3700–3703.

74. E. Carosati, R. Mannhold, P. Whal, J.B. Hansen, T. Fremming, I. Zamora, G. Cianchetta, and M. Baroni, *Virtual screening for novel openers of pancreatic K_{ATP} channels*, J. Med. Chem. 50 (2007), pp. 2117–2126.

75. Q.Y. Zhang, J. Wan, X. Xu, G.F. Yang, Y.L. Ren, J.J. Liu, H. Wang, and Y. Guo, *Structure-based rational quest for potential novel inhibitors of human HMG-CoA reductase by combining CoMFA 3D QSAR modeling and virtual screening*, J. Comb. Chem. 9 (2007), pp. 131–138.

76. S. Oloff, R.B. Mailman, and A. Tropsha, *Application of validated QSAR models of D_1 dopaminergic antagonists for database mining*, J. Med. Chem. 48 (2005), pp. 7322–7332.

77. J.G. Topliss and R.J. Costello, *Chance correlations in structure–activity studies using multiple regression analysis*, J. Med. Chem. 15 (1972), pp. 1066–1068.

78. S. Wold, C. Albano, W.J. Dunn III, U. Edlund, K. Esbensen, P. Geladi, S. Hellberg, E. Johansson, W. Lindberg, and M. Sjostrom, *Multivariate data analysis in chemistry*, in *Chemometrics: Mathematics and Statistics in Chemistry*, B. Kowalski, Ed., Reidel, Dordrecht, the Netherlands, 1984, pp. 17–95.

79. J.E. Dubois, J.J. Aaron, P. Alcais, J.P. Doucet, F. Rothenberg, and R. Uzan, *Quantitative study of substituent interactions in aromatic electrophilic substitution. I. Bromination of polysubstituted benzenes*, J. Am. Chem. Soc. 94 (1972), pp. 6823–6828.

80. G. Goethals, F. Membrey, B. Ancian, and J.P. Doucet, *Variable substituent response to electron demand and substituent-substituent interactions in protonation equilibriums of 1,1 diphenylethylenes*, J. Org. Chem. 43 (1978), pp. 4944–4947.

81. B. Ancian, F. Membrey, and J.P. Doucet, *Carbon-13 chemical shift response to substituent effects in arylmethyl and arylhydroxy carbenium ions. Evidence for substituent interaction in disubstituted ions depending upon the carbenium-like character at the trigonal carbon*, J. Org. Chem. 43 (1978), pp. 1509–1518.

82. J.W. McFarland, *Comparative molecular field analysis of anticoccidial triazines*, J. Med. Chem. 35 (1992), pp. 2543–2550.

83. J. Devillers, Ed., *Neural Networks in QSAR and Drug Design*, Academic Press, San Diego, CA, 1996.

84. J. Devillers, *Genetic algorithms in computer-aided molecular design*, in *Genetic Algorithms in Molecular Modeling*, J. Devillers, Ed., Academic Press, San Diego, CA, 1996, pp. 1–34.

85. B.T. Luke, *Evolutionary programming applied to the development of quantitative structure activity relationships and quantitative structure–property relationships*, J. Chem. Inf. Comput. Sci. 34 (1994), pp. 1279–1287.

86. B.T. Luke, *An overview of genetic methods*, in *Genetic Algorithms in Molecular Modeling*, J. Devillers, Ed., Academic Press, San Diego, CA, 1996, pp. 35–66.

87. J. Zupan and J. Gasteiger, *Neural networks: A new method for solving chemical problems or just a passing phase?* Anal. Chim. Acta 248 (1991), pp. 1–30.

88. J. Gasteiger and J. Zupan, *Neural networks in chemistry*, Angew. Chem. Int. Ed. Engl. 32 (1993), pp. 503–527.

89. D.J. Livingstone and D.T. Manallack, *Neural networks in 3D QSAR*, QSAR Comb. Sci. 22 (2003), pp. 510–518.

90. S.P. Niculescu, *Artificial neural networks and genetic algorithms in QSAR*, J. Mol. Struct. (Theochem) 622 (2003), pp. 71–83.

91. K.L.E. Kaiser, *The use of neural networks in QSARs for acute aquatic toxicological endpoints*, J. Mol. Struct. (Theochem) 622 (2003), pp. 85–95.

92. D.T. Manallack and D.J. Livingstone, *Neural networks in drug discovery: Have they lived up to their promise?* Eur. J. Med. Chem. 34 (1999), pp. 195–208.

93. T. Aoyama, Y. Suzuki, and H. Ichikawa, *Neural networks applied to pharmaceutical problems. III. Neural networks applied to quantitative structure–activity relationship (QSAR) analysis*, J. Med. Chem. 33 (1990), pp. 2583–2590.

94. T.A. Andrea and H. Kalayeh, *Applications of neural networks in quantitative structure–activity relationships of dihydrofolate reductase inhibitors*, J. Med. Chem. 34 (1991), pp. 2824–2836.

95. J. Polanski and B. Walczak, *The comparative molecular surface analysis (COMSA): A novel tool for molecular design*, Comput. Chem. 24 (2000), pp. 615–625.

96. B. Walczack and D.L. Massart, *Local modelling with radial basis function networks*, Chemom. Intell. Lab. Syst. 50 (2000), pp. 179–198.

97. X.J. Yao, A. Panaye, J.P. Doucet, R.S. Zhang, H.F. Chen, M.C. Liu, Z.D. Hu, and B.T. Fan, *Comparative study of QSAR/QSPR correlations using support vector machines, radial basis function neural network and multiple linear regression*, J. Chem. Inf. Comput. Sci. 44 (2004), pp. 1257–1266.

98. T. Niwa, *Using general regression and probabilistic neural netwotks to predict human intestinal absorption with topological descriptors derived from two-dimensional chemical structures*, J. Chem. Inf. Comput. Sci. 43 (2003), pp. 113–119.

99. A. Panaye, B.T. Fan, J.P. Doucet, X.J. Yao, R.S. Zhang, M.C. Liu, and Z.D. Hu, *Quantitative structure–toxicity relationships (QSTRs): A comparative study of various non linear methods. General regression neural network, radial basis function neural network and support vector machine in predicting toxicity of nitro- and cyano-aromatics to Tetrahymena pyriformis*, SAR QSAR Environ. Res. 17 (2006), pp. 75–91.

100. V.N. Vapnik, *The Nature of Statistical Learning Theory*, Springer-Verlag, Berlin, Germany, 1995.

101. J.P. Doucet, F. Barbault, H. Xia, A. Panaye, and B.T. Fan, *Nonlinear SVM approaches to QSPR/QSAR studies and drug design*, Curr. Comput.-Aided Drug Des. 3 (2007), pp. 263–289.

102. R. Hu, J.P. Doucet, M. Delamar, and R. Zhang, *QSAR models for 2-amino-6-arylsulfonylbenzonitriles and congeners HIV-1 reverse transcriptase inhibitors based on linear and non linear regression methods*, Eur. J. Med. Chem. 44 (2009), pp. 2158–2171.

103. S. Ren, *Modeling the toxicity of aromatic compounds to Tetrahymena pyriformis: The response surface methodology with nonlinear methods*, J. Chem. Inf. Comput. Sci. 43 (2003), pp. 1679–1687.

104. A. Yasri and D.J. Hartsough, *Toward an optimal procedure for variable selection and QSAR model building*, J. Chem. Inf. Comput. Sci. 41 (2001), pp. 1218–1227.

105. R. Leardi, *Genetic algorithms in feature selection*, in *Genetic Algorithms in Molecular Modeling*, J. Devillers, Ed., Academic Press, San Diego, CA, 1996, pp. 67–86.

106. H. Kubinyi, *Variable selection in QSAR studies. I. An evolutionary algorithm*, Quant. Struct. Act. Relat. 13 (1994), pp. 285–294.

107. H. Kubinyi, *Variable selection in QSAR studies. II. A highly efficient combination of systematic search and evolution*, Quant. Struct. Act. Relat. 13 (1994), pp. 393–401.

108. D. Rogers and A.J. Hopfinger, *Application of genetic function approximation to quantitative structure–activity relationships and quantitative structure–property relationships*, J. Chem. Inf. Comput. Sci. 34 (1994), pp. 854–866.

109. D. Rogers, *Some theory and examples of genetic function approximation with comparison to evolutionary techniques*, in *Genetic Algorithms in Molecular Modeling*, J. Devillers, Ed., Academic Press, San Diego, CA, 1996, pp. 87–107.

110. J. Polanski and R. Gieleciak, *The comparative molecular surface analysis (CoMSA) with modified uninformative variable elimination-PLS (UVE-PLS) method: Application to the steroids binding the aromatase enzyme*, J. Chem. Inf. Comput. Sci. 43 (2003), pp. 656–666.

111. R. Gieleciak and J. Polanski, *Modeling robust QSAR. 2. Iterative variable elimination schemes for CoMSA: Application for modeling benzoic acid pK_a values*, J. Chem. Inf. Model. 47 (2007), pp. 547–556.

112. J.M. Sutter, S.L. Dixon, and P.C. Jurs, *Automated descriptor selection for quantitative structure–activity relationships using generalized simulated annealing*, J. Chem. Inf. Comput. Sci. 35 (1995), pp. 77–84.

113. A. Golbraikh and A. Tropsha, *Beware of q2!*, J. Mol. Graph. Model. 20 (2002), pp. 269–276.

114. A.O. Aptula, N.G. Jeliazkova, T.W. Schultz, and M.T.D. Cronin, *The better predictive model: High q2 for the training set or low root mean square error of prediction for the test set?* QSAR Comb. Sci. 24 (2005), pp. 385–396.

115. H. Kubinyi, F.A. Hamprecht, and T. Mietzner, *Three-dimensional quantitative similarity–activity relationships (3D QSiAR) from SEAL similarity matrices*, J. Med. Chem. 41 (1998), pp. 2553–2564.

116. K. Baumann, *Distance profiles (DiP): A translationally and rotationally invariant 3D structure descriptor capturing steric properties of molecules*, Quant. Struct. Act. Relat. 21 (2002), pp. 507–519.

117. W. Tong, H. Fang, H. Hong, Q. Xie, R. Perkins, J. Anson, and D.M. Sheehan, *Regulatory application of SAR/QSAR for priority setting of endocrine disruptors: A perspective*, Pure Appl. Chem. 75 (2003), pp. 2375–2388.

118. W. Tong, Q. Xie, H. Hong, L. Shi, H. Fang, and R. Perkins, *Assessment of prediction confidence and domain extrapolation of two structure–activity relationship models for predicting estrogen receptor binding activity*, Environ. Health Perspect. 112 (2004), pp. 1249–1254.

119. H. Kubinyi, *Comparative molecular field analysis (CoMFA)*, in *The Encyclopedia of Computational Chemistry*, Vol. 1, P.v.R Schleyer, N.L. Allinger, T. Clark, J. Gasteiger, P.A. Kollman, H.F Schaefer III, and P.R. Schreiner, Eds., John Wiley & Sons, Chichester, U.K., 1998, pp. 448–460.

120. W.J. Dunn and D. Rogers, *Genetic partial least squares in QSAR*, in *Genetic Algorithms in Molecular Modeling*, J. Devillers, Ed., Academic Press, San Diego, CA, 1996, pp. 109–130.

121. C.L. Waller, B.W. Juma, L.E. Gray Jr., and W.R. Kelce, *Three-dimensional quantitative structure–activity relationships for androgen receptor ligands*, Toxicol. Appl. Pharmacol. 137 (1996), pp. 219–227.

122. P.M. Burden, T.H. Ai, H.Q. Lin, M. Akinci, M. Costandi, T.M. Hambley, and G.A.R. Johnston, *Chiral derivatives of 2-cyclohexylideneperhydro-4,7-methanoindenes, a novel class of nonsteroidal androgen receptor ligand: Synthesis, X-ray analysis and biological activity*, J. Med. Chem. 43 (2000), pp. 4629–4635.

123. C.L. Waller, D.L. Minor, and J.D. McKinney, *Using three-dimensional quantitative structure–activity relationships to examine estrogen receptor binding affinities of polychlorinated hydroxybiphenyls*, Environ. Health Perspect. 103 (1995), pp. 702–707.

124. S.H. Unger and C. Hansch, *Model building in structure–activity relations. Reexamination of adrenergic blocking activity of β-halo-β-arylalkylamines*, J. Med. Chem. 16 (1973), pp. 745–749.

125. SYBYL. Program available from Tripos, Inc., St. Louis, MO.

126. E.A. Coats, *The CoMFA steroids as a benchmark dataset for development of 3D QSAR methods*, Perspect. Drug Discov. Des. 12–14 (1998), pp. 199–213.

127. U. Norinder, *Recent progress in CoMFA methodology and related techniques*, Perspect. Drug Discov. Des. 12–14 (1998), pp. 25–39.

128. R.D. Cramer III, S.A. DePriest, D.E. Patterson, and P.E. Hecht, *The developing practice of comparative molecular field analysis*, in *3D QSAR in Drug Design-Theory, Methods and Applications*, H. Kubinyi, Ed., ESCOM, Leiden, the Netherlands, 1993, pp. 443–485.

129. G. Folkers, A. Merz, and D. Rognan, *CoMFA: Scope and limitations*, in *3D QSAR in Drug Design, Theory, Methods and Applications*, H. Kubinyi, Ed., ESCOM Science Publishers B.V., Leiden, the Netherlands, 1993, pp. 583–618.

130. R.T. Kroemer, P. Hecht, S. Guessregen, and K.R. Liedl, *Improving the predictive quality of CoMFA models*, Perspect. Drug Discov. Des. 12–14 (1998), pp. 41–56.

131. C.L. Waller and G.E. Kellogg, *Adding chemical information to CoMFA models with alternative 3D QSAR fields*. NetSci. January 1996. http://www.awod.com/netsci/Science/Compchem/feature10.html

132. H. Kubinyi, Ed., *3D QSAR in Drug Design, Theory, Methods and Applications*, ESCOM Science Publishers B.V., Leiden, the Netherlands, 1993.

133. H. Kubinyi, G. Folkers, and Y.C. Martin, Eds., *3D QSAR in Drug Design.*, Vol. 2 *Ligand–Protein Interactions and Molecular Similarity*, Kluwer/ESCOM, Dordrecht, the Netherlands, 1998; also published in Perspect. Drug Discov. Des. 9–11 (1998).

134. H. Kubinyi, G. Folkers, and Y.C. Martin, Eds., *3D QSAR in Drug Design.*, Vol. 3 *Recent Advances*, Kluwer/ESCOM, Dordrecht, the Netherlands, 1998; also published in Perspect. Drug Discov. Des. 12–14 (1998).

135. Y.C. Martin, K.H. Kim, and C.T. Lin, *Comparative molecular field analysis: CoMFA*, in *Advances in Quantitative Structure–Property Relationships*, Vol. 1, M. Charton, Ed., JAI Press, Greenwich, CT, 1996, pp. 1–52.

136. K.H. Kim, *List of CoMFA references 1993–1996 and 1997*, Perspect. Drug Discov. Des. 12–14 (1998), pp. 317–333, 334–338.

137. K.H. Kim, G. Greco, and E. Novellino, *A critical review of recent CoMFA applications*, Perspect. Drug. Discov. Des. 12–14 (1998), pp. 257–315.

138. H. Kubinyi, *Comparative molecular field analysis (CoMFA)*, in *Handbook of Chemoinformatics. From Data to Knowledge*, Vol. 4, J. Gasteiger, Ed., Wiley-VCH, Weinheim, Germany, 2003, pp. 1555–1574.

139. U. Thibaut, G. Folkers, G. Klebe, H. Kubinyi, A. Merz, and D. Rognan, *Recommendations for CoMFA studies and 3D QSAR publications*, Quant. Struct. Act. Relat. 13 (1994), pp. 1–3.

140. Y.C. Martin, *3D-QSAR: Current state, scope and limitations*, Perspect. Drug Discov. Des. 12–14 (1998), pp. 3–23.

141. S. Peterson, *Improved CoMFA modeling by optimization of settings: Toward the design of inhibitors of the HCV NS3 protease*, PhD thesis, Faculty of Pharmacy, Uppsala University, Uppsala, Sweden, 2007.

142. S.P. Korhonen, *FLUFF-BALL, a fuzzy superposition and QSAR technique: Towards an automated computational detection of biologically active compounds using multivariate methods*, PhD thesis, Kuopio University, Kuopio, Finland, 2007.

143. S. de Jong, *SIMPLS: An alternative approach to partial least squares regression*, Chemom. Intell. Lab. Syst. 18 (1993), pp. 251–263.

144. B.L. Bush and R.B. Nachbar Jr., *Sample-distance partial least squares: PLS optimized for many variables, with application to CoMFA*, J. Comput.-Aided Mol. Des. 7 (1993), pp. 587–619.

145. K. Hasegawa, T. Kimura, Y. Miyashita, and K. Funatsu, *Nonlinear partial least squares modeling of phenyl alkylamines with the monoamine oxidase inhibitory activities*, J. Chem. Inf. Comput. Sci. 36 (1996), pp. 1025–1029.

146. M. Wagener, J. Sadowski, and J. Gasteiger, *Autocorrelation of molecular surface properties for modeling corticostreroid binding globulin and cytosolic Ah receptor activity by neural networks*, J. Am. Chem. Soc. 117 (1995), pp. 7769–7775.

147. W. Kabsch, *A solution for the best rotation to relate two sets of vectors*, Acta Crystallogr. A 32 (1976), pp. 922–923.

148. C. Lemmen and T. Lengauer, *Computational methods for the structural alignment of molecules*, J. Comput.-Aided Mol. Des. 14 (2000), pp. 215–232.

149. B.A. Bhongade and A.K. Gadad, *3D-QSAR CoMFA/CoMSIA studies on Urokinase plasminogen activator (uPA) inhibitors: A strategic design in novel anticancer agents*, Bioorg. Med. Chem. 12 (2004), pp. 2797–2805.

150. S. Dove and A. Buschauer, *Improved alignment by weighted field fit in CoMFA of Histamine H_2 receptor agonistic imidazolylpropylguanidines*, Quant. Struct. Act. Relat. 18 (1999), pp. 329–341.

151. S.K. Kearsley and G.M. Smith, *An alternative method for the alignment of molecular structures: Maximizing electrostatic and steric overlap*, Tetrahedron Comput. Methodol. 3 (1990), pp. 615–633.

152. G. Klebe, T. Mietzner, and F. Weber, *Different approaches toward an automatic structural alignment of drug molecules: Applications to sterol mimics, thrombin and thermolysin inhibitors*, J. Comput.-Aided Mol. Des. 8 (1994), pp. 751–778.

153. M. Feher and J.M. Schmidt, *Multiple flexible alignment with SEAL: A study of molecules acting on the colchicine binding site*, J. Chem. Inf. Comput. Sci. 40 (2000), pp. 495–502.

154. D.M. Ferguson and D.J. Raber, *A new approach to probing conformational space with molecular mechanics: Random incremental pulse search*, J. Am. Chem. Soc. 111 (1989), pp. 4371–4378.

155. G. Jones, P. Willett, and R.C. Glen, *A genetic algorithm for flexible molecular overlay and pharmacophore elucidation*, J. Comput.-Aided Mol. Des. 9 (1995), pp. 532–549.

156. P. Labute, C. Williams, M. Feher, E. Sourial, and J.M. Schmidt, *Flexible alignment of small molecules*, J. Med. Chem. 44 (2001), pp. 1483–1490.

157. J.T. Bolin, D.J. Filman, D.A. Matthews, R.C. Hamlin, and J. Kraut, *Crystal structures of Escherichia coli and Lactobacillus casei dihydrofolate reductase refined at 1.7 Å resolution. I. General features and binding of methotrexate*, J. Biol. Chem. 257 (1982), pp. 13650–13662.

158. J. Gasteiger and M. Marsili, *Iterative partial equalization of orbital electronegativity—A rapid access to atomic charges*, Tetrahedron 36 (1980), pp. 3219–3228.

159. J. Gasteiger and H. Saller, *Calculation of the charge distribution in conjugated systems by a quantification of the resonance concept*, Angew. Chem. Int. Ed. Engl. 24 (1985), pp. 687–689.

160. M. Marsili and J. Gasteiger, *Pi-Charge distribution from molecular topology and pi-orbital electronegativity*, Croat. Chem. Acta 53 (1980), pp. 601–614.

161. J. Gasteiger and M. Marsili, *Prediction of proton magnetic resonance shifts: The dependence on hydrogen charges obtained by iterative partial equalization of orbital electronegativity*, Organ. Magn. Reson. 15 (1981), pp. 353–360.

162. M.J.S. Dewar, E.G. Zoebisch, E.F. Healy, and J.J.P. Stewart, *Development and use of quantum mechanical molecular models. 76. AM1: A new general purpose quantum mechanical molecular model*, J. Am. Chem. Soc. 107 (1985), pp. 3902–3909.

163. J.J.P. Stewart, *Optimization of parameters for semiempirical methods I. Method*, J. Comput. Chem. 10 (1989), pp. 209–220.

164. S. Hellberg, S. Wold, W.J. Dunn III, J. Gasteiger, and M.G. Hutchings, *The anaesthetic activity and toxicity of halogenated ethyl methyl ethers, a multivariate QSAR modelled by PLS*, Quant. Struct. Act. Relat. 4 (1985), pp. 1–11.

165. P. Geladi and B.R. Kowalski, *Partial least-squares regression: A tutorial*, Anal. Chim. Acta 185 (1986), pp. 1–17.

166. M.M.C. Ferreira, *Multivariate QSAR*, J. Braz. Chem. Soc. 13 (2002), pp. 742–753.

167. R.D. Cramer III, J.D. Bunce, D.E. Patterson, and I.E. Frank, *Cross-validation, bootstrapping, and partial least squares compared with multiple regression in conventional QSAR studies*, Quant. Struct. Act. Relat. 7 (1988), pp. 18–25.

168. J.G. Topliss and R.P. Edwards, *Chance factors in studies of quantitative structure–activity relationships*, J. Med. Chem. 22 (1979), pp. 1238–1244.

169. T.W. Schultz, T.I. Netzeva, and M.T.D Cronin, *Evaluation of QSARs for ecotoxicity: A method for assigning quality and confidence*, SAR QSAR Environ. Res. 15 (2004), pp. 385–397.

170. K. Hasegawa, T. Kimura, and K. Funatsu, *Nonlinear CoMFA using QPLS as a novel 3D-QSAR approach*, Quant. Struct. Act. Relat. 16 (1997), pp. 219–223.

171. S. Ajmani, K. Jadhav, and S.A. Kulkarni, *Three-dimensional QSAR using the k-nearest neighbor method and its interpretation*, J. Chem. Inf. Model. 46 (2006), pp. 24–31.

172. C. Teixeira, N. Serradji, F. Maurel, and F. Barbault, *Docking and 3D-QSAR studies of BMS-806 analogs as HIV-1 gp120 entry inhibitors*, Eur. J. Med. Chem. 44 (2009), pp. 3524–3532.

173. C.L. Waller, T.I. Oprea, K. Chae, H.K. Park, K.S. Korach, S.C. Laws, T.E. Wiese, W.R. Kelce, and L.E. Gray Jr., *Ligand-based identification of environmental estrogens*, Chem. Res. Toxicol. 9 (1996), pp. 1240–1248.

174. R. Bursi and M.B. Groen, *Application of (quantitative) structure–activity relationships to progestagens: From serendipity to structure-based design*, Eur. J. Med. Chem. 35 (2000), pp. 787–796.

175. I.K. Pajeva and M. Wiese, *Interpretation of CoMFA results—A probe set study using hydrophobic fields*, Quant. Struct. Act. Relat. 18 (1999), pp. 369–379.

176. GRID v. 20. Molecular Discovery Ltd, West Way House, Elms Parade, Oxford, U.K., 2002.

177. M. Baroni, G. Costantino, G. Cruciani, D. Riganelli, R. Valigi, and S. Clementi, *Generating optimal linear PLS estimations (GOLPE): An advanced chemometric tool for handling 3D-QSAR problems*, Quant. Struct. Act. Relat. 12 (1993), pp. 9–20.

178. G. Cruciani, P. Crivori, P.-A. Carrupt, and B. Testa, *Molecular fields in quantitative structure–permeation relationships: The VolSurf approach*, J. Mol. Struct. (Theochem) 503 (2000), pp. 17–30.

179. P. Crivori, G. Cruciani, P.A. Carrupt, and B. Testa, *Predicting blood–brain barrier permeation from three-dimensional molecular structure*, J. Med. Chem. 43 (2000), pp. 2204–2216.

180. R. Ragno, S. Simeoni, S. Valente, S. Massa, and A. Mai, *3D-QSAR studies on histone deacetylase inhibitors. A GOLPE/GRID approach on different series of compounds*, J. Chem. Inf. Model. 46 (2006), pp. 1420–1430.

181. J.L. Melville and J.D. Hirst, *On the stability of CoMFA models*, J. Chem. Inf. Comput. Sci. 44 (2004), pp. 1294–1300.

182. S.D. Peterson, W. Schaal, and A. Karlén, *Improved CoMFA modeling by optimization of settings*, J. Chem. Inf. Model. 46 (2006), pp. 355–364.

183. E. Morgan, *Chemometrics: Experimental Design*, Chadwick N., Ed., John Wiley & Sons Ltd, Chichester, U.K., 1991.

184. S.J. Cho and A. Tropsha, *Cross-validated R^2-guided region selection for comparative molecular field analysis: A simple method to achieve consistent results*, J. Med. Chem. 38 (1995), pp. 1060–1066.

185. B.R. Sadler, S.J. Cho, K.S. Ishaq, K. Chae, and K.S. Korach, *Three-dimensional quantitative structure–activity relationship study of nonsteroidal estrogen receptor ligands using the comparative molecular field analysis/cross-validated r^2-guided region selection approach*, J. Med. Chem. 41 (1998), pp. 2261–2267.

186. R.X. Wang, Y. Gao, L. Liu, and L.H. Lai, *All-orientation search and all-placement search in comparative molecular field analysis*, J. Mol. Model. 4 (1998), pp. 276–283.

187. H.F. Chen, Q. Li, X.J. Yao, B.T. Fan, S.G. Yuan, A. Panaye, and J.P. Doucet, *CoMFA/CoMSIA/HQSAR and docking study of the binding mode of selective cyclooxygenase (COX-2) inhibitors*, QSAR Comb. Sci. 23 (2004), pp. 36–55.

188. L.M. Shi, H. Fang, W. Tong, J. Wu, R. Perkins, R.M. Blair, W.S. Branham, S.L. Dial, C.L. Moland, and D.M. Sheehan, *QSAR models using a large diverse set of estrogens*, J. Chem. Inf. Comput. Sci. 41 (2001), pp. 186–195.

189. M. Pastor, G. Cruciani, and S. Clementi, *Smart region definition: A new way to improve the predictive ability and interpretability of three-dimensional quantitative structure–activity relationships*, J. Med. Chem. 40 (1997), pp. 1455–1464.

190. G. Cruciani, S. Clementi, and M. Pastor, *GOLPE-guided region selection*, Perspect. Drug Discov. Des. 12–14 (1998), pp. 71–86.

191. U. Norinder, *Single and domain mode variable selection in 3D QSAR applications*, J. Chemom. 10 (1996), pp. 95–105.

192. G.E. Kellogg, S.F. Semus, and D.J. Abraham, *HINT: A new method of empirical hydrophobic field calculation for CoMFA*, J. Comput.-Aided Mol. Des. 5 (1991), pp. 545–552.

193. D.J. Abraham and A.J. Leo, *Extension of the fragment method to calculate amino acid zwitterion and side chain partition coefficients*, Proteins Struct. Funct. Genet. 2 (1987), pp. 130–152.

194. F.C. Wireko, G.E. Kellogg, and D.J. Abraham, *Allosteric modifiers of hemoglobin. 2. Crystallographically determined binding sites and hydrophobic binding/interaction analysis of novel hemoglobin oxygen effectors*, J. Med. Chem. 34 (1991), pp. 758–767.

195. U.J. Norinder, *Experimental design based 3-D QSAR analysis of steroid–protein-interactions: Application to human CBG complexes*, J. Comput.-Aided Mol. Des. 4 (1990), pp. 381–389.

196. K.P. Coleman, W.A. Toscano Jr., and T.E. Wiese, *QSAR models of the in vitro estrogen activity of bisphenol A analogs*, QSAR Comb. Sci. 22 (2003), pp. 78–88.

197. J.L. Fauchére, P. Quarendon, and L. Kaetterer, *Estimating and representing hydrophobicity potential*, J. Mol. Graph. 6 (1988), pp. 203–206.

198. B. Testa, P.A. Carrupt, P. Gaillard, F. Billois, and P. Weber, *Lipophilicity in molecular modeling*, Pharm. Res. 13 (1996), pp. 335–343.

199. P. Gaillard, P.A. Carrupt, B. Testa, and A. Boudon, *Molecular lipophilicity potential, a tool in 3D QSAR: Method and applications*, J. Comput.-Aided Mol. Des. 8 (1994), pp. 83–96.

200. P. Broto, G. Moreau, and C. Vandycke, *Molecular structures: Perception, autocorrelation descriptor and SAR studies. System of atomic contribution for the calculation of the n-octanol/water partition coefficients*, Eur. J. Med. Chem. 19 (1984), pp. 71–78.

201. A.K. Ghose and G.M. Crippen, *Atomic physicochemical parameters for three-dimensional structure-directed quantitative structure–activity relationships I. Partition coefficients as a measure of hydrophobicity*, J. Comput. Chem. 7 (1986), pp. 565–577.

202. P. Gaillard, P.A. Carrupt, B. Testa, and P. Schambel, *Binding of arylpiperazines, (aryloxy) propanolamines and tetrahydropyridyl-indoles to the 5-HT$_{1A}$ receptor: Contribution of the molecular lipophilicity potential to three-dimensional quantitative structure–affinity relationships models*, J. Med. Chem. 39 (1996), pp. 126–134.

203. T. Masuda, K. Nakamura, T. Jikihara, F. Kasuya, K. Igarashi, M. Fukui, T. Takagi, and H. Fujiwara, *3D-quantitative structure–activity relationships for hydrophobic interactions: Comparative molecular field analysis (CoMFA) including molecular lipophilicity potentials as applied to the glycine conjugation of aromatic as well as aliphatic carboxylic acids*, Quant. Struct. Act. Relat. 15 (1996), pp. 194–200.

204. S. Kneubüehler, U. Thull, C. Altomare, V. Carta, P. Gaillard, P.A. Carrupt, A. Carotti, and B. Testa, *Inhibition of monoamine oxidase-B by 5H-indeno[1,2-c]pyridazines: Biological activities, quantitative structure-activity relationships (QSARs) and 3D-QSARs*, J. Med. Chem. 38 (1995), pp. 3874–3883.

205. U. Thull, S. Kneubühler, P. Gaillard, P.A. Carrupt, B. Testa, C. Altomare, A. Carotti, P. Jenner, and K.S. McNaught, *Inhibition of monoamine oxidase by isoquinoline derivatives. Qualitative and 3D-quantitative structure–activity relationships*, Biochem. Pharmacol. 50 (1995), pp. 869–877.

206. U. Norinder, *3-D QSAR analysis of steroid/protein interactions: The use of different maps*, J. Comput.-Aided Mol. Des. 5 (1991), pp. 419–426.

207. C. Altomare, S. Cellamare, A. Carotti, G. Casini, M. Ferrapi, E. Gavuzzo, F. Mazza, P.A. Carrupt, P. Gaillard, and B Testa, *X-ray crystal structure, partitioning behavior, and molecular modeling study of piracetam-type nootropics: Insights into the pharmacophore*, J. Med. Chem. 38 (1995), pp. 170–179.

208. M. Bradley and C.L. Waller, *Polarizability fields for use in three-dimensional quantitative structure activity relationships (3D-QSAR)*, J. Chem. Inf. Comput. Sci. 41 (2001), pp. 1301–1307.

209. S.P. Korhonen, K. Tuppurainen, A. Asikainen, R. Laatikainen, and M. Peräkylä, *SOMFA on large diverse xenoestrogen dataset: The effect of superposition algorithms and external regression tools*, QSAR Comb. Sci. 26 (2007), pp. 809–819.

210. C.L. Waller and J.D. McKinney, *Three-dimensional quantitative structure–activity relationships of dioxins and dioxin-like compounds: Model validation and Ah receptor characterization*, Chem. Res. Toxicol. 8 (1995), pp. 847–858.

211. K.H. Kim, *3D-quantitative structure–activity relationships: Describing hydrophobic interactions directly from 3D structures using a comparative molecular field analysis CoMFA approach*, Quant. Struct. Act. Relat. 12 (1993), pp. 232–238.

212. K.H. Kim, G. Greco, E. Novellino, C. Silipo, and A. Vittoria, *Use of the hydrogen bond potential function in a comparative molecular field analysis (CoMFA) on a set of benzo-diazepines*, J. Comput.-Aided Mol. Des. 7 (1993), pp. 263–280.

213. C.E. Bohl, C. Chang, M.L. Mohler, J. Chen, D.D. Miller, P.W. Swaan, and J.T. Dalton, *A ligand-based approach to identify quantitative structure–activity relationships for the androgen receptor*, J. Med. Chem. 47 (2004), pp. 3765–3776.

214. E. Carosati, S. Sciabola, and G. Cruciani, *Hydrogen bonding interactions of covalently bonded fluorine atoms: From crystallographic data to a new angular function in the GRID force field*, J. Med. Chem. 47 (2004), pp. 5114–5125.

215. R.S. Bohacek and C. McMartin, *Definition and display of steric, hydrophobic, and hydrogen-bonding properties of ligand binding sites in proteins using Lee and Richards accessible surface: Validation of a high-resolution graphical tool for drug design*, J. Med. Chem. 35 (1992), pp. 1671–1684.

216. B. Lee and F.M. Richards, *The interpretation of protein structures: Estimation of static accessibility*, J. Mol. Biol. 55 (1971), pp. 379–400.

217. C.L. Waller and G.R. Marshall, *Three-dimensional quantitative structure–activity relationship of angiotensin-converting enzyme and thermolysin inhibitors. II. A comparison of CoMFA models incorporating molecular orbital fields and desolvation free energies based on active-analog and complementary-receptor-field alignment rules*, J. Med. Chem. 36 (1993), pp. 2390–2403.

218. T. Sulea and E.O. Purisima, *Desolvation free-energy field derived from boundary element continuum dielectric calculations*, Quant. Struct. Act. Relat. 18 (1999), pp. 154–158.

219. P.-A. Carrupt, B. Testa, and P. Gaillard, *Computational approaches to lipophilicity: Methods and applications*, in *Reviews in Computational Chemistry*, Vol. 11, K.B. Lipkowitz and D.B. Boyd, Eds., Wiley-VCH, New York, 1997, pp. 241–315.

220. K. Gohda, I. Mori, D. Ohta, and T. Kikuchi, *A CoMFA analysis with conformational propensity: An attempt to analyze the SAR of a set of molecules with different conformational flexibility using a 3D-QSAR method*, J. Comput.-Aided Mol. Des. 14 (2000), pp. 265–275.

221. G.E. Kellogg, L.B. Kier, P. Gaillard, and L.H. Hall, *The E-state fields: Application to 3D QSAR*, J. Comput.-Aided Mol. Des. 10 (1996), pp. 513–520.

222. L.B. Kier and L.H. Hall, *An atom-centered index for drug QSAR models*, in *Advances in Drug Design*, Vol. 22, B. Testa, Ed., Academic Press, London, U.K., 1992, pp. 1–38.

223. L.B. Kier and L.H. Hall, *The electrotopological state structure modeling for QSAR and database analysis*, in *Topological Indices and Related Descriptors in QSAR and QSPR*, J. Devillers and A.T. Balaban, Eds., Gordon & Breach Science Publishers, Amsterdam, the Netherlands, 1999, pp. 491–562.

224. C. Oostenbrink and W.F. van Gunsteren, *Free energies of binding of polychlorinated biphenyls to the estrogen receptor from a single simulation*, Proteins Struct. Funct. Bioinform. 54 (2004), pp. 237–246.

225. R.T. Kroemer and P. Hecht, *Replacement of steric 6-12 potential-derived interaction energies by atom-based indicator variables in CoMFA leads to models of higher consistency*, J. Comput.-Aided Mol. Des. 9 (1995), pp. 205–212.

226. P. Floersheim, J. Nozulak, and H.P. Weber, *Experience with comparative molecular-field analysis*, in *Trends in QSAR and Molecular Modelling 92*, C.G. Wermuth, Ed., ESCOM, Leiden, the Netherlands, 1993, pp. 227–232.

227. T. Sulea, T.I. Oprea, S. Muresan, and S.L. Chan, *A different method for steric field evaluation in CoMFA improves model robustness*, J. Chem. Inf. Comput. Sci. 37 (1997), pp. 1162–1170.

228. S. Muresan, T.Sulea, D. Ciubotariu, L.Kurunczi, and Z. Simon, *van der Waals intersection envelop volumes as a possible basis for steric interaction in CoMFA*, Quant. Struct. Act. Relat. 15 (1996), pp. 31–32.

229. A. Poso, K. Tuppurainen, and J. Gynther, *Modelling of molecular mutagenicity with comparative molecular field analysis (CoMFA): Structural and electronic properties of MX compounds related to TA 100 mutagenicity*, J. Mol. Struct. (Theochem) 304 (1994), pp. 255–260.

230. C. Navajas, A. Poso, K. Tuppurainen, and J. Gynther, *Comparative molecular field analysis (CoMFa) of MX compounds using different semi-empirical methods: LUMO field and its correlation with mutagenic activity*, Quant. Struct. Act. Relat. 15 (1996), pp. 189–193.

231. C.L. Waller, M.V. Evans, and J.D. McKinney, *Modeling the cytochrome P450-mediated metabolism of chlorinated volatil organic compounds*, Drug Metab. Dispos. 24 (1996), pp. 203–210.

232. T.G. Gantchev, H. Ali, and J.E. van Lier, *Quantitative structure–activity relationships/ comparative molecular field analysis (QSAR/CoMFA) for receptor-binding properties of halogenated estradiol derivatives*, J. Med. Chem. 37 (1994), pp. 4164–4176.

233. H.F. Chen, X.J. Yao, M. Petitjean, H.R. Xia, J.H. Yao, A. Panaye, J.P. Doucet, and B.T. Fan, *Insight into the bioactivity and metabolism of human glucagon receptor antagonists from 3D-QSAR analyses*, QSAR Comb. Sci. 23 (2004), pp. 603–620.

234. S. Funar-Timofei and G. Schüürmann, *Comparative molecular field analysis (CoMFA) of anionic azo dye-fiber affinities I: Gas-phase molecular orbital descriptors*, J. Chem. Inf. Comput. Sci. 42 (2002), pp. 788–795.

235. G. Schüürmann and S. Funar-Timofei, *Multilinear regression and comparative molecular field analysis (CoMFA) of azo dye-fiber affinities. 2. Inclusion of solution-phase molecular orbital descriptors*, J. Chem. Inf. Comput. Sci. 43 (2003), pp. 1502–1512.

236. K.H. Kim and Y.C. Martin, *Direct prediction of dissociation constants (pK_a's) of clonidine-like imidazolines, 2-substituted imidazoles, and 1-methyl-2-substituted-imidazoles from 3D structures using a comparative molecular field analysis (CoMFA) approach*, J. Med. Chem. 34 (1991), pp. 2056–2060.

237. K.H. Kim and Y.C. Martin, *Direct prediction of linear free energy substituent effects from 3D structures using comparative molecular field analysis. 1. Electronic effects of substituted benzoic acids*, J. Org. Chem. 56 (1991), pp. 2723–2729.

238. R.T. Kroemer, P. Hecht, and K.R. Liedl, *Different electrostatic descriptors in comparative molecular field analysis: A comparison of molecular electrostatic and Coulomb potentials*, J. Comput. Chem. 17 (1996), pp. 1296–1308.

239. T. Hou, L. Zhu, L. Chen, and X. Xu, *Mapping the binding site of a large set of quinazoline type EGF-R inhibitors using molecular field analyses and molecular docking studies*, J. Chem. Inf. Comput. Sci. 43 (2003), pp. 273–287.

240. S. Hannongbua, K. Nivesanond, L. Lawtrakul, P. Pungpo, and P. Wolschann, *3D-quantitative structure–activity relationships of HEPT derivatives as HIV-1 reverse transcriptase inhibitors, based on ab initio calculations*, J. Chem. Inf. Comput. Sci. 41 (2001), pp. 848–855.

241. M.J. Frisch, M. Head-Gordon, H.B. Schlegel, K. Raghavachari, J.S. Binkley, C. Gonzalez, D.J. Defrees et al., *Gaussian 88*, Gaussian Inc., Pittsburg, PA, 1988.

242. T.A. Halgren, *Merck molecular force field. I. Basis, form, scope, parametrization and performance of MMFF94*, J. Comput. Chem. 17 (1996), pp. 490–519.

243. R.R. Mittal, L. Harris, R.A. McKinnon, and M.J. Sorich, *Partial charge calculation method affects CoMFA QSAR prediction accuracy*, J. Chem. Inf. Model. 49 (2009), pp. 704–709.

244. R.T. Kroemer, P. Ettmayer, and P. Hecht, *3D-quantitative structure–activity relationships of human immunodeficiency virus type-1 proteinase inhibitors: Comparative molecular field analysis of 2-heterosubstituted statine derivatives—Implication for the design of novel inhibitors*, J. Med. Chem. 38 (1995), pp. 4917–4928.

245. H.B. Broughton, M. Gordaliza, M.-A. Castro, J.M. Miguel del Corral, and A. San Feliciano, *Modified CoMFA methods for the analysis of antineoplastic effects of lignan analogues*, J. Mol. Struct. (Theochem) 504 (2000), pp. 287–294.

246. A.R. Ortiz, M. Pastor, A. Palomer, G. Cruciani, F. Gago, and R.C. Wade, *Reliability of comparative molecular field analysis models: Effects of data scaling and variable selection using a set of human synovial fluid phospholipase A_2 inhibitors*, J. Med. Chem. 40 (1997), pp. 1136–1148.

247. A.R. Ortiz, M.T. Pisabarro, F. Gago, and R.C. Wade, *Prediction of drug binding affinities by comparative binding energy analysis*, J. Med. Chem. 38 (1995), pp. 2681–2691.

248. M. Murcia and A.R. Ortiz, *Virtual screening with flexible docking and COMBINE-based models. Application to a series of factor Xa inhibitors*, J. Med. Chem. 47 (2004), pp. 805–820.

249. S.J. Cho, M.L. Serrano Garsia, J. Bier, and A. Tropsha, *Structure-based alignment and comparative molecular field analysis of acetylcholinesterase inhibitors*, J. Med. Chem. 39 (1996), pp. 5064–5071.

250. M. Pastor, G. Cruciani, and K.A. Watson, *A strategy for the incorporation of water molecules present in a ligand binding site into a three-dimensional quantitative structure-activity relationship analysis*, J. Med. Chem. 40 (1997), pp. 4089–4102.

251. Y.C. Martin, M.G. Bures, E.A. Danaher, J. DeLazzer, I. Lico, and P. Pavlik, *A fast new approach to pharmacophore mapping and its application to dopaminergic and benzodiazepine agonists*, J. Comput.-Aided Mol. Des. 7 (1993), pp. 83–102.

252. G.R. Marshall, C.D. Barry, H.E. Bosshard, R.A. Dammkoehler, and D.A. Dunn, *The conformational parameter in drug design: The active analogue approach*, in *Computer-Assisted Drug Design*, E.C. Olson and R.E. Christoffersen, Eds., *ACS Symp.* Series 112, Vol. 112 , American Chemical Society, Washington, DC, 1979, pp. 205–226.

253. H.F. Chen, X.C. Dong, B.S. Zen, K. Gao, S.G. Yuan, A. Panaye, J.P. Doucet, and B.T. Fan, *Virtual screening and rational drug design method using structure generation system based on 3D-QSAR and docking*, SAR QSAR Environ. Res. 14 (2003), pp. 251–264.

254. G. Wolber, A.A. Dornhofer, and T. Langer, *Efficient overlay of small organic molecules using 3D pharmacophores*, J. Comput. Aided Mol. Des. 20 (2006), pp. 773–788.

255. J. Greene, S. Kahn, H. Savoj, P. Sprague, and S. Teig, *Chemical function queries for 3D database search*, J. Chem. Inf. Comput. Sci. 34 (1994), pp. 1297–1308.

256. D. Barnum, J. Greene, A. Smellie, and P. Sprague, *Identification of common functional configutarations among molecules*, J. Chem. Inf. Comput. Sci. 36 (1996), pp. 563–571.

257. U. Norinder, *Refinement of catalyst hypotheses using simplex optimisation*, J. Comput. Aided Mol. Des. 14 (2000), pp. 545–557.

258. T. Langer and R.D. Hoffmann, *On the use of chemical function-based alignments as input for 3D-QSAR*, J. Chem. Inf. Comput. Sci. 38 (1998), pp. 325–330.

259. W.M. Suhre, S. Ekins, C. Chang, P.W. Swaan, and S.H. Wright, *Molecular determinants of substrate/inhibitor binding to the human and rabbit renal organic cation transporters hOCT2 and rbOCT2*, Mol. Pharmacol. 67 (2005), pp. 1067–1077.

260. R. Bureau, C. Daveu, I. Baglin, J. Sopkova-De Oliveira Santos, J.C. Lancelot, and S. Rault, *Association of two 3D QSAR analyses. Application to the study of partial agonist serotonin-3 ligands*, J. Chem. Inf. Comput. Sci. 41 (2001), pp. 815–823.

261. F. Briens, R. Bureau, and S. Rault, *Applicability of CATALYST in ecotoxicology, a new promising tool for 3D-QSAR; study of chlorophenols*, Ecotoxicol. Environ. Saf. 43 (1999), pp. 241–251.

262. S. Dixon, A.M. Smondyrev, E.H. Knoll, S.N. Rao, D.E. Shaw, and R.A. Friesner, *PHASE: A new engine for pharmacophore perception, 3D QSAR model development and 3D database screening: 1. Methodology and preliminary results*, J. Comput.-Aided Mol. Des. 20 (2006), pp. 647–671.

263. D.A. Evans, T.N. Doman, D.A. Thorner, and M.J. Bodkin, *3D QSAR methods: Phase and catalyst compared*, J. Chem. Inf. Model. 47 (2007), pp. 1248–1257.

264. K. Iwase and S. Hirono, *Estimation of active conformations of drugs by a new molecular superposing procedure*, J. Comput.-Aided Mol. Des. 13 (1999), pp. 499–512.

265. G. Klebe and U. Abraham, *On the prediction of binding properties of drug molecules by comparative molecular field analysis*, J. Med. Chem. 36 (1993), pp. 70–80.

266. C.A. Marhefka, B.M. Moore II, T.C. Bishop, L. Kirkovsky, A. Mukherjee, J.T. Dalton, and D.D. Miller, *Homology modeling using multiple molecular dynamics simulations and docking studies of the human androgen receptor ligand binding domain bound to testosterone and non steroidal ligands*, J. Med. Chem. 44 (2001), pp. 1729–1740.

267. C.L. Waller, T.I. Oprea, A. Giolitti, and G.R. Marshall, *Three-dimensional QSAR of human immunodeficiency virus (1) protease inhibitors. 1. A CoMFA study employing experimentally-determined alignment rules*, J. Med. Chem. 36 (1993), pp. 4152–4160.

268. T.I. Oprea, C.L. Waller, and G.R. Marshall, *Three-dimensional quantitative structure-activity relationship of human immunodeficiency virus (I) protease inhibitors. 2. Predictive power using limited exploration of alternate binding modes*, J. Med. Chem. 37 (1994), pp. 2206–2215.

269. S.A. DePriest, D. Mayer, C.B. Naylor, and G.R. Marshall, *3D-QSAR of angiotensin-converting enzyme and thermolysin inhibitors: A comparison of CoMFA models based on deduced and experimentally determined active site geometries*, J. Am. Chem. Soc. 115 (1993), pp. 5372–5384.

270. P. Bernard, D.B. Kireev, J.R. Chrétien, P.L. Fortier, and L. Coppet, *Automated docking of 82 N-benzylpiperidine derivatives to mouse acetylcholinesterase and comparative molecular field analysis with "natural" alignment*, J. Comput.-Aided Mol. Des. 13 (1999), pp. 355–371.

271. A. Golbraikh, P. Bernard, and J.R. Chretien, *Validation of protein-based alignment in 3D quantitative structure–activity relationships with CoMFA models*, Eur. J. Med. Chem. 35 (2000), pp. 123–136.

272. A.M. Gamper, R.H. Winger, K.R. Liedl, C.A. Sotriffer, J.M. Varga, R.T. Kroemer, and B.M. Rode, *Comparative molecular field analysis of haptens docked to the multispecific antibody IgE(Lb4)*, J. Med. Chem. 39 (1996), pp. 3882–3888.

273. D.S. Goodsell and A. Olson, *Automated docking of substrates to proteins by simulated annealing*, Proteins Struct. Funct. Genet. 8 (1990), pp. 195–202.

274. W. Sippl, *Binding affinity prediction of novel estrogen receptor ligands using receptor-based 3-D QSAR methods*, Bioorg. Med. Chem. 10 (2002), pp. 3741–3755.

275. H.F. Chen, Q. Li, X.J. Yao, B.T. Fan, S.G. Yuan, A. Panaye, and J.P. Doucet, *3D-QSAR and docking study of the binding mode of steroids to progesterone receptor in active site*, QSAR Comb. Sci. 22 (2003), pp. 604–613.

276. J. Mestres, D.C. Rohrer, and G.M. Maggiora, *A molecular-field–based similarity study of non nucleoside HIV-1 reverse transcriptase inhibitors. 2. The relationship between alignment solutions obtained from conformationally rigid and flexible matching*, J. Comput.-Aided Mol. Des. 14 (2000), pp. 39–51.

277. S. Hannongbua, P. Pungpo, J. Limtrakul, and P. Wolschann, *Quantitative structure–activity relationships and comparative molecular field analysis of TIBO derivatised HIV-1 reverse transcriptase inhibitors*, J. Comput.-Aided Mol. Des. 13 (1999), pp. 563–577.

278. R.V.C. Guido, G. Oliva, C.A. Montanari, and A.D. Andricopulo, *Structural basis for selective inhibition of trypanosomatid glyceraldehyde-3-phosphate dehydrogenase: Molecular docking and 3D QSAR studies*, J. Chem. Inf. Model. 48 (2008), pp. 918–929.

279. G. Jones, P. Willett, R.C. Glen, A.R. Leach, and R. Taylor, *Development and validation of a genetic algorithm for flexible docking*, J. Mol. Biol. 267 (1997), pp. 727–748.

280. B. Kramer, M. Rarey, and T. Lengauer, *Evaluation of the FlexX incremental construction algorithm for protein–ligand docking*, Proteins Struct. Funct. Genet. 37 (1999), pp. 228–241.

281. M. Jalaie and J.A. Erickson, *Homology model directed alignment selection for comparative molecular field analysis: Application to photosystem II inhibitors*, J. Comput. Aided Mol. Des. 14 (2000), pp. 181–197.

282. I.D. Kuntz, J.M. Blaney, S.J. Oatley, R. Langridge, and T.E. Ferrin, *A geometric approach to macromolecule–ligand interactions*, J. Mol. Biol. 161 (1982), pp. 269–288.

283. R.M.A. Knegtel, I.D. Kuntz, and C.M. Oshiro, *Molecular docking to ensembles of protein structures*, J. Mol. Biol. 266 (1997), pp. 424–440.

284. J. Pan, G.Y. Liu, J. Cheng, X.J. Chen, and X.L. Ju, *CoMFA and molecular docking studies of benzoxazoles and benzothiazoles as CYP450 1A1 inhibitors*, Eur. J. Med. Chem. 45 (2010), pp. 967–972.

285. A.R. Leach, *Ligand docking to proteins with discrete side-chain flexibility*, J. Mol. Biol. 235 (1994), pp. 345–356.

286. E.C. Meng, B.K. Schoichet, and I.D. Kuntz, *Automated docking with grid-based energy evaluation*, J. Comput. Chem. 13 (1992), pp. 505–524.

287. M. Arakawa, K. Hasegawa, and K. Funatsu, *Novel alignment method of small molecules using the Hopfield neural network*, J. Chem. Inf. Comput. Sci. 43 (2003), pp. 1390–1395.

288. M. Arakawa, K. Hasegawa, and K. Funatsu, *Application of the novel molecular alignment method using the Hopfield neural network to 3D-QSAR*, J. Chem. Inf. Comput. Sci. 43 (2003), pp. 1396–1402.

289. J.P. Doucet and A. Panaye, *3D Structural information: From property prediction to substructure recognition with neural networks*, SAR QSAR Environ. Res. 8 (1998), pp. 249–272.

290. J.J. Hopfield and D.W. Tank, *Neural computation of decisions in optimization problems*, Biol. Cybern. 52 (1985), pp. 141–152.

291. J.J. Hopfield, *Neural networks and physical systems with emergent collective computational abilities*, Proc. Natl Acad. Sci. U.S.A. 79 (1982), pp. 2554–2558.

292. J.L. McClelland and D.E. Rumelhart, *Explorations in Parallel Distributed Processing. A Handbook of Models, Programs and Exercises*, MIT Press, Cambridge, MA, 1988.

293. H. Abdi, *Les Réseaux de Neurones*, Presses universitaires de Grenoble, Grenoble, France, 1994, pp. 105, 123.

294. E. Feuilleaubois, V. Fabart, and J.P. Doucet, *Implementation of the three-dimensional pattern search problem on Hopfield-like neural networks*, SAR QSAR Environ. Res. 1 (1993), pp. 97–114.

295. G.L. Bilbro, R. Mann, T.K. Miller, W.E. Snyder, D.E. van den Bout, and M. White, *Optimization by mean field annealing*, in *Advances in Neural Information Processing Systems*, D.S. Touretzky, Ed., Morgan Kaufman, San Mateo, CA, 1989, pp. 91–98.

296. M. Martín-Valdivia, A. Ruiz-Sepúlveda, and F. Triguero-Ruiz, *Improving local minima of Hopfield networks with augmented Lagrange multipliers for large scale TSPs*, Neural Netw. 13 (2000), pp. 283–285.

297. C. Marot, P. Chavatte, and D. Lesieur, *Comparative molecular field analysis of selective cyclooxygenase-2 (COX-2) inhibitors*, Quant. Struct. Act. Relat. 19 (2000), pp. 127–134.

298. N.J. Richmond, C.A. Abrams, P.R.N. Wolohan, E. Abrahamian, P. Willett, and R.D. Clark, *GALAHAD: 1 Pharmacophore identification by hypermolecular alignment of ligands in 3D*, J. Comput. Aided Mol. Des. 20 (2006), pp. 567–587.

299. N.J. Richmond, P. Willett, and R.D. Clark, *Alignment of three-dimensional molecules using an image recognition algorithm*, J. Mol. Graph. Model. 23 (2004), pp. 199–209.

300. J.K. Shepphird and R.D. Clark, *A marriage made in torsional space: Using GALAHAD models to drive pharmacophore multiplet searches*, J. Comput.-Aided. Mol. Des. 20 (2006), pp. 763–771.

301. W. Long, P. Liu, Q. Li, Y. Xu, and J. Gao, *3D-QSAR studies on a class of IKK-2 inhibitors with GALAHAD used to develop molecular alignment models*, QSAR Comb. Sci. 9 (2008), pp. 1113–1119.

302. A.V. Anghelescu, R.K. DeLisle, J.F. Lowrie, A.E. Klon, X. Xie, and D.J. Diller, *Technique for generating three-dimensional alignments of multiple ligands from one-dimensional alignments*, J. Chem. Inf. Model. 48 (2008), pp. 1041–1054.

303. S.L. Dixon and K.M. Merz Jr., *One-dimensional molecular representations and similarity calculations: Methodology and validation*, J. Med. Chem. 44 (2001), pp. 3795–3809.

304. N.P. Todorov, I.L. Alberts, I.J.P. de Esch, and P.M. Dean, *QUASI: A novel method for simultaneous superposition of multiple flexible ligands and virtual screening using partial similarity*, J. Chem. Inf. Model. 47 (2007), pp. 1007–1020.

305. T.D.J. Perkins and P.M. Dean, *An exploration of a novel strategy for superposing several flexible molecules*, J. Comput.-Aided Mol. Des. 7 (1993), pp. 155–172.

306. D.A. Cosgrove, D.M. Bayada, and A.P. Johnson, *A novel method of aligning molecules by local surface shape similarity*, J. Comput.-Aided Mol. Des. 14 (2000), pp. 573–591.

307. J.E.J. Mills, I.J.P. de Esch, T.D.J. Perkins, and P.M. Dean, *Slate: A method for the superposition of flexible ligands*, J. Comput.-Aided Mol. Des. 15 (2001), pp. 81–96.

308. T.D.J. Perkins, J.E.J. Mills, and P.M. Dean, *Molecular surface-volume and property matching to superpose flexible dissimilar molecules*, J. Comput. Aided Mol. Des. 9 (1995), pp. 479–490.

309. J. Nilsson, S. de Jong, and A.K. Smilde, *Multiway calibration in 3D QSAR*, J. Chemom. 11 (1997), pp. 511–524.

310. W.J. Dunn III, A.J. Hopfinger, C. Catana, and C. Duraiswami, *Solution of the conformation and alignment tensors for the binding of trimethoprim and its analogs to dihydrofolate reductase: 3D-quantitative structure–activity relationships study using molecular shape analysis, 3-way partial least-squares regression, and 3-way factor analysis*, J. Med. Chem. 39 (1996), pp. 4825–4832.

311. K. Hasegawa, M. Arakawa, and K. Funatsu, *Simultaneous determination of bioactive conformations and alignment rules by multi-way PLS modeling*, Comput. Biol. Chem. 27 (2003), pp. 211–216.

312. K. Hasegawa, M. Arakawa, and K. Funatsu, *Rational choice of bioactive conformations through use of conformation analysis and 3-way partial least squares modeling*, Chemom. Intell. Lab. Syst. 50 (2000), pp. 253–261.

313. K. Hasegawa, S. Matsuoka, M. Arakawa, and K. Funatsu, *New molecular surface-based 3D-QSAR method using Kohonen neural network and 3-Way PLS*, Comput. Chem. 26 (2002), pp. 583–589.

314. R. Bro, *Multiway calibration. Multilinear PLS*, J. Chemom. 10 (1996), pp. 47–61.

315. R.D. Cramer, *Topomer CoMFA: A design methodology for rapid lead optimization*, J. Med. Chem. 46 (2003), pp. 374–388.

316. R.D. Cramer, R.D. Clark, D.E. Patterson, and A.M. Ferguson, *Bioisosterism as a molecular diversity descriptor: Steric fields of single "topomeric" conformers*, J. Med. Chem. 39 (1996), pp. 3060–3069.

317. R.S. Pearlman, 3D molecular structures: Generation and use in 3D searching, in *3D QSAR in Drug Design, Theory, Methods and Applications*, H. Kubinyi, Ed., ESCOM Science Publishers B.V., Leiden, the Netherlands, 1993, pp. 41–79.

318. CONCORD, Program available from Tripos, Inc., St. Louis, MO.

319. R.D. Cramer, R.J. Jilek, and K.M. Andrews, *dbtop: Topomer similarity searching of conventional structure databases*, J. Mol. Graph. Model. 20 (2002), pp. 447–462.

320. T. Ohgaru, R. Shimizu, K. Okamoto, M. Kawase, Y. Shirakuni, R. Nishikiori, and T. Takagi, *Ordinal classification using comparative molecular field analysis*, J. Chem. Inf. Model. 48 (2008), pp. 207–212.

321. T. Ohgaru, R. Shimizu, K. Okamoto, N. Kawashita, M. Kawase, Y. Shirakuni, R. Nishikiori, and T. Takagi, *Enhancement of ordinal CoMFA by ridge logistic partial least squares*, J. Chem. Inf. Model. 48 (2008), pp. 910–917.

322. S. Clementi, G. Cruciani, P. Fifi, D. Riganelli, R. Valigi, and G. Musumarra, *A new set of principal properties for heteroaromatics obtained by GRID*, Quant. Struct. Act. Relat. 15 (1996), pp. 108–120.

323. T.C. Lin, P.A. Pavlik, and Y.C. Martin, *Use of molecular fields to compare series of potentially bioactive molecules designed by scientists or by computer*, Tetrahedron Comput. Methodol. 3 (1990), pp. 723–738.

324. M. Cocchi and E. Johansson, *Amino acids characterization by GRID and multivariate data analysis*, Quant. Struct. Act. Relat. 12 (1993), pp. 1–8.

325. T. Langer, *Molecular similarity characterization using CoMFA*, Perspect. Drug Discov. Des. 12–14 (1998), pp. 215–231.

326. U. Norinder, *Theoretical amino-acid descriptors. Application to bradykinin potentiating peptides*, Peptides 12 (1991), pp. 1223–1227.

327. A. Ashek, C. Lee, H. Park, and S.J. Cho, *3D QSAR studies of dioxins and dioxin-like compounds using CoMFA and CoMSIA*, Chemosphere 65 (2006), pp. 521–529.

328. J.P. Doucet and A. Panaye, *Molecular field analysis methods for modeling endocrine disruptors*, in *Endocrine Disruption Modeling*, J. Devillers, Ed., Taylor & Francis, Boca Raton, FL, 2009, pp. 295–335.

329. G. Klebe, *Comparative molecular similarity indices analysis: CoMSIA*, Perspect. Drug Discov. Des. 12–14 (1998), pp. 87–104.

330. G. Klebe, *The use of composite crystal-field environments in molecular recognition and the de novo design of protein ligands*, J. Mol. Biol. 237 (1994), pp. 212–235.

331. M. Böhm, J. Stürzebecher, and G. Klebe, *Three-dimensional quantitative structure-activity relationship analyses using comparative molecular field analysis and comparative molecular similarity indices analysis to elucidate selectivity differences of inhibitors binding to trypsin, thrombin, and factor Xα*, J. Med. Chem. 42 (1999), pp. 458–477.

332. J. Boström, M. Böhm, K. Gundertofte, and G. Klebe, *A 3D QSAR study on a set of dopamine D_4 receptor antagonists*, J. Chem. Inf. Comput. Sci. 43 (2003), pp. 1020–1027.

333. D. Robert, L. Amat, and R. Carbó-Dorca, *Quantum similarity QSAR: Study of inhibitors binding to thrombin, trypsin, and factor Xα, including a comparison with CoMFA and CoMSIA methods*, Int. J. Quantum Chem. 80 (2000), pp. 265–282.

334. R.J. Hu, F. Barbault, M. Delamar, and R.S. Zhang, *Receptor- and ligand-based 3D-QSAR study for a series of non-nucleoside HIV-1 reverse transcriptase inhibitors*, Bioorg. Med. Chem. 17 (2009), pp. 2400–2409.

335. M.M. Dias, R.R. Mittal, R.A. McKinnon, and M.J. Sorich, *Systematic statistical comparison of comparative molecular similarity indices analysis molecular fields for computer-aided lead optimization*, J. Chem. Inf. Model. 46 (2006), pp. 2015–2021.

336. M. Böhm and G. Klebe, *Development of new hydrogen-bond descriptors and their application to comparative molecular field analyses*, J. Med. Chem. 45 (2002), pp. 1585–1597.

337. M.L. Verdonk, J.C. Cole, and R. Taylor, *SuperStar: A knowledge-based approach for identifying interaction sites in proteins*, J. Mol. Biol. 289 (1999), pp. 1093–1108.

338. H. Lanig, W. Utz, and P. Gmeiner, *Comparative molecular field analysis of dopamine D4 receptor antagonists including 3-[4-(4-chlorophenyl)piperazin-1-ylmethyl]pyrazolo[1,5-a]pyridine (FAUC 113), 3-[4-(4-chlorophenyl)piperazin-1-ylmethyl]-1H-pyrrolo-[2,3-b] pyridine (L-745,870), and clozapine*, J. Med. Chem. 44 (2001), pp. 1151–1157.

339. J. Boström, K. Gundertofte, and T. Liljefors, *A pharmacophore model for dopamine D4 receptor antagonists*, J. Comput.-Aided Mol. Des. 14 (2000), pp. 769–786.

340. V.N. Viswanadhan, A.K. Ghose, G.R. Revankar, and R.K. Robins, *Atomic physicochemical parameters for three-dimensional structure directed quantitative structure–activity relationships. 4. Additional parameters for hydrophobic and dispersive interactions and their application for an automated superposition of certain naturally occurring nucleoside antibiotics*, J. Chem. Inf. Comput. Sci. 29 (1989), pp. 163–172.

341. G. Klebe, T. Mietzner, and F. Weber, *Methodological developments and strategies for a fast flexible superposition of drug-size molecules*, J. Comput. Aided Mol. Des. 13 (1999), pp. 35–49.

342. A. Weber, M. Böhm, C.T. Supuran, A. Scozzafava, C.A. Sotriffer, and G. Klebe, *3D QSAR selectivity analyses of carbonic anhydrase inhibitors: Insights fort the design of isozyme selective inhibitors*, J. Chem. Inf. Model. 46 (2006), pp. 2737–2760.

343. G.M. Morris, D.S. Goodsell, R.S. Halliday, R. Huey, W.E. Hart, R.K. Belew, and A.J. Olson, *Automated docking using a Lamarkian genetic algorithm and an empirical binding free energy function*, J. Comput. Chem. 19 (1998), pp. 1639–1662.

344. D.D. Robinson, P.J. Winn, P.D. Lyne, and W.G. Richards, *Self-organizing molecular field analysis: A tool for structure-activity studies*, J. Med. Chem. 42 (1999), pp. 573–583.

345. M. Li, L. Du, B. Wu, and L. Xia, *Self-organizing molecular field analysis on α_{1a}-adrenoceptor dihydropyridine antagonists*, Bioorg. Med. Chem. 11 (2003), pp. 3945–3951.

346. S.R. Krystek Jr., J.T. Hunt, P.D. Stein, and T.R. Stouch, *Three-dimensional quantitative structure–activity relationships of sulfonamide endothelin inhibitors*, J. Med. Chem. 38 (1995), pp. 659–668.

347. A. Asikainen, J. Ruuskanen, and K. Tuppurainen, *Spectroscopic QSAR methods and self-organizing molecular field analysis for relating molecular structure and estrogenic activity*, J. Chem. Inf. Comput. Sci. 43 (2003), pp. 1974–1981.

348. S.P. Korhonen, K. Tuppurainen, R. Laatikainen, and M. Peräkylä, *Improving the performance of SOMFA by use of standard multivariate methods*, SAR QSAR Environ. Res. 16 (2005), pp. 567–579.

349. M.Y. Li, H. Fang, and L. Xia, *Pharmacophore-based design, synthesis, biological evaluation, and 3D-QSAR studies of aryl-piperazines as α_1-adrenoceptor antagonists*, Bioorg. Med. Chem. Lett. 15 (2005), pp. 3216–3219.

350. S. Li and Y. Zheng, *Self-organizing molecular field analysis on a new series of COX-2 selective inhibitors: 1,5-Diarylimidazoles*, Int. J. Mol. Sci. 7 (2006), pp. 220–229.

351. H. Liu, X. Huang, J. Shen, X. Luo, M. Li, B. Xiong, G. Chen et al., *Inhibitory mode of 1,5-diarylpyrazole derivatives against cyclooxygenase-2 and cyclooxygenase-1: Molecular docking and 3D QSAR analyses*, J. Med. Chem. 45 (2002), pp. 4816–4827.

352. Y. Ren, G. Chen, Z. Hu, X. Chen, and B. Yan, *Applying novel three-dimensional holographic vector of atomic interaction field to QSAR studies of artemisinin derivatives*, QSAR Comb. Sci. 27 (2008), pp. 198–207.

353. T.A. Martinek, F. Ötvös, G. Dervarics, G. Tóth, and F. Fülöp, *Ligand-based prediction of active conformation by 3D-QSAR flexibility descriptors and their application in 3+3D-QSAR models*, J. Med. Chem. 48 (2005), pp. 3239–3250.

354. P.A. Smith, M.J. Sorich, R.A. McKinnon, and J.O. Miners, *Pharmacophore and quantitative structure-activity relationship modeling: Complementary approaches for the rationalization and prediction of UDP-glucoronosyltransferase 1A4 substrate selectivity*, J. Med. Chem. 46 (2003), pp. 1617–1626.

355. M. Esposito, N. Müller, and A. Hemphill, *Structure–activity relationships from in vitro efficacies of the thiazolide series against the intracellular apicomplexan protozoan Neospora caninum*, Int. J. Parasitol. 37 (2007), pp. 183–190.

356. S.P. Korhonen, K. Tuppurainen, R. Laatikainen, and M. Peräkylä, *FLUFF-BALL, a template-based grid-independent superposition and QSAR Technique: Validation using a benchmark steroid data set*, J. Chem. Inf. Comput. Sci. 43 (2003), pp. 1780–1793.

357. S.P. Korhonen, K. Tuppurainen, R. Laatikainen, and M. Peräkylä, *Comparing the performance of FLUFF-BALL to SEAL-CoMFA with a large diverse estrogen data set: From relevant superpositions to solid predictions*, J. Chem. Inf. Model. 45 (2005), pp. 1874–1883.

358. A.H. Asikainen, J. Ruuskanen, and K.A. Tuppurainen, *Consensus kNN QSAR: A versatile method for predicting the estrogenic activity of organic compounds in silico. A comparative study with five estrogen receptors and a large, diverse set of ligands*, Environ. Sci. Technol. 38 (2004), pp. 6724–6729.

359. W. Tong, R. Perkins, R. Strelitz, E.R. Collantes, S. Keenan, W.J. Welsh, W.S. Branham, and D.M. Sheehan, *Quantitative structure–activity relationships (QSARs) for estrogen binding to the estrogen receptor: Predictions across species*, Environ. Health Perspect. 105 (1997), pp. 1116–1124.

360. J. Manchester and R. Czermínski, *SAMFA: Simplifying molecular description for 3D-QSAR*, J. Chem. Inf. Model. 48 (2008), pp. 1167–1173.

361. V. Svetnik, A. Liaw, C. Tong, J.C. Culberson, R.P. Sheridan, and B.P. Feuston, *Random forest: A classification and regression tool for compound classification and QSAR modeling*, J. Chem. Inf. Comput. Sci. 43 (2003), pp. 1947–1958.

362. Q.-S. Xu, Y.-Z. Liang, and Y.-P. Du, *Monte Carlo cross-validation for selecting a model and estimating the prediction error in multivariate calibration*, J. Chemom. 18 (2004), pp. 112–120.

363. J. Polanski, R. Gieleciak, T. Magdziarz, and A. Bak, *GRID formalism for the comparative molecular surface analysis: Application to the CoMFA Benchmark steroids, azo dyes and HEPT derivatives*, J. Chem. Inf. Comput. Sci. 44 (2004), pp. 1423–1435.

364. H. Verli, M. Girão Albuquerque, R. Bicca de Alencastro, and E.J. Barreiro, *Local intersection volume: A new 3D descriptor applied to develop a 3D-QSAR pharmacophore model for benzodiazepine receptor ligands*, Eur. J. Med. Chem. 37 (2002), pp. 219–229.

365. Y. Tominaga and I. Fujiwara, *Novel 3D descriptors using excluded volume: Application to 3D quantitative structure–activity relationships*, J. Chem. Inf. Comput. Sci. 37 (1997), pp. 1158–1161.

366. M. Hahn and D. Rogers, *Receptor surface models*, Perspect. Drug Discov. Des. 12–14 (1998), pp. 117–133.

367. A.N. Jain, K. Koile, and D. Chapman, *COMPASS: Predicting biological activities from molecular surface properties. Performance comparisons on a steroid benchmark*, J. Med. Chem. 37 (1994), pp. 2315–2327.

368. B.J. Burke and A.J. Hopfinger, *1-(Substituted-benzyl)imidazole-2(3H)-thiones inhibitors of dopamine β-hydroxylase*, J. Med. Chem. 33 (1990), pp. 274–281.

369. C. Silipo and C. Hansch, *Correlation analysis. Its application to the structure-activity relation of triazines inhibiting dihydrofolate reductase*, J. Am. Chem. Soc. 97 (1975), pp. 6849–6861.

370. L.I. Kruse, C. Kaiser, W.E. DeWolf Jr., J.S. Frazee, S.T. Ross, J. Wawro, M. Wise et al., *Multisubstrate inhibitors of dopamine β-hydroxylase. 2. Structure–activity relationships* at the phenethylamine binding site, J. Med. Chem. 30 (1987), pp. 486–494.

371. A.J. Hopfinger, *Theory and application of molecular potential energy fields in molecular shape analysis: A quantitative structure-activity relationship study of 2,4-diamino-5-benzylpyrimidines as dihydrofolate reductase inhibitors*, J. Med. Chem. 26 (1983), pp. 990–996.

372. A.J. Hopfinger, *Inhibition of dihydrofolate reductase: Structure–activity correlations of 2,4-diamino-5-benzylpyrimides based upon molecular shape analysis*, J. Med. Chem. 24 (1981), pp. 818–822.

373. J.M. Blaney, S.W. Dietrich, M.A. Reynolds, and C. Hansch, *Quantitative structure-activity relationship of 5-(X-benzyl)-2,4-diaminopyrimidines inhibiting bovine liver dihydrofolate reductase*, J. Med. Chem. 22 (1979), pp. 614–717.

374. J.S. Tokarski and A.J. Hopfinger, *Three-dimensional molecular shape analysis— Quantitative structure–activity relationship of a series of cholecystokinin—A receptor antagonists*, J. Med. Chem. 37 (1994), pp. 3639–3654.

375. A.J. Hopfinger, B.J. Burke, and W.J. Dunn III, *A generalized formalism of three-dimensional quantitative structure–property relationship analysis for flexible molecules using tensor representation*, J. Med. Chem. 37 (1994), pp. 3768–3774.

376. B.J. Burke, W.J. Dunn III, and A.J. Hopfinger, *Construction of a molecular shape analysis – Three-dimensional quantitative structure–analysis relationship for an analog series of pyridobenzodiazepinone inhibitors of muscarinic 2 and 3 receptors*, J. Med. Chem. 37 (1994), pp. 3775–3788.

377. M.G. Koehler and A.J. Hopfinger, *Molecular modelling of polymers. 5. Inclusion of intermolecular energetics in estimating glass and crystal–melt transition temperatures*, Polymer 30 (1989), pp. 116–126.

378. U. Holzgrabe and A.J. Hopfinger, *Conformational analysis, molecular shape comparison, and pharmacophore identification of different allosteric modulators of muscarinic receptors*, J. Chem. Inf. Comput. Sci. 36 (1996), pp. 1018–1024.

379. D.E. Patterson, R.D. Cramer, A.M. Ferguson, R.D. Clark, and L.E. Weinberger, *Neighborhood behavior: A useful concept for validation of "Molecular Diversity" descriptors*, J. Med. Chem. 39 (1996), pp. 3049–3059.

380. A.C. Good, E.E. Hodgkin, and W.G. Richards, *Utilization of Gaussian functions for the rapid evaluation of molecular similarity*, J. Chem. Inf. Comput. Sci. 32 (1992), pp. 188–191.

381. A.C. Good, *The calculation of molecular similarity: Alternative formulas, data manipulation and graphical display*, J. Mol. Graph. 10 (1992), pp. 144–151.

382. Q. Huang, X. He, C. Ma, R. Liu, S. Yu, C.A. Dayer, G.R. Wenger, R. McKernan, and J.M. Cook, *Pharmacophore/receptor models for GABA$_A$/BzR subtypes ($\alpha1\beta3\gamma2$, $\alpha5\beta3\gamma2$, and $\alpha6\beta3\gamma2$) via a comprehensive ligand-mapping approach*, J. Med. Chem. 43 (2000), pp. 71–95.

383. R.C.A. Martins, M.G. Albuquerque, and R.B. Alencastro, *Local intersection volume (LIV) descriptors: 3D-QSAR models for PGI$_2$ receptor ligands*, J. Braz. Chem. Soc. 13 (2002), pp. 816–821.

384. N.A. Meanwell, M.J. Rosenfeld, A.K. Trehan, J.L. Romine, J.J.K. Wright, C.L. Brassard, J.O. Buchanan et al., *Nonprostanoid prostacyclin mimetics. 3. Structural variations of the diphenyl heterocycle moiety*, J. Med. Chem. 35 (1992), pp. 3498–3512.

385. A.M. Doweyko, *Three dimensional pharmacophores from binding data*, J. Med. Chem. 37 (1994), pp. 1769–1778.

386. J.J. Kaminski and A.M. Doweyko, *Antiulcer agents. 6. Analysis of the in vitro biochemical and in vivo gastric antisecretory activity of substituted imidazo[1,2-a] pyridines and related analogues using comparative molecular field analysis and hypothetical active site lattice methodologies*, J. Med. Chem. 40 (1997), pp. 427–436.

387. S. Dastmalchi and M. Hamzeh-Mivehrod, *Molecular modelling of human aldehyde oxidase and identification of the key interactions in the enzyme–substrate complex*, DARU 13 (2005), pp. 82–93.

388. S. Guccione, A.M. Doweyko, H. Chen, G. Uccello Barretta, and F. Balzano, *3D-QSAR using "multiconformer" alignment: The use of HASL in the analysis of 5-HT$_{1A}$ thieno-pyrimidinone ligands*, J. Comput.-Aided Mol. Des. 14 (2000), pp. 647–657.

389. M. Hahn, *Receptor surface models. 1. Definition and construction*, J. Med. Chem. 38 (1995), pp. 2080–2090.

390. M. Hahn and D. Rogers, *Receptor surface models. 2. Application to quantitative structure–activity relationships studies*, J. Med. Chem. 38 (1995), pp. 2091–2102.

391. A. Hirashima, E. Kuwano, and M. Eto, *Three dimensional receptor surface model of octopaminergic agonists for the locust neuronal octopamine receptor*, Internet Electron. J. Mol. Des. 2 (2003), pp. 274–287. http://www.biochempress.com

392. A.R. Shaikh, M. Ismael, C.A. Del Carpio, H. Tsuboi, M. Koyama, A. Endou, M. Kubo, E. Broclawik, and A. Miyamoto, *Three-dimensional quantitative structure–activity relationship (3D-QSAR) and docking studies on (benzothiazole-2-yl) acetonitrile derivatives as c-Jun N-terminal kinase-3 (JNK3) inhibitors*, Bioorg. Med. Chem. Lett. 16 (2006), pp. 5917–5925.

393. D.E. Walters and R.M. Hinds, *Genetically evolved receptor models: A computational approach to construction of receptor models*, J. Med. Chem. 37 (1994), pp. 2527–2536.

394. D.E Walters, *Genetically evolved receptor models (GERM) as a 3D QSAR tool*, Perspect. Drug Discov. Des. 12–14 (1998), pp. 159–166.

395. D.E. Walters and T.D. Muhammad, *Genetically evolved receptor models (GERM): A procedure for construction of atomic-level receptor site models in the absence of a receptor crystal structure*, in *Genetic Algorithms in Molecular Modeling*, J. Devillers, Ed., Academic Press, London, U.K., 1996, pp. 193–210.

396. A.C. Good, S.S. So, and W.G. Richards, *Structure–activity relationships from molecular similarity matrices*, J. Med. Chem. 36 (1993), pp. 433–438.

397. A.N. Jain, N.L. Harris, and J.Y. Park, *Quantitative binding site model generation: Compass applied to multiple chemotypes targeting the 5-HT$_{1A}$ receptor*, J. Med. Chem. 38 (1995), pp. 1295–1308.

398. U. Norinder, *3D-QSAR investigation of the Tripos benchmark steroids and some protein-tyrosine kinase inhibitors of styrene type using the TDQ approach*, J. Chemom. 10 (1996), pp. 533–545.

399. J. Polanski, R. Gieleciak, and A. Bak, *The comparative molecular surface analysis (COMSA)—A nongrid 3D QSAR method by a coupled neural network and PLS system: Predicting pK$_a$ values of benzoic and alkanoic acids*, J. Chem. Inf. Comput. Sci. 42 (2002), pp. 184–191.

400. J. Gasteiger, X. Li, C. Rudolph, J. Sadowski, and J. Zupan, *Representation of molecular electrostatic potentials by topological feature maps*, J. Am. Chem. Soc. 116 (1994), pp. 4608–4620.

401. J. Zupan and J. Gasteiger, *Neural Networks in Chemistry and Drug Design*, 2nd edn, Wiley-VCH, BRD, Weinheim, Germany, 1999.

402. S. Anzali, G. Barnickel, M. Krug, J. Sadowski, M. Wagener, J. Gasteiger, and J. Polanski, *The comparison of geometric and electronic properties of molecular surfaces by neural networks: Application to the analysis of corticosteroid-binding globulin activity of steroids*, J. Comput.-Aided Mol. Des. 10 (1996), pp. 521–534.

403. J. Polanski, *The receptor-like neural network for modeling corticosteroid and testosterone binding globulins*, J. Chem. Inf. Comput. Sci. 37 (1997), pp. 553–561.

404. J. Polanski, *Self-organizing neural network for modeling 3D QSAR of colchicinoids*, Acta Biochim. Pol. 47 (2000), pp. 37–45.

405. T.I. Oprea and A.E. Garcia, *Three-dimensional quantitative structure–activity relationships of steroid aromatase inhibitors*, J. Comput.-Aided Mol. Des. 10 (1996), pp. 186–200.

406. R.D. Beger, D.A. Buzatu, J.G. Wilkes, and J.O. Lay, *^{13}C NMR quantitative spectrometric data activity relationship (QSDAR) models of steroids binding the aromatase enzyme*, J. Chem. Inf. Comput. Sci. 41 (2001), pp. 1360–1366.

407. R.D. Beger and J.G. Wilkes, *Comparative structural connectivity spectra analysis (CoSCoSA) models of steroids binding to the aromatase enzyme*, J. Mol. Recognit. 15 (2002), pp. 154–162.

408. V. Centner, D.L. Massart, O.E. de Noord, S. de Jong, B.M. Vandeginste, and C. Sterna, *Elimination of uninformative variables for multivariate calibration*, Anal. Chem. 68 (1996), pp. 3851–3858.

409. D.M. Tanenbaum, Y. Wang, S.P. Williams, and P.B. Sigler, *Crystallographic comparison of the estrogen and progesterone receptor's ligand binding domains*, Proc. Natl Acad. Sci. U.S.A. 95 (1998), pp. 5998–6003.

410. A.M.A. Hammad and M.O. Taha, *Pharmacophore modeling, quantitative structure–activity relationship analysis and shape-complemented in silico screening allow access to novel influenza neuraminidase inhibitors*, J. Chem. Inf. Model. 49 (2009), pp. 978–996.

411. A. Tromelin and E. Guichard, *Use of Catalyst in a 3D-QSAR study of the interactions between flavor compounds and β-lactoglobulin*, J. Agric. Food Chem. 51 (2003), pp. 1977–1983.

412. A.K. Debnath, *Generation of predictive pharmacophore models for CCR5 antagonists: Study with piperidine- and piperazine-based compounds as a new class of HIV-1 entry inhibitors*, J. Med. Chem. 46 (2003), pp. 4501–4515.

413. Z. Zhu and J.K. Buolamwini, *Constrained NBMPR analogue synthesis, pharmacophore mapping and 3D-QSAR modeling of equilibrative nucleoside transporter 1 (ENT1) inhibitory activity*, Bioorg. Med. Chem. 16 (2008), pp. 3848–3865.

414. G. Moreau and P. Broto, *The autocorrelation of a topological structure. A new molecular descriptor*, Nouv. J. Chim. 4 (1980), pp. 359–360.

415. G. Moreau and P. Broto, *Autocorrelation of molecular structures: Application to SAR studies*, Nouv. J. Chim. 4 (1980), pp. 757–764.

416. P. Broto, G. Moreau, and C. Vandycke, *Molecular structures: Perception, autocorrelation descriptor and SAR studies. Use of the autocorrelation descriptor in the QSAR study of two non-narcotic analgesic studies*, Eur. J. Med. Chem. 19 (1984), pp. 79–84.

417. P. Broto, G. Moreau, and C. Vandycke, *Molecular structures: Perception, autocorrelation descriptor and SAR studies. Autocorrelation descriptor*, Eur. J. Med. Chem. 19 (1984), pp. 66–70.

418. M.K. Gupta, R. Sagar, A.K. Shaw, and Y.S. Prabhakar, *CP-MLR directed QSAR studies on the antimycobacterial activity of functionalized alkenols—Topological descriptors in modeling the activity*, Bioorg. Med. Chem. 13 (2005), pp. 343–351.

419. M. Chastrette and D. Crétin, *Structure–property relationships—Determination of the vapor pressure of hydrocarbons and oxygenated compounds using multifunctional autocorrelation mehtod (MAM)*, SAR QSAR Environ. Res. 3 (1995), pp. 131–149.

420. J. Devillers, *EVA/PLS versus autocorrelation/neural network estimation of partition coefficients*, Perspect. Drug Discov. Des. 19 (2000), pp. 117–131.

421. J. Devillers, *Autocorrelation descriptors for modeling (eco)toxiological endpoints*, in *Topological Indices and Related Descriptors in QSAR and QSPR*, J. Devillers and A.T. Balaban, Eds., Gordon & Breach Science Publishers, New York, 1999, pp. 595–612.

422. C.T. Klein, D. Kaiser, and G. Ecker, *Topological distance based 3D descriptors for use in QSAR and diversity analysis*, J. Chem. Inf. Comput. Sci. 44 (2004), pp. 200–209.

423. C.L. Waller and J.D. McKinney, *Comparative molecular field analysis of polyhalogenated dibenzo-p-dioxins, dibenzofurans and biphenyls*, J. Med. Chem. 35 (1992), pp. 3660–3666.

424. S. Moro, M. Bacilieri, B. Cacciari, and G. Spalluto, *Autocorrelation of molecular electrostatic potential surface properties combined with partial least squares analysis as new strategy for the prediction of the activity of human A_3 adenosine receptor antagonists*, J. Med. Chem. 48 (2005), pp. 5698–5704.

425. S. Moro, P. Braiuca, F. Deflorian, C. Ferrari, G. Pastorin, B. Cacciari, P.G. Baraldi, K. Varani, P.A. Borea, and G. Spalluto, *Combined target-based and ligand-based drug design approach as a tool to define a novel 3D-pharmacophore model of human A_3 adenosine receptor antagonists: Pyrazolo[4,3-e]1,2,4-triazolo[1,5-c]pyrimidine derivatives as a key study*, J. Med. Chem. 48 (2005), pp. 152–162.

426. S. Moro, M. Bacilieri, B. Cacciari, C. Bolcato, C. Cusan, G. Pastorin, K.N. Klotz, and G. Spalluto, *The application of a 3D-QSAR (autoMEP/PLS) approach as an efficient pharmacodynamic-driven filtering method for small-sized virtual library: Application to a lead optimization of a human A_3 adenosine receptor antagonist*, Bioorg. Med. Chem. 14 (2006), pp. 4923–4932.

427. J. Hinze and H.H. Jaffé, *Electronegativity. I. Orbital electronegativity of neutral atoms*, J. Am. Chem. Soc. 84 (1962), pp. 540–546.

428. TSAR, Oxford Molecular Ltd., The Medawar Centre, Oxford Science, Park, Oxford. Stanford-on-Thames, Oxford, U.K.

429. K. Baumann, *An alignment-independent versatile structure descriptor for QSAR and QSPR based on the distribution of molecular features*, J. Chem. Inf. Comput. Sci. 42 (2002), pp. 26–35.

430. J.T. Clerc and A.L. Terkovics, *Versatile topological structure descriptor for quantitative structure/property studies*, Anal. Chim. Acta 235 (1990), pp. 93–102.

431. G. Greco, E. Novellino, C. Silipo, and A. Vittoria, *Comparative molecular field analysis on a set of muscarinic agonists*, Quant. Struct. Act. Relat. 10 (1991), pp. 289–299.

432. A.M. Ferguson, T. Heritage, P. Jonathon, S.E. Pack, L. Phillips, J. Rogan, and P.J. Snaith, *EVA: A new theoretically based molecular descriptor for use in QSAR/QSPR analysis*, J. Comput.-Aided Mol. Des. 11 (1997), pp. 143–152.

433. K. Baumann, H. Albert, and M. von Korff, *A systematic evaluation of the benefits and hazards of variable selection in latent variable regression. Part I. Search algorithm, theory and simulations*, J. Chemom. 16 (2002), pp. 339–350.

434. K. Baumann, M. von Korff, and H. Albert, *A systematic evaluation of the benefits and hazards of variable selection in latent variable regression. Part II. Practical applications*, J. Chemom. 16 (2002), pp. 351–360.

435. E. Gancia, G. Bravi, P. Mascagni, and A. Zaliani, *Global 3D-QSAR methods: MS-WHIM and autocorrelation*, J. Comput-Aided Mol. Des. 14 (2000), pp. 293–306.

436. G. Cruciani and K.A. Watson, *Comparative molecular field analysis using GRID force-field and GOLPE variable selection methods in a study of inhibitors of glycogen phosphorylase b*, J. Med. Chem. 37 (1994), pp. 2589–2601.

437. F. Fontaine, M. Pastor, and F. Sanz, *Incorporating molecular shape into the alignment-free GRid INdependent Descriptors*, J. Med. Chem. 47 (2004), pp. 2805–2815.

438. J. Brea, C.F. Masaguer, M. Villazón, M.I. Cadavid, E. Raviña, F. Fontaine, C. Dezi, M. Pastor, F. Sanz, and M.I. Loza, *Conformationally constrained butyrophenones as new pharmacological tools to study 5-HT$_{2A}$ and 5-HT$_{2C}$ receptor behaviours*, Eur. J. Med. Chem. 38 (2003), pp. 433–440.

439. E. Raviña, J. Negreira, J. Cid, C.F. Masaguer, E. Rosa, M.E. Rivas, J.A. Fontenla et al., *Conformationally constrained butyrophenones with mixed dopaminergic (D$_2$) and serotoninergic (5-HT$_{2A}$, 5-HT$_{2C}$) affinities: Synthesis, pharmacology, 3D-QSAR, and molecular modeling of (aminoalkyl)benzo- and -thienocycloalkanones as putative atypical antipsychotics*, J. Med. Chem. 42 (1999), pp. 2774–2797.

440. S. Huo, J. Wang, P. Cieplak, P.A. Kollman, and I.D. Kuntz, *Molecular dynamics and free energy analyses of cathepsin D-inhibitor interactions: Insight into structure-based ligand design*, J. Med. Chem. 45 (2002), pp. 1412–1419.

441. P. Benedetti, R. Mannhold, G. Cruciani, and M. Pastor, *GBR compounds and mepyramines as cocaine abuse therapeutics: Chemometric studies on selectivity using grid independent descriptors (GRIND)*, J. Med. Chem. 45 (2002), pp. 1577–1584.

442. G. Caron and G. Ermondi, *Influence of conformation on GRIND-based three-dimensional quantitative structure–activity relationship (3D-QSAR)*, J. Med. Chem. 50 (2007), pp. 5039–5042.

443. R. Ragno, S. Simeoni, D. Rotili, A. Caroli, G. Botta, G. Brosch, S. Massa, and A. Mai, *Class II-selective histone deacetylase inhibitors. Part 2: Alignment-independent GRIND 3-D QSAR, homology and docking studies*, Eur. J. Med. Chem. 43 (2008), pp. 621–632.

444. D.S. Goodsell, G.M. Morris, and A.J. Olson, *Automated docking of flexible ligands: Applications of autodock*, J. Mol. Recognit. 9 (1996), pp. 1–5.

445. B. Bertoša, B. Kojić-Prodić, R.C. Wade, M. Ramek, S. Piperaki, A. Tsantili-Kakoulidou, and S. Tomić, *A new approach to predict the biological activity of molecules based on similarity of their interaction fields and the logP and logD values: Application to auxins*, J. Chem. Inf. Comput. Sci. 43 (2003), pp. 1532–1541.

446. S. Tomić, R.R. Gabdoulline, B. Kojić-Prodić, and R.C. Wade, *Classification of auxin plant hormones by interaction property similarity indices*, J. Comput.-Aided Mol. Des. 12 (1998), pp. 63–79.

447. S. Tomić, R.R. Gabdoulline, B. Kojić-Prodić, and R.C. Wade, *Classification of auxin related compounds based on similarity of their interaction fields: Extension to a new set of compounds*, Internet J. Chem. 26 (1998), p. 1. <http://www.ijc.com/articles/1998v1/26/>ISSN 1099.

448. S. Antolić, E. Dolusǐć, E.K. Kozǐć, B. Kojić-Prodić, V. Magnus, M. Ramek, and S. Tomić, *Auxin activity and molecular structure of 2-alkylindole-3-acetic acids*, Plant Growth Regul. 39 (2003), pp. 235–252.

449. G. Cruciani, M. Pastor, and W. Guba, *VolSurf: A new tool for the pharmacokinetic optimization of lead compounds*, Eur. J. Pharm. Sci. 11(suppl 2) (2000), pp. S29–S39.

450. G. Cruciani, S. Clementi, and M. Baroni, *Variable selection in PLS analysis*, in *3D QSAR in Drug Design, Theory, Methods and Applications*, H. Kubinyi, Ed., ESCOM Science Publishers B.V., Leiden, the Netherlands, 1993, pp. 551–564.

451. E.-J. Woo, J. Marshall, J. Bauly, J.-G. Chen, M. Venis, R.M. Napier, and R.W. Pickersgill, *Crystal structure of auxin-binding protein 1 in complex with auxin*, EMBO J. 21 (2002), pp. 2877–2885.

452. UNITY Chemical Information Software, TRIPOS, St. Louis, MO.

453. ISIS keys, ISIS/Base Molecular design Ltd, 14600 Catalina Street, Irvine, CA.

454. G. Cruciani, M. Pastor, and R. Mannhold, *Suitability of molecular descriptors for database mining. A comparative analysis*, J. Med. Chem. 45 (2002), pp. 2685–2694.

455. M. Baroni, G. Cruciani, S. Sciabola, F. Perruccio, and J.S. Mason, *A common reference framework for analyzing/comparing proteins and ligands. Fingerprints for ligands and proteins (FLAP): Theory and application*, J. Chem. Inf. Model. 47 (2007), pp. 279–294.

456. S. Sciabola, I. Morao, and M.J. de Groot, *Pharmacophoric fingerprint method (TOPP) for 3D-QSAR modeling: Application to CYP2D6 metabolic stability*, J. Chem. Inf. Model. 47 (2007), pp. 76–84.

457. F. Fontaine, M. Pastor, H. Guitiérrez-De-Terán, J.J. Lozano, and F. Sanz, *Use of alignment-free molecular descriptors in diversity analysis and optimal sampling of molecular libraries*, Mol. Div. 6 (2003), pp. 135–147.

458. N. Stiefl and K. Baumann, *Mapping property distributions of molecular surfaces: Algorithm and evaluation of a novel 3D quantitative structure–activity relationship technique*, J. Med. Chem. 46 (2003), pp. 1390–1407.

459. F. Glover, *Tabu search—Part 1*, ORSA J. Comput. 1 (1989), pp. 190–206.

460. F. Glover, *Tabu search—Part 2*, ORSA J. Comput. 2 (1990), pp. 4–32.

461. J.P. Doucet and J. Weber, *Drug receptor interactions: Receptor mapping and pharmacophore approach*, in *Computer-Aided Molecular Design. Theory and Applications*, Academic Press, San Diego, CA, 1996, pp. 363–404.

462. A.G. Maldonado, M. Petitjean, J.P. Doucet, A. Panaye, and B.T. Fan, *MolDIA: XML based system of molecular diversity analysis towards virtual screening and QSPR*, SAR QSAR Environ. Res. 17 (2006), pp. 11–23.

463. M.A. Johnson and G.M. Maggiora, Eds., *Concepts and Applications of Molecular Similarity*, John Wiley & Sons, New York, 1990.

464. P.M. Dean, Ed., *Molecular Similarity in Drug Design*, Chapman & Hall, Blackie Publishers, Glasgow, U.K., 1995.

465. P. Willett, *Similarity and Clustering in Chemical Information Systems*, Research Studies Press, Letchworth, Herfordshire, U.K., 1987.

466. P. Gund, J.D. Andose, J.B. Rhodes, and G.M. Smith, *Three-dimensional molecular modeling and drug design*, Science 208 (1980), pp. 1425–1431.

467. N. Nikolova and J. Jaworska, *Approaches to measure chemical similarity—A review*, QSAR Comb. Sci. 22 (2003), pp. 1006–1026.

468. G. Rum and W.C. Herndon, *Molecular similarity concepts. 5. Analysis of steroid–protein binding constants*, J. Am. Chem. Soc. 113 (1991), pp. 9055–9060.

469. H. Kubinyi, *QSAR. Hansch Analysis and Related Approaches (Methods and Principles in Medicinal Chemistry)*, Vol. 1. R. Mannhold, P. Kroogsgard-Larsen, and H. Timmerman, Eds., VCH, Weinheim, Germany, 1993.

470. Y.C. Martin, C.T. Lin, C. Hetti, and J. DeLazzer, *PLS analysis of distance matrices to detect nonlinear relationships between biological potency and molecular properties*, J. Med. Chem. 38 (1995), pp. 3009–3015.

471. R. Carbó-Dorca, D. Robert, L. Amat, X. Gironés, and E. Besalú, *Molecular Quantum Similarity in QSAR and Drug Design, Lecture Notes in Chemistry*, Vol. 73, Springer Verlag, Berlin, Germany, 2000.

472. X. Fradera, L. Amat, E. Besalú, and R. Carbó-Dorca, *Application of molecular quantum similarity to QSAR*, Quant. Struct. Act. Relat. 16 (1997), pp. 25–32.

473. R. Carbó-Dorca, *Stochastic transformation of quantum similarity matrices and their use in quantum QSAR (QQSAR) models*, Intern. J. Quantum Chem. 79 (2000), pp. 163–177.

474. E.E. Hodgkin and W.G. Richards, *Molecular similarity based on electrostatic potential and electric field*, Int. J. Quantum Chem. Quantum Biol. Quantum Pharma. Symp. 32–S14 (1987), pp. 105–110.

475. M. Manaut, F. Sanz, J. José, and M. Milesi, *Automatic search for maximum similarity between molecular electrostatic potential distributions*, J. Comput.-Aided Mol. Des. 5 (1991), pp. 371–380.

476. C. Burt, W.G. Richards, and P. Huxley, *The application of molecular similarity calculations*, J. Comput. Chem. 11 (1990), pp. 1139–1146.

477. A.C. Good and W.G. Richards, *Rapid evaluation of shape similarity using Gaussian functions*, J. Chem. Inf. Comput. Sci. 33 (1993), pp. 112–116.

478. D.J. Livingstone, G. Hesketh, and D. Clayworth, *Novel method for the display of multivariate data using neural networks*, J. Mol. Graph. 9 (1991), pp. 115–118.

479. A.Y. Meyer and W.G. Richards, *Similarity of molecular shape*, J. Comput.-Aided Mol. Des. 5 (1991), pp. 427–439.

480. M.S. Allen, Y.C. Tan, M.L. Trudell, K. Narayanan, L.R. Schindler, M.J. Martin, C. Schultz et al. *Synthetic and computer-assisted analyses of the pharmacophore for the benzodiazepine receptor inverse agonist site*, J. Med. Chem. 33 (1990), pp. 2343–2357.

481. F.I. Carroll, Y. Gao, M.A. Rahman, P. Abraham, K. Parham, A.H. Lewin, J.W. Boja, and M.J. Kuhar, *Synthesis, ligand binding, QSAR and CoMFA study of 3β-(p-substituted phenyl)tropane-2β-carboxylic acid methyl esters*, J. Med. Chem. 34 (1991), pp. 2719–2725.

482. G. Maret, N. El Tayar, P.A. Carrupt, B. Testa, P. Jenner, and M. Baird, *Toxication of MPTP (1-methyl-4-phenyl-1,2,3,6-tetrahydropyridine) and analogs by monoamine oxidase. A structure–reactivity relationship study*, Biochem. Pharm. 40 (1990), pp. 783–792.

483. D.C. Horwell, W. Howson, M. Higginbottom, D. Naylor, G.S. Ratcliffe, and S. Williams, *Quantitative structure–activity relationships (QSARs) of N-terminus fragments of NK1 Tachykinin antagonists: A comparison of classical QSARs and three-dimensional QSARs from similarity matrices*, J. Med. Chem. 38 (1995) pp. 4454–4462.

484. C.A. Montanari, M.S. Tute, A.E. Beezer, and J.C. Mitchell, *Determination of receptor-bound drug conformations by QSAR using flexible fitting to derive a molecular similarity index*, J. Comput. Aided Mol. Des. 10 (1996), pp. 67–73.

485. R.J. Abraham and G.H. Grant, *Charge calculations in molecular mechanics. IX. A general parametrisation of the scheme for saturated halogen, oxygen and nitrogen compounds*, J. Comput.-Aided Mol. Des. 6 (1992), pp. 273–286 (and preceding papers).

486. A. Agarwal, P.P. Pearson, E.W. Taylor, H.B. Li, T. Dahlgren, M. Herslof, Y. Yang, G. Lambert, D.L. Nelson, J.W. Regan, and A.R. Martin, *Three-dimensional quantitative structure-activity relationshps of 5-HT receptor binding data for tetrahydropyridinylindole derivatives: A comparison of the Hansch and CoMFA methods*, J. Med. Chem. 36 (1993), pp. 4006–4014.

487. S.S. So and M. Karplus, *Evolutionary optimization in quantitative-structure–activity relationship: An application of genetic neural networks*, J. Med. Chem. 39 (1996), pp. 1521–1530.

488. S.S. So and M. Karplus, *Genetic neural networks for quantitative structure–activity relationships: Improvements and application of benzodiazepine affinity for benzodiazepine/GABA_A receptors*, J. Med. Chem. 39 (1996), pp. 5246–5256.

489. B.D. Silverman and D.E. Platt, *Comparative molecular moment analysis (CoMMA): 3D QSAR without molecular superposition*, J. Med. Chem. 39 (1996), pp. 2129–2140.

490. M.S. Allen, A.J. LaLoggia, L.J. Dorn, M.J. Martin, G. Costantino, T.J. Hagen, K.F. Koehler, P. Skolnick, and J.M. Cook, *Predictive binding of β-carboline inverse agonists and antagonists via the CoMFA/GOLPE approach*, J. Med. Chem. 35 (1992), pp. 4001–4010.

491. S.J. Cho, A. Tropsha, M. Suffness, Y.C. Cheng, and H. Lee, *Antitumor agents. 163. Three-dimensional quantitative structure–activity relationship study of 4'-O-demethylepipodophillotoxin analogues using the modified CoMFA/q2-GRS approach*, J. Med. Chem. 39 (1996), pp. 1383–1395.

492. K.A. Watson, E.P. Mitchell, L.N. Johnson, G. Cruciani, J.C. Son, C.J.F. Bichard, G.W.J. Fleet, N.G. Oikonomakos, M. Kontou, and S.E. Zographos, *Glucose analogue inhibitors of glycogen phosphorylase: From crystallographic analysis to drug prediction using GRID force-field and GOLPE variable selection*, Acta Crystallogr. D Biol. Crystallogr. 51 (1995), pp. 458–472.

493. K.A. Watson, E.P. Mitchell, L.N. Johnson, C.J.F. Bichard, M.G. Orchard, G.W.J. Fleet, N.G. Oikonomakos, D.D. Leonidas, and J.C. Son, *Design of inhibitors of glycogene phosphorylase: A study of α- and β-C-glucosides and 1-thio-β-D-glucose compounds*, Biochemistry 33 (1994), pp. 5745–5758.

494. N.G. Oikonomakos, M. Kontou, S.E. Zographos, K.A. Watson, L.N. Johnson, C.J.F. Bichard, G.W.J. Fleet, and K.R. Acharya, *N-Acetyl-β-D-glucopyranosylamine: A potent T-state inhibitor of glycogen phophorylase. A comparison with α-D-glucose*, Protein Sci. 4 (1995), pp. 2469–2477.

495. R. Benigni, M. Cotta-Ramusino, F. Giorgi, and G. Gallo, *Molecular similarity matrices and quantitative structure-activity relationships: A case study with methodological implications*, J. Med. Chem. 38 (1995), pp. 629–635.

496. ASP program, Oxford Molecular LTD, The Madawar Center, Oxford Science Park. Oxford, U.K.

497. L. Amat, D. Robert, E. Besalú, and R. Carbó-Dorca, *Molecular quantum similarity measures tuned 3D QSAR: An antitumoral family validation study*, J. Chem. Inf. Comput. Sci. 38 (1998), pp. 624–631.

498. X. Gironés and R. Carbó-Dorca, *Modelling toxicity using molecular quantum similarity measures*, QSAR Comb. Sci. 25 (2006), pp. 579–589.

499. L. Amat, E. Besalú, R. Carbó-Dorca, and R. Ponec, *Identification of active molecular sites using quantum-self-similarity measures*, J. Chem. Inf. Comput. Sci. 41 (2001), pp. 978–991.

500. P. Bultinck, T. Kuppens, X. Gironés, and R. Carbó-Dorca, *Quantum similarity superposition algorithm (QSSA): A consistent scheme for molecular alignment and molecular similarity based on quantum chemistry*, J. Chem. Inf. Comput. Sci. 43 (2003), pp. 1143–1150.

501. P. Constans, L. Amat, and R. Carbó-Dorca, *Toward a global maximization of the molecular similarity function: Superposition of two molecules*, J. Comput. Chem. 18 (1997), pp. 826–846.

502. X. Gironés, D. Robert, and R. Carbó-Dorca, *TGSA: A molecular superposition program based on topo-geometrical considerations*, J. Comput. Chem. 22 (2001), pp. 255–263.

503. L. Amat and R. Carbó-Dorca, *Quantum similarity measures under atomic shell approximation: First order density fitting using elementary Jacobi rotations*, J. Comput. Chem. 18 (1997), pp. 2023–2039.

504. L. Amat and R. Carbó-Dorca, *Fitted electronic density functions from H to Rn for use in quantum similarity measures: cis-Diamminedichloroplatinum(II) complex as an application example*, J. Comput. Chem. 20 (1999), pp. 911–920.

505. L. Amat ASA repository for different basis sets at the Institute of Computational, Chemistry, http://stark.udg.es/cat/similarity/ASA/funcset.html (accessed July 2004).

506. D. Robert, L. Amat, and R. Carbó-Dorca, *Three-dimensional quantitative structure-activity relationships from tuned molecular quantum similarity measures: Prediction of the corticosteroid-binding globulin binding affinity for a steroid family*, J. Chem. Inf. Comput. Sci. 39 (1999), pp. 333–344.

507. D. Hadjipavlou-Litina and C. Hansch, *Quantitative structure–activity relationships of the benzodiazepines. A review and reevaluation*, Chem. Rev. 94 (1994), pp. 1483–1505.

508. L. Amat, R. Carbó-Dorca, and R. Ponec, *Molecular quantum similarity measures as an alternative to logP values in QSAR studies*, J. Comput. Chem. 19 (1998), pp. 1575–1583.

509. L. Amat, R. Carbó-Dorca, and R. Ponec, *Simple linear QSAR models based on quantum similarity measures*, J. Med. Chem. 42 (1999), pp. 5169–5180.

510. A. Gallegos, R. Carbó-Dorca, R. Ponec, and K. Waisser, *Similarity approach to QSAR. Application to antimycobacterial benzoxazines*, Int. J. Pharm. 269 (2004), pp. 51–60.

511. T. Kotani and K. Higashiura, *Rapid evaluation of molecular shape similarity index using pairwise calculation of the nearest atomic distances*, J. Chem. Inf. Comput. Sci. 42 (2002), pp. 58–63.

512. T.I. Oprea, D. Ciubotariu, T.I. Sulea, and Z. Simon, *Comparison of the minimal steric difference (MTD) and comparative molecular field analysis (CoMFA) methods for analysis of binding of steroids to carrier proteins*, Quant. Struct. Act. Relat. 12 (1993), pp. 21–26.

513. C. Lemmen, T. Lengauer, and G. Klebe, *FlexS: A method for fast flexible ligand superposition*, J. Med. Chem. 41 (1998), pp. 4502–4520.

514. W. Zheng and A. Tropsha, *Novel variable selection quantitative structure-property relationship approach based on the k-nearest-neighbor principle*, J. Chem. Inf. Comput. Sci. 40 (2000), pp. 185–194.

515. N. Metropolis, A.W. Rosenbluth, M.N. Rosenbluth, A.H. Teller, and E. Teller, *Equation of state calculations by fast computing machines*, J. Chem. Phys. 21 (1953), pp. 1087–1092.

516. G.R. Desiraju, B. Gopalakrishnan, R.K.R. Jetti, D. Raveendra, J.A.R.P. Sarma, and H.S. Subramanya, *Three-dimensional quantitative structure activity relationship (3D-QSAR) studies of some 1,5-diarylpyrazoles: Analogue based design of selective cyclooxygenase-2 inhibitors*, Molecules 5 (2000), pp. 945–955.

517. M.E. Suh, S.Y. Park, and H.J. Lee, *Comparison of QSAR methods (CoMFA, CoMSIA, HQSAR) of anticancer 1-N-substituted imidazoquinoline-4,9-dione derivatives*, Bull. Kor. Chem. Soc. 23 (2002), pp. 417–422.

518. A. Golbraikh and A. Tropsha, *QSAR modeling using chirality descriptors derived from molecular topology*, J. Chem. Inf. Comput. Sci. 43 (2003), pp. 144–154.

519. J. Zupan, M. Novič, and J. Gasteiger, *Neural networks with counter-propagation learning strategy used for modelling*, Chemom. Intell. Lab. Syst. 27 (1995), pp. 175–187.

520. M. Vračko, M. Novič, and J. Zupan, *Study of structure-toxicity relationship by a counterpropagation neural network*, Anal. Chim. Acta 384 (1999), pp. 319–332.

521. M. Vračko, D. Mills, and S.C. Basak, *Structure-mutagenicity modelling using counter propagation neural networks*, Environ. Toxicol. Pharmacol. 16 (2004), pp. 25–36.

522. S.C. Basak, D.R. Mills, A.T. Balaban, and B.D. Gute, *Prediction of mutagenicity of aromatic and heteroaromatic amines from structure: A hierarchical QSAR approach*, J. Chem. Inf. Comput. Sci. 41 (2001), pp. 671–678.

523. M. Novič and M. Vračko, *Kohonen and counterpropagation neural networks employed for modeling endocrine disruptors*, in *Endocrine Disruption Modeling*, J. Devillers, Ed., Taylor & Francis, Boca Raton, FL, 2009, pp. 199–234.

524. T. Ghafourian and M.T.D. Cronin, *The effect of variable selection on the non-linear modelling of oestrogen receptor binding*, QSAR Comb. Sci. 25 (2006), pp. 824–835.
525. I. Kuzmanovski and M. Novič, *Counter-propagation neural networks in Matlab*, Chemom. Intell. Lab. Syst. 90 (2008), pp. 84–91.
526. M. Arakawa, K. Hasegawa, and K. Funatsu, *QSAR study of anti-HIV HEPT analogues based on multi-objective genetic programming and counter-propagation neural network*, Chemom. Intell. Lab. Syst. 83 (2006), pp. 91–98.
527. H.F. Chen, X.J. Yao, Q. Li, S.G. Yuan, A. Panaye, J.P. Doucet, and B.T. Fan, *Comparative study of non-nucleoside inhibitors with HIV-1 reverse transcriptase based on 3D-QSAR and docking*, SAR QSAR Environ. Res. 14 (2003), pp. 455–474.
528. R. Garg, S.P. Gupta, H. Gao, M.S. Babu, A.K. Debnath, and C. Hansch, *Comparative quantitative structure–activity relationship studies on anti-HIV drugs,* Chem. Rev. 99 (1999), pp. 3525–3601.
529. C.M.R. Ginn, D.B. Turner, P. Willett, A.M. Ferguson, and T.W. Heritage, *Similarity searching in files of three-dimensional chemical structures: Evaluation of the EVA descriptor and combination of rankings using data fusion*, J. Chem. Inf. Comput. Sci. 37 (1997), pp. 23–37.
530. D.B. Turner, P. Willett, A.M. Ferguson, and T.W. Heritage, *Evaluation of a novel molecular vibration-based descriptor (EVA) for QSAR studies: 2. Model validation using a benchmark steroid dataset*, J. Comput.-Aided Mol. Des. 13 (1999), pp. 271–296.
531. K. Tuppurainen, *EEVA(electronic eigenvalues): A new QSAR/QSPR descriptor for electronic substituent effects based on molecular orbital energies*, SAR QSAR Environ. Res. 10 (1999), pp. 39–46.
532. K. Wolinski, J.F. Hinton, and P. Pulay, *Efficient implementation of the gauge-independent atomic orbital method for NMR chemical shift calculations*, J. Am. Chem. Soc. 112 (1990), pp. 8251–8260.
533. ACD/LabsCNMR software, version 4.0. Toronto, Canada.
534. W. Bremser, *HOSE—A novel substructure code*, Anal. Chim. Acta 103 (1978), pp. 355–365.
535. M.C. Hemmer, V. Steinhauer, and J. Gasteiger, *Deriving the 3D structure of organic molecules from their infrared spectra*, Vib. Spectrosc. 19 (1999), pp. 151–164.
536. R. Benigni, L. Passerini, D.J. Livingstone, M.A. Johnson, and A. Giuliani, *Infrared spectra information and their correlation with QSAR descriptors*, J. Chem. Inf. Comput. Sci. 39 (1999), pp. 558–562.
537. R. Benigni, A. Guiliani, and L. Passerini, *Infrared spectra as chemical descriptors for QSAR models*, J. Chem. Inf. Comput. Sci. 41 (2001), pp. 727–730.
538. F.R. Burden, *A chemically intuitive molecular index based on the eigenvalues of a modified adjacency matrix*, Quant. Struct. Act. Relat. 16 (1997), pp. 309–314.
539. D.B. Turner and P. Willett, *Evaluation of the EVA descriptor for QSAR studies: 3. The use of a genetic algorithm to search for models with enhanced predictive properties (EVA_GA)*, J. Comput.-Aided Mol. Des. 14 (2000), pp. 1–21.
540. E.W. Berking, F. van Meel, E.W. Turpijn, and J. van der Vies, *Binding of progestagens to receptor proteins in MCF-7 cells*, J. Steroid Biochem. 19 (1983), pp. 1563–1570.
541. R.D. Beger and J.G. Wilkes, *Models of polychlorinated dibenzodioxins, dibenzofurans, and biphenyls binding affinity to the aryl hydrocarbon receptor developed using ^{13}C NMR data*, J. Chem. Inf. Comput. Sci. 41 (2001), pp. 1322–1329.
542. R.D. Beger, J.P. Freeman, J.O. Lay Jr., J.G. Wilkes, and D.W. Miller, *^{13}C NMR and electron ionization mass spectrometric data–activity relationship model of estrogen receptor binding*, Toxicol. Appl. Pharmacol. 169 (2000), pp. 17–25.
543. R.D. Beger, J.P. Freeman, J.O. Lay Jr., J.G. Wilkes, and D.W. Miller, *Use of ^{13}C NMR spectrometric data to produce a predictive model of estrogen receptor binding activity*, J. Chem. Inf. Comput. Sci. 41 (2001), pp. 219–224.

544. R.D. Beger, K.J. Holm, D.A. Buzatu, and J.G. Wilkes, *Using simulated 2D ^{13}C NMR nearest neighbor connectivity spectral data patterns to model a diverse set of estrogens*, Internet Electron. J. Mol. Des. 2 (2003), pp. 435–453.

545. W.S. Branham, S.L. Dial, C.L. Moland, B.S. Hass, R.M. Blair, H. Fang, L. Shi, W. Tong, R.G. Perkins, and D.M. Sheehan, *Phytoestrogens and mycoestrogens bind to the rat uterine estrogen receptor*, J. Nutr. 132 (2002), pp. 658–664.

546. R.M. Blair, H. Fang, W.S. Branham, B.S. Hass, S.L. Dial, C.L. Moland, W. Tong, L. Shi, R. Perkins, and D.M. Sheehan, *The estrogen receptor relative binding affinities of 188 natural and xenochemicals: Structural diversity of ligands*, Toxicol. Sci. 54 (2000), pp. 138–153.

547. W. Tong, D.R. Lowis, R. Perkins, Y. Chen, W.J. Welsh, D.W. Goddette, T.W. Heritage, and D.M. Sheehan, *Evaluation of quantitative structure–activity relationship methods for large-scale prediction of chemicals binding to the estrogen receptor*, J. Chem. Inf. Comput. Sci. 38 (1998), pp. 669–677.

548. G.G. Kuiper, J.G. Lemmen, B. Carlsson, J.C. Corton, S.H. Safe, P.T. van der Saag, B. van der Burg, and J.A. Gustafsson, *Interaction of estrogenic chemicals and phytoestrogens with estrogen receptor β*, Endocrinology 139 (1998), pp. 4252–4263.

549. T. Ghafourian and M.T.D. Cronin, *The impact of variable selection on the modelling of oestrogenicity*, SAR QSAR Environ. Res. 16 (2005), pp. 171–190.

550. R.D. Beger and J.G. Wilkes, *Developing ^{13}C NMR quantitative spectrometric data-activity relationship (QSDAR) models of steroid binding to the corticosteroid binding globulin*, J. Comput.-Aided Mol. Des. 15 (2001), pp. 659–669.

551. R.D. Beger, D.A. Buzatu, J.G. Wilkes, and J.O. Lay Jr., *Comparative structural connectivity spectra analysis (CoSCoSA) models of steroid binding to the corticosteroid binding globulin*, J. Chem. Inf. Comput. Sci. 42 (2002), pp. 1123–1131.

552. C. De Gregorio, L.B. Kier, and L.H. Hall, *QSAR modeling with the electrotopological state indices: Corticosteroids*, J. Comput.-Aided Mol. Des. 12 (1998), pp. 557–561.

553. K. Tuppurainen and J. Ruuskanen, *Electronic eigenvalue (EEVA): A new QSAR/QSPR descriptor for electronic substituent effect based on molecular orbital energies. A QSAR approach to the Ah receptor binding affinity of polychlorinated biphenyls (PCBs), dibenzo-p-dioxins (PCDDs) and dibenzofurans (PCDFs)*, Chemosphere 41 (2000), pp. 843–848.

554. K. Tuppurainen, M. Viisas, R. Laatikainen, and M. Peräkylä, *Evaluation of a novel electronic eigenvalue (EEVA) molecular descriptor for QSAR/QSPR studies: Validation using a benchmark steroid data set*, J. Chem. Inf. Comput. Sci. 42 (2002), pp. 607–613.

555. V.A. Palyulin, E.V. Radchenko, and N.S. Zefirov, *Molecular field topology analysis method in QSAR studies of organic compouds*, J. Chem. Inf. Comput. Sci. 40 (2000), pp. 659–667.

556. A.H. Asikainen, J. Ruuskanen, and K.A. Tuppurainen, *Alternative QSAR models for selected estradiol and cytochrome P450 ligands: Comparison between classical, spectroscopic, CoMFA and GRID/GOLPE methods*, SAR QSAR Environ. Res. 16 (2005), pp. 555–565.

557. E. Napolitano, R. Fiaschi, K.E. Carlson, and J.A. Katzenellenbogen, *11β-Substituted estradiol derivatives, potential high-affinity carbon-11-labeled probes for the estrogen receptor: A structure-affinity relationship study*, J. Med. Chem. 38 (1995), pp. 429–434.

558. H. Gao, J.A. Katzenellenbogen, R. Garg, and C. Hansch, *Comparative QSAR analysis of estrogen receptor ligands*, Chem. Rev. 99 (1999), pp. 723–724.

559. L. Sun, Y. Zhou, L. Genrong, and S.Z. Li, *Molecular electronegativity-distance vector (MEDV-4): A two-dimensional QSAR method for the estimation and prediction of biological activities of estradiol derivatives*, J. Mol. Struct. (Theochem) 679 (2004), pp. 107–113.

560. A. Poso, J. Gynther, and R. Juvonen, *A comparative molecular field analysis of cytochrome P450 2A5 and 2A6 inhibitors*, J. Comput. Aided Mol. Des. 15 (2001), pp. 195–202.

561. C. Hansch, S.B. Mekapati, A. Kurup, and R.P. Verma, *QSAR of cytochrome P450*, Drug Metabol. Rev. 36 (2004), pp. 105–156.

562. J. Schuur and J. Gasteiger, *Infrared spectra simulation of substituted benzene derivatives on the basis of a novel 3D structure representation*, Anal. Chem. 69 (1997), pp. 2398–2405.

563. J. Gasteiger, J. Schuur, P. Selzer, L. Steinhauer, and V. Steinhauer, *Finding the 3D structure of a molecule in its IR spectrum*, Fresenius J. Anal. Chem. 359 (1997), pp. 50–55.

564. J.H. Schuur, P. Selzer, and J. Gasteiger, *The coding of the three-dimensional structure of molecules by molecular transforms and its application to structure-spectra correlations and studies of biological activity*, J. Chem. Inf. Comput. Sci. 36 (1996), pp. 334–344.

565. J. Gasteiger, J. Sadowski, J. Schuur, P. Selzer, L. Steinhauer, and V. Steinhauer, *Chemical information in 3D space*, J. Chem. Inf. Comput. Sci. 36 (1996), pp. 1030–1037.

566. C. Epouhe, B.T. Fan, S.G. Yuan, A. Panaye, and J.P. Doucet, *Contribution to structural elucidation: Behaviours of substructures partially defined from 2D-NMR*, Chin. J. Chem. 21 (2003), pp. 1268–1274.

567. A.T. Balaban, Ed., *From Chemical Topology to Three-Dimensional Geometry*, Plenum Press, New York, 1997.

568. E. Estrada and E. Molina, *3D connectivity indices in QSPR/QSAR studies*, J. Chem. Inf. Comput. Sci. 41 (2001), pp. 791–797.

569. W. Zheng, S.J. Cho, and A. Tropsha, *Rational combinatorial library design. 1. Focus-2D: A new approach to the design of targeted combinatorial chemical libraries*, J. Chem. Inf. Comput. Sci. 38 (1998), pp. 251–258.

570. S.J. Cho, W. Zheng, and A. Tropsha, *Rational combinatorial library design. 2. Rational design of targeted combinatorial peptide libraries using chemical similarity probe and the inverse QSAR approaches*, J. Chem. Inf. Comput. Sci. 38 (1998), pp. 259–268.

571. J.V. de Julián-Ortiz, J. Gálvez, C. Muñoz-Collado, R. Garciá-Domenech, and C. Gimeno-Cardona, *Virtual combinatorial syntheses and computational screening of new potential anti-herpes compounds*, J. Med. Chem. 42 (1999), pp. 3308–3314.

572. J. Galvez, R. Garcia-Domenech, J.V. de Julian-Ortiz, and R. Soler, *Topological approach to drug design*, J. Chem. Inf. Comput. Sci. 35 (1995), pp. 272–284.

573. D.T. Stanton and P.C. Jurs, *Development and use of charged partial surface area structural descriptors in computer-assisted quantitative structure–property relationship studies*, Anal. Chem. 62 (1990), pp. 2323–2329.

574. R.J. Abraham and P.E. Smith, *Charge calculation in molecular mechanics IV: A general method for conjugated systems*, J. Comput. Chem. 9 (1988), pp. 288–297.

575. D.T. Stanton, L.M. Egolf, P.C. Jurs, and M.G. Hicks, *Computer-assisted prediction of normal boiling points of pyrans and pyrroles*, J. Chem. Inf. Comput. Sci. 32 (1992), pp. 306–316.

576. D.T. Stanton, S. Dimitrov, V. Grancharov, and O.G. Mekenyan, *Charged partial, surface area (CPSA) descriptors QSAR applications*, SAR QSAR Environ. Res. 13 (2002), pp. 341–351.

577. P. Schmieder, Y. Koleva, and O. Mekenyan, *A reactivity pattern for discrimination of ER agonism and antagonism based on 3-D molecular attributes*, SAR QSAR Environ. Res. 13 (2002), pp. 353–364.

578. M. Karelson, V.S. Lobanov, and A.R. Katritzky, *Quantum-chemical descriptors in QSAR/QSPR studies*, Chem. Rev. 96 (1996), pp. 1027–1043.

579. E. Clementi, *Computational Aspects for Large Chemical Systems*, Springer Verlag, Berlin, Germany, 1980.

580. R. Franke, *Theoretical Drug Design Methods*, Elsevier Science Publishers, Amsterdam, the Netherlands, 1984.

581. M.T.D. Cronin and T.W. Schultz, *Development of quantitative structure–activity relationships for the toxicity of aromatic compounds to Tetrahymena pyriformis: Comparative assessment of the methodologies*, Chem. Res. Toxicol. 14 (2001), pp. 1284–1295.

582. M.T.D. Cronin, N. Manga, J.R. Seward, G.D. Sinks, and T.W. Schultz, *Parametrization of electrophilicity for the prediction of the toxicity of aromatic compounds*, Chem. Res. Toxicol. 14 (2001), pp. 1498–1505.

583. J.R. Seward, M.T.D. Cronin, and T.W. Schultz, *The effect of precision of molecular orbital descriptors on toxicity modeling of selected pyridines*, SAR QSAR Environ. Res. 13 (2002), pp. 325–340.

584. J. Mullay, *A method for calculating atomic charges in large molecules*, J. Comput. Chem. 9 (1988), pp. 399–405.

585. J. Mullay, *A simple method for calculating atomic charge in molecules*, J. Am. Chem. Soc. 108 (1986), pp. 1770–1775.

586. S.J. Wiener, P.A. Kollman, D.A. Case, U.C. Singh, C. Ghio, G. Alagona, S. Profeta, and P. Wiener, *A new force field for molecular mechanical simulation of nucleic acids and proteins*, J. Am. Chem. Soc. 106 (1984), pp. 765–784.

587. A.R. Katritzky and E.V. Gordeeva, *Traditional topological indexes vs electonic, geometrical, and combined molecular descriptors in QSAR/QSPR research*, J. Chem. Inf. Comput. Sci. 33 (1993), pp. 835–857.

588. O. Ivanciuc, T. Ivanciuc, and A.T. Balaban, *Vertex- and edge-weighted molecular graphs and derived structural descriptors*, in *Topological Indices and Related Descriptors in QSAR and QSPR*, J. Devillers and A.T. Balaban, Eds., Gordon & Breach Science Publishers, Amsterdam, the Netherlands, 1999, pp. 169–220.

589. H. Hosoya, *Topological index, a newly proposed quantity characterizing the topological nature of structural isomers of saturated hydrocarbons*, Bull. Chem. Soc. Jpn 44 (1971), pp. 2332–2339.

590. M. Randić, *On characterization of molecular branching*, J. Am. Chem. Soc. 97 (1975), pp. 6609–6615.

591. L.B. Kier, L.H. Hall, W.J. Murray, and M. Randić, *Molecular connectivity. I. Relationship to nonspecific local anesthesia*, J. Pharm. Sci. 64 (1975), pp. 1971–1974.

592. M. Randić, *Characterization of atoms, molecules and classes of molecules based on path enumeration*, MATCH 7 (1979), pp. 5–64.

593. M Randić, *On history of the Randić index and emerging hostility toward chemical graph theory*, Commun. Math. Comput. Chem./MATCH 59 (2008), pp. 5–124.

594. M. Randić, *Generalized molecular descriptors*, J. Math. Chem. 7 (1991), pp. 155–168.

595. A. Gallegos and X. Gironés, *Topological quantum similarity indices based on fitted densities: Theoretical background and QSPR applications*, J. Chem. Inf. Model. 45 (2005), pp. 321–326.

596. A. Gallegos Saliner, *Molecular quantum similarity in QSAR; applications in computer-aided molecular design*, PhD thesis, Girona University, Girona, Spain, 2004.

597. M. Randić, *On characterization of three-dimensional structures*, Int. J. Quantum Chem. Quantum Biol. Quantum Pharma. Symp. 34-S15 (1988), pp. 201–208.

598. M. Randić, B. Jerman-Blažić, and N. Trinajstić, *Development of 3-dimensional molecular descriptors*, Comput. Chem. 14 (1990), pp. 237–246.

599. M. Randić and M. Razinger, *Molecular topographic indices*, J. Chem. Inf. 35 (1995), pp. 140–147.

600. B. Bogdanov, S. Nikolić, and N. Trinajstić, *On the three-dimensional Wiener number*, J. Math. Chem. 3 (1989), pp. 299–309.

601. M.V. Diudea, D. Horvath, and A. Graovac, *Molecular topology. 15. 3D distance matrices and related topological indices*, J. Chem. Inf. Comput. Sci. 35 (1995), pp. 129–135.

602. M. Randić, *Molecular shape profiles*, J. Chem. Inf. Comput. Sci. 35 (1995), pp. 373–382.

603. M. Randić, *Molecular bonding profiles*, J. Math. Chem. 19 (1996), pp. 375–392.

604. A.T. Balaban, *From chemical topology to 3D geometry*, J. Chem. Inf. Comput. Sci. 37 (1997), pp. 645–650.

605. S.C. Basak, C.D. Grunwald, and G.J. Niemi, *Use of graph theoretic and geometric molecular descriptors in structure–activity relationships*, in *Topological Indices and Related Descriptors in QSAR and QSPR*, J. Devillers and A.T. Balaban (Eds.), Gordon and Breach Science Publishers, Amsterdam, The Netherlands (1999), pp. 73–116.

606. S.C. Basak, B.D. Gute, and G.D. Grunwald, *A hierarchical approach to the development of QSAR models using topological, geometrical and quantum chemical parameters*, in *Topological Indices and Related Descriptors in QSAR and QSPR*, J. Devillers and A.T. Balaban, Eds., Gordon & Breach Science Publishers, Amsterdam, the Netherlands, 1999, pp. 675–696.

607. E. Estrada, *Spectral moments of the edge-adjacency matrix of molecular graphs. 2. Molecules contaning heteroatoms and QSAR applications*, J. Chem. Inf. Comput. Sci. 37 (1997), pp. 320–328.

608. S.P. van Helden and H. Hamersma, *3D-QSAR of the receptor binding of steroids. A comparison of multiple regression neural networks and comparative molecular field analysis*, in *QSAR and Molecular Modeling: Concepts, Computational Tools and Biological Applications, Proceedings of the 10th European Symposium on Structure–Activity Relationships: QSAR and Molecular Modeling*, Barcelona, Spain, F. Sanz, J. Giraldo and D F. Manaut, Eds., J.R. Prous Science Publishers, Barcelona, Spain, 1995, pp. 481–483.

609. S.P. van Helden, H. Hamersma, and V.J. van Geerestein, *Prediction of the progesterone receptor binding of steroids using a combination of genetic algorithms and neural networks*, in *Genetic Algorithms in Molecular Modeling*, J. Devillers, Ed., Academic Press, San Diego, CA, 1996, pp. 159–192.

610. S.S. So, S.P. van Helden, V.J. van Geerestein, and M. Karplus, *Quantitative structure–activity relationship studies of progesterone receptor binding steroids*, J. Chem. Inf. Comput. Sci. 40 (2000), pp. 762–772.

611. S.P. Niculescu and K.L.E. Kaiser, *Modeling the relative binding affinity of steroids to the progesterone receptor with probabilistic neural networks*, Quant. Struct. Act. Relat. 20 (2001), pp. 223–226.

612. J. Sun, H.F. Chen, H.R. Xia, J.H. Yao, and B.T. Fan, *Comparative study of factor Xa inhibitors using molecular docking/SVM/HQSAR/3D-QSAR methods*, QSAR Comb. Sci. 25 (2006), pp. 25–45.

613. L.J. Tang, Y.P. Zhou, J.H. Jiang, H.Y. Zou, H.L. Wu, G.L. Shen, and R.Q. Yu, *Radial basis function network-based transform for a nonlinear support vector machine as optimized by a particle swarm optimization algorithm with application to QSAR studies*, J. Chem. Inf. Model. 47 (2007), pp. 1438–1445.

614. E. Estrada and L.A. Montero, *Bond order weighted graphs in molecules as structure-property indices*, Mol. Eng. 2 (1993), pp. 363–373.

615. E. Estrada, *Three-dimensional molecular descriptors based on electron charge density weighted graphs*, J. Chem. Inf. Comput. Sci. 35 (1995), pp. 708–713.

616. M. Lobato, L. Amat, E. Besalú, and R. Carbó-Dorca, *Structure–activity relationships of a steroid family using quantum similarity measures and topological quantum similarity indices*, Quant. Struct. Act. Relat. 16 (1997), pp. 465–472.

617. E. Estrada, *Edge adjacency relationships and a novel topological index related to molecular volume*, J. Chem. Inf. Comput. Sci. 35 (1995), pp. 31–33.

618. E. Estrada, *Edge adjacency relationships in molecular graphs containing heteroatoms: A new topological index related to molar volume*, J. Chem. Inf. Comput. Sci. 35 (1995), pp. 701–707.

619. E. Estrada and L. Rodríguez, *Edge-connectivity indices in QSPR/QSAR studies. 1. Comparison to other topological indices in QSPR studies*, J. Chem. Inf. Comput. Sci. 39 (1999), pp. 1037–1041.

620. E. Estrada and A. Ramírez, *Edge adjacency relationships and molecular topographic descriptors. Definition and QSAR applications*, J. Chem. Inf. 36 (1996), pp. 837–843.

621. E. Estrada, *Spectral moments of the edge adjacency matrix in molecular graphs. 1. Definition and applications to the prediction of physical properties of alkanes*, J. Chem. Inf. Comput. Sci. 36 (1996), pp. 844–849.

622. S. Vilar, E. Estrada, E. Uriarte, L. Santana, and Y. Gutierrez, *In silico studies toward the discovery of new anti-HIV nucleoside compounds through the use of TOPS-MODE and 2D/3D connectivity indices. 2. Purine derivatives*, J. Chem. Inf. Model. 45 (2005), pp. 502–514.

623. E. Estrada, E.J. Delgado, J.B. Alderete, and G.A. Jaña, *Quantum-connectivity descriptors in modeling solubility of environmentally important organic compounds*, J. Comput. Chem. 25 (2004), pp. 1787–1796.

624. E. Estrada and I. Gutman, *A topological index based on distances of edges of molecular graphs*, J. Chem. Inf. Comput. Sci. 36 (1996), pp. 850–853.

625. H.P. Schultz, *Topological organic chemistry. 1. Graph theory and topological indices of alkanes*, J. Chem. Inf. Comput. Sci. 29 (1989), pp. 227–228.

626. E. Estrada, *Novel strategies in the search of topological indices*, in *Topological Indices and Related Descriptors in QSAR and QSPR*, J. Devillers and A.T. Balaban, Eds., Gordon & Breach Science Publishers, Amsterdam, the Netherlands, 1999, pp. 403–453.

627. P.L.A. Popelier and P.J. Smith, *QSAR models based on quantum topological molecular similarity*, Eur. J. Med. Chem. 41 (2006), pp. 862–873.

628. R.F.W. Bader, *Atoms in Molecules: A Quantum Theory*, Oxford University Press, Oxford, U.K., 1990.

629. A. Thakur, M. Thakur, P.V. Khadikar, C.T. Supuran, and P. Sudele, *QSAR study on benzenesulphonamide carbonic anhydrase inhibitors: Topological approach using Balaban index*, Bioorg. Med. Chem. 12 (2004), pp. 789–793.

630. T.W. Schultz, D.T. Lin, T.S. Wilke, and M. Arnold, *Quantitative structure-activity relationships for the Tetrahymena pyriformis population growth endpoint: A mechanism of action approach*, in *Practical Applications of Quantitative Structure–Activity Relationships (QSAR) in Environmental Chemistry and Toxicology*, W. Karcher and J. Devillers, Eds., ECSC, EEC, EAEC, Brussels and Luxembourg, Kluwer Acad. Dordrecht, 1990, pp. 241–262.

631. K. Roy and G. Ghosh, *QSTR with extended topochemical atom indices. 2. Fish toxicity of substituted benzenes*, J. Chem. Inf. Comput. Sci. 44 (2004), pp. 559–567.

632. T.A. Roy, A.J. Krueger, C.R. Mackerer, W. Neil, A.M. Arroyo, and J.J. Yang, *SAR models for estimating the percutaneous absorption of polynuclear aromatic hydrocarbons*, SAR QSAR Environ. Res. 9 (1998), pp. 171–185.

633. A. Gallegos, D. Robert, X. Gironés, and R. Carbó-Dorca, *Structure-toxicity relationships of polycyclic aromatic hydrocarbons using molecular quantum similarity*, J. Comput. Aided Mol. Des. 15 (2001), pp. 67–80.

634. X. Gironés, A. Gallegos, and R. Carbó-Dorca, *Modeling antimalarial activity: Application of kinetic energy density quantum similarity measures as descriptors in QSAR*, J. Chem. Inf. Comput. Sci. 40 (2000), pp. 1400–1407.

635. X. Gironés, A. Gallegos, and R. Carbó-Dorca, *Antimalarial activity of synthetic 1,2,4-trioxanes and cyclic peroxy ketals, a quantum similarity study*, J. Comput. Aided Mol. Des. 15 (2001), pp. 1053–1063.

636. J. Cioslowski and E.D. Fleischmann, *Assessing molecular similarity from results of ab initio electronic structure calculation*, J. Am. Chem. Soc. 113 (1991), pp. 64–67.

637. M.A. Avery, S. Mehrotra, T.L. Jonhson, J.D. Bonk, J.A. Vroman, and R. Miller, *Structure-activity relationships of the antimalarial agent Artemisinin. 5. Analogs of 10-deoxoartemisinin substituted at C-3 and C-9*, J. Med. Chem. 39 (1996), pp. 4149–4155.

638. E. Besalú, A. Gallegos, and R. Carbó-Dorca, *Topological quantum similatity indices and their use in QSAR. Application to several families of antimalarial compounds, MATCH-Communications in Mathematical and in Computer Chemistry* (Special issue dedicated to Prof. Balaban) M. Diudea and O. Ivanciuc, Eds., *MATCH-Commun. Math.Comput. Chem.* 44 (2001), pp. 41–64.

639. A. Gallegos-Saliner, L. Amat, R. Carbó-Dorca, T.W. Schultz, and M.T.D. Cronin, *Molecular quantum similarity analysis of estrogenic activity*, J. Chem. Inf. Comput. Sci. 43 (2003), pp. 1166–1176.

640. N. Brown, B. McKay, and J. Gasteiger, *Fingal: A novel approach to geometric finger-printing and a comparative study of its applications to 3D-QSAR modelling*, QSAR Comb. Sci. 24 (2005), pp. 480–484.

641. C.L. Waller, *A comparative QSAR study using CoMFA, HQSAR, and FRED/SKEYS paradigms for estrogen receptor binding affinities of structurally diverse compounds*, J. Chem. Inf. Comput. Sci. 44 (2004), pp. 758–765.

642. S. Sciabola, E. Carosati, L. Cucurull-Sanchez, M. Baroni, and R. Mannhold, *Novel TOPP descriptors in 3D-QSAR analysis of apoptosis inducing 4-aryl-4H-chromenes: Comparison versus other 2D- and 3D-descriptors*, Bioorg. Med. Chem. 15 (2007), pp. 6450–6462.

643. MDL Information Systems, Inc. 14600 Catalina Street, San Leandro, CA.

644. J.L. Durant, B.A. Leland, D.R. Henry, and J.G. Nourse, *Reoptimization of MDL keys for use in drug discovery*, J. Chem. Inf. Comput. Sci. 42 (2002), pp. 1273–1280.

645. M.J. McGregor and P.V. Pallai, *Clustering of large databases of compounds: Using the MDL "keys" as structural descriptors*, J. Chem. Inf. Comput. Sci. 37 (1997), pp. 443–448.

646. J.M. Barnard and G.M. Downs, *Chemical fragment generation and clustering software*, J. Chem. Inf. Comput. Sci. 37 (1997), pp. 141–142.

647. PipeLine Pilot 4.5.2, 2005, version 4.5.2. Scitegic Inc., 9665 Chesapeake Dr., Suite 401, San Diego, CA.

648. S.E. O'Brien and P.L.A. Popelier, *Quantum molecular similarity. 3. QTMS descriptors*, J. Chem. Inf. Comput. Sci. 41 (2001), pp. 764–775.

649. A. Da Settimo, G. Primofiore, F. Da Settimo, A.M. Marini, E. Novellino, G. Greco, C. Martini, G. Giannaccini, and A. Lucacchini, *Synthesis, structure–activity relationships, and molecular modeling studies of N-(indol-3-ylglyoxylyl)benzylamine derivatives acting as the benzodiazepine receptor*, J. Med. Chem. 39 (1996), pp. 5083–5091.

650. L.H. Hall, B. Mohney, and L.B. Kier, *The electrotopological state: Structure information at the atomic level for molecular graphs*, J. Chem. Inf. Comput. Sci. 31 (1991), pp. 76–82.

651. C. Hansch, J. Schaeffer, and R. Kerley, *Alcohol dehydrogenase structure–activity relationships*, J. Biol. Chem. 247 (1972), pp. 4703–4710.

652. N. Kakeya, N. Yata, A. Kamada, and M. Aoki, *Biological activities of drugs. VII. Structure-activity relationship of sulfonamide carbonic anhydrase inhibitors,* Chem. Pharm. Bull. (Tokyo) 17 (1969), pp. 2000–2007.

653. E. Argese, C. Bettiol, G. Giurin, and P. Miana, *Quantitative structure–activity relationships for the toxicity of chlorophenols to mammalian submitochondrial particles*, Chemosphere 38 (1999), pp. 2281–2292.

654. P.L.A. Popelier, U.A. Chaudry, and P.J. Smith, *Quantum topological molecular similarity. Part 5. Further development with an application to the toxicity of polychlorinated dibenzo-p-dioxins (PCDDs)*, J. Chem. Soc. Perkin Trans. 2 (2002), pp. 1231–1237.

655. X. Gironés, L. Amat, D. Robert, and R. Carbó-Dorca, *Use of electron–electron repulsion energy as a molecular descriptor in QSAR and QSPR studies*, J. Comput. Aided Mol. Des. 14 (2000), pp. 477–485.

656. K. Roy and P.L.A. Popelier, *Exploring predictive QSAR models using quantum topological molecular similarity (QTMS) descriptors for toxicity of nitroaromatics to Saccharomyces cerevisiae*, QSAR Comb. Sci. 27 (2008), pp. 1006–1012.

657. B.D. Silverman, D.E. Platt, M. Pitman, and I. Rigoutsos, *Comparative molecular momment analysis (CoMMA)*, Perspect. Drug Discov. Des. 12–14 (1998), pp. 183–196.

658. H.J. Breslin, M.J. Kukla, D.W. Ludovici, R. Mohrbacher, W. Ho, M. Miranda, J.D. Rodgers et al., *Synthesis and anti-HIV-1 activity of 4,5,6,7-tetrahydro-5-methylimidazo[4,5,1-jk][1,4]benzodiazepin-2(1H)-one (TIBO) derivatives. 3*, J. Med. Chem. 38 (1995), pp. 771–793.

659. L. Saíz-Urra, M. Pérez González, and M. Teijeira, *QSAR studies about cytotoxicity of benzophenazines with dual inhibition toward both topoisomerases I and II: 3D-MoRSE descriptors and statistical considerations about variable selection*, Bioorg. Med. Chem. 14 (2006), pp. 7347–7358.

660. R.A. Gupta, A.K. Gupta, L.K. Soni, and S.G. Kaskhedikar, *Rationalization of physicochemical characters of oxazolyl thiosemicarbazone analogs towards multidrug resistant tuberculosis: A QSAR approach*, Eur. J. Med. Chem. 42 (2007), pp. 1109–1116.

661. M. Pérez González, C. Terán, M. Teijeira, and A. Morales Helguera, *QSAR studies using radial distribution function for predicting A_1 adenosine receptors agonists*, Bull. Math. Biol. 69 (2007), pp. 347–359.

662. R. Todeschini, M. Lasagni, and E. Marengo, *New molecular descriptors for 2D- and 3D-structures theory*, J. Chemom. 8 (1994), pp. 263–272.

663. R. Todeschini, P. Gramatica, R. Provenzani, and E. Marengo, *Weighted holistic invariant molecular descriptors. Part 2. Theory developmemt and application on modeling physico-chemical properties of polyaromatic hydrocarbons*, Chemom. Intell. Lab. Syst. 27 (1995), pp. 221–229.

664. G. Bravi, E. Gancia, P. Mascagni, M. Pegna, R. Todeschini, and A. Zaliani, *MS-WHIM, new 3D theorethical descriptors derived from molecular surface properties: A comparative 3D QSAR study in a series of steroids*, J. Comput. Aided Mol. Des. 11 (1997), pp. 79–92.

665. G. Bravi and J.H. Wikel, *Application of MS-WHIM descriptors: 1. Introduction of new molecular surface properties and 2. Prediction of binding affinity data*, Quant. Struct. Act. Relat. 19 (2000), pp. 29–38.

666. G. Bravi and J.H. Wikel, *Application of MS-WHIM descriptors: 3. Prediction of molecular properties*, Quant. Struct. Act. Relat. 19 (2000), pp. 39–49.

667. R. Todeschini and P. Gramatica, *3D-modelling and prediction by WHIM descriptors. Part 5. Theory development and chemical meaning of WHIM descriptors*, Quant. Struct. Act. Relat. 16 (1997), pp. 113–119.

668. R. Leardi, R. Boggia, and M. Terrile, *Genetic algorithms as a stategy for feature selection*, J. Chemom. 6 (1992), pp. 267–281.

669. P. Gramatica, *WHIM descriptors of shape*, QSAR Comb. Sci. 25 (2006), pp. 327–332.

670. R. Todeschini, C. Bettiol, G. Giurin, P. Gramatica, P. Miana, and E. Argese, *Modeling and prediction by using WHIM descriptors in QSAR studies: Submitochondrial particles (SMP) as toxicity biosensors of chlorophenols*, Chemosphere 33 (1996), pp. 71–79.

671. R. Todeschini, M. Vighi, R. Provenzani, A. Finizio, and P. Gramatica, *Modeling and prediction by using WHIM descriptors in QSAR studies: Toxicity of heterogeneous chemicals on Daphnia magna*, Chemosphere 32 (1996), pp. 1527–1545.

672. R. Todeschini and P. Gramatica, *3D-modelling and prediction by WHIM descriptors. Part 6. Application of WHIM descriptors in QSAR studies*, Quant. Struct. Act. Relat. 16 (1997), pp. 120–125.

673. E. Papa, F. Villa, and P. Gramatica, *Statistically validated QSARs, based on theoretical descriptors, for modeling aquatic toxicity of organic chemicals in Pimephales promelas (fathead minnow)*, J. Chem. Inf. 45 (2005), pp. 1256–1266.

674. P. Gramatica, F. Consolaro, and S. Pozzi, *QSAR approach to POPs screening for atmospheric persistence*, Chemosphere 43 (2001), pp. 655–664.

675. P. Gramatica, N. Navas, and R. Todeschini, *3D-modelling and prediction by WHIM descriptors. Part 9. Chromatographic relative retention time and physico-chemical properties of polychlorinated biphenyls (PCBs)*, Chemom. Intell. Lab. Syst. 40 (1998), pp. 53–63.

676. M.L. Connolly, *Analytical molecular surface calculation*, J. Appl. Cryst. 16 (1983), pp. 548–558.

677. A. Poso, R. Juvonen, and J. Gynther, *Comparative molecular field analysis of compounds with CYP2A5 binding affinity*, Quant. Struct. Act. Relat. 14 (1995), pp. 507–511.

678. S. Ekins, G. Bravi, S. Binkley, J.S. Gillepsie, B.J. Ring, J.H. Wikel, and S.A. Wrighton, *Three- and four-dimensional-quantitative structure activity relationship (3D/4D-QSAR) analyses of CYP2C9 inhibitors*, Drug Metab. Dispos. 28 (2000), pp. 994–1002.

679. K.J. Schaper and M.L.R. Samitier, *Calculation of octanol/water partition coefficients (logP) using artificial neural networks and connection matrices*, Quant. Struct. Act. Relat. 16 (1997), pp. 224–230.

680. H. van De Waterbeemd, G. Camenisch, G. Folkers, and O.A. Raevsky, *Estimation of Caco-2 cell permeability using calculated molecular descriptors*, Quant. Struct. Act. Relat. 15 (1996), pp. 480–490.

681. VolSurf program, version 4.1.4. Molecular Discovery Ltd. http://www.moldiscovery.com

682. C.J. Cramer and D.G. Truhlar, *AM1-SM2 and PM3-SM3 parameterized SCF solvation models for free energies in aqueous solutions*, J. Comput.-Aided Mol. Des. 6 (1992), pp. 629–666.

683. E. Lo Piparo, K. Koehler, A. Chana, and E. Benfenati, *Virtual screening for aryl hydrocarbon receptor binding prediction*, J. Med. Chem. 49 (2006), pp. 5702–5709.

684. E. Filipponi, G. Cruciani, O. Tabarrini, V. Ceccheti, and A. Fravolini, *QSAR study and VolSurf characterization of anti-HIV quinolone library*, J. Comput.-Aided Mol. Des. 15 (2001), pp. 203–217.

685. S. Liu, C. Cao, and Z. Li, *Approach to estimation and prediction for normal boiling point (NBP) of alkanes based on a novel molecular distance-edge (MDE)vector, λ*, J. Chem. Inf. Comput. Sci. 38 (1998), pp. 387–394.

686. S.S. Liu, C.S. Yin, Z.L. Li, and S.X. Cai, *QSAR study of steroid benchmark and dipeptides based on MEDV-13*, J. Chem. Inf. Comput. Sci. 41 (2001), pp. 321–329.

687. S.S. Liu, C.S. Yin, and L.S. Wang, *Combined MEDV-GA-MLR method for QSAR of three panels of steroids, dipeptides, and COX-2 inhibitors*, J. Chem. Inf. Comput. Sci. 42 (2002), pp. 749–756.

688. P. Zhou, H. Zeng, F.F. Tian, B. Li, and Z.L. Li, *Applying novel molecular electronegativity-interaction vector (MEIV) to QSPR study on collision cross section of singly protonated peptides*, QSAR Comb. Sci. 26 (2007), pp. 117–121.

689. R.S. Pearlman and K.M. Smith, *Novel software tools for chemical diversity*, Perspect. Drug Discov. Des. 9–11 (1998), pp. 339–353.

690. R.S. Pearlman and K.M. Smith, *Metric validation and the receptor-relevant subspace concept*, J. Chem. Inf. Comput. Sci. 39 (1999), pp. 28–35.

691. R.S. Pearlman. *Novel software for addressing chemical diversity*. http://www.netsci.org/Science/Combichem/feature08.html

692. F.R. Burden, *Molecular identification number for substructure searches*, J. Chem. Inf. Comput. Sci. 29 (1989), pp. 225–227.

693. J.S. Mason and B.R. Beno, *Library design using BCUT chemistry-space descriptors and multiple four-point pharmacophore fingerprints: Simultaneaous optimization and structure-based diversity*, J. Mol. Graph. Mod. 18 (2000), pp. 438–451.

694. D.T. Stanton, *Evaluation and use of BCUT descriptors in QSAR and QSPR studies*, J. Chem. Inf. Comput. Sci. 39 (1999), pp. 11–20.

695. M. Pérez González, C. Terán, M. Teijeira, P. Besada, and M.J. González-Moa, *BCUT descriptors for predicting affinity toward A_3 adenosine receptors*, Bioorg. Med. Chem. Lett. 15 (2005), pp. 3491–3495.

696. V. Consonni, R. Todeschini, and M. Pavan, *Structure/response correlations and similarity/diversity analysis by GETAWAY descriptors. 1. Theory of the novel 3D molecular descriptors*, J. Chem. Inf. Comput. Sci. 42 (2002), pp. 682–692.

697. V. Consonni, R. Todeschini, M. Pavela, and P. Gramatica, *Structure/response correlations and similarity/diversity analysis by GETAWAY descriptors. 2. Application of the novel 3D molecular descriptors to QSAR/QSPR studies*, J. Chem. Inf. Comput. Sci. 42 (2002), pp. 693–705.

698. H. Zhang, H. Li, and C. Liu, *CoMFA, CoMSIA, and molecular hologram QSAR studies of novel neuronal nAChRs ligands-open ring analogues of 3-pyridyl ether*, J. Chem. Inf. Model. 45 (2005), pp. 440–448.

699. H.B. Zhang, C.P. Liu, and H. Li, *CoMFA and CoMSIA studies of nAChRs ligands: Epibatidine analogues*, QSAR Comb. Sci. 23 (2004), pp. 80–88.

700. A.J. Hopfinger, A. Reaka, P. Venkatarangan, J.S. Duca, and S. Wang, *Construction of a virtual high throughput screen by 4D-QSAR analysis: Application to a combinatorial library of glucose inhibitors of glycogen phosphorylase b*, J. Chem. Inf. Comput. Sci. 39 (1999), pp. 1151–1160.

701. M. Shen, C. Béguin, A. Golbraikh, J.P. Stables, H. Kohn, and A. Tropsha, *Application of predictive QSAR models to database mining: Identification and experimental validation of novel anticonvulsant compounds*, J. Med. Chem. 47 (2004), pp. 2356–2364.

702. A. Hillebrecht and G. Klebe, *Use of 3D QSAR models for database screening: A feasability study*, J. Chem. Inf. Model. 48 (2008), pp. 384–396.

703. P. Benedetti, R. Mannhold, G. Cruciani, and G. Ottaviani, *GRIND/ALMOND investigations on $CysLT_1$ receptor antagonists of the quinolinyl(bridged)aryl type*, Bioorg. Med. Chem. 12 (2004), pp. 3607–3617.

704. P. Labute, *A widely applicable set of descriptors*, J. Mol. Graph. Model. 18 (2000), pp. 464–477.

705. E.S. Istvan and J. Deisenhofer, *Structural mechanism for statin inhibition of HMG-CoA reductase*, Science 292 (2001), pp. 1160–1170.

706. H. Claussen, C. Buning, M. Rarey, and T. Lengauer, *FlexE: Efficient molecular docking considering protein structure variations*, J. Mol. Biol. 308 (2001), pp. 377–395.

707. E. Abrahamian, P.C. Fox, L. Nærum, I.T. Christensen, H. Thøgersen, and R.D. Clark, *Efficient generation, storage, and manipulation of fully flexible pharmacophore multiplets and their use in 3D-similarity searching*, J. Chem. Inf. Comput. Sci. 43 (2003), pp. 458–468.

708. S.A. Hindle, M. Rarey, C. Buning, and T.J. Lengauer, *Flexible docking under pharmacophore type constraints*, J. Comput.-Aided Mol. Des. 16 (2002), pp. 129–149.

709. J.J. Irwin and B.K. Shoichet, *ZINC—A Free database of commercially available compounds for virtual screening*, J. Chem. Inf. Model. 45 (2005), pp. 177–182. http://zinc.docking.org

710. A. Palomer, J. Pascual, F. Cabré, M.L. García, and D. Mauleón, *Derivation of pharmacophore and CoMFA models for leukotriene D_4 receptor antagonists of the quinolinyl(bridged)aryl series*, J. Med. Chem. 43 (2000), pp. 392–400.

711. M.A. Kastenholz, M. Pastor, G. Cruciani, E.E.J. Haaksma, and T. Fox, *GRID/CPCA: A new computational tool to design selective ligands*, J. Med. Chem. 43 (2000), pp. 3033–3044.

712. J.J. Lozano, M. Pastor, G. Cruciani, K. Gaedt, N.B. Centeno, F. Gago, and F. Sanz, *3D-QSAR methods on the basis of ligand-receptor complexes. Application of COMBINE and GRID/GOLPE methodologies to a series of CYP1A2 ligands*, J. Comput. Aided Mol. Des. 14 (2000), pp. 341–353.

713. MOLCONN-Z software. Hall Associated Consulting, Quincy, MA. http://www.edusoft-lc.com/molconn

714. A. Vedani, D.R. McMasters, and M. Dobler, *Multi-conformational ligand representation in 4D-QSAR: Reducing the bias associated with ligand alignement*, Quant. Struct. Act. Relat. 19 (2000), pp. 149–161.

715. O.A. Santos-Filho and A.J. Hopfinger, *The 4D-QSAR paradigm: Application to a novel set of non-peptidic HIV protease inhibitors*, Quant. Struct. Act. Relat. 21 (2002), pp. 369–381.

716. 4D-QSAR user's manual, version 3.0, October 2001. The Chem21 Group, Inc. 1780 Wilson Drive, Lake Forest, IL.

717. T. Huber, A.E. Torda, and W.F. van Gunsteren, *Optimization methods for conformational sampling using a Boltzmann-weighted mean field approach*, Biopolymers 39 (1996), pp. 103–114.

718. J. Gálvez, *On a topological interpretation of electronic and vibrational molecular energies*, J. Mol. Struct. (Theochem) 429 (1998), pp. 255–264.

719. M. Ravi, A.J. Hopfinger, R.E. Hormann, and L. Dinan, *4D-QSAR analysis of a set of ecdysteroids and a comparison to CoMFA modeling*, J. Chem. Inf. Comput. Sci. 41 (2001), pp. 1587–1604.

720. L. Dinan, R.E. Hormann, and T. Fujimoto, *An extensive ecdysteroid CoMFA*, J. Comput. Aided Mol. Des. 13 (1999), pp. 185–207.

721. S.E. Hagen, J. Domagala, C. Gajda, M. Lovdahl, B.D. Tait, E. Wise, T. Holler et al., *4-Hydroxy-5,6-dihydropyrones as inhibitors of HIV protease: The effect of heterocyclic substituents at C-6 on antiviral potency and pharmacokinetic parameters*, J. Med. Chem. 44 (2001), pp. 2319–2332.

722. M.G. Albuquerque, A.J. Hopfinger, E.J. Barreiro, and R.B. de Alencastro, *Four-dimensional quantitative structure–activity relationship analysis of a series of interphenylene 7-oxabicycloheptane oxazole thromboxane A_2 receptor antagonists*, J. Chem. Inf. Comput. Sci. 38 (1998), pp. 925–938.

723. O.A. Santos-Filho and A.J. Hopfinger, *A search for sources of drug resistance by 4D-QSAR analysis of a set of antimalarian dihydrofolate reductase inhibitors*, J. Comput. Aided Mol. Des. 15 (2001), pp. 1–12.

724. J.S. Duca and A.J. Hopfinger, *Estimation of molecular similarity based on 4D-QSAR analysis: Formalism and validation*, J. Chem. Inf. Comput. Sci. 41 (2001), pp. 1367–1387.

725. M. Iyer, T. Zheng, A.J. Hopfinger, and Y.J. Tseng, *QSAR analyses of skin penetration enhancers*, J. Chem. Inf. Model. 47 (2007), pp. 1130–1149.

726. D. Pan, Y. Tseng, and A.J. Hopfinger, *Quantitative structure-based design: Formalism and application of receptor-dependent RD-4D-QSAR analysis to a set of glucose analogue inhibitors of glycogen phosphorylase*, J. Chem. Inf. Comput. Sci. 43 (2003), pp. 1591–1607.

727. X. Huang, T. Liu, J. Gu, X. Luo, R. Ji, Y. Cao, H. Xue et al., *3D-QSAR model of flavonoids binding at benzodiazepine site in $GABA_A$ receptors*, J. Med. Chem. 44 (2001), pp. 1883–1891.

728. X. Hong and A.J. Hopfinger, *3D-pharmacophores of flavonoid binding at the benzodiazepine $GABA_A$ receptor site using 4D-QSAR analysis*, J. Chem. Inf. Comput. Sci. 43 (2003), pp. 324–336.

729. X. Huang, L. Xu, X. Luo, K. Fan, R. Ji, G. Pei, K. Chen, and H. Jiang, *Elucidating the inhibiting mode of AHPBA derivatives against HIV-1 protease and building predictive 3D-QSAR models*, J. Med. Chem. 45 (2002), pp. 333–343.

730. C.L. Senese and A.J. Hopfinger, *Receptor-independent 4D-QSAR analysis of a set of norstatine derived inhibitors of HIV-1 protease*, J. Chem. Inf. Comput. Sci. 43 (2003), pp. 1297–1307.

731. A.C. Nair, P. Jayatilleke, X. Wang, S. Miertus, and W.J. Welsh, *Computational studies on tetrahydropyrimidine-2-one HIV-1 protease inhibitors: Improving three-dimensional quantitative structure–activity relationship comparative molecular field analysis models by inclusion of calculated inhibitor- and receptor-based properties*, J. Med. Chem. 45 (2002), pp. 973–983.

732. P. Thipnate, J. Liu, S. Hannongbua, and A.J. Hopfinger, *3D pharmacophore mapping using 4D QSAR analysis for the cytotoxicity of lamellarins against human hormone-dependent T47D breast cancer cells*, J. Chem. Inf. Model. 49 (2009), pp. 2312–2322.

733. D. Pan, M. Iyer, J. Liu, Y. Li, and A.J. Hopfinger, *Constructing optimum blood brain barrier QSAR models using a combination of 4D-molecular similarity measures and cluster analysis*, J. Chem. Inf. Comput. Sci. 44 (2004), pp. 2083–2098.

734. A. Kulkarni, Y. Han, and A.J. Hopfinger, *Predicting Caco-2 cell permeation coefficients of organic molecules using membrane-interaction QSAR analysis*, J. Chem. Inf. Comput. Sci. 42 (2002), pp. 331–342.

735. J. Polanski and A. Bak, *Modeling steric and electronic effects in 3D- and 4D-QSAR schemes: Predicting benzoic pK_a values and steroid CGB binding affinities*, J. Chem. Inf. Comput. Sci. 43 (2003), pp. 2081–2092.

736. J. Polanski, A. Bak, R. Gieleciak, and T. Magdziarz, *Self-organizing neural networks for modeling robust 3D and 4D QSAR: Application to dihydrofolate reductase inhibitors*, Molecules 9 (2004), pp. 1148–1159.

737. A. Bak and J. Polanski, *Modeling robust QSAR 3: SOM-4D-QSAR with iterative variable elimination IVE-PLS: Application to steroid, azo dye, and benzoic acid series*, J. Chem. Inf. Mod. 47 (2007), pp. 1469–1480.

738. J. Polanski, R. Gieleciak, and A. Bak, *Probability issues in molecular design: Predictive and modeling ability in 3D-QSAR schemes*, Comb. Chem. High Throughput Screen. 7 (2004), pp. 793–807.

739. User and reference manual. Quasar. http://www.biograf.ch

740. A. Bak and J. Polanski, *A 4D-QSAR study on anti-HIV HEPT analogues*, Bioorg. Med. Chem. 14 (2006), pp. 273–279.

741. S. Timofei and W.M.F. Fabian, *Comparative molecular field analysis of heterocyclic monoazo dye-fiber affinities*, J. Chem. Inf. Comput. Sci. 38 (1998), pp. 1218–1222.

742. J. Polanski, R. Gieleciak, and M. Wyszomirski, *Comparative molecular surface analysis (CoMSA) for modeling dye-fiber affinities of the azo and anthraquinone dyes*, J. Chem. Inf. Comput. Sci. 43 (2003), pp. 1754–1762.

743. V.E. Kuz'min, A.G. Artemenko, R.N. Lozytska, A.S. Fedtchouk, V.P. Lozitsky, E.N. Muratov, and A.K. Mescheriakov, *Investigation of anticancer activity of macrocyclic Schiff bases by means of 4D-QSAR based on simplex representation of molecular structure*, SAR QSAR Environ. Res. 16 (2005), pp. 219–230.

744. O. Mekenyan, N. Nikolova, and P. Schmieder, *Dynamic 3D QSAR techniques: Applications in toxicology*, J. Mol. Struct. (Theochem) 622 (2003), pp. 147–165.

745. J. Ivanov, S. Karabunarliev, and O. Mekenyan, *3DGEN: A system for exhaustive 3D molecular design proceeding from molecular topology*, J. Chem. Inf. Comput. Sci. 34 (1994), pp. 234–243.

746. O. Mekenyan, D. Dimitrov, N. Nikolova, and S. Karaburnarliev, *Conformational coverage by a genetic algorithm*, J. Chem. Inf. Comput. Sci. 39 (1999), pp. 997–1016.

747. O.G. Mekenyan, S. Karaburnarliev, J.M. Ivanov, and D.N. Dimitrov, *A new development of the oasis computer system for modeling molecular properties*, Comput. Chem. 18 (1994), pp. 173–187.

748. R. Serafimova, M. Todorov, T. Pavlov, S. Kotov, E. Jacob, A. Aptula, and O. Mekenyan, *Identification of the structural requirements for mutagenicity, by incorporating molecular flexibility and metabolic activation of chemicals II. General Ames mutagenicity model*, Chem. Res. Toxicol. 20 (2007), pp. 662–676.

749. M.T.D. Cronin, J.D. Walker, J.S. Jaworska, M.H.I. Comber, C.D. Watts, and A.P. Worth, *Use of QSARs in international decision-making frameworks to predict ecologic effects and environmental fate of chemical substances*, Environ. Health Perspect. 111 (2003), pp. 1376–1390.

750. M.A. Lill, M. Dobler, and A. Vedani, *In silico prediction of receptor-mediated environmental toxic phenomena—Application to endocrine disruption*, SAR QSAR Environ. Res. 16 (2005), pp. 149–169.

751. M.A. Lill, A. Vedani, and M. Dobler, *Raptor: Combining dual-shell representation, induced-fit simulation, and hydrophobicity scoring in receptor modeling: Application toward the simulation of structurally diverse ligand sets*, J. Med. Chem. 47 (2004), pp. 6174–6186.

752. M.A. Lill, F. Winiger, A. Vedani, and B. Ernst, *Impact of induced fit on ligand binding to the androgen receptor: A multidimensional QSAR study to predict endocrine-disrupting effects of environmental chemicals*, J. Med. Chem. 48 (2005), pp. 5666–5674.

753. A. Vedani, M. Dobler, and P. Zbinden, *Quasi-atomistic receptor surface models: A bridge between 3-D QSAR and receptor modeling*, J. Am. Chem. Soc. 120 (1998), pp. 4471–4477.

754. S. Ducki, G. Mackenzie, N.J. Lawrence, and J.P. Snyder, *Quantitative structure-activity relationship (5D-QSAR) study of combretastatin-like analogues as inhibitors of tubulin assembly*, J. Med. Chem. 48 (2005), pp. 457–465.

755. M.A. Lill and A. Vedani, *Combining 4D pharmacophore generation and multidimensional QSAR: Modeling ligand binding to the bradykinin B_2 receptor*, J. Chem. Inf. Model. 46 (2006), pp. 2135–2145.

756. A. Vedani, M. Dobler, H. Dollinger, K.-M. Hasselbach, F. Birke, and M.A Lill, *Novel ligands for the chemokine receptor-3 (CCR3): A receptor-modeling study based on 5D-QSAR*, J. Med. Chem. 48 (2005), pp. 1515–1527.

757. R. Wang, Y. Lu, X. Fang, and S. Wang, *An extensive test of 14 scoring functions using the PDBbind refined set of 800 protein–ligand complexes*, J. Chem. Inf. Comput. Sci. 44 (2004), pp. 2114–2125.

758. H. Chen, P.D. Lyne, F. Giordanetto, T. Lovell, and J. Li, *On evaluating molecular-docking methods for pose prediction and enrichment factors*, J. Chem. Inf. Model. 46 (2006), pp. 401–415.

759. G.L. Warren, C.W. Andrew, A.M. Capelli, B. Clarke, J. LaLonde, M.H. Lambert, M. Lindvall et al., *A critical assessment of docking programs and scoring functions*, J. Med. Chem. 49 (2006), pp. 5912–5931.

760. I.J. Enyedy and W.J. Egan, *Can we use docking and scoring for hit-to-lead optimization?* J. Comput. Aided Mol. Des. 22 (2008), pp. 161–168.

761. J. Carlsson, L. Boukharta, and J. Åqvist, *Combining docking, molecular dynamics and the linear interaction energy method to predict binding modes and affinities for non-nucleoside inhibitors to HIV-1 reverse transcriptase*, J. Med. Chem. 51 (2008), pp. 2648–2656.

762. A.R. Leach, B.K. Shoichet, and C.E. Peishoff, *Prediction of protein–ligand interactions. Docking and scoring: Successes and gaps*, J. Med. Chem. 49 (2006), pp. 5851–5855.

763. A.R. Leach, *Molecular Modelling: Principles and Applications*, Addison Wesley Longman, Essex, U.K., 1996.

764. E. Stjernschantz, J. Marelius, C. Medina, M. Jacobsson, N.P.E. Vermeulen, and C. Oostenbrink, *Are automated molecular dynamics simulations and binding free energy calculations realistic tools in lead optimization? An evaluation of the linear interaction energy (LIE) method*, J. Chem. Inf. Model. 46 (2006), pp. 1972–1983.

765. R. Wang, L. Lai, and S. Wang, *Further development and validation of empirical scoring functions for structure-based binding affinity prediction*, J. Comput. Aided Mol. Des. 16 (2002), pp. 11–26.

766. M. Kontoyianni, L.M. McClellan, and G.S. Sokol, *Evaluation of docking performance: Comparative data on docking algorithms*, J. Med. Chem. 47 (2004), pp. 558–565.

767. J. Tirado-Rives and W.L. Jorgensen, *Contribution of conformer focusing to the uncertainty in predicting free energies for protein–ligand binding*, J. Med. Chem. 49 (2006), pp. 5880–5884.

768. M. Rarey, B. Kramer, T. Lengauer, and G. Klebe, *A fast flexible docking method using an incremental construction algorithm*, J. Mol. Biol. 261 (1996), pp. 470–489.

769. C.A. Baxter, C.W. Murray, D.E. Clark, D.R. Westhead, and M.D. Eldridge, *Flexible docking using Tabu search and an empirical estimate of binding affinity*, Proteins Struct. Funct. Genet. 33 (1998), pp. 367–382.

770. T. Hou, J. Wang, L. Chen, and X. Xu, *Automated docking of peptides and proteins by using a genetic algorithm combined with a Tabu search*, Protein Eng. 12 (1999), pp. 639–647.

771. D.K. Gehlhaar, G.M. Verkhivker, P.A. Rejto, C.J. Sherman, D.B. Fogel, L.J. Fogel, and S.T. Freer, *Molecular recognition of the inhibitor AG-1343 by HIV-1 protease: Conformationally flexible docking by evolutionary programming*, Chem. Biol. 2 (1995), pp. 317–324.

772. R. Abagyan, M. Totrov, and D. Kuznetzov, *ICM—A new method for protein modeling and design: Applications to docking and structure prediction from the distorted native conformation*, J. Comput. Chem. 15 (1994), pp. 488–506.

773. R.A. Friesner, J.L Banks, R.B. Murphy, T.A Halgren, J.J. Klicic, D.T. Mainz, M.P. Repasky et al., *Glide: A new approach for rapid, accurate docking and scoring. 1. Method and assessment of docking accuracy*, J. Med. Chem. 47 (2004), pp. 1739–1749.

774. T.A. Halgren, R.B. Murphy, R.A. Friesner, H.S. Beard, L.L. Frye, W.T. Pollard, and J.L. Banks, *Glide: A new approach for rapid, accurate docking and scoring. 2. Enrichment factors in database screening*, J. Med. Chem. 47 (2004), pp. 1750–1759.

775. M.D. Eldridge, C.W. Murray, T.R. Auton, G.V. Paolini, and R.P. Mee, *Empirical scoring functions: I. The development of a fast empirical scoring function to estimate the binding affinity of ligands in receptor complexes*, J. Comput.-Aided Mol. Des. 11 (1997), pp. 425–445.

776. R. Wang, L. Liu, L. Lai, and Y. Tang, *SCORE: A new empirical method for estimating the binding affinity of a protein–ligand complex*, J. Mol. Model. 4 (1998), pp. 379–384.

777. H.J. Böhm, *The computer program LUDI: A new method for the de novo design of enzyme inhibitors*, J. Comput. Aided Mol. Des. 6 (1992), pp. 61–78.

778. H.-J. Böhm, *The development of a simple empirical scoring function to estimate the binding constant for a protein–ligand complex of known three-dimensional structure*, J. Comput. Aided Mol. Des. 8 (1994), pp. 243–256.

779. I. Muegge and Y.C. Martin, *A general and fast scoring function for protein–ligand interactions: A simplified potential approach*, J. Med. Chem. 42 (1999), pp. 791–804.

780. I. Muegge, *PMF scoring revisited*, J. Med. Chem. 49 (2006), pp. 5895–5902.

781. R.S. DeWitte and E.I. Shakhnovich, *SMoG: De novo design method based on simple, fast and accurate free energy estimates. 1. Methodology and supporting evidence*, J. Am. Chem. Soc. 118 (1996), pp. 11733–11744.

782. C. Bissantz, G. Folkers, and D. Rognan, *Protein-based virtual screening of chemical databases. 1. Evaluation of different docking/scoring combinations*, J. Med. Chem. 43 (2000), pp. 4759–4767.

783. J.C. Cole, C.W. Murray, J.W.M. Nissink, R.D. Taylor, and R. Taylor, *Comparing protein–ligand docking programs is difficult*, Proteins Struct. Funct. Bioinform. 60 (2005), pp. 325–332.

784. M.D. Cummings, R.L. DesJarlais, A.C. Gibbs, V. Mohan, and E.P. Jaeger, *Comparison of automated docking programs as virtual screening tools*, J. Med. Chem. 48 (2005), pp. 962–976.

785. T.J.A. Ewing, S. Makino, A.G. Skillman, and I.D. Kuntz, *DOCK 4.0: Search strategies for automated molecular docking of flexible molecule databases*, J. Comput. Aided Mol. Des. 15 (2001), pp. 411–428.

786. R.L. DesJarlais, R.P. Sheridan, G.L. Seibel, J.S. Dixon, I.D. Kuntz, and E. Venkataraghavan, *Using shape complementarity as an initial screen in designing ligands for a receptor binding site of known three-dimensional structure*, J. Med. Chem. 31 (1988), pp. 722–729.

787. A.M. Lesk, *Detection of three-dimensional patterns of atoms in chemical structures*, Commun. ACM 22 (1979), pp. 219–224.

788. D.A. Case, T.A. Darden, T.E. Cheatham III, C.L. Simmerling, J. Wang, R.E. Duke, R. Luo et al., AMBER 10, University of California, San Francisco, CA, 2008.

789. DOCK. http://dock.compbio.ucsf.edu/DOC-6/doc6_manual.htm

790. K.D. Stewart, J.A. Bentley, and M. Cory, *Docking ligands into receptors: The test case of α-chimotrypsin*, Tetrahedron Comput. Method. 3 (1990), pp. 713–722.

791. R.L. DesJarlais, R.P. Sheridan, J.S. Dixon, I.D. Kuntz, R. Venkataraghavan, *Docking flexible ligands to macromolecular receptors by molecular shape*, J. Med. Chem. 29 (1986), pp. 2149–2153.

792. A.R. Leach and I.D. Kuntz, *Conformational analysis of flexible ligands in macromolecular receptor sites*, J. Comput. Chem. 13 (1992), pp. 730–748.

793. S. Makino and I.D. Kuntz, *Automated flexible ligand docking method and its application for database search*, J. Comput. Chem. 18 (1997), pp. 1812–1825.

794. H. Gohlke, M. Hendlich, and G. Klebe, *Knowledge-based scoring function to predict protein–ligand interactions*, J. Mol. Biol. 295 (2000), pp. 337–356.

795. AUTODOCK. http://autodock.scripts.edu/downloads

796. O. Trott and A.J. Olson, *Software news and update: AutoDock Vina: Improving the speed and accuracy of docking with a new scoring function, efficient optimization, and multithreading*, J. Comput. Chem. 31 (2010), pp. 455–461.

797. G.M. Morris, R. Huey, W. Lindstrom, M.F. Sanner, R.K. Belew, D.S. Goodsell, and A.J. Olson, *AutoDock4 and AutoDockTools4: Automated docking with selective receptor flexibility*, J. Comput. Chem. 30 (2009), pp. 2785–2791.

798. N. Marchand-Geneste, M. Cazaunau, A.J.M. Carpy, M. Laguerre, J.M. Porcher, and J. Devillers, *Homology model of the rainbow trout estrogen receptor (rtERα) and docking of endocrine disrupting chemicals (EDCs)*, SAR QSAR Environ. Res. 17 (2006), pp. 93–106.

799. G. Klebe and T. Mietzner, *A fast and efficient method to generate biologically relevant conformations*, J. Comput. Aided Mol. Des. 8 (1994), pp. 583–606.

800. *DAYLIGHT Software Manual*. DAYLIGHT Inc., Daylight Chem. Inf. Softwares, Laguna Niguel, CA. http://www.daylight.com

801. I.L. Alberts, N.P. Todorov, P. Källblad, and P.M. Dean, *Ligand docking and design in a flexible receptor site*, QSAR Comb. Sci. 24 (2005), pp. 503–507.

802. G. Jones, P. Willett, and R.C. Glen, *Molecular recognition of receptor sites using a genetic algorithm with a description of desolvation*, J. Mol. Biol. 245 (1995), pp. 43–53.

803. W.L. Jorgensen, D.S. Maxwell, and J. Tirado-Rives, *Development and testing of the OPLS all-atom force field on conformational energetics and properties of organic liquids*, J. Am. Chem. Soc. 118 (1996), pp. 11225–11236.

804. A.N. Jain, *Surflex: Fully automatic flexible molecular docking using a molecular-similarity-based search engine*, J. Med. Chem. 46 (2003), pp. 499–511.

805. R.A. Friesner, R.B. Murphy, M.P. Repasky, L.L. Frye, J.R. Greenwood, T.A. Halgren, P.C. Sanschagrin, and D.T. Mainz, *Extra precision Glide: Docking and scoring incorporating a model of hydrophobic enclosure for protein–ligand complexes*, J. Med. Chem. 49 (2006), pp. 6177–6196.

806. G. Nemethy, K.D. Gibson, K.A. Palmer, C.N. Yoon, G. Paterlini, A. Zagari, S. Rumsey, and H.A. Scheraga, *Energy parameters in polypeptides. 10. Improved geometrical parameters and nonbonded interactions for use in the ECEPP/3 algorithm, with application to proline-containing peptides*, J. Phys. Chem. 96 (1992), pp. 6472–6484.

807. R. Abagyan and M. Totrov, *Biased probability Monte Carlo conformational searches and electrostatic calculations for peptides and proteins*, J. Mol. Biol. 235 (1994), pp. 983–1002.

808. M. Stahl, N.P. Todorov, T. James, H. Mauser, H.-J. Boehm, and P.M. Dean, *A validation study on the practical use of automated de novo design*, J. Comput. Aided Mol. Des. 16 (2002), pp. 459–478.

809. C.M. Venkatachalam, X. Jiang, T. Oldfield, and M. Waldman, *Ligand fit: A novel method for the shape-directed rapid docking of ligands to protein active sites*, J. Mol. Graph. Model. 21 (2003), pp. 289–307.

810. C. McMartin and R.S. Bohacek, *QXP: Powerful, rapid computer algorithm for structure-based drug design*, J. Comput. Aided Mol. Des. 11 (1997), pp. 333–344.

811. I. Bytheway and S. Cochran, *Validation of molecular docking calculations involving FGF-1 and FGF-2*, J. Med. Chem. 47 (2004), pp. 1683–1693.

812. P. Cozzini, M. Fornabaio, A. Marabotti, D.J. Abraham, G.E. Kellogg, and A. Mozzarelli, *Simple, intuitive calculations of free energy of binding for protein–ligand complexes. 1. Models without explicit constrained water*, J. Med. Chem. 45 (2002), pp. 2469–2483.

813. H. Gohlke, M. Hendlich, and G. Klebe, *Predicting binding modes, binding affinities and "hot spots" for protein–ligand complexes using a knowledge-based scoring function*, Perspect. Drug Discov. Des. 20 (2000), pp. 115–144.

814. T.J.A. Ewing and I.D. Kuntz, *Critical evaluation of search algorithms for automated molecular docking and database screening*, J. Comput. Chem. 18 (1997), pp. 1175–1189.

815. M. Stahl and M. Rarey, *Detailed analysis of scoring functions for virtual screening*, J. Med. Chem. 44 (2001), pp. 1035–1042.

816. E. Kellenberger, J. Rodrigo, P. Muller, and D. Rognan, *Comparative evaluation of eight docking tools for docking and virtual screening accuracy*, Proteins Struct. Funct. Bioinform. 57 (2004), pp. 225–242.

817. R. Wang, Y. Lu, and S. Wang, *Comparative evaluation of 11 scoring functions for molecular docking*, J. Med. Chem. 46 (2003), pp. 2287–2303.

818. P. Ferrara, H. Gohlke, D.J. Price, G. Klebe, and C.L. Brooks III, *Assessing scoring functions for protein–ligand interactions*, J. Med. Chem. 47 (2004), pp. 3032–3047.

819. A. Vedani, *YETI: An interactive molecular mechanics program for small-molecule protein complexes*, J. Comput. Chem. 9 (1988), pp. 269–280.

820. A. Vedani and D.W. Huhta, *A new force field for modeling metalloproteins*, J. Am. Chem. Soc. 112 (1990), pp. 4759–4767.

821. A.J. Tervo, T.H. Nyrönen, T. Rönkkö, and A. Poso, *Comparing the quality and predictiveness between 3D QSAR models obtained from manual and automated alignment*, J. Chem. Inf. Comput. Sci. 44 (2004), pp. 807–816.

822. H.J. Böhm, *Prediction of binding constants of protein ligands: A fast method for the prioritization of hits obtained from de novo design or 3D database search programs*, J. Comput. Aided Mol. Des. 12 (1998), pp. 309–323.

823. A.N. Jain, *Scoring noncovalent protein-ligand interactions: A continuous differentiable function tuned to compute binding affinities*, J. Comput. Aided Mol. Des. 10 (1996), pp. 427–440.

824. J.S. Tokarski and A.J. Hopfinger, *Prediction of ligand–receptor binding thermodynamics by free energy force field (FEFF) 3D-QSAR analysis: Application to a set of peptidomimetic renin inhibitors*, J. Chem. Inf. Comput. Sci. 37 (1997), pp. 792–811.

825. J.L. Martin, L.N. Johnson, and S.G. Withers, *Comparison of the binding of glucose and glucose 1-phosphate derivatives to T-state glycogen phosphorylase b*, Biochemistry 29 (1990), pp. 10745–10757.

826. O.A. Santos-Filho and A.J. Hopfinger, *Structure-based QSAR analysis of a set of 4-hydroxy-5,6-dihydropyrones as inhibitors of HIV-1 protease: An application of the receptor-dependent (RD) 4D-QSAR formalism*, J. Chem. Inf. Model. 46 (2006), pp. 345–354.

827. V. Zoete, O. Michielin, and M. Karplus, *Protein-ligand binding free energy estimation using molecular mechanics and continuum electrostatics. Application to HIV-1 protease inhibitors*, J. Comput. Aided Mol. Des. 17 (2003), pp. 861–880.

828. R.W. Zwanzig, *High-temperature equation of state by a perturbation method. I. Nonpolar gases*, J. Chem. Phys. 22 (1954), pp. 1420–1426.

829. J. Pitera and P. Kollman, *Designing an optimum guest for a host using multimolecule free energy calculations: Predicting the best ligand for Rebek's "Tennis Ball,"* J. Am. Chem. Soc. 120 (1998), pp. 7557–7567.

830. B.L. Tembe and J.A. McCammon, *Ligand-receptor interactions*, J. Comput. Chem. 8 (1984), pp. 281–283.

831. C. de Graaf, C. Oostenbrink, P.H.J. Keizers, B.M.A. van Vugt-Lussenburg, J.N.M. Commandeur, and N.P.E. Vermeulen, *Free energies of binding of R- and S- propanolol to wild-type and F483A mutant cytochrome P450 2D6 from molecular dynamics simulations*, Eur. Biophys. J. 36 (2007), pp. 589–599.

832. J.W. Pitera and W.F. van Gunsteren, *A comparison of non-bonded scaling approaches for free energy calculations*, Mol. Simul. 28 (2002), pp. 45–65.

833. B.C. Oostenbrink, J.W. Pitera, M.M.H. van Lipzig, J.H.N. Meerman, and W.F. van Gunsteren, *Simulations of the estrogen receptor ligand-binding domain: Affinity of natural ligands and xenoestrogens*, J. Med. Chem. 43 (2000), pp. 4594–4605.

834. H. Liu, A.E. Mark, and W.F. van Gunsteren, *Estimating the relative free energy of different molecular states with respect to a single reference state*, J. Phys. Chem. 100 (1996), pp. 9485–9494.

835. J. Åqvist, C. Medina, and J.E. Samuelsson, *A new method for predicting binding affinity in computer-aided drug design*, Protein Eng. 7 (1994), pp. 385–391.

836. J. Åqvist, V.B Luzhkov, and B.O. Brandsdal, *Ligand binding affinities from MD simulations*, Acc. Chem. Res. 35 (2002), pp. 358–365.

837. P.E. Smith and W.F. van Gunsteren, *Predictions of free energy differences from a single simulation of the initial state*, J. Chem. Phys. 100 (1994), pp. 577–585.

838. P.R. Gerber, A.E. Mark, and W.F. van Gunsteren, *An approximate but efficient method to calculate free energy trends by computer simulation: Application to dihydrofolate reductase-inhibitor complexes*, J. Comput.-Aided Mol. Des. 7 (1993), pp. 305–323.

839. T. Hansson, J. Marelius, and J. Åqvist, *Ligand binding affinity prediction by linear interaction energy methods*, J. Comput.-Aided Mol. Des. 12 (1998), pp. 27–35.

840. K.B. Ljungberg, J. Marelius, D. Musil, P. Svensson, B. Norden, and J. Åqvist, *Computational modelling of inhibitor binding to human thrombin*, Eur. J. Pharm. Sci. 12 (2001), pp. 441–446.

841. R.C. Rizzo, M. Udier-Blagović, D.P. Wang, E.K. Watkins, M.B. Kroeger Smith, R.H. Smith Jr., J. Tirado-Rives, and W.L. Jorgensen, *Prediction of activity for nonnucleoside inhibitors with HIV-1 reverse transcriptase based on Monte Carlo simulations*, J. Med. Chem. 45 (2002), pp. 2970–2987.

842. J. Åqvist and T. Hansson, *On the validity of electrostatic linear response in polar solvents*, J. Phys. Chem. 100 (1996), pp. 9512–9521.

843. W. Wang, J. Wang, and P.A. Kollman, *What determines the van der Waals coefficient β in the LIE (linear interaction energy) method to estimate binding free energies using molecular dynamics simulations?*, Proteins Struct. Funct. Genet. 34 (1999), pp. 395–402.

844. Schrödinger, LLC, New York. http://schrodinger.com

845. R.D. Clark, A. Strizhev, J.M. Leonard, J.F. Blake, and J.B. Matthew, *Consensus scoring for ligand/protein interactions*, J. Mol. Graph. Model. 20 (2002), pp. 281–295.

846. D.K. Jones-Hertzog and W.L. Jorgensen, *Binding affinities for sulfonamide inhibitors with human thrombin using Monte Carlo simulations with a linear response method*, J. Med. Chem. 40 (1997), pp. 1539–1549.

847. D.Z. Huang, U. Lüthi, P. Kolb, M. Cecchini, A. Barberis, and A. Caflisch, *In silico discovery of β-secretase inhibitors*, J. Am. Chem. Soc. 128 (2006), pp. 5436–5443.

848. M. Nervall, P. Hanspers, J. Carlsson, L. Boukharta, and J. Åqvist, *Predicting binding modes from free energy calculations*, J. Med. Chem. 51 (2008), pp. 2657–2667.

849. H. Guitiérrez-de-Terán, M. Nervall, K. Ersmark, P. Liu, L.K. Janka, B.M. Dunn, A. Hallberg, and J. Åqvist, *Inhibitor binding to the plasmepsin IV aspartic protease from Plasmodium falciparum*, Biochemistry 45 (2006), pp. 10529–10541.

850. J. Srinivasan, T.E. Cheatham III, P. Cieplak, P.A. Kollman, and D.A. Case, *Continuum solvent studies of the stability of DNA, RNA and phosphoramidate-DNA helices*, J. Am. Chem. Soc. 120 (1998), pp. 9401–9409.

851. D. Sitkoff, K.A. Sharp, and B. Honig, *Accurate calculation of hydration free energies using macroscopic solvent models*, J. Phys. Chem. 98 (1994), pp. 1978–1988.

852. B.R. Brooks, D. Janežič, and M. Karplus, *Harmonic analysis of large system. I. Methodology*, J. Comput. Chem. 16 (1995), pp. 1522–1542.

853. D. Janežič and B.R. Brooks, *Harmonic analysis of large system. II. Comparison of different protein models*, J. Comput. Chem. 16 (1995), pp. 1543–1553.

854. D. Janežič, R.M. Venable, and B.R. Brooks, *Harmonic analysis of large system. III. Comparison with molecular dynamics*, J. Comput. Chem. 16 (1995), pp. 1554–1566.

855. J. Wang, P. Morin, W. Wang, and P.A. Kollman, *Use of MM-PBSA in reproducing the binding free energies to HIV-1RT of TIBO derivatives and predicting the binding mode to HIV-1 RT of Efavirenz by docking and MM-PBSA*, J. Am. Chem. Soc. 123 (2001), pp. 5221–5230.

856. D. Roccatano, A. Amadei, M.E.F. Apol, A. Di Nola, and H.J.C. Berendsen, *Application of the quasi-Gaussian entropy theory to molecular dynamics simulations of Lennard-Jones fluids*, J. Chem. Phys. 109 (1998), pp. 6358–6363.

857. M.K. Gilson and B. Honig, *Calculation of the total electrostatic energy of a macromolecular system: Solvation energies, binding energies and conformational analysis*, Proteins 4 (1988), pp. 7–18.

858. Y.Y. Sham, Z.T. Chu, H. Tao, and A. Warshel, *Examining methods for calculations of binding free energies: LRA, LIE, PDLD-LRA, and PDLD/S-LRA calculations of ligands binding to an HIV protease*, Proteins Struct. Funct. Genet. 39 (2000), pp. 393–407.

859. B. Kuhn and P.A. Kollman, *Binding of a diverse set of ligands to avidin and streptavidin: An accurate quantitative prediction of their relative affinities by a combination of molecular mechanics and continuum solvent models*, J. Med. Chem. 43 (2000), pp. 3786–3791.

860. G.M. Torrie and J.P. Valleau, *Nonphysical sampling distribution in Monte Carlo free energy estimation: Umbrella sampling*, J. Comput. Phys. 12 (1977), pp. 187–199.

861. X.J. Kong and C.L. Brooks III, *λ-Dynamics: A new approach to free energy calculations*, J. Chem. Phys. 105 (1996), pp. 2414–2423.

862. R.J. Radmer and P.A. Kollman, *The application of three approximative free energy calculation methods to structure based ligand design: Trypsin and its complex with inhibitors*, J. Comput.-Aided Mol. Des. 12 (1998), pp. 215–227.

863. M.A.L. Eriksson, J. Pitera, and P.A. Kollman, *Prediction of the binding free energies of new TIBO-like HIV-1 reverse transcriptase inhibitors using a combination of PROFEC, PB/SA, CMC/MD and free energy calculations*, J. Med. Chem. 42 (1999), pp. 868–881.

864. R.H. Smith, W.L. Jorgensen, J. Tirado-Rives, M.L. Lamb, P.A.J. Janssen, C.J. Michejda, and M.B. K. Smith, *Prediction of binding affinities for TIBO inhibitors of HIV-1 reverse transcriptase using Monte Carlo simulations in a linear response method*, J. Med. Chem. 41 (1998), pp. 5272–5286.

865. B. Kuhn, P. Gerber, T. Schulz-Gasch, and M. Stahl, *Validation and use of the MM-PBSA approach for drug discovery*, J. Med. Chem. 48 (2005), pp. 4040–4048.

866. C. Obiol-Pardo and J. Rubio-Martinez, *Comparative evaluation of MMPBSA and XSCORE to compute binding free energy in XIAP-peptide complexes*, J. Chem. Inf. Model. 47 (2007), pp. 134–142.

867. M. Born, *Volumen und hydratationswärme der Ionen*, Zeitschrift für Physik 1 (1920), pp. 45–48.

868. R. Constanciel and R. Contreras, *Self consistent field theory of solvent effects representation by continuum models: Introduction of desolvation contribution*, Theor. Chim. Acta 65 (1984), pp. 1–11.

869. W.C. Still, A. Tempczyk, R.C. Hawley, and T. Hendrickson, *Semianalytical treatment of solvation for molecular mechanics and dynamics*, J. Am. Chem. Soc. 112 (1990), pp. 6127–6129.

870. C.R.W. Guimarães and M. Cardozo, *MM-GB/SA rescoring of docking poses in structure-based lead optimization*, J. Chem. Inf. Model. 48 (2008), pp. 958–970.

871. S.J. Wodak and J. Janin, *Analytical approximation to the accessible surface area of proteins*, Proc. Natl Acad. Sci. U.S.A. 77 (1980), pp. 1736–1740.

872. X. Zou, Y. Sun, and I.D. Kuntz, *Inclusion of solvation in ligand binding free energy calculations using the generalized-Born model*, J. Am. Chem. Soc. 121 (1999), pp. 8033–8043.

873. P. Koehl, *Implicit solvent models for protein simulations.* http://nook.cs.udavis.edu/ékoehl/ProShape/born.html (accessed May 20, 2009).

874. Z. Yu, M.P. Jacobson, and R.A. Friesner, *What role do surfaces play in GB models? A new-generation of surface-generalized Born model based on a novel Gaussian surface for biomolecules*, J. Comput. Chem. 27 (2006), pp. 72–89.

875. P.D. Lyne, M.L. Lamb, and J.C. Saeh, *Accurate prediction of the relative potencies of members of a series of kinase inhibitors using molecular docking and MM-GBSA scoring*, J. Med. Chem. 49 (2006), pp. 4805–4808.

876. N. Huang, C. Kalyanaraman, J.J. Irwin, and M.P. Jacobson, *Physics-based scoring of protein-ligand complexes: Enrichment of known inhibitors in large-scale virtual screening*, J. Chem. Inf. Model. 46 (2006), pp. 243–253.

877. R.J. Hu, *Molecular modeling studies of HIV-1 reverse transcriptase and some of its inhibitors*, PhD thesis, Lanzhou and Paris-Denis-Diderot Universities, Lanzhou, China, 2009.

878. K. Roy and J.T. Leonard, *QSAR modeling of HIV-1 reverse transcriptase inhibitor 2-amino-6-arylsulfonylbenzonitriles and congeners using molecular connectivity and E-state parameters*, Bioorg. Med. Chem. 12 (2004), pp. 745–754.

879. J.T. Leonard and K. Roy, *Classical QSAR modeling of HIV-1 reverse transcriptase inhibitor 2-amino-6-arylsulfonylbenzonitriles and congeners*, QSAR Comb. Sci. 23 (2004), pp. 23–35.

880. M.P. Freitas, *MIA-QSAR modelling of anti HIV-1 activities of some 2-amino-6-arylsulfonylbenzonitriles and their thio and sulfinylcongeners*, Org. Biomol. Chem. 4 (2006), pp. 1154–1159.

881. P.P. Roy and K. Roy, *On some aspects of variable selection for partial least squares regression models*, QSAR Comb. Sci. 27 (2008), pp. 302–313.

882. S.R. Johnson, *The trouble with QSAR (or How I learned to stop worrying and embrace fallacy)*, J. Chem. Inf. Model. 48 (2008), pp. 25–26.

883. A.M. Doweyko, *3D-QSAR illusions*, J. Comput. Aided Mol. Des. 18 (2004), pp. 587–596.

884. M.T.D. Cronin and T.W. Schultz, *Pitfalls in QSAR*, J. Mol. Struct. (Theochem) 622 (2003), pp. 39–51.

885. G.M. Maggiora, *On outliers and activity cliffs—Why QSAR often disappoints*, J. Chem. Inf. Model. 46 (2006), pp. 1535.

886. J. Manchester and R. Czermínski, *Caution: Popular BENCHMARK data sets do not distinguish the merits of 3D QSAR methods*, J. Chem. Inf. Model. 49 (2009), pp. 1449–1454.

887. W.H. Press, B.P. Flannery, S.A. Teukolsky, and W.T. Vetterling, *Numerical Recipes. The Art of Scientific Computing*, Cambridge University Press, Cambridge, U.K., 1989, pp. 274–334.

888. E.M. Engler, J.D. Andose, and P.v.R. Schleyer, *Critical evaluation of molecular mechanics*, J. Am. Chem. Soc. 95 (1973), pp. 8005–8025.

889. J.F. Stanton and D.E. Bernholdt, *An empirically adjusted Newton–Raphson algorithm for finding local minima on molecular potential energy surfaces*, J. Comput. Chem. 11 (1990), pp. 58–63.

890. K. Zimmmermann, *ORAL: All purpose molecular mechanics simulator and energy minimizer*, J. Comput. Chem. 12 (1991), pp. 310–319.

891. J.L.M. Dillen, *PEFF: A program for the development of empirical force fields*, J. Comput. Chem. 13 (1992), pp. 257–267.

892. J.H. Holland, *Adaptation in Natural and Artificial Systems*, University of Michigan Press, Ann Arbor, MI, 1975.

893. J.H. Holland, *Les Algorithmes Génétiques*, revue Pour Sci. 179 (1992), pp. 44–51.

894. B.T. Fan and R.S. Zhang, Introduction to Computer Chemistry, Lanzhou University Press, Lanzhou, China, 1999, pp. 183–199.

895. J.H Friedman and B.W. Silverman, *Flexible parsimonious smoothing and additive modeling*, Technometrics 31 (1989), pp. 3–39.

896. H. Kubinyi, *Evolutionary variable selection in regression and PLS analyses*, J. Chemom. 10 (1996), pp. 119–133.

897. D.L. Selwood, D.J. Livingstone, J.C.W. Comley, A.B. O'Dowd, A.T. Hudson, P. Jackson, K.S. Jandu, V.S. Rose, and J.N. Stables, *Structure–activity relationships of antifilarial antimycin analogs: A multivariate pattern recognition study*, J. Med. Chem. 33 (1990), pp 136–142.

898. J.H. Wikel and E.R. Dow, *The use of neural networks for variable selection in QSAR*, Bioorg. Med. Chem. Lett. 3 (1993), pp. 645–651.

899. M. Fodslette Møller, *A scaled conjugate gradient algorithm for fast supervised learning*, Neural Netw. 6 (1993), pp. 525–533.

900. D.J. Maddalena and G.A.R. Johnston, *Prediction of receptor properties and binding affinity of ligands to benzodiazepines/GABA$_A$ receptors using artificial neural networks*, J. Med. Chem. 38 (1995), pp. 715–724.

901. D.H. Andrews, *The relation between the Raman spectra and the structure of organic molecules*, Phys. Rev. 36 (1930), pp. 544–554.

902. J.R. Maple, M.J. Hwang, T.P. Stockfish, U. Dinur, M. Waldman, C.S. Ewig, and A.T. Hagler, *Derivation of class II force fields. I. Methodology and quantum force field for the alkyl functional group and alkane molecules*, J. Comput. Chem. 15 (1994), pp. 162–182.

903. S. Lifson and A. Warshel, *Consistent force field for calculations of conformations, vibrational spectra, and enthalpies of cycloalkane and n-alkane molecules*, J. Chem. Phys. 49 (1968), pp. 5116–5129.

904. A.K. Rappe, C.J. Casewit, K.S. Colwell, W.A. Goddard III, and W.M. Skiff, *UFF, a full periodic table force field for molecular mechanics and molecular dynamics simulations*, J. Am. Chem. Soc. 114 (1992), pp. 10024–10035.

905. N.L. Allinger, *Conformational analysis. 130. MM2. A hydrocarbon force field utilizing V1 and V2 torsional terms*, J. Am. Chem. Soc. 99 (1977), pp. 8127–8134.

906. N.L. Allinger, *Calculation of molecular structure and energy by force-field methods in Advances in Physical Organic Chemistry*, Vol. 13, V. Gold and J. Bethell, Eds., Academic Press, London, U.K., 1976, pp. 1–82.

907. N.L. Allinger, K.H. Chen, and J.H. Lii, *An improved force field (MM4) for saturated hydrocarbons*, J. Comput. Chem. 17 (1996), pp. 642–668.

908. N. Nevins, C. Knohsiang, and N.L. Allinger, *Molecular mechanics (MM4) calculations on alkenes*, J. Comput. Chem. 17 (1996), pp. 669–694.

909. N. Nevins, J.H. Lii, and N.L. Allinger, *Molecular mechanics (MM4) calculations on conjugated hydrocarbons*, J. Comput. Chem. 17 (1996), pp. 695–729.

910. N. Nevins and N.L. Allinger, *Molecular mechanics (MM4) vibrational frequency calculations for alkenes and conjugated hydrocarbons*, J. Comput. Chem. 17 (1996), pp. 730–746.

911. A. Hocquet and M. Langgard, *An evaluation of the MM+ force field*, J. Mol. Model. 4 (1998), pp. 94–112.

912. HyperChem 7.0, Hypercube Inc., Gainesville, FL.

913. M. Clark, R.D. Cramer III, and N. Van Opdenbosch, *Validation of the general purpose tripos 5.2 force field*, J. Comput. Chem. 10 (1989), pp. 982–1012.

914. A.D. Mackerell Jr., J. Wiorkiewicz-Kuczera, and M. Karplus, *An all-atom empirical energy function for the simulation of nucleic acids*, J. Am. Chem. Soc. 117 (1995), pp. 11946–11975.

915. T.A. Halgren, *Merck molecular force field. II. MMFF94 van der Waals and electrostatic parameters for intermolecular interactions*, J. Comput. Chem. 17 (1996), pp. 520–552.

916. T.A. Halgren, *Merck molecular force field. III. Molecular geometries and vibrational frequencies for MMFF94*, J. Comput. Chem. 17 (1996), pp. 553–586.

917. T.A. Halgren and R.B. Nachbar, *Merck molecular force field. IV. Conformational energies and geometries for MMFF94*, J. Comput. Chem. 17 (1996), pp. 587–615.

918. T.A. Halgren, *Merck molecular force field. V. Extension of MMFF94 using experimental data, additional computational data, and empirical rules*, J. Comput. Chem. 17 (1996), pp. 616–641.

919. R. Lavery, K. Zakrzewska, and H. Sklenar, *JUMNA, junction minimisation of nucleic acids*, Comput. Phys. Commun. 91 (1995), pp. 135–158.

920. C.K. Varma, *Molecular mechanical force fields, review and critical analysis of modern day force fields with application to protein and nucleic acid structures*, December 2001, Biochemistry 218, Stanford University.

921. J. Wang, R.M. Wolf, J.W. Caldwell, P.A. Kollman, and D.A. Case, *Development and testing of a general Amber force field*, J. Comput. Chem. 25 (2004), pp. 1157–1174.

922. Y. Duan, C. Wu, S. Chowdhury, M.C. Lee, G. Xiong, W. Zhang, R. Yang et al., *A point-charge force field for molecular mechanics simulations of proteins based on condensed-phase quantum mechanical calculations*, J. Comput. Chem. 24 (2003), pp. 1999–2012.

923. J.W. Ponder and D.A. Case, *Force fields for protein simulation*, Adv. Protein Chem. 66 (2003), pp. 27–85.

924. T.E. Cheatham III and M.A. Young, *Molecular dynamics simulations of nucleic acids: Successes, limitations and promise*, Biopolymers (Nucleic Acid Sci.) 56 (2001), pp. 232–256.

925. V. Hornak, R. Abel, A. Okur, B. Strockbine, A. Roitberg, and C. Simmerling, *Comparison of multiple Amber force fields and development of improved protein backbone parameters*, Proteins Struct. Funct. Bioinform. 65 (2006), pp. 712–725.

926. P. Altoè, M. Stenta, A. Bottoni, and M. Garavelli, *A tunable QM/MM approach to chemical reactivity, structure and physico-chemical properties prediction*, Theor. Chem. Acc. 118 (2007), pp. 219–240.

927. H. Lin, Y. Zhang, and D.G. Truhlar. http://comp.chem.umn.edu/qmmm

928. C.I. Bayly, P. Cieplak, W.D. Cornell, and P.A. Kollman, *A well-behaved electrostatic potential based method using charge restraints for deriving atomic charges: The RESP model*, J. Phys. Chem. 97 (1993), pp. 10269–10280.

929. W.D. Cornell, P. Cieplak, C.I. Bayly, and P.A. Kollman, *Application of RESP charges to calculate conformational energies, hydrogen bond energies, and free energies of solvation*, J. Am. Chem. Soc. 115 (1993), pp. 9620–9631.

930. C.A. Reynolds, J.W. Essex, and W.G. Richards, *Atomic charges for variable molecular conformations*, J. Am. Chem. Soc. 114 (1992), pp. 9075–9079.

931. A. Jakalian, B.L. Bush, D.B. Jack, and C.I. Bayly, *Fast, efficient generation of high-quality atomic charges AM1-BCC model: I. Method*, J. Comput. Chem. 21 (2000), pp. 132–146.

932. A. Jakalian, D.B. Jack, and C.I. Bayly, *Fast, efficient generation of high-quality atomic charges AM1-BCC model: II. Parameterization and validation*, J. Comput. Chem. 23 (2002), pp. 1623–1641.

933. W.F. Van Gunsteren and H.J.C. Berendsen, *Computer simulation of molecular dynamics: Methodology, applications and perspectives in Chemistry*, Angew. Chem. Int. Ed. Engl. 29 (1990), pp. 992–1023.

934. D. Rognan, *Molecular dynamics simulations: A tool for drug design*, Persepect. Drug Discov. Des. 9–11 (1998), pp. 181–209.

935. L. Verlet, *Computer "experiments" on classical fluids. I. Thermodynamical properties of Lennard Jones molecules*, Phys. Rev. 159 (1967), pp. 98–103.

936. R.W. Hockney, *The potential calculation and some applications*, Methods Comput. Phys. 9 (1970), pp. 136–211.

937. W.C. Swope, H.C. Andersen, P.H. Berens, and K.R. Wilson, *A computer simulation method for the calculation of equilibrium constants for the formation of physical clusters of molecules: Application to small water clusters*, J. Chem. Phys. 76 (1982), pp. 637–649.

938. J.P. Ryckaert, G. Cicotti, and H.J.C. Berendsen, *Numerical integration of the cartesian equations of motion of a system with constraints: Molecular dynamics of n-alkanes*, J. Comput. Phys. 23 (1977), pp. 327–342.

939. W.L. Jorgensen, J. Chandrasekhar, J.D. Madura, R.W. Impey, and M.L. Klein, *Comparison of simple potential functions for simulating liquid water*, J. Chem. Phys. 79 (1983), pp. 926–935.

940. H.J.C. Berendsen, J.R. Grigera, and T.P. Straatsma, *The missing term in effective pair potentials*, J. Phys. Chem. 91 (1987), pp. 6269–6271.

941. T. Darden, D. York, and L. Pedersen, *Particle mesh Ewald an N-log(N) method for Ewald sums in large systems*, J. Chem. Phys. 98 (1993), pp. 10089–10092.

942. T. Darden, L. Perera, L. Li, and L. Pedersen, *New tricks for modelers from the crystallography toolkit: The particle mesh Ewald algorithm and its use in nucleic acid simulations*, Structure 7 (1999), pp. R55–R60.

943. E. Kolossov and R. Stanforth, *The quality of QSAR models: Problems and solutions*, SAR QSAR Environ. Res. 18 (2007), pp. 89–100.

944. H. Akaike, *Information theory and an extension of the maximum likehood principle*, in *Second International Symposium on Information Theory*, B.N. Petrov and F. Csaki, Eds., Akademiai Kiado, Budapest, Hungary, 1973, pp. 267–281.

945. H. Akaike, *A new look at the statistical model identification*, IEEE Trans. Automat. Contr. AC-19 (1974), pp. 716–723.

946. D.M. Hawkins, S.C. Basak, and D. Mills, *Assessing model fit by cross-validation*, J. Chem. Inf. Comput. Sci. 43 (2003), pp. 579–586.

947. K.B. Wiberg, W.E. Pratt, and W.F. Bailey, *Nature of substituent effects in nuclear magnetic resonance spectroscopy. 1. Factor analysis of carbon-13 chemical shifts in aliphatic halides*, J. Org. Chem. 45 (1980), pp. 4936–4947.

948. H. Kaneko, M. Arakawa, and K. Funatsu, *Development of a new regression analysis method using independent component analysis*, J. Chem. Inf. Model. 48 (2008), pp. 534–541.

949. A.K. Smilde, *Comments on multilinear PLS*, J. Chemom. 11 (1997), pp. 367–377.

950. F. Lindgren and S. Rännar, *Alternative partial least squares (PLS) algorithms*, Perspect. Drug Discov. Des. 12–14 (1998), pp. 105–113.

951. U. Indahl, *A twist to partial least squares regression*, J. Chemom. 19 (2005), pp. 32–44.

952. H. Martens and T. Naes, *Multivariate Calibration*, John Wiley & Sons, Chichester, U.K., 1989.

953. P. Bastien, *Régression PLS et données censurées*, PhD Thesis, Conservatoire National Arts et Metiers, Paris, France, 2008.

954. W.G. Glen, W.J. Dunn III, and D.R. Scott, *Principal components analysis and partial least squares regression*, Tetrahedron Comput. Methodol. 2 (1989), pp. 349–376.

955. W.G. Glen, M. Sarker, W.J. Dunn III, and D.R. Scott, *UNIPALS; Software for principal components analysis and partial least squares regression*, Tetrahedron Comput. Methodol. 2 (1989), pp. 377–396.

956. A. Berglund and S. Wold, *INLR, implicit non-linear latent variable regression*, J. Chemom. 11 (1997), pp. 141–156.

957. A. Berglund and S. Wold, *A serial extension of multiblock PLS*, J. Chemom. 13 (1999), pp. 461–471.

958. L. Eriksson, E. Johansson, F. Lindgren, and S. Wold, *GIFI-PLS: Modeling of nonlinearities and discontinuities in QSAR*, Quant. Struct. Act. Relat. 19 (2000), pp. 345–355.

959. K. Hasegawa, Y. Miyashita, and K. Funatsu, *GA strategy for variable selection in QSAR studies: GA-based PLS analysis of calcium channel antagonists*, J. Chem. Inf. Comput. Sci. 37 (1997), pp. 306–310.

960. K. Hasegawa and K. Funatsu, *GA strategy for variable selection in QSAR studies: GAPLS and D-optimal designs for predictive QSAR model*, J. Mol. Struct. Theochem 425 (1998), pp. 255–262.

961. K. Tuppurainen, S.P. Korhonen, and J. Ruuskanen, *Performance of multicomponent self-organizing regression (MCSOR) in QSAR, QSPR, and multivariate calibration: Comparison with partial least-squares (PLS) and validation with large external data sets*, SAR QSAR Environ. Res. 17 (2006), pp. 549–561.

962. J. Chen and X.Z. Wang, *A new approach to near-infrared spectral data analysis using independent component analysis*, J. Chem. Inf. Comput. Sci. 41 (2001), pp. 992–1001.

963. X. Shao, W. Wang, Z. Hou, and W. Cai, *A new regression method based on independent component analysis*, Talanta 69 (2006), pp. 676–680.

964. I.V. Tetko, *Neural networks studies. 4. Introduction to associative neural networks*, J. Chem. Inf. Comput. Sci. 42 (2002), pp. 717–728.

965. A.F. Duprat, T. Huynh, and G. Dreyfus, *Toward a principled methodology for neural network design and performance evaluation in QSAR. Application to the prediction of logP*, J. Chem. Inf. Comput. Sci. 38 (1998), pp. 586–594.

966. I.V. Tetko, D.J. Livingstone, and A.I. Luik, *Neural network studies 1. Comparison of overfitting and overtraining*, J. Chem. Inf. Comput. Sci. 35 (1995), pp. 826–833.

967. T. Aoyama, Y. Suzuki, and H. Ichikawa, *Neural networks applied to quantitative structre-activity relationships*, J. Med. Chem. 33 (1990), pp. 905–908.

968. I.V. Tetko, V.P. Solov'ev, A.V. Antonov, X.J. Yao, J.P. Doucet, B.T. Fan, F. Hoonakker et al., *Benchmarking of linear and nonlinear approaches for quantitative structure–property relationship studies of metal complexation with ionophores*, J. Chem. Inf. Model. 46 (2006), pp. 808–819.

969. I.V. Tetko, V.V. Kovalishyn, and D.J. Livingstone, *Volume learning algorithm artificial neural networks for 3D QSAR studies*, J. Med. Chem. 44 (2001), pp. 2411–2420.

970. M.J.L. Orr, *Introduction to Radial Basis Function Networks*, Centre for Cognitive Science, Edinburgh University, Edinburgh, U.K., 1996.

971. M.J.L. Orr, *MATLAB Routines for Subset Selection and Ridge Regression in Linear Neural Networks*, Centre for Cognitive Science, Edinburgh University, Edinburgh, U.K., 1996.

972. X.J. Yao, B.T. Fan, J.P. Doucet, A. Panaye, M.C. Liu, R.S. Zhang, X.Y. Zhang, and Z.D. Hu, *Quantitative structure property relationship models for the prediction of liquid heat capacity*, QSAR Comb. Sci. 22 (2003), pp. 29–48.

973. D.F. Specht, *A general regression neural network*, IEEE Trans. Neural Netw. 2 (1991), pp. 568–576.

974. E. Parzen, *On estimation of a probability density function and mode*, Ann. Math. Stat. 3 (1962), pp. 1065–1076.

975. T. Kohonen, *Self-Organization and Associative Memory*, 3rd edn, Springer Verlag, Berlin, Germany, 1989.

976. W.J. Melssen, J.R.M. Smits, L.M.C. Buydens, and G. Kateman, *Tutorial: Using artificial neural networks for solving chemical problems. Part II. Kohonen self-organising feature maps and Hopfield networks*, Chemom. Intell. Lab. Syst. 23 (1994), pp. 267–291.

977. T. Kohonen, *The self-organizing map (SOM)*, Neurocomputing 21 (1998), pp. 1–6. http://www.cis.hut.fi/projects

978. S. Anzali, J. Gasteiger, U. Holzgrabe, J. Polanski, J. Sadowski, A. Teckentrup, and M. Wagener, *The use of self-organizing neural networks in drug design*, Perspect. Drug Discov. Des. 9–11 (1998), pp. 273–299.

979. M.S. Friedrichs and P.G. Wolynes, *Toward protein tertiary structure recognition by means of associative memory Hamiltonians*, Science 246 (1989), pp. 371–373.

980. J.H. Friedman and J. Tukey, *Projection Pursuit algorithm for exploratory data analysis*, IEEE Trans. Comp. 9 (1974), pp. 881–889.

981. J.H. Friedman and W. Stuetzle, *Projection pursuit regression*, J. Am. Stat. Assoc. 76 (1981), pp. 817–823.

982. J.H. Friedman, *Classification and multiple regression through projection pursuit*, Technical Report LCS 12, Laboratory for Computational Statistics, Department of Statistics, Stanford University, Stanford, CA, 1985.

983. J.H. Friedman, *Exploratory projection pursuit*, J. Am. Stat. Assoc. 82 (1987), pp. 249–266.

984. The comprehensive R Archive Network, http://cran.r-project.org

985. J. Shine, *Projection pursuit regression and its relationship to neural networks*, Data Mining Seminar, George Mason University, Fairfax, Virginia, October 24, 2003.

986. V. Nguyen-Cong and B.M. Rode, *Quantitative electronic structure–activity relationships of pyridinium cephalosporins using nonparametric regression methods*, Eur. J. Med. Chem. 31 (1996), pp. 479–484.

987. J.H. Friedman, *Multivariate adaptive regression splines*, Ann. Stat. 19 (1991), pp. 1–67.

988. R.D. De Veaux, D.C. Psichogios, and L.H. Ungar, *A comparison of two nonparametric estimation schemes: MARS and neural networks*, Comput. Chem. Eng. 17 (1993), pp. 819–837.

989. Q.S. Xu, M. Daszykowski, B. Walczak, F. Daeyaert, M.R. de Jonge, J. Heeres, L.M.H. Koymans et al., *Multivariate adaptive regression-splines—Studies of HIV reverse transcriptase inhibitors*, Chemom. Intell. Lab. Syst. 72 (2004), pp. 27–34.

990. W.S. Cleveland, *Robust locally weighted regression and smoothing scatterplots*, J. Am. Stat. Assoc. 74 (1979), pp. 829–836.

991. W.S. Cleveland and S.J. Devlin, *Locally weighted regression: An approach to regression analysis by local fitting*, J. Am. Stat. Assoc. 83 (1988), pp. 596–610.

992. A.Buja, T. Hastie, and R. Tibshirani, *Linear smoothers and additive models*, Ann. Stat. 17 (1989), pp. 453–555.

993. T. Hastie and R. Tibshirani, *GAM: Generalized additive models*. http://www-stat.stanford.edu/software/gam/index.html

994. R. Tibshirani, *Regression shrinkage and selection via the Lasso*, J. R. Stat. Soc. Series B Stat. Methodol. 58 (1996), pp. 267–288.

995. V.V. Zernov, K.V. Balakin, A.A. Ivaschenko, N.P. Savchuk, and I.V. Pletnev, *Drug discovery using support vector machines. The case studies of drug-likeness, agrochemical-likeness, and enzyme inhibition predictions*, J. Chem. Inf. Comput. Sci. 43 (2003), pp. 2048–2056.

996. C. Cortes and V. Vapnik, *Support vector networks*, in *Machine Learning*, Vol. 20, Kluwer Academic Publishers, Boston, MA, pp. 273–297, 1995.

997. N. Cristianini and J. Shawe-Taylor, *An Introduction to Support Vector Machines and Other Kernel-Based Learning Methods*, Cambridge University Press, Cambridge, U.K., 2000.

998. B. Schölkopf and A.J. Smola, *Learning with Kernels. Support Vector Machines, Regularization, Optimization and Beyond*, The MIT Press, Cambridge, MA, London U.K., 2002.

999. A.J. Smola and B. Schölkopf, *A tutorial on support vector regression*, Statist. Comput. 14 (2004), pp. 199–222.

1000. U. Thissen, M. Pepers, B. Üstün, W.J Melssen, and L.M.C. Buydens, *Comparing support vector machine to PLS for spectral regression applications*, Chemom. Intell. Lab. Syst. 73 (2004), pp. 169–179.

1001. V.D. Sánchez, *Advanced support vector machines and kernel methods*, Neurocomputing 55 (2003), pp. 5–20.

1002. C.C. Chang and C.J. Lin, *A library for support vector machines*. LIBSVM, version 2-31. http://csie.ntu.edu.tw/~cjlin/libsvm/#nuandone

1003. R. Czermiñski, A.Yasri, and D. Hartsough, *Use of support vector machine in pattern classification: Application to QSAR studies*, Quant. Struct. Act. Relat. 20 (2001), pp. 227–240.

1004. B. Üstün, W.J. Melssen, and L.M.C. Buydens, *Faciliting the application of Support Vector Regression by using a universal Pearson VII function based kernel*, Chemom. Intell. Lab. Syst. 81 (2006), pp. 29–40.

1005. P. Lind and T. Maltseva, *Support vector machines for the estimation of aqueous solubility*, J. Chem. Inf. Comput. Sci. 43(2003), pp. 1855–1859.

1006. P. Mahé, N. Ueda, T. Akutsu, J.L. Perret, and J.P. Vert, *Graph kernels for molecular structure–activity relationship analysis with support vector machines*, J. Chem. Inf. Model. 45 (2005), pp. 939–951.

1007. H. Fröhlich, J.K. Wegner, F. Sieker, and A. Zell, *Kernel functions for attributed molecular graphs—A new similarity-based approach to ADME prediction in classification and regression*, QSAR Comb. Sci. 25 (2006), pp. 317–326.

1008. M. Song, C.M. Breneman, J. Bi, N. Sukumar, K.P. Bennett, S. Cramer, and N. Tugcu, *Prediction of protein retention times in anion-exchange chromatography systems using support vector regression*, J. Chem. Inf. Comput. Sci. 42 (2002), pp. 1347–1357.

1009. J.A.K. Suykens and J. Vandewalle, *Chaos control using least-squares support vector machines*, Int. J. Circ. Theor. Appl. 27 (1999), pp. 605–615.

1010. J.A.K. Suykens, J. Vandewalle, and B. De Moor, *Optimal control by least squares support vector machines*, Neural Netw. 14 (2001), pp. 23–35.

1011. X.J. Yao, H.X. Liu, R.S. Zhang, M.C. Liu, Z.D. Hu, A.Panaye, J.P. Doucet, and B.T. Fan, *QSAR and classification study of 1,4-dihydropyridine calcium channel antagonists based on least squares support vector machines*, Mol. Pharm. 2 (2005), pp. 348–356.

1012. T.W. Schultz, T.I. Netzeva, and M.T.D. Cronin, *The use of diversity versus representivity in the training and validation of quantitative structure–activity relationships*, SAR QSAR Environ. Res. 14 (2003), pp. 59–81.

1013. M.F. Parretti, R.T. Kroemer, J.H. Rothman, and W.G. Richards, *Alignment of molecules by the Monte Carlo optimization of molecular similarity indices*, J. Comput. Chem. 18 (1997), pp. 1344–1353.

1014. H. Chen, J. Zhou, and G. Xie, *PARM: A genetic evolved algorithm to predict bioactivity*, J. Chem. Inf. Model. 38 (1998), pp. 243–250.

1015. A. Cherkasov, F. Ban, O. Santos-Filho, N. Thorsteinson, M. Fallahi, and G.L. Hammond, *An updated steriod benchmark set and its application in the discovery of novel nanomolar ligands of steroid sex harmone-binding globulin*, J. Med. Chem. 51 (2008), pp. 2047–2056.

Index

A

ACE (angiotensin-converting enzyme), 25, 27, 31, 36; *see also* Sutherland data set
Acetylcholinesterase, *see* AChE
AChE (acetylcholinesterase); *see also* Sutherland data set
  by CoMFA, 31, 36, 38
  flexibility in docking, 326
  by MQSM, 176–177
  by SM-GNN, 167
Active analogue approach, 34
ADAPT (automated data analysis using pattern recognition techniques) program, 212, 244
Adenosine A1 receptor binding, 146, 234
Adenosine A3 receptor binding, 135, 245, 263
Adenosine binding GADPH, 38–40
α1–Adreno receptor antagonist, 74–75
AhR receptor ligand
  by autocorrelation, 135
  by BCP, 228
  by CoMFA, 23, 56
  by CoSA, 197
  by 5D-QSAR, 300
  by EEVA, 200–201
  by free energy calculation, 350
  by GETAWAY, 249
  by MS-WHIM, 240
  by Quasar, 299
  by SESP, 139
  by SM-GNN, 166
  by VolSurf, 242
Aldehyde oxidase, 103
Aldose reductase, 323
Alignment in CoMFA; *see also* SEAL
  field fit, 7
  multifit, 7
  property density, 9–11
  rigid atom-atom fit, 6
Alignment–refinement; *see also* Pharmacophore analysis; SEAL
  1D to 3D, 46
  via Hopfield network, 41–44
  integrating docking, 37–41, 317 (*see also* Docking)
  QUASI method, 47
  QSSA, 173, 218
  TGSA, 172–173, 218

Alkanoic acid, 118
ALMOND program, 143
AMBER (assisted molecular building with energy refinement) program, 357, 391, 397
Amide, 226–227
Amidinophenylalanine, 63–65
Amine mutagenicity, 191
2-Amino-6-arylsulfonylbenzonitrile, 366–370
Analogy in model building and energy refinement, *see* AMBER
Androgen receptor, *see* AR binding
Angiotensin-converting enzyme, *see* ACE
ANN (artificial neural network)
  associative NN, 430–431
  BNN, 425–430
  CPANN, 188–192
  GNN, 164–169, 214, 388–390
  GRNN, 434–435, 452–453
  Hopfield NN, 41–44, 438–442
  KNN, 115–117, 435–438
  RBFNN, 431–434, 452–453
Antiasthmatic activity, 260–262
Anti-HIV activity, *see also* Pyridinone; TIBO derivative
  2-amino-6-arylsulfonylbenzonitrile, 366–370
  by CoMFA, 14–15, 192
  dihydropyrone, 275–276, 341
  Efavirenz, 359–360
  HEPT, 192, 288, 290
  HIV protease inhibitor, 101–102
  by LIE treatment, 354
  NNRTI, 353, 359
  nucleosides, 216
  phenylbutanoic acid, 280
  quinolone, 242
  statin, 101–102
Antimalarial activity, *see Plasmodium falciparum*
AR (androgen receptor) binding, 295, 305–306
Aromatic diols, 349
Artemisinin, 75, 222, 243
Artificial neural network, *see* ANN
Aryl-chromene, 224
Aryl-piperazines, 22, 74
ASA (atomic shell approximation), 174, 218
Atomic shell approximation, *see* ASA
Autocorrelation method, 239
Autocorrelation model electrostatic potential, *see* AutoMEP
Autocorrelation vector, 133–136

521

T - #0308 - 071024 - C8 - 234/156/25 - PB - 9780367383169 - Gloss Lamination